Electrical Machine Drives Control

Electrical Machine Drives Control

An Introduction

Juha Pyrhönen

Department of Electrical Engineering
Lappeenranta University of Technology, Finland

Valéria Hrabovcová

Faculty of Electrical Engineering
University of Žilina, Slovakia

R. Scott Semken

Department of Mechanical Engineering
Lappeenranta University of Technology, Finland

This edition first published 2016

© 2016 John Wiley & Sons Ltd

Registered office
John Wiley & Sons Ltd, The Atrium, Southern Gate, Chichester, West Sussex, PO19 8SQ, United Kingdom

For details of our global editorial offices, for customer services and for information about how to apply for permission to reuse the copyright material in this book please see our website at www.wiley.com.

Library of Congress Cataloging-in-Publication Data

Names: Pyrhönen, Juha, author. | Hrabovcová, Valéria, author. | Semken, R. Scott, author.
Title: Electrical machine drives control : An introduction / Juha Pyrhönen,
    Valéria Hrabovcová, R. Scott Semken.
Description: Chichester, West Sussex, United Kingdom : John Wiley & Sons,
    Inc., [2016] | Includes bibliographical references and index.
Identifiers: LCCN 2016015388 | ISBN 9781119260455 (cloth) | ISBN 9781119260400
    (epub) | ISBN 9781119260448 (epdf)
Subjects: LCSH: Electric driving. | Electric motors–Electronic control.
Classification: LCC TK4058 .P89 2016 | DDC 621.46–dc23 LC record available at
https://lccn.loc.gov/2016015388

A catalogue record for this book is available from the British Library.

ISBN: 9781119260455

Set in 10/12 pt TimesLTStd-Roman by Thomson Digital, Noida, India
Printed and bound in Malaysia by Vivar Printing Sdn Bhd

10 9 8 7 6 5 4 3 2 1

Contents

Preface

A basic study of electrical drives is fundamental to an electrical engineering curriculum, and, today, gaining a better academic understanding of the theory and application of controlled-velocity electrical drive technologies is increasingly important. Electrical drives provide superior control properties for a wide variety of processes, and the number of applications for precision-controlled motor drives is increasing. A modern electrical drive accurately controls motor torque and speed with relatively high electromechanical conversion efficiencies, making it possible to considerably reduce energy consumption. Because of the present pervasive use of electric machinery and the associated large energy flows, the introduction of more effective and efficient electrical drives promises significant environmental benefit, and electrical engineers are responding by introducing new and more efficient electrical drives to a myriad of industrial processes.

A controlled-velocity electrical drive combines power electronics, *electric machinery*, a *control system*, and drive mechanisms to apply force or torque to execute any number of desired functions. The term *electric machinery* refers primarily to the electromagnetic mechanical devices that convert electricity to mechanical power or mechanical power to electricity—that is, to electric motors or generators. The term *control system* refers to the control electronics, instrumentation, and coding that monitor the condition of the electric machinery and adjust operating speed and/or match force or torque to load.

With a rigorous introduction to theoretical principles and techniques, this academic reference and research book offers the master of science or doctoral student in electrical engineering a textbook that provides the background needed to carry out detailed analyses with respect to controlled-velocity electrical drives. At the same time, for engineers in general, the text can serve as a guide to understanding the main phenomena associated with electrical machine drives. The edition includes up-to-date theory and design guidelines, taking into account the most recent advances in the field. The years of scientific research activity and the extensive pedagogical skill of the authors have combined to produce this comprehensive approach to the subject matter. The considered electric machinery consists of not only classic rotating machines, such as direct current, asynchronous, and synchronous motors and generators, but also new electric machine architectures that have resulted as the controller and power electronics have continued to develop and as new materials, such as permanent magnets, have been introduced. Examples covered include permanent magnet synchronous machines, switched reluctance machines, and synchronous reluctance machines.

The text is comprehensive in its analysis of existing and emerging electrical drive technologies, and it thoroughly covers the variety of drive control methods. In comparison to other books in the field, this treatment is unique. The authors are experts in the theory and design of electric machinery. They clearly define the most basic electrical drive concepts and go on to explain the critical details while maintaining a solid connection to theory and design of the associated electric machinery. Addressing a number of industrial applications, the authors take their investigation of electrical drives beyond theory to examine a number of practical aspects of control and application. Scalar, vector, and direct torque control methods are thoroughly covered with the nonidealities of direct torque control being given particular focus.

The expert body of knowledge that makes up this book has been built up over a number of years with contributions from numerous colleagues from both the Lappeenranta University of Technology and the University of Žilina in Slovakia. The authors are grateful for their help.

In particular, the authors would like to thank Professor Tapani Jokinen for his extensive contributions in general, Professor Olli Pyrhönen for his expert guidance on the control of synchronous electrical machine drives, Dr. Pasi Peltoniemi for the detailed and valuable example on tuning the control of an electrically excited synchronous machine, and M.Sc. Juho Montonen for his permanent magnet machine analysis. The authors would also like to specifically thank Dr. Hanna Niemelä, who translated some of the included text from its original Finnish. Finally, we give our warmest thanks to our families, who accommodated our long hours of writing, editing, and manuscript preparation.

This academic reference and research book uniquely provides comprehensive materials concerning all aspects of controlled-velocity electrical drive technology including control and operation. The treatise is based on the authors' extensive expertise in the theory and design of electric machinery, and in contrast to existing publications, its handling of electrical drives is solidly linked to the theory and design of the associated electric machinery.

Abbreviations and Symbols

A	magnetic vector potential [Vs/m], linear current density [A/m]
AC	alternating current
AM	asynchronous machine
ASIC	application-specific integrated circuit
A1–A2	armature winding terminals of a DC machine
AlNiCo	aluminium nickel cobalt permanent magnet
A	transmission ratio
B	magnetic flux density, vector [T] [Vs/m^2]
B	magnetic flux density, scalar [T] [Vs/m^2]
BLDC	brushless DC motor
B1–B2	commutating pole winding of a DC machine
C	capacitance [F], machine constant, speed of light [m/s]
C_E	constant, function of machine construction
C_T	torque producing dimensionless factor
C1–C2	compensating winding of a DC machine
$C_{io,i}$	outer or inner capacitance between the ball and the race in the ball bearing [F]
C_g	capacitance between the races of the ball bearing [F]
C_{01}, C_{02}	capacitance of the filter [F]
C_{wf}	capacitance between the stator winding and the stator frame [F]
C_{wr}	capacitance between the stator winding and the rotor core [F]
C_{sr}	capacitance between the stator and rotor cores [F]
c	experimentally determined coefficient, distributed capacitance [F/m]
c/h	duty cycles per hour
c'	capacitance per unit length [F/m]
CENELEC	Comité Européen de Normalisation Electrotechnique
CHP	combined heat and power
CSI	current source inverter
D	diameter [m], friction coefficient, code (drive end)
D1, D2	diode 1, diode 2
$D\Omega$	viscous friction, frictional torque
DC	direct current
DFIG	doubly fed induction generator
DFIM	doubly fed induction motor
DFLC	direct flux linkage control
DTC	direct torque control
D1–D2	series magnetizing winding terminals of a DC machine

d	thickness [m], axis
DOL	Direct On Line
DSC	Direct Self Control
E	electromotive force (emf) [V], RMS, electric field strength [V/m], scalar
E_{PMph}	phase value of emf induced by PM [V]
emf	electromotive force [V]
$\boldsymbol{E}$	electric field strength, vector [V/m]
ESR	equivalent series resistance [Ω]
e	electromotive force [V], instantaneous value $e(t)$ or per-unit value
$\boldsymbol{e}_{\text{s}}$	back electromotive force vector induced by the stator flux linkage ψ_{s} [V] or per-unit value
e_{m}	back electromotive force induced by the air gap flux linkage ψ_{m} [V] or per-unit value
F	force [N], scalar
$\boldsymbol{F}$	force [N], vector
F1, F2	terminals of field winding
FEA	finite element analysis
FLC	flux linkage control
FOC	field oriented control
FPGA	field programmable gate array
F_{m}	magnetomotive force $\oint \boldsymbol{H} \cdot d\boldsymbol{l}$ [A], (mmf)
FPGA	field-programmable gate array
f	frequency, characteristic oscillation frequency [Hz], or per-unit value
f_{sw}	switching frequency [Hz] or per-unit value
g	distributed conductance [S/m]
G_{m}	transfer function
G_{ce}	closed loop transfer function
GTO	gate turn-off thyristor
H	magnetic field strength [A/m]
h_{PM}	height of permanent magnet material [m]
I	electric current [A], RMS
IE1, 2, 3, 4	efficiency classes
IC	cooling methods
IGBT	insulated-gate bipolar transistor
IGCT	integrated gate-commutated thyristor, integrated gate controlled thyristor
IEC	International Electrotechnical Commission
IEEE	Institute of Electrical and Electronics Engineers
IM	induction motor
Im	imaginary part
IP	enclosure class
$i(t)$	instantaneous value of current [A]
i_{B}	base value for current [A]
$I_{\text{k}}, I_{\text{s}}$	starting current [A]
I_{st}	locked rotor current (starting) [A]
I_{ef}	effective load current [A]
i_{a}	armature current [A]
i_{com}	common current linkage [A]

i_f	field current [A]
$\boldsymbol{i_m}$	magnetizing current space vector [A] or per-unit value
i_{PM}	PM represented by a current source in the rotor [A] or per-unit value
i_{PE}	current in the protective earth wire of the motor cable [A] or per-unit value
i_{mPE}	earthing current [A] or per-unit value
J	moment of inertia [kgm^2], inertia, current density [A/m^2], magnetic polarization [Vs/m^2]
J_m	moment of inertia of the motor [kgm^2]
J_{load}	load moment of inertia [kgm^2]
J_{tot}	total moment of inertia [kgm^2]
j	imaginary unit
K	kelvin, transformation ratio, constant
K_p	amplification
k	coupling factor
k_C	Carter factor
k_d	distribution factor
k_{gain}	gain coefficient
k_p	pitch factor
k_{ri}	reduction factor (current ratio of synchronous machine)
k_{riav}	ratio of magnitudes of the current space vectors
k_{rs}	transformation ratio between stator and rotor
k_{sq}, k_{sk}	skewing factor
k_w	winding factor
$k_w N$	effective number of turns
L	inductance [H]
L_c	choke
LCI	load commutated inverter
L_D	total inductance of the direct damper winding [H] or per-unit value
$L_{D\sigma}$	leakage inductance of the direct damper winding [H] or per-unit value
L_d	direct axis synchronous inductance [H] or per-unit value
L_{dD}	mutual inductance between the stator equivalent winding on the d-axis and the direct equivalent damper winding [H] or per-unit value
L_{dF}	mutual inductance between the stator equivalent winding on the d-axis and the field winding (in practice L_{md}) [H] or per-unit value
L_s'	transient inductance [H] or per-unit value
L_d'	direct axis transient inductance [H] or per-unit value
L_d''	direct axis subtransient inductance [H] or per-unit value
L_f	total inductance of the field winding [H] or per-unit value
L_F	inductance of the DC field winding [H] or per-unit value
$L_{f\sigma}$	leakage inductance of the field winding [H] or per-unit value
L_k	short-circuit inductance [H] or per-unit value
$L_{k\sigma}$	mutual leakage inductance between the field winding and the direct damper winding, i.e., the Canay inductance [H] or per-unit value
L_m	magnetizing inductance [H] or per-unit value
L_{md}	magnetizing inductance of an m-phase synchronous machine, in d-axis [H] or per-unit value
L_{mn}	mutual inductance [H] or per-unit value

L_{mq}	quadrature magnetizing inductance [H] or per-unit value
L_Q	total inductance of the quadrature damper winding [H] or per-unit value
$L_{Q\sigma}$	leakage inductance of the quadrature damper winding [H] or per-unit value
L_{pd}	main inductance of a single phase [H] or per-unit value
L_p	main inductance of a single phase [H] or per-unit value
L_q	quadrature axis synchronous inductance [H] or per-unit value
L_q''	quadrature axis subtransient inductance [H] or per-unit value
L_{qQ}	mutual inductance between the stator equivalent winding on the q-axis and the quadrature equivalent damper winding (in practice L_{mq}) [H] or per-unit value
L_0	equivalent inductance [H] or per-unit value
L_{m0}	magnetizing inductance at no load at the rated stator flux level or per-unit value
LSRM	linear switched reluctance machine
L_s	stator synchronous inductance [H] or per-unit value
$L_{s\sigma}$	stator leakage inductance [H] or per-unit value
L_{01}, L_{02}	inductance of the filter [H] or per-unit value
L'	transient inductance [H] or per-unit value
L''	subtransient inductance [H] or per-unit value
L1, L2, L3	network phases
l	length [m], magnetizing route [m], distance [m], relative inductance, distributed inductance [H/m]
l_{cr}	critical cable length [m]
l_e	effective core length [m]
l'	equivalent core length, effective machine length [m], inductance per unit length [H/m]
M	mutual inductance [H], or per-unit value
M	modulation index
MMF	magnetomotive force [A]
m	number of phases, mass [kg]
m_a	amplitude modulation ratio
m_f	frequency modulation ratio
m'	phase number of the reduced system
m_0	constant
N	number of turns in a winding, magnetic north pole, code (nondrive end)
N_p	number of turns of one pole pair
NdFeB	neodymium iron boron permanent magnet
NEMA	National Electrical Manufacturers Association
NPC	neutral point clamped (inverter)
$\boldsymbol{n}$	normal unit vector of the surface
n	number of teeth, number of units determined by the subscript
pu	per unit
P	power, losses [W] or per-unit value
P_{ef}	effective power [W]
P_e	electrical power [W] or per-unit value
P_{el}	electrical power [W] or per-unit value
P_{in}	input power [W] or per-unit value

P_{mec}	mechanical power [W] or per-unit value
PE	protective earth wire of the motor cable
PID	proportional-integrating-differentiating controller
PMSM	permanent magnet synchronous motor (or machine)
PWM	pulse-width-modulation
PM	permanent magnet
PMaSynRM	permanent magnet assisted synchronous reluctance motor
MTPV	maximum torque per volt
MTPA	maximum torque per ampere
P_ρ	friction loss [W] or per-unit value
p	number of pole pairs
Q	electric charge [C], number of slots
q	number of slots per pole and phase, instantaneous charge, $q(t)$ [C]
R	resistance [Ω] or per-unit value
R_{ball}	resistance of the ball of the ball bearing [Ω]
R_{ri}	resistance of the inner race of the ball bearing [Ω]
R_{ro}	resistance of the outer race of the ball bearing [Ω]
R_D	resistance of the direct damper winding [Ω] or per-unit value
R_E'	representing the part of mechanical power associated with R_r
R_f	resistance of the field winding [Ω] or per-unit value
R_F	resistance of the field winding [Ω] or per-unit value
RM	reluctance machine
RMS	root mean square
R_s	stator resistance [Ω]
R_Q	resistance of the quadrature damper winding [Ω] or per-unit value
R_{01}, R_{02}	resistance of the filter [Ω] or per-unit value
r	radius [m], distributed resistance [Ω/m]
$\boldsymbol{r}$	radius unit vector
S1–S9	duty types of electrical machines
S	apparent power [VA], or per-unit value, surface [m²]
S	switch, magnetic south pole
SM	synchronous motor
SR	switched reluctance
SRM	switched reluctance motor
SynRM	synchronous reluctance motor
SVM	space vector modulated inverters
S_{st}	maximum permitted starting apparent power [VA] or per-unit value
S_{PE}	power processing ability required by power electronics [VA] or per-unit value
S_U, S_V, S_W	switching function variables
SmCo	samarium cobalt permanent magnet
SynRM	synchronous reluctance machine
$s,$	slip, Laplace domains operator
s_b	slip at T_b
s_0	base slip value
T	temperature [K] [°C], duration [s], torque [Nm], cycle time of the oscillation [s]
T1, T2	transistor 1, transistor 2

TEFC	totally enclosed fan cooled
T_{sub}	duration of the subsequent of the modulation [s]
ΔT	temperature rise [K] [°C]
T	torque space vector [Nm] or per-unit value
T_b	pull out torque, breakdown torque [Nm] or per-unit value
T_{em}	electromagnetic torque [Nm] or per-unit value
T_e	electromagnetic torque [Nm] or per-unit value
T_L	load torque [Nm] or per-unit value
T_{max}	maximal torque [Nm] or per-unit value
T_N	nominal, rated torque [Nm] or per-unit value
$T_{pull\text{-}in}$	synchronizing torque [Nm] or per-unit value
T_s	starting torque [Nm] or per-unit value
T_{wL}	working torque of the load [Nm] or per-unit value
t_0	operating period [s]
t_c	commutation period [s]
t_{cef}	effective cooling period [s]
T_1	starting torque, locked rotor torque [Nm] or per-unit value
T_u	minimum torque [Nm] or per-unit value
T_I	integrating time constant [s]
T_D	differentiating time constant [s]
t	time [s]
t	tangential unit vector
t_j	cycle time [s]
t_p	time of pulse propagation (wave propagation time) [s]
t_r	rise time, duration of converter pulse [s]
U	voltage [V], RMS
U_d	supply voltage [V]
$U_{DC,meas}$	measured intermediate voltage [V]
U	depiction of a phase
u	voltage, instantaneous value $u(t)$, incoming voltage [V] or per-unit value
u_{cm}	common mode voltage (star point voltage) [V] or per-unit value
u_r	reflected voltage [V] or per-unit value
u_{drop}	voltage drop estimation error [V] or per-unit value
u_2	forward travelling voltage [V] or per-unit value
u_{DCmE}	voltage from DC link midpoint to PE [V] or per-unit value
Δu	voltage drop [V] or per-unit value
$\Delta U_{DC,offs}$	offset voltage [V] or per-unit value
V	depiction of a phase
VDE	Verband Deutscher Elektroingenieure
VRM	variable reluctance motor
VSI	voltage source inverter
v	speed, velocity, wave velocity, propagation speed of the voltage pulse [m/s]
v	vector
W	energy [J], coil span (width) [m]
W	depiction of a phase
W^*, W^x	magnetic co-energy [J]
W_e	magnetic energy (energy stored in magnetic field) [J]

W_{mec}	mechanical work [J]
W_{mt}	energy converted into mechanical work when the transistor is conducting [J]
W_{md}	mechanical work when the diode is conducting [J]
W_{fc}	energy stored in the magnetic field [J]
W_R	energy returning to the voltage source [J]
W_d	energy returned through the diode to the voltage source [J]
w_{ins}	thickness of the insulation layer,[m]
w_{Fe}	thickness of the iron layer,[m]
X	reactance [Ω]
x	coordinate, axis
Y	admittance [S]
y	axis
Z	impedance, nonlinear impedance of the ball bearing [Ω]
Z_m	characteristic impedance of the motor cable [Ω]
Z_s	characteristic impedance of the filter [Ω]
Z_{s01}, Z_{s02}	impedance of the filter [Ω]
Z_0	characteristic impedance [Ω]
z	coordinate, length [m]
z_Q	number of conductors in a slot
α	angle [rad] [°], coefficient, temperature coefficient, relative pole width of the pole shoe
α_i	factor of the arithmetical average of the flux density
α_{PM}	relative permanent magnet width
α_{SM}	relative pole width coefficient for synchronous machines
β	angle [rad] [°]
Γ	energy ratio, cylinder that confines the rotor, integration route
γ	angle, rotor angle [rad] [°], coefficient
γ_c	commutation angle [rad] [°]
γ_D	switch conducting angle, dwelling angle [rad] [°]
γ_0	turn on switching angle [rad] [°]
δ	air gap (length) [m], load angle [rad] [°]
δ_{de}	equivalent air gap (slotting taken into account) in the d-axis [m]
δ_e	equivalent air gap (slotting taken into account) [m]
δ_{ef}	effective air gap (influence of iron taken into account) [m]
δ', δ_0	minimum air gap, (air gap in the middle of the pole shoe) [m]
δ'_0	minimum air gap, influence of slotting is taken into account [m]
δ'_d	equivalent direct axis air gap [m]
δ'_q	equivalent quadrature axis air gap [m]
δ_s	load angle [rad] [°]
δ_m	load angle [rad] [°]
$\Delta\delta_{ef}$	additional effective air gap caused by PM [m]
ε	permittivity [F/m], stroke angle [rad] [°], angle, correction term
ε_r	relative permittivity
ε_0	permittivity of vacuum $8.854 \cdot 10^{-12}$ [F/m]
η	efficiency
$\boldsymbol{\Theta}$	current linkage vector [A] or per-unit value
Θ	current linkage [A], angle [rad] [°]

θ	angle [rad] [°]
ϑ	angle [rad] [°]
κ	angle, current angle [rad] [°], vector position in a sector
λ	angle [rad] [°],
μ	permeability [Vs/Am],
μ_r	relative permeability
μ_0	permeability of vacuum, $4 \cdot \pi \cdot 10^{-7}$ [Vs/Am] [H/m]
ν	pulse velocity [m/s], ordinal of harmonic
Π	surface [m^2]
ρ	resistivity [Ωm], reflection factor (coefficient)
ρ_ν	transformation ratio for IM impedance, resistance, inductance
σ	leakage factor, ratio of the leakage flux to the main flux, Maxwell stress [N/m^2]
σ_F	tension, tension force [Pa]
σ_{Fn}	normal tension [Pa]
σ_{Ftan}	tangential tension [Pa]
σ_{mec}	mechanical stress [Pa]
τ	relative time, transmission coefficient, control bit (torque or flux linkage)
τ'_{d0}	direct axis transient time constant with an open-circuit stator winding [s]
τ'_d	direct axis transient time constant [s]
τ''_d	direct axis subtransient time constant [s]
τ''_q	quadrature axis subtransient time constant [s]
τ''_{q0}	quadrature axis subtransient time constant with open-circuit stator winding [s]
τ_p	pole pitch [m]
τ_v	phase zone distribution
τ_A	armature time constant [s]
τ_{mec}	mechanical time constant [s]
$\boldsymbol{\Phi}$	magnetic flux space vector [Wb] [Vs] or per-unit value
Φ	magnetic flux [Wb] [Vs]
Φ_δ	air gap flux [Wb] [Vs]
Φ_h	main magnetic flux [Wb] [Vs]
ϕ	magnetic flux, instantaneous value $\phi(t)$ [Wb] [Vs],
φ	phase shift angle, power factor angle [rad] [°]
ψ	magnetic flux linkage [Vs] or per-unit value
ψ_h	flux linkage of a single phase [Vs] or per-unit value
ψ_m	air-gap flux linkage [Vs] or per-unit value
$\psi_{s,u}$	stator flux linkage integrated from the converter voltages [Vs] or per-unit value
$\psi_{s,i}$	stator flux linkage calculated from the current model [Vs] or per-unit value
ψ_{s0}	initial flux linkage ($\sim\psi_{PM}$) [Vs] or per-unit value
ψ_A	armature reaction flux linkage [Vs] or per-unit value
ψ_C	compensating winding flux linkage [Vs] or per-unit value
ψ_B	commutating pole winding flux linkage [Vs] or per-unit value
ψ_F	field winding flux linkage [Vs] or per-unit value
ψ_{PM}	permanent magnet flux linkage [Vs] or per-unit value
ψ_{tot}	total flux linkage of the machine [Vs] or per-unit value
Ω	mechanical angular speed [rad/s] or per-unit value

Ω_{hs}	speed of high speed area starts at Ω_{hs} or per-unit value
ω	electric angular velocity [rad/s], angular frequency [rad/s] or per-unit value

Subscripts

0	section
1	primary, fundamental component, beginning of a phase, locked rotor torque, phase number
2	secondary, end of a phase, phase number
3	phase number
a	armature, shaft
A	armature
arm	armouring
ad	additional
av	average
act	actual
b	base value, peak value of torque, blocking
bar	concerning bar
bearing	concerning bearing
C	capacitor
c	conductor, commutation
cp	constant power
calc	calculated
corr	correction
cr, crit	critical
D	direct, damper
d	direct, direct axis, distribution
DC	direct current
E	back emf (electromotive force)
e	electrical, electric
eff	effective
el	electric, electrical
em	electromagnetic
err	error
est	estimate
ext	external
f,	field, filter
filt	filtered
F	force, field
Fe	iron
grid	concerning a grid
i	internal
k	short circuit, ordinal
L	load
LL	line to line
m	mutual, main, motor, mechanical
M	motor

mag	magnetizing, magnetic
max	maximum
mec	mechanical
meas	measured
min	minimum
N	rated
n	nominal, normal, normalized, normalization, orthogonal component
non-sal	nonsaliency
new	new value
old	old value
p	pole, pitch
ph	phase
pu	per unit value
PM	permanent magnet
q	quadrature, quadrature axis, zone
r	rotor, rotor reference frame
ref	reference
res	reserve
s	stator
sal	saliency
sk	skewing
slipring	concerning a slip ring
sub	subtransient
sum	vector sum of currents
syn	synchronous
sw	switching
t	tangential
tan	tangential
tot	total
tr	transient
triangle	triangle waveform
u	pull-up torque
v	zone, coil
w	end winding leakage flux
x	x-direction, axis
y	y-*direction, axis*
z	z-direction, phasor of voltage phasor graph
δ	air gap
Φ	flux
ν	harmonic
σ	leakage

Superscripts

^,	peak/maximum value, amplitude
'	imaginary, apparent, reduced, referred, virtual, transient
*	base winding, complex conjugate

s   stator reference frame
r   rotor reference frame
g   general reference frame

Boldface symbols are used for space vectors

i current space vector, $\boldsymbol{i} = i_x + j i_y$, $\boldsymbol{i} = i e^{j\theta}$ [A] or per-unit value
i absolute or per-unit value of current space vector
I complex RMS phasor of the current

1

Introduction to electrical machine drives control

Few technologies are more important to our collective quality of life than electrical drive technology. One could say that electric motors drive and electric generators power the world. Further, power electronics offers an indefatigable tool for accurate power conversion. And it seems the importance of the technology is poised to rise to even greater heights in the course of the next few decades as more reliable, more cost effective, and more flexible electrical drive systems become available.

For more than a century, electrical machine drives have been powering production processes for numerous industries. Applications include pumping, ventilation, compression, milling, crushing, grinding, conveying, and transporting. In modern robot-dependent manufacturing systems, electrical drives are responsible for precise position control of various robot arms and end effectors.

Concerns about air quality in cities and the increasing demand for improvements in energy efficiency favour using even more electric or hybrid vehicles for transportation needs. The current rate of change toward even more electromobility is limited only by today's high price of electric storage technology. The electrical drives themselves, that is, the motors and converters, are more than sufficient to serve as a replacement for the existing internal combustion engines in cars and buses.

Today, more than 50% of the world population lives in urban areas, and that percentage is growing. This growth in population powers increasing demand for more and better methods of moving people, materials, and things. Electrical machine drives are becoming an increasingly essential element of these transportation applications. Globalization, the accelerating process of international integration, puts added demand on sea and air transport, and ships and even aircraft are relying more and more on the most up-to-date electrical drive systems.

Electrical Machine Drives Control: An Introduction, First Edition. Juha Pyrhönen, Valéria Hrabovcová and R. Scott Semken.
© 2016 John Wiley & Sons, Ltd. Published 2016 by John Wiley & Sons, Ltd.

In addition, the average age of the world population is advancing at a rate unparalleled in human history. By 2050, the elderly will account for 16% of the global population. Caring for these 1.5 billion senior citizens over the age of 65 will strain the world's existing healthcare infrastructure. Fortunately, intelligent machinery has the potential to address the needs of the ageing population and to ease this demographic challenge. As the sinew of intelligent machinery, the increasing importance of electrical machines drives again seems to be clear.

Climate change is also bringing about ever more troubling environmental challenges. Permafrost in Siberia is melting and releasing methane into the atmosphere, there are stronger and increasingly damaging storms, and many drought areas are experiencing unprecedented levels of dryness. The burning of carbon-based fossil fuels to produce both electrical and motive power has been identified as a major contributor to climate change, and moving toward electrical power production technologies that do not burn fuels is a possible solution. Electrical generator drives are essential components of several of the more climate-friendly power production options currently available such as hydro, wind, and geothermal. Moreover, electric vehicles, a green alternative to fuel-burning cars, buses, and trucks, also rely on electrical motor drives.

At present, electric motors are the world's single biggest consumer of electricity, accounting for about 70% of industrial power consumption and nearly 45% of total global electricity consumption. Most in service are polyphase current (AC) induction motors, which are inexpensive and easy to maintain and can be directly connected to an AC power source. However, the majority of these AC induction motors lack flexible speed control, so they are not being used as efficiently as possible. Modern electrical drive technology is beginning to offer more cost-effective solutions with excellent speed control, making it possible to significantly improve efficiencies and minimize power consumption. These developments will encourage the replacement of AC motor systems in existing applications and the implementation of modern electrical drives for any new ones.

1.1   What is an electrical machine drive?

The word *drive* comes from the Anglo-Saxon word *dríf-an*, which was a verb meaning *to urge (an animal or person) to move*. It is used as a noun here that can be defined as *the means for giving motion to a machine or machine part*. Therefore, an *electrical drive* can be defined as an electrical means of imparting motion. When an electrical drive is operated in reverse, it becomes a means of harnessing motion to generate electricity. To be more specific, when an electrical drive is driving, it can be referred to as an *electrical motor drive*. When it is driven, it can be referred to as an *electrical generator drive*.

Depending on the application, electric machines often operate in both motoring and generating modes. And, often, there is no technology difference between an electrical motor drive and an electrical generator drive. For example, the electric drive motor that propels an electric train or automobile—referred to as a traction motor—must run forward and backward and brake in both directions.

Electrical machine drives can be categorized as either noncontrolled or controlled motor or generator drives. Most motor drives working in industrial applications are noncontrolled. Almost exclusively, these are three-phase AC induction motors with direct on line (DOL) or across the line starting. Large-scale power generation mostly uses DOL drives based on synchronous generator drives.

To improve performance and efficiency, many applications are making use of controlled electrical drives. Controlled electrical motor drives are starting to become more popular in cases where the drives are tied into an industrial automation system. Distributed generation is driving demand in electrical power industries for speed-controlled electrical generator drives. In wind power, for example, so-called full power converters are becoming more common where both the generator and the network connection are fully controlled via power electronics.

1.2   Controlled variable speed drives

The primary function of any variable speed drive is to control speed, force production, acceleration, deceleration, and direction of movement, whether it be rotary or linear. Unlike constant speed electric machines, variable speed drives can smoothly change speed to anywhere within their design operating range, and this adjustability makes it possible to optimize production processes for improved product quality, production speed, or safety.

Electrical variable speed drives are offered in a number of basic types, but the two most versatile for general purpose applications, and therefore the most common, are direct current (DC) drives and adjustable frequency AC drives. An electrical variable speed drive typically includes the following three principle elements.

The *high-level controller* enables (a) the operator to start, stop, and change speed via a human-machine interface (HMI) using buttons, switches, and potentiometers or (b) a plant control and set point master computer to send similar commands.

The *drive controller* converts the fixed voltage and frequency of an AC power source into adjustable power output to control the electric drive motor over its range of speeds.

The *drive motor* transforms electrical energy into motor movement. Shaft rotation or linear actuator movement speed varies with power applied by the drive controller.

1.2.1   DC variable speed drives

DC drives are motor speed control systems based on DC motors or generators.

In a traditional rotary DC motor, the rotor (armature) spins inside a magnetic field that is initially produced either electromagnetically or via attached permanent magnets (PMs). The most common electromagnetic approach is to supply the field and armature windings separately. The result is referred to as a separately excited DC motor. If, instead, the no-load magnetic field is produced using PMs, the result is referred to as a PMDC motor. Separately excited and PMDC represent two of the more important and commonly used DC motor types.

In a separately excited and compensated DC motor, speed is directly proportional to the voltage applied to the armature and inversely proportional to motor flux, which is a function of field current. As a result, speed can be controlled via either armature voltage or field current. In a PMDC motor, speed is also directly proportional to the applied voltage. However, since the PMDC magnetic field remains constant, PMDC motor speed cannot be increased beyond the rated speed by reducing armature field current.

DC drive control

The speed and torque of a DC motor are independent. Speed is proportional to the applied voltage, and torque is proportional to the applied current.

As in all drives, power varies in direct proportion to speed. That is, 100% rated power is developed only at 100% rated motor speed with rated torque. Constant power over a specified speed range is needed for some applications. An armature-controlled DC drive can deliver less-than-maximum nearly constant power over a portion of its operating speed range. Because it is a function of speed, the level of power available depends on where in the speed range it is needed. For example, a particular drive might be capable of delivering 50% of its maximum power from 50% to 100% of its rated speed, so if 4 kW was needed over the upper half of the drives speed range, an armature-voltage–controlled drive rated for 8 kW would be required.

In addition to being armature-voltage controllable, the performance of separately excited DC drives can be influenced by changes in field current. Normally, they operate using a constant field excitation, but they can be pushed over their rated speed by reducing field flux beyond the rated speed point. This is called field weakening.

The advantages of the DC drive

Brushed DC motors are more complicated than AC motors and require more maintenance. Their most vulnerable component is the mechanical commutator, which acts as a mechanical inverter in a motor or a mechanical rectifier in a generator. The maximum speed of a DC motor depends on its mechanical endurance, which may be limited because of the commutator and brushes. Some of the disadvantages of the traditional DC motor can be overcome with a brushless DC motor architecture. The brushless DC motor moves the armature to the stator side and uses power-electronic commutation. Its architecture is similar to that of a PM synchronous AC motor.

The primary advantages can be summarized as follows.

- DC drives can be less complex and less expensive for most power ratings.

- DC drives can provide starting and accelerating torques exceeding 400% of rated (Sowmya, 2014).

- DC drives are able to control speed over a wide range (above and below rated speed).

- DC drives can be quick starting, stopping, reversing, and accelerating.

- DC drives offer accurate speed control and a linear speed-torque curve.

- DC drives dominate in sub-kilowatt power applications.

- DC drives are easier to understand for maintenance and operations personnel.

1.2.2   AC variable speed drives

AC drives are machine speed control systems based on AC motors or generators. AC motors typically operate using three-phase AC. Single-phase supplied AC induction motors are also widely used for lighter duty applications. The motors can be rotary or linear. In general, the controller characteristics are the same for either. For clarity, the following discussion focuses on rotary AC motor drives.

A rotary AC motor has a stationary stator and a spinning rotor. The stator is wound with a circular array of conductor coils (the windings) that produces static lines of current and a rotating magnetic field. The rotor carries lines of current that also produce a magnetic field. Both rotate as the rotor spins. The interaction between the rotor or stator currents and the common rotating magnetic field is responsible for the force production (torque) of the motor. Depending on motor type, the rotor currents may be produced via electromagnetic induction or via an active set of rotor windings. In a PM machine, the function of the stator is the same. However, the PM rotor lacks the lines of current and only contributes a spinning magnetic field. In analysis, the PM can be replaced by an equivalent current, if needed. The stator currents and the common rotating magnetic field are responsible for force production in a PM machine.

The two most common AC motor types are induction motors and synchronous motors, each with a number of variations.

The induction motor

An induction motor (also called an asynchronous motor) relies on a slight difference in speed between the rotating magnetic field of the stator and the rotating speed of the rotor to induce current in the rotor's AC windings or integral conductive squirrel cage. This difference in speed is referred to as *slip*.

Single-phase supplied AC induction motors are often two-phase capacitor-run motors. They can exhibit good performance properties for a particular working condition. Because induced coil currents produce a virtual second phase during operation, shaded-pole induction motors act like two-phase motors with their virtual second phase working as a short-circuit winding that produces a rotating field component in the air gap to start the motor. The single-phase motor types are not excellent performers. In general, they are not as efficient as multiple-phase induction motors; however, they are ubiquitous in both industrial and household settings, because of their simple construction, low cost, and reliability and because single-phase voltage sources are readily available. Single-phase frequency converters and small three-phase motors are available if speed-controllable single-phase motor drives are required. Naturally, this should be the trend to enhance energy efficiency.

Three-phase induction motors are the workhorses of industry. The two most common types use either active rotor windings or a rotor squirrel-cage architecture. Because in the first type AC current is transmitted to the active rotor windings via slip rings, it is commonly referred to as a slip-ring induction motor. The second type is referred to as a squirrel-cage induction motor.

Slip-ring induction motors equipped with external rotor resistors have high starting torque, smooth acceleration under heavy loads, adjustable speed, and good running characteristics. Traditionally, they have been used in applications such as lifts, cranes, and conveyors. More recently, their general popularity and market share have dropped off significantly. However, the doubly fed induction generator, a slip-ring machine, remains the most popular generator type for wind turbines. Squirrel-cage induction motors are simpler and rugged in construction. They are relatively inexpensive and require little maintenance. They are the preferred choice for lathes, drilling machines, pumps, and compressors, among other applications.

The synchronous motor

In contrast to the induction motor, a synchronous motor does not rely on slip induction. The magnetic poles of its rotor remain magnetically locked with the rotating air-gap magnetic field, which is synchronous with the frequency of the AC supply current. In a synchronous motor, the rotor poles are produced via an active set of windings or a circular array of PMs.

Synchronous motors are available with power ratings from less than 1 kW to tens of megawatts. Typically, sub-kilowatt synchronous motors are used in applications where a precise constant speed is needed, such as in clocks, timers, and tape players. Above 10 kW, the main benefits of synchronous motors are their high efficiency and an ability to provide power-factor correction. Larger synchronous motors can be found in higher-powered fans and blowers. Three-phase synchronous motors are also being used as traction motors for electric vehicles. The best-known example of synchronous traction motor use in an electric vehicle is France's high-speed TGV trains (for *Train à Grande Vitesse*, which is French for "high-speed train"). The largest synchronous motor drive is the ABB-supplied 101-MW wind tunnel drive motor owned and operated by the USA National Aeronautics and Space Administration (NASA).

AC drive control

There are various methods used to transform incoming AC power into the adjustable form of input needed to control AC motor speed and torque. Two well-known examples are pulse-width modulation (PWM) and six-step or trapezoidal waveform conversion.

PWM varies the average value of voltage (and current) by rapidly switching (on and off) the voltage input to the motor. The relative duration of the on and off periods, referred to as the duty cycle, determines the amount of voltage supplied. Long on periods and short off periods correspond to high voltage, and short on periods and long off periods correspond to low voltage. Duty cycle is expressed in percentage with 100% being maximum voltage. PWM assumes inductive loads. With inductance, energy can be stored within the magnetic circuit to maintain relatively smooth current in response to the PWM supply.

For an AC drive to operate smoothly in response to PWM, the motor must receive on-off switching pulses that are short relative to the time it takes for the load to respond. The PWM resultant waveform must appear smooth to the load. Typically, AC drive switching frequencies can be from a few to tens of kHz.

Power electronics (switching devices)

To implement the appropriate power conversion method (to vary frequency and voltage of the motor input power supply), both AC induction motors and synchronous motors are increasingly being coupled with power electronics switching systems to produce variable-speed AC drives. For induction motors in variable-torque fan, pump, and compressor applications, the result offers significant and important energy savings opportunities. For large synchronous motors, the result also makes it easier to get the massive rotors moving.

These power electronics can be classified based on the different topologies. By far the most common AC drives are voltage-source inverter (VSI) drives. Other topologies include the current-source inverter (CSI), the load commutated inverter (LCI), and the cycloconverter (CCV).

In a VSI, the DC output of a diode-bridge converter stores energy in a capacitor bus to supply voltage input to an inverter. In place of a diode rectifier, AC conversion can also be accomplished using a more complicated active-switch converter capable of four-quadrant operation and two-directional power flow. Most AC drives are VSI with PWM voltage output. In a CSI, the DC output of an SCR-bridge (silicon controlled rectifier, i.e., thyristor) converter stores energy in a series-reactor connection to supply current input to an inverter. CSIs typically output PWM or a six-step waveform. In an LCI drive, the DC output of an SCR-bridge converter stores energy via a DC-link inductor circuit to supply a second SCR-bridge inverter. LCIs output quasi-sinusoidal six-step current. The CCV is a direct frequency converter approach that transforms an incoming AC waveform of constant frequency and voltage into an outgoing AC waveform of varying frequency and varying voltage.

The advantages of the AC drive

The primary advantages of the speed-controlled AC drive can be summarized as follows.

- AC drives can be smaller and lighter for higher power ratings.

- AC drives can accommodate widely varying loads or extended low-load operation.

- AC drives can operate more easily at higher speeds (over 3000 or 3600 rpm).

- AC drives need less apparent starting power and offer lower-cost electronic motor reversing.

- AC drives offer better speed and torque control.

- AC drives are more common in high power applications.

- AC drives require less maintenance are a better choice when access is limited.

1.3 Electrical machine drive implementation

Figure 1.1 compares typical configurations for a controlled and noncontrolled drive. The shaded areas represent the core electrical drive components. In each case, the primary inputs are power, either electrical or mechanical, and the control signal, which could be speed, torque, or position for the controlled drive or simply an on/off command for the noncontrolled drive. Output is the mechanical power of the output shaft for a motor drive or the generated electricity for a generator drive.

Instrumentation monitors the state of the motor by measuring critical parameters such as rotation speed, electric current, and operating temperature. The measurement signals are transmitted to the controller, which issues control signals to the power electronics based on the input references and transducer signals. Based on signals from the controller, the power

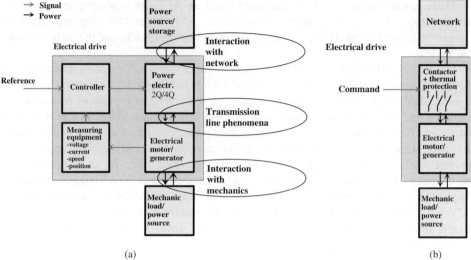

(a)   (b)

Figure 1.1 Block diagrams illustrating the principles of a controlled and noncontrolled electrical drive where (a) is a drive with a two- or four-quadrant 2Q/4Q power electronic converter and (b) is a direct-on-line drive. The input of an electrical drive consists of electrical or mechanical power and an operational reference. Output is the mechanical or electrical power transmitted either to a mechanical or electric load. In principle, a controlled electrical drive consists of power electronics, an electric motor or generator, monitoring instrumentation, and a controller. A noncontrolled electrical drive consists of a contactor with protective thermal relays and the electric motor or generator. In either case, there are interactions between the drive system and the network, between the power electronic converter and the electric machine (transmission line phenomena), and between the electric machine and the mechanical system.

electronics either converts incoming electrical power to supply motors that produce mechanical power or converts outgoing electrical power to supply a storage system or network in response to incoming mechanical power.

Present-day converters all apply some kind of a modulation technology to convert electrical power from one form to another. The most common is PWM, with which, for example, incoming DC voltage can be converted to a suitable AC voltage to accommodate an AC motor. PWM applies constant voltage in pulses of a given duty cycle (pulses/cycle) and frequency (cycles/second). For electrical drives, PWM depends on the system having a suitable level of inductance to filter the PWM input and to achieve smooth electric current curves despite the chopped voltage.

As implied by the blocks shown in Figure 1.2, the expert in electrical drives should exhibit competence in several technical areas. Obviously, expertise in energy technology, electric machines technology, electrical power network technology, measurement technology, mechanics, control engineering, thermodynamics, and telecommunications technology is required. How the electrical drive system interacts with its mechanical and electrical

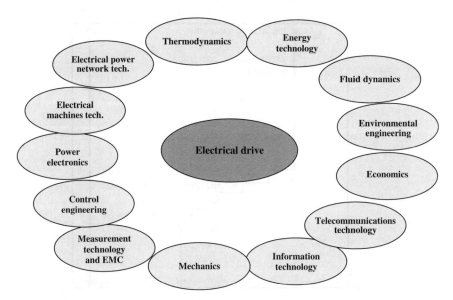

Figure 1.2 Fields of operation of an electrical drives expert. EMC, electromagnetic compatibility.

interfaces also must be understood. Figure 1.2 illustrates the main fields of operation and the areas of expertise of an electrical drives expert.

Telecommunications technology is perhaps not the first thing that comes to mind when considering electrical drives; however, present-day drives such as those used for ship propulsion and wind power are operated and monitored remotely. For example, the condition of electrical drives in ships is monitored in land-based centres via telecommunication protocols utilizing satellites. Similarly, machine systems such as harbour cranes are monitored via the Internet by the manufacturer for maintenance and product development purposes.

1.4   Controlled electrical drives and energy efficiency

The growing use of controlled electrical drives significantly affects overall energy use and energy technology itself. They play an increasingly significant role in the management of global energy consumption. Controlled electrical drives improve energy efficiencies by more accurately controlling process energy flow. The following paragraphs will discuss some of these aspects in more detail.

A great many human activities depend on electromechanical energy conversion. It is particularly important to many industrial processes and for the housing, commerce, and transportation sectors as well. In the early 21st century, global electricity consumption reached 18,500 TWh/a, and current forecasts suggest it will reach 30,000 TWh/a in the near future. Electrical drives will represent 50 to 60% of this total, so it is easy to understand that

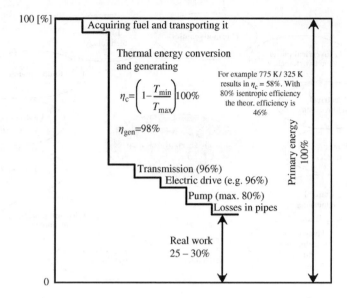

Figure 1.3 Chain of efficiencies in electricity production (from chemical or nuclear energy), transmission, and final use indicating low total efficiency. System performance can be considered acceptable if 25% to 30% of the primary energy can be used to perform useful work; however, actual performance can be significantly lower.

improving electrical drive efficiency is a key factor in reducing consumption and the associated emissions of carbon dioxide (CO_2).

Electricity is still produced mostly by rotary electric machines driven by incoming mechanical power provided by turbines, internal combustion engines, and the like. Electricity production from solar cells and fuel cells is increasing, but so far, generators remain the most important means of energy conversion. Furthermore, the strong future predicted for wind power ensures a continuing important role for rotary electric machines in power generation.

Since most electricity is still produced by burning fossil fuels or by nuclear power, it is important to understand the relationship between primary energy demand and final process efficiency. Figure 1.3 illustrates how the efficiency of each link in a chain of energy conversions leads to a surprising low resultant total efficiency for the electrical energy system of a condensing power plant.

The biggest losses are due to the inherently low efficiencies of thermal cycles producing mechanical work from thermal energy. The maximum theoretical efficiency η_c of a thermomechanical energy conversion is known as the Carnot efficiency and is defined as follows.

$$\eta_C = 1 - \frac{T_{\min}}{T_{\max}} \tag{1.1}$$

Absolute temperatures T_{min} and T_{max} are the minimum and maximum of the thermal cycle. Using a steam power plant as an example, the temperature of the condensed water supplied to the boiler would be T_{min}, and the temperature of the supercharged steam leaving the boiler would be T_{max}. In Figure 1.1, these temperatures are 325 K and 775 K, respectively. The maximum corresponds to the practical upper limit of present-day steam power plants. Cooling the condensed water to the minimum 325 K also is within typical capabilities. For these temperatures, however, Equation (1.1) yields a maximum theoretical efficiency of 58%. When all efficiency-limiting factors are taken into account (e.g., the isentropic efficiency of the turbine (typically 80%) and the power needed to run the power plants components), the actual energy conversion efficiency drops to within the range of 40% to 45%. In combined cycles (e.g., a gas turbine system with waste heat boiler plus steam turbine), it is possible to reach 60% electricity production efficiency, and in places where the heat of the process can be utilized, it is possible to reach 80% total efficiency. Plants that utilize process heat in this way are referred to as combined heat and power (CHP) plants. CHP technology is prevalent and most efficient in northern countries like Finland where water and housing must be heated year round.

The chain of efficiencies illustrated in Figure 1.3 suggests that electricity, which is a highly refined energy carrier, should be reserved for applications that cannot be served using other forms of energy transmission. An electrical drive is one such application. With this in mind, some countries prohibit the use of electricity for direct resistive heating, for example.

A system using a heat pump driven by a PM synchronous motor drive is capable of producing as much as seven units of heat output with just one unit of electricity input (the performance coefficient of the heat pump is typically 3 to 7). Looking again at Figure 1.3 reveals that a heat pump system based on an electrical drive may produce more heat than can be produced by burning fuel directly. If 30% of the original primary energy goes to driving the heat pump and if the performance coefficient is seven, then $0.3 \times 7 = 2.1$, or 210% goes to heat the process. Therefore, if electricity is to be the energy transfer medium used for heating, using an electrical drive to run a heat pump to produce heat energy that is 90 to 210% of the primary energy is better than using direct resistive heating to produce heat energy that is just 30% of the primary energy.

The final process in the chain, pumps and pipelines for the system in Figure 1.3, should have as high efficiency as possible to avoid excessive primary energy consumption. Therefore, the electrical drives expert and the pump specialist should collaborate to select the best possible technology for the final drive. The motor drive, the pump, and the pipelines must work together to achieve optimal final drive efficiencies.

Because electrical drives are more controllable than mechanical or hydraulic systems, their use in many fields of human activity is increasing. Moreover, since electrical drives are typically more energy efficient, they are becoming important from the point of view of energy efficiency.

The latest developments in electrical drives are in mobile equipment such as electric or hybrid vehicles and heavy working machines. Careful integration of electrical drive technologies into mobile equipment systems is resulting in significantly improved energy efficiency, better emission control, and better performance. Hybrid mobile equipment uses electrical motor drives to power the subsystems or the drive wheels. These motor drives are powered by an electrical generator drive, which in turn is powered by an internal combustion engine. Given the nearly constant speed and load required by an electrical generator drive, the internal combustion engine of a hybrid can be sized optimally and designed to operate cleanly at peak efficiency. For example, the efficiency of a modern low emission diesel engine, designed to power the generator of a hybrid excavator (50 to 500 kW), can easily reach an efficiency of 40%.

Electrical power is being produced more efficiently and cleanly, and modern frequency converters are improving the efficiencies of process control as well. Efficient process control can save energy and reduce emissions. To illustrate, one manufacturer of electrical machine drives, ABB, has reported that their installed base accounts for 445 TWh in energy savings; equivalent to the annual production of 56 nuclear reactors or the yearly consumption of more than 110 million households (ABB, 2015). Compared with fossil fuel–generated electricity, this represents a reduction in CO_2 emissions of approximately 336 billion kg, which is the amount emitted in a year by more than 90 million cars. ABB enjoys approximately 20% market share, so real benefits associated with worldwide electrical machine drive usage can be estimated to be roughly 8 EJ in energy savings and 1.7 trillion kg in CO_2 reduction.

1.5 The electrical drive as an element of a controlled industrial process

In industrial processes, effective speed or torque control is necessary to achieve reasonable energy consumption and high quality. This requirement becomes clear if the particular process requirements are analysed. Industrial processes can be divided into two main categories: material transformation and material transportation. In each case, being able to adapt to changing process requirements can lead to improvements in efficiency and quality. The precise control of speed in response to varying load conditions offered by controllable electrical drives provides this ability.

Presently, electrical drives are essential process components in a number of industrial areas, including the chemical industry; fabrication workshops; the plastics industry; the pulp, paper, and printing industries; the food and soft drink industries; mining; metals; and power plants. In addition, many electrical drives are being used for the heating, plumbing, and ventilation of buildings.

All industrial processes require both material and energy. Figure 1.4 shows the energy and material flows for a typical manufacturing process. Energy and material are consumed to output a product. The figure illustrates the energy and material inputs and depicts consumption as energy and material loss outputs. Processing can involve mechanical power, the electromagnetic effect, heating or cooling, chemical reactions, or biological reactions. Mechanical power is applied most effectively via speed-controlled electrical motor drives. With precision speed control available, processes can be developed that produce high-quality results with minimal materials and energy consumption.

Process equipment can be categorized as transport devices, equipment that moves materials from place to place, or processing devices, equipment that modifies the geometries or properties of materials.

Transport devices include various systems to move solid materials and devices for controlled delivery of liquids or gases. Device construction varies depending on

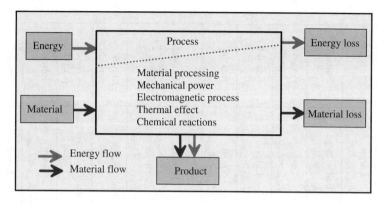

Figure 1.4 Energy and material flows in a typical production process.

material to be transported. Solids such as containers, metals, wood, minerals, or even human beings are moved by hoists, conveyors, lifts, and vehicles. Fluids such as water, oil, or liquid chemicals are transported by pumps through pipes or tubes. Gases are forced by blowers, compressors, or fans through ducts or tubes. A particular application of this kind is air conditioning.

Processing devices carry out a nearly unlimited variety of material modifications. Every electrical process device integrates energy control, an electric motor, mechanical power transmission, and the material modification machinery. An electrical drive provides the first three: energy control, motor, and transmission. It converts the electrical energy fed to the system into the required mechanical energy needed to carry out the processing function of the operating machine. In the most demanding processing tasks, servo drives are needed for precise position control. For example, electrical component placement machines used in circuit card assembly in electronics industry must be controlled precisely to place the components on a circuit card with a high degree of accuracy.

Speed control is an important element of electrical drives. It can be accomplished, for instance, by using a frequency converter to control energy flow or by using mechanical gears as to control power transmission. Frequency converters are popular, because they make it possible to control ordinary noncontrolled electric motors. Newer PM motors, which have greater torque density than induction motors, often can be applied directly without the need for power transmission gearing. Electrical drives are referred to as high-speed drives when the frequency converter feeds the electric motor at a frequency that is notably higher than the line frequency. These high-speed drives are gradually becoming more common.

Table 1.1 reviews some general industrial electrical drive applications and lists some of their more important behavioural characteristics. Table 1.2 lists traditional and more modern methods of achieving different speeds in electrical drives.

Electrical machine drives are the essential workhorses of modern industry. Their successful development and implementation demand expertise from multiple engineering disciplines. When the most appropriate electrical drive is applied correctly to carry out its targeted process function, energy efficiency can be optimized, and the cumulative effects throughout industry can bring substantial environmental benefit.

Table 1.1 Typical industrial (or alike) electrical drive applications

Application	Power [MW]	Typical Speed [min^{-1}]	Starting Torque T_s/T_N	Synchronizing Torque Demand (DOL) $T_{pull-in}/T_N$	Maximum Load Torque T_{max}/T_N	Moment of Inertia J_{load}/J_{motor}	Relative Smoothness of Torque
Grinder	2–20	1000–1800	Small <20%	Small <10%	Normal 1.5	Normal $J_{load} \sim j_{motor}$	Smooth
Chopping Machine	0.1–3	250–500	Small <20%	Small <10%	Large 2.5–3	Large $J_{load} > J_{motor}$	Large variations
Centrifugal compressor	…20	1000–1800 10,000–100,000	Normal <40%	Normal <40%	Normal <1.5	Fairly large $J_{load} > J_{motor}$	Smooth
Reciprocating compressor	…20	150–500	Normal <40%	Normal <40%	Fairly large 2	Large $J_{load} > J_{motor}$	Periodic variation
Blower	…15	300–1800	Small <20%	Normal <40%	Normal 1.5	Large $J_{load} > J_{motor}$	Smooth
Centrifugal pump	…10	500–1800	Normal <40%	Normal <40%	Normal 1.5	Small $J_{load} < J_{motor}$	Smooth
Vacuum pump - Nash - Sulzer	…3 …3	200–400 1000–1800 reduction gear	Small <20% Small <20%	Normal <40% Large ~100%	Normal 1.5 Normal 1.5	Normal $J_{load} \sim J_{motor}$ Fairly large $J_{load} > J_{motor}$	Smooth Smooth
Ore or cement mill	…10	150–500	Large >100%	Large ~100%	Fairly large 2	Normal, varies under load	May vary periodically
Position-controlled drives	…0.2	0–6000	High	—	Large	Normal, varies under load	Varies periodically
Propeller	…50	70–300	Small <20%	Normal <40%	Normal 1.5	Small $J_{load} < J_{motor}$	Smooth
Wind turbine	…10	10–1800	—	—	Normal 1.5	Large $J_{load} > J_{generator}$	Smooth
Mobile machine or train traction	0.05–5	1000–6000 Reduction gear	High	—	Normal 1.5	large $J_{load} > J_{motor}$	Varies under load
Ship propulsion	1–30	100–	Low	—	Normal	Normal $J_{load} \sim J_{motor}$	Smooth

Table 1.2 Traditional and modern speed control means

Control Method	Mechanical Variator	Hydraulic Coupling	Motor Pole Number Switching	Voltage Control	Slip-Ring Machine Cascade	Power-Electronics–Based Control
Power range	0–75 kW	15–12,000 kW	0–5000 kW	0–5 kW	100–10,000 kW	0–100,000 kW
Max speed	$4000\,\text{min}^{-1}$	$2900\,\text{min}^{-1}$	$3000\,\text{min}^{-1}$	$3000\,\text{min}^{-1}$	$3000\,\text{min}^{-1}$	$100,000\,\text{min}^{-1}$
Speed control range [%]	8–100	25–100	25–100 stepped	60–100	50–100	0–100
Speed of torque control	Slow	Slow	Slow	Slow	Fast	Very fast
Speed of speed control	Slow	Slow	Slow	Fairly fast	Fast	Fast

References

ABB. (2015). Energy efficiency. A solution. Retrieved from http://www.slideshare.net/ABBDrivesandControls/energy-efficiency-a-solution-48833156

Sowmya, M., et al. (2014). Closed loop speed regulation of DC motor using phase controlled thyristor converter. *International Journal of Engineering Sciences Research*, 5, article 06368. ISSN: 2230–8504; e-ISSN-2230-8512.

2

Aspects common to all controlled electrical machine drive types

This book examines existing and emerging electrical drives, offering thorough coverage of the variety of drive control methods. The electric machinery considered includes not only classic rotating machines, such as direct current (DC), asynchronous, and synchronous motors and generators, but also new electric machine architectures that have resulted from the simultaneous development of the electromagnetic machinery, the controller, and the power electronics combined with the introduction of new materials, such as permanent magnets. This chapter covers a number of aspects that are common to electric machine control in general.

2.1 Pulse width modulation converter electrical motor drive

AC drive control is based on frequency and voltage regulation. DC drive control is based on regulating voltage. In both cases, state-of-the art electrical drives make use of a pulse width modulation (PWM) principle. PWM is widely applied to convert electrical power from one form to another, for example, from DC to AC or vice versa. When applying PWM, two fundamental principles must be understood: a voltage source may not be short circuited, and a current source may not be open circuited. Furthermore, a good electric circuit design will have its input and output sides designed so that one is a voltage source and the other is a current source.

Electric machines are inductive. Inductance works to oppose changes in current by inducing instantaneous proportional opposing voltages. In effect, the inductance of an electric machine applies opposing induced voltages as needed to maintain constant

Electrical Machine Drives Control: An Introduction, First Edition. Juha Pyrhönen, Valéria Hrabovcová and R. Scott Semken.
© 2016 John Wiley & Sons, Ltd. Published 2016 by John Wiley & Sons, Ltd.

machine current, so it can be regarded as a current source. When PWM pulses are supplied to the terminals of an electric machine, its construction determines how it reacts to the pulses with its transient or subtransient inductance. Subtransient inductance manifests in synchronous machines with multiple rotor windings. These transient inductances must be at a suitable level when the PWM switching frequency is selected. Magnetic flux leakage resulting in leakage inductance is often regarded as an adverse phenomenon, but in reality, for example, an induction machine could not be supplied via a PWM pulse pattern in the absence of leakage inductance. Machine inductive behaviour will be discussed in more detail in later chapters.

The current-source nature of electric machinery suggests that power should be supplied as a voltage source. This is why voltage-source converters have become so popular. However, the current-source nature for some electric machines is unclear. One example is a rotor surface magnet servomotor with low synchronous inductance. A current-source converter may be a better solution to drive these machines. In practice, however, voltage-source converters dominate the market, and current-source converters are seen only in special cases.

In a voltage-source converter, the DC link voltage is converted into a three-phase AC voltage having the desired frequency and amplitude. The output of a PWM converter consists of sharp voltage pulses of varying durations. The phrase "instantaneous average" describes the idea of producing nearly sinusoidal currents by modulating the converter power switches so the instantaneously average output voltages follow sinusoidal patterns. There must be a suitable switching frequency, so the current-source nature of the motor can filter the sharp voltage pulses to produce acceptable current waveforms.

Induction motors are always high inductance. As a result, it works well with a voltage-source PWM converter. The induction motor remains the workhorse of the industry. It is overwhelmingly dominant in terms of number of units installed and in terms of total installed power. Moreover, it can be integrated with various electrical drive types. However, other motor types are increasing in importance for electrical drive applications. Various synchronous machines, in particular, have become more popular lately. Industrial DC-motor drives are losing popularity.

Figure 2.1 illustrates a simplified schematic for a squirrel-cage motor drive supplied by a voltage-source vector-controlled frequency converter. Vector control refers to the use of space vectors which were developed in the 20th century to model the electrical machine transients. In this material, the terms "vector" and "space vector" are used synonymously even though space vectors are not vectors in the classic sense. In electrical machine modelling and control, space vectors are frequently used to clearly and concisely represent measured three-dimensional quantities. This is especially true for three-phase systems, where space-vector representation makes it easier to recognize and interpret electromagnetic behaviours. The following text introduces the voltage space vector. Chapter 4 offers a more rigorous definition of space vectors and their application.

Switches k1–k6 carry out the pulse width modulation of the drive. In practice, these switching functions can be provided by controllable switches, such as metal oxide semiconductor field-effect (MOSFE) or insulated-gate bipolar (IGB) transistors, or by gate turn-off (GTO) or integrated gate-commutated (IGC) thyristors and an anti–parallel freewheeling diode to ensure that the current source, which might be a current-maintaining inductive component, never becomes open circuited. The switching commands given by k1–k6 are based on motor control. The frequency converter comes embedded with software that

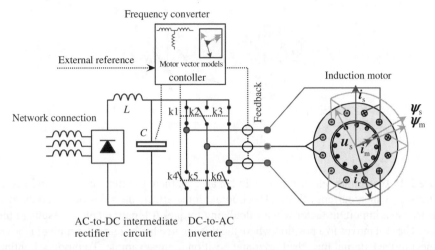

Figure 2.1 Induction motor drive equipped with a vector-control voltage-source frequency converter. The stator current vector, the rotor current vector, and the magnetizing current vector are denoted i_s, i_r, and i_m, respectively. The stator voltage is u_s, and the vectors for stator flux linkage and air-gap flux linkage are ψ_s and ψ_m. The current and flux linkage vectors are in the directions of the corresponding magnetic effects. The drive controller has one or more motor models and rotates an image of the machine to emulate its behaviour and to be able to make control decisions.

includes a motor model. In operation, the converter models motor behaviour, and then rotates actual motor space vectors as the model prediction dictates. The converter supplies motor inputs in a controlled manner responding to incoming commands, for example, a speed command from an upper-level controller.

The ability of motors to convert electrical energy into mechanical energy is based on electrical and magnetic force effects. The Lorentz force equation can be expressed verbally as follows. A force dF acts on a charge dQ moving at velocity v through a current-carrying conductor of length dl subjected to a magnetic flux density B. In equation form,

$$dF = dQv \times B = dQ \frac{dl}{dt} \times B = \frac{dQ}{dt} dl \times B = idl \times B \qquad (2.1)$$

Figure 2.2 illustrates how the Lorentz force results in DC motor torque production. The current-carrying motor-armature winding experiences torque-producing forces when subject to magnetic flux.

The Lorentz force may be used in torque calculations for all machines with conductors in a magnetic flux field.

In AC motors, according to the law of induction, the voltage space vector u_s fed to an AC motor stator winding is integrated into the flux linkage vector of the stator as follows.

$$\psi_s \approx \int u_s dt \qquad (2.2)$$

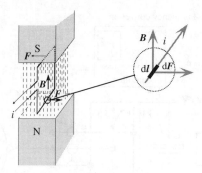

Figure 2.2 Elementary electric motor. In the air gap, there is a flux density B and a rotating coil through which current i flows. The Lorentz force affects the sides of the armature coil. Coil size is an important factor when calculating torque. A bigger diameter results in bigger torque. The coil moves to a position where the flux created by the current running in the coil is collinear to the external flux, the horizontal position in this example. To produce continuous torque, several coils are needed at different armature angles.

The absolute value of the flux linkage space vector ψ is directly proportional to the flux density B, the effective number of turns $k_w N$, and the surface S through which the flux travels.

$$\psi = k_w N \int B \cdot dS \qquad (2.3)$$

When voltage is applied, corresponding currents begin to flow. Considering electromechanical energy conversion, the flux linkage of the air gap, which is the most important of the flux linkage components, can be expressed with this equation.

$$\boldsymbol{\psi}_m = L_m \boldsymbol{i}_m \qquad (2.4)$$

Flux linkage in the air gap is in the direction of the magnetizing current vector $\boldsymbol{i}_m$ and is associated with the magnetizing inductance L_m. A vector equation can be determined for the electromagnetic torque vector $\boldsymbol{T}_e$ from the Lorentz force.

$$\boldsymbol{T}_e = \frac{3}{2} p (\boldsymbol{\psi}_s \times \boldsymbol{i}_s) = \frac{3}{2} p (\boldsymbol{\psi}_m \times \boldsymbol{i}_s) \qquad (2.5)$$

The length of the current-linkage vectors correspond to magnitude, that is, the maximum value of the respective sinusoidal current waveforms. The right-hand rule can be applied to determine current-linkage vector directions. The direction of the *current space vector* is the direction of the current linkage of the winding. Definitions for *current space vector* and *voltage space vector* are introduced later in Chapter 4. Because flux linkage and voltage are related according to the law of induction, changing the direction of the voltage vector in the stator winding also changes flux linkage direction. If the direction of the voltage vector in the winding of a three-phase motor changes continually in the right sequence, the flux linkage of the motor begins to rotate. The moving rotor of an induction motor can lag behind the driving flux rotation in the air gap. This is referred to as slip. For a synchronous machine, the rotor has no permanent slip. A synchronous machine has temporary slip during load transients. After reaching a certain load angle, the rotor rotates at the same speed as the air-gap flux.

Motor control can be implemented using a frequency converter, which varies voltage amplitude and AC frequency. A common voltage-source converter consists of three parts. The 50 or 60 Hz three-phase line voltage is fed to an AC-to-DC rectifier unit, which converts it to DC voltage. This DC voltage is fed to a DC intermediate circuit, which filters the voltage output of the rectifier. Next, the DC-to-AC inverter switches each motor phase to either the negative or the positive bus bar of the DC intermediate circuit in the appropriate sequence. The previous Figure 2.1 illustrated a circuit for a frequency converter–supplied induction motor drive showing switches k1–k5. In the figure, switches k1, k3, and k5 are conductive. By switching on only k1, k5, and k6, the flux linkage is made to rotate counterclockwise.

The same frequency converter technology described here can be applied to all rotating field machines and, in case of network converter, the electrical network. Currently, synchronous motors are becoming more popular for many electrical drive applications. As a result, there is active development in electrical drives based on synchronous motors.

Drives based on DC motors are currently less popular in industrial settings, and AC motor–based drives are gaining ground. DC motor–based drives, however, remain popular for small motor applications with less demanding performance requirements. One typical area of application includes the numerous auxiliary drives present in modern vehicles—as many as 100 in a passenger car. Electrical drive applications in automobiles include the fuel pump, the starter, the generator, various cooling fans, the windshield wipers, the mirror-turning motors, and the seat-adjustment motors. Many of these are ON/OFF drives, but larger DC motor drives can also use PWM choppers to adequately control current.

Electrical drives based on small series DC motors and AC power supplies are built in extremely large quantities to operate all kinds of hand tools, including drills, grinders, dryers, and kitchen appliances. In this book, however, the focus is mostly on higher-power precision drives used mainly in industry, distributed generation, commerce, and housing and increasingly often in mobile applications.

2.2 Converter interface to power source

Most industrial drives are connected to an electrical utility grid. Electricity network technology becomes essential when considering the interface of the electricity network and power electronics. Generally, just knowing the voltage of the connection point and the short-circuit power is sufficient. However, it is often necessary to analyse resonances that may occur in the network and to understand issues related to the quality of the electricity. The short-circuit power at the point of common coupling can usually be defined from the properties of the transformer supplying the point of common coupling. The transformer nameplate gives the rated voltage U, rated apparent power S_N, rated frequency, and short-circuit voltage u_k or impedance z_k in percent. The rated impedance of the transformer is expressed according to the following equation.

$$Z_N = \frac{U^2}{S_N} \qquad (2.6)$$

If the resistance is neglected, the corresponding rated inductance is

$$L_N = \frac{Z_N}{\omega} = \frac{U^2}{2\pi f S_N} \qquad (2.7)$$

and the corresponding short-circuit inductance and short-circuit apparent power of the transformer are

$$L_k = u_k L_N \tag{2.8}$$

$$S_k = \frac{S_N}{u_k} \tag{2.9}$$

EXAMPLE 2.1: The rated apparent power of a 50 Hz, 400 V transformer is 3000 kVA. Its short-circuit voltage is $u_k = 6\%$. Calculate the short-circuit inductance of the transformer and the short-circuit power at the transformer terminals assuming an infinite supplying network.

SOLUTION: The line-to-line voltage on the secondary side is 400 V. If resistance is neglected, the short-circuit inductance of the transformer seen from the secondary side is

$$L_k = u_k \frac{U^2}{2\pi f S_N} = 0.06 \frac{3\left(\frac{400}{\sqrt{3}} \text{V}\right)^2}{2\pi \frac{50}{\text{s}} \cdot 3000000 \text{ VA}} = 10.2 \cdot 10^{-6} \text{H} = 10.2 \, \mu\text{H}$$

The short-circuit apparent power of the transformer is

$$S_k = \frac{S_N}{u_k} = \frac{3000000 \text{ VA}}{0.06} = 50 \text{ MVA}$$

The transformer forms the limiting factor for the short-circuit inductance and power at the transformer terminals. Because the short-circuit apparent power supplying the transformer is usually so large, there is no need to account for it in the network inductance determination.

If the electrical drive is supplied via a cable of significant length, its effect on the inductance of the point of common coupling must be added to the transformer inductance. Cable manufacturers give the inductance per metre, from which it is easy to calculate cable inductance.

Short-circuit apparent power and short-circuit inductance must be known to connect an electrical drive to a common coupling point. For example, when connecting a thyristor bridge, the bridge manufacturer's guidelines about the minimum commutation inductance in front of the bridge must be followed and a suitable inductance must be added if the short-circuit inductance of the network is not sufficient to establish the commutation inductance specified by the manufacturer.

For direct network drives based on squirrel-cage induction motors, the short-circuit power of the supplying network must be large enough to avoid overlarge voltage drops during start-up. Typically, about 7 times its rated current surges through an induction motor during a DOL start, so the available short-circuit power should be approximately 10 times the rated power of the motor.

If different filters are used at the network interface, it is possible for oscillations due develop because of the network parasitic capacitances and inductances. This must be monitored to avoid resonances between the network and the filter.

EXAMPLE 2.2: Calculate the starting voltage drop across the terminals of the transformer from the previous example during an induction motor start-up given the following rated values: 200 kW, 400 V, 400 A. Suppose that the starting current is 7 times the rated current.

SOLUTION: The rated current of the machine is 400 A. Starting current is, for example, 7 times the rated current, or $I_{kIM} = 2800$ A. The inductance of the transformer is $L_k = 10.2$ µH, and the 50 Hz impedance is 0.0032 Ω. Therefore, the phase voltage drop on the transformer secondary side is about 9 V (15.5 V line to line) during machine start-up, which is 3.84% of the rated voltage. The terminal phase voltage on the transformer secondary side is 221 V, and the line-to-line terminal voltage is 385 V.

Even with a 3000 kVA/200 kW power ratio, the voltage drop caused by the starting current is significant and can manifest as, for example, a blinking of lighting.

Overall, at least some basic knowledge about the network supplying the electrical drive must be known to design an effective system.

In place of a supplying network, electric drives may be operated directly from a voltage source. The frequency converter DC link may be directly connected to an accumulator in mobile drives or to a supercapacitor in hybrid systems. A battery or a supercapacitor cannot replace the DC-link capacitor shown in Figure 2.1. It is instead connected in parallel with it. Naturally, the voltage sources connected in parallel with the converter DC link must be modelled somehow. Typically, batteries can be replaced by a voltage source and an equivalent series resistance (ESR) and possibly a series inductance.

A supercapacitor introduces large capacitance. Its parasitic elements can also be easily replaced with an ESR and series inductance. For a supercapacitor-supported DC link, the problems related to highly variable supercapacitor voltage may be mitigated using a bidirectional DC-DC converter positioned between the DC-link capacitor and the super-capacitor to keep the DC-link voltage of the motor converter constant. Figure 2.3 illustrates a

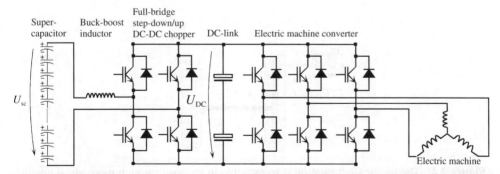

Figure 2.3 Supercapacitor system of a parallel hybrid vehicle machine drive converter. The full-bridge converter can transfer energy to and from the DC link of the electric machine converter. The electric machine works in both generating and motoring modes, charging and discharging the supercapacitor.

power electronic system with an electric machine converter and a full-bridge DC-DC converter capable of keeping the DC-link voltage U_{DC} constant as the supercapacitor voltage U_{sc} varies.

2.3 Fundamental mechanics

The fundamental mechanics relevant to electrical drives can be expressed, in principle and within set boundaries or limits, as two basic equations that describe the mechanics of a rotating object. The first is an expression of electrical torque.

$$T_e = T_L + J\frac{d\Omega}{dt} + D\Omega \tag{2.10}$$

T_e is the electrical torque produced by the motor, which will be consumed to compensate for the load torque T_L, the frictional torque $D\Omega$, the inertia J, and the dynamic countertorque caused by the angular acceleration $d\Omega/dt$.

The second equation relates mechanical torque to load torque for electrical drives with gearing. If the torque efficiency of the gearing is assumed to be 100%, the following relationship can be expressed. The number of teeth for the two gears are n_L and n_m, and a is transmission ratio.

$$\frac{T_m}{T_L} = \frac{\Omega_L}{\Omega_m} = \frac{\theta_L}{\theta_m} = \frac{n_m}{n_L} = a \tag{2.11}$$

Figure 2.4 illustrates a geared motor drive system.

Currently, linear- motion drives are also important. Linear movement can be achieved using a rotating machine coupled to an appropriate mechanism or by using a linear electric motor. The linear movement corresponding differential equation for electrical force F_e is dependent on the load force F_L, mass m, speed v, and drag coefficient D as

$$F_e = F_L + m\frac{dv}{dt} + Dv \tag{2.12}$$

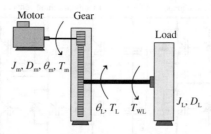

Figure 2.4 The motor shaft rotation is transmitted to the load through mechanical gearing. J_m is the inertia of the motor, D_m is the friction term of the motor, θ_m is the rotation angle of the motor, and T_m is the motor torque. On the load side, T_L is the torque produced by the motor to the load, and T_{WL} is the working torque of the load. The efficiency of a gear is often given as 99% per cogwheel contact.

Interaction between the drive system and the mechanics can be complicated by the presence of mechanical resonance. Mechanical vibration, in direct response to the combined torsional spring, mass, and damping components of the system, also can be present.

Peak performance is needed for servo-based electrical drives. The general equation for an electric circuit with inductance and resistance is as follows.

$$U = E + L\frac{di}{dt} + Ri \tag{2.13}$$

Comparing Equation (2.13) with Equation (2.10) for electrical torque shows that they are analogous. Electrical resistance R corresponds to friction D in mechanics, and inductance L corresponds to inertia J. The back emf E is analogous to load T_L.

The time constant of the LR circuit is $\tau_{el} = L/R$. The corresponding mechanical time constant should be $\tau_{mech} = J/D$. In engineering, however, the mechanical time constant of an electrical drive is defined normally as the time at rated motor torque that is needed to accelerate the inertia J of the drive to the rated speed.

2.4   Basic mechanical load types

Based on their principal behaviour, loads can be categorized into basic types. The most important examples of load torque as a function of mechanical rotation speed are depicted in Figure 2.5.

Mathematically, the countering torques can be divided into the following categories. Static friction can be described with

$$T_L \cong \Omega_L^{-1} \tag{2.14}$$

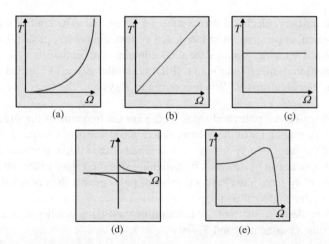

Figure 2.5   Principal dependence of load torque on mechanical rotation speed for (a) the blower $\sim\Omega^2$, (b) the piston pump $\sim\Omega^1$, (c) lift $\sim\Omega^0$, (d) the influence of static friction $\sim\Omega^{-1}$, and (e) the flywheel as a load for an induction motor connected directly to the network.

Constant torque, produced at constant speed in hoists, lifts, etc., is naturally

$$T_L \cong \Omega_L^0 \qquad (2.15)$$

Torque can be directly proportional to angular speed. This happens in general, for example, in reciprocating (piston) pumps.

$$T_L \cong \Omega_L^1 \qquad (2.16)$$

Torque is proportional to the square of angular speed in centrifugal pumps and blowers.

$$T_L \cong \Omega_L^2 \qquad (2.17)$$

There are torque relationships based on higher powers of Ω. However, these are not considered basic load types. Equation (2.17), which relates torque to the square of angular velocity, is encountered most commonly, because most electrical drives in use today are for pumping and blowing.

Analogous equations and load types may be found for linear movements. Especially important is, for example, the air friction for vehicles. This friction may be described using an equation analogous to Equation (2.17). The force needed to compensate for the air displacement caused by a moving vehicle is proportional to the square of the speed v.

$$F_L \cong v^2 \qquad (2.18)$$

2.5   Proportional-integral-derivative controller in electrical drives

To maximize efficiency, electrical drives must be controlled. The control variable may be motor torque, speed, or position. In an integrated system, almost any physical variable can be used as the control variable, such as the CO_2 content in a ventilation system. Commonly today, the proportional-integral-derivative (PID) controller is used to provide generic feedback control in electrical drives. When controlling a system, the transfer function of it is needed.

A transfer function has poles and zeros. These are the frequencies for which the value of the denominator or numerator of the transfer function becomes zero, respectively. The values of the poles and the zeros of a system determine whether the system is stable. In the simplest sense, a control system can be designed by assigning specific values to the poles and zeros of the system. Control systems must have a number of poles greater than or equal to the number of zeros. Such systems are called proper.

In the Laplace domain, the transfer function of the PID controller with integrating and differentiating time constants T_I and T_D is

$$G_{PID} = K_P \left(1 + \frac{1}{sT_I} + sT_D \right) = K_P \frac{s^2 T_D T_I + sT_I + 1}{sT_I} \qquad (2.19)$$

Controller amplification can be written as

$$K_P \frac{T_I + T_D}{T_I} \qquad (2.20)$$

Combining the expressions from Equations (2.19) and (2.20), the transfer function can be expressed

$$G_{PID} = K_P \left(\frac{T_I + T_D}{T_I} + \frac{1}{sT_I} + sT_D \right) = K_P \frac{(1 + sT_I)(1 + sT_D)}{sT_I} \qquad (2.21)$$

The PID controller keeps the variable being controlled close to its reference value. PID control is applied as follows.

1. There is a reference value, for example, a speed reference coming from an external source such as an upper-level controller.

2. A feedback device is used to measure the actual value of the controlled variable and produce a feedback signal.

3. The feedback signal is subtracted from the reference value to size the error remaining in the controlled variable.

4. The PID controller manipulates this error signal to produce an internal reference for the converter.

Figure 2.6 illustrates the placement of a PID controller in a speed-controlled drive. The speed controller accepts the speed error as an input and produces a torque reference for the motor drive system. If the speed is too low, more torque is needed and vice versa. The proportional controller multiplies the error, the integrating controller integrates the error signal ensuring that no stable error is possible, and the differentiator attempts to suppress fast changes.

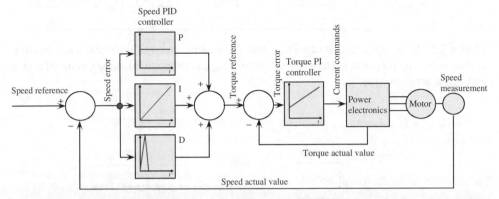

Figure 2.6 A PID controller used to control the speed of an electrical drive. The speed controller provides the torque controller a reference value. The torque controller may also be a PI or PID controller depending on the case. This simple approach is valid for DC motor drives. For AC drives, the torque controller is more complicated but works on the same principles.

2.6 The speed, torque, or position control of an electrical drive

A modern electrical drive consists of controllers, converters, the electric machine, the mechanical load, and measurement instrumentation. Figure 2.7 illustrates the basic construction of a DC motor drive with motor speed as the reference. The simplified system assumes a fully compensated motor with armature current directly representing drive torque and constant magnetization. In AC electrical drives, the relationship between torque and motor current is more complicated, but the basic construction of Figure 2.7 still serves as a useful example.

Figure 2.8 shows the basic construction for an electrical drive system in which torque is the reference.

Speed and torque reference control electrical drives can be found, for example, in paper machinery where paper roller pairs must synchronize speed and operate at the same torque. One roll drive operates according to the principle illustrated in Figure 2.7, and the second operates according to Figure 2.8. The torque reference for the second drive comes directly from the speed controller output of the first. In general, the torque reference configuration is dangerous and additional precautions must be built in. For example, a speed limiter might be added to prohibit overspeed in the event of load loss.

Figure 2.9 illustrates a servo drive, which is designed to control rotor position. Servo drives are widely used in different automated piece goods–handling processes.

Figure 2.10 offers a principal example of behaviour for a position-controlled electrical drive. Here, torque rises quickly to the maximum rated value as the drive motor accelerates to maximum rated speed. The drive continues at speed until signalled to stop. The electric drive

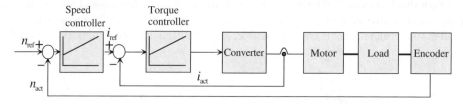

Figure 2.7 Basic construction of a DC motor drive using speed as the reference. Armature current is directly proportional to torque; therefore, i_{ref} can be regarded as the torque reference and i_{act} as actual torque.

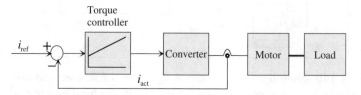

Figure 2.8 Basic construction of an electrical drive with torque as reference. Torque is assumed to be proportional to current.

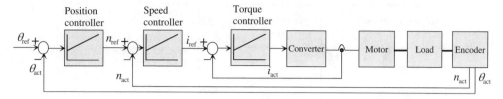

Figure 2.9 Position-controlled servo drive.

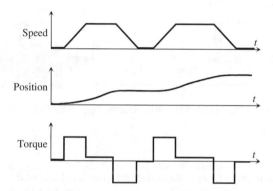

Figure 2.10 Servo-drive speed, position, and torque behaviour at maximum performance.

motor then ramps up braking torque to stop the motor shaft precisely at the programmed destination. Torque is always ramped up to avoid problems associated with abrupt starts and stops such as jerkiness. See Figure 2.10. A position-controlled drive, such as that depicted in Figure 2.9, cannot brake strongly as shown in the speed curve and must approach its destination slowly.

Examples of embedded PID controllers dedicated to different motor type controls will be discussed in later chapters.

2.7 Control time rates and embedded system principles

Today, information technology is a necessary component of electrical drive system implementation. Modern processors are capable of handling the high-accuracy control algorithms, and they adjust electrical drive physical behaviour to meet set targets. In an ideal case, the integrated circuit coding and the related measurement data combine to produce an almost-complete representation of the physical world electrical drive.

Efficient data transmission between the various components of the electrical drive becomes possible with information technology. This exchange of data can be wireless, or the data may be transmitted along electrical or optical cables. It may also be possible to eliminate data transmission cables altogether and transmit data along power cables using a carrier wave.

Figure 2.11 illustrates typical time rates and the signal-processing tasks related to the control of an electrical drive.

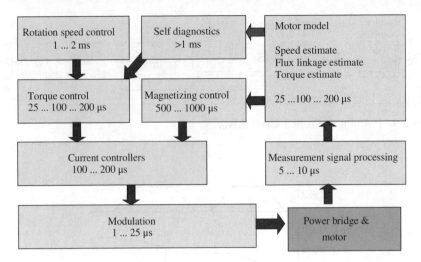

Figure 2.11 Time-level requirements and sequence of tasks for a motor drive.

An electrical motor drive system has different time-level demands. The electromagnetic phenomena are fast, but the mechanical phenomena are significantly slower. In Figure 2.11, the fastest times relate to the modulation of the PWM signal. The shortest PWM pulse durations may typically be in the range of 1 µs.

Motor currents must be measured at high enough rates (every 5 to 10 µs) to get acceptable feedback information for control. The higher rate gives a clear image of electric current behaviours and makes it possible to appropriately filter the actual current value before performing any current control actions at the 100 to 200 µs level.

The motor model is typically calculated every 25 to 200 µs to evaluate the electromagnetic state of the motor. In classic direct torque control (DTC), for example, the stator flux linkage is estimated every 25 µs by integrating stator voltage. Moreover, various correction actions must take place at the same rate.

There is substantial magnetic energy in the magnetic circuit of a motor, and therefore changes to its magnetic state may take place at a slower rate, such as every millisecond.

Because torque control is essential and because torque can change rapidly, torque calculations must be carried out at high speed. In the classic DTC, torque is updated every 25 µs. In vector-controlled drives, the rate may be slightly lower.

Because of its high mechanical inertia, system speed changes at a significantly slower rate than torque. Therefore, it is sufficient to update the speed controller algorithms every 1 to 2 ms.

Powerful data-processing capability is needed to carry out the aforementioned signal-processing tasks. At the beginning of the 21st century, the tendency still has been to use signal processors and related auxiliary circuits, such as application-specific integrated circuits (ASICs), to control the power electronics of electrical drives. However, the development trend seems to be leading toward packing all the control electronics and required code for the power electronics onto a single field-programmable gate array (FPGA) circuit. See Figure 2.12.

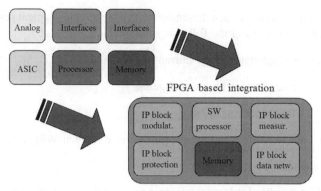

Source, the Laboratory of Control Engineering, LUT

Figure 2.12 Development trend in control electronics toward FPGAs and away from ASICs. With FPGAs, all the necessary functions, including modulators, measurements, protection algorithms, data systems, and memories, are included in the one device. A software processor uploaded in the FPGA runs the necessary programmes.

The most efficient electrical drives are capable of extremely accurate control of rotation speed and torque. Offering fast and precise position control, these are referred to as servo drives.

Traditionally, frequency converters operate with control cycle durations of a few dozen microseconds. This speed can be increased significantly, when information is not transmitted between different units, using an FPGA-based integration. For example, a welding machine control was implemented at the Laboratory of Control Engineering at the Lappeenranta University of Technology with a control cycle of 64 ns.

The continuing development of power electronics and microprocessors is improving the characteristics of electrical drives. Markets for intelligent electrical drives will expand as the awareness of these improved drives increases. They offer compelling competitive advantages such as low maintenance, energy efficiency, and sophisticated process control.

2.8   Per-unit values

Normally, electrical drives are analysed using per-unit (pu) values. Per-unit values reveal directly a parameter's relative magnitude. For instance, a relative magnetizing inductance of $L_m = 3$ pu for an asynchronous machine is considered quite high. On the other hand, a value of $L_m = 1.5$ pu is low. The stator leakage inductance $L_{s\sigma}$ per-unit values typically vary in the range of 0.1 to 0.2. Focusing on per-unit values in this way makes it possible to compare electric machines that have different ratings. For example, their relative magnetizing inductances can be compared. Per-unit values are especially important when machine control is implemented in an embedded system. By coding the control in terms of per-unit values, a common software program can be used to control motors of different sizes and ratings.

In practice, there are two per-unit value types. One type is used for direct motor control, and the other is used for embedded programming using fixed-point numbers. The per-unit values are obtained by dividing each dimension with a base value. Typical per-unit values for motor control are derived using the following base values:

- peak value for rated stator phase current $\hat{i}_N$

- peak value for rated stator phase voltage $\hat{u}_N$

- rated angular frequency $\omega_N = 2\pi f_{sN}$

- rated flux linkage, corresponding also to the rated angular velocity $\hat{\psi}_N$

- rated impedance Z_N

- time in which 1 radian in electrical degrees $t_N = 1 \, \text{rad}/\omega_N$ is travelled at a rated angular frequency. Relative time τ is thus measured as an angle $\tau = \omega_N t$

- apparent power S_N corresponding to rated current and voltage

- rated torque T_N corresponding to rated power and frequency

When operating sinusoidally, the rated RMS current of the electric machine is I_N and the line-to-line RMS voltage is U_N. Therefore, the base value for current is

$$I_b = \hat{i}_N = \sqrt{2} I_N \tag{2.22}$$

The base value for voltage is

$$U_b = \hat{u}_N = \sqrt{2} \frac{U_N}{\sqrt{3}} \tag{2.23}$$

Angular frequency is

$$\omega_N = 2\pi f_{sN} \tag{2.24}$$

The base value for flux linkage is

$$\psi_b = \hat{\psi}_N = \frac{\hat{u}_N}{\omega_N} \tag{2.25}$$

The impedance base value is

$$Z_b = Z_N = \frac{\hat{u}_N}{\hat{i}_N} \tag{2.26}$$

The base value for inductance is

$$L_b = L_N = \frac{\hat{u}_N}{\omega_N \hat{i}_N} \tag{2.27}$$

The base value for capacitance is

$$C_b = C_N = \frac{\hat{i}_N}{\omega_N \hat{u}_N} \tag{2.28}$$

The apparent power base value is

$$S_b = S_N = \frac{3}{2} \hat{i}_N \hat{u}_N = \sqrt{3} U_N I_N \tag{2.29}$$

And, finally, the base value for torque is

$$T_b = T_N = \frac{3}{2\omega_N} \hat{i}_N \hat{u}_N \cos \varphi_N = \frac{\sqrt{3} U_N I_N}{\omega_N} \cos \varphi_N \tag{2.30}$$

The appropriate per-unit values are

$$u_{s,pu} = \frac{u_s}{\hat{u}_N} \tag{2.31}$$

$$i_{s,pu} = \frac{i_s}{\hat{i}_N} \tag{2.32}$$

Equations (2.31) and (2.32) also show that i_s and u_s represent the amplitudes of space vectors $\boldsymbol{i}_s$ and $\boldsymbol{u}_s$.

$$R_{s,pu} = \frac{R_s \hat{i}_N}{\hat{u}_N} \tag{2.33}$$

The relative inductance values are the same as the relative reactance values. Therefore, the following expression, for example, for magnetizing inductance L_m can be formulated.

$$L_{m,pu} = \frac{L_m}{L_b} = \frac{L_m}{\frac{\hat{u}_N}{\omega_N \hat{i}_N}} = \frac{\hat{i}_N}{\hat{u}_N} X_m = X_{m,pu} \tag{2.34}$$

The per-unit value of stator flux linkage is

$$\psi_{s,pu} = \frac{\omega_N \psi_s}{\hat{u}_N} \tag{2.35}$$

The per-unit angular velocity and speed are as follows.

$$\omega_{pu} = \frac{\omega}{\omega_N} = \frac{n}{f_N} = n_{pu} \tag{2.36}$$

Given that n is rotational speed in 1/s. The relative time gets the form

$$\tau = \omega_N t \tag{2.37}$$

Also, according to Equation (2.37), relative time replaces real time when using per-unit equations. Real time, however, can be used in per-unit equations. This is accomplished by replacing relative time τ with $\omega_N t$.

In general, the equations introduced in this text are written using absolute vector values and absolute time. However, most of the included examples employ a per-unit presentation. The reader should take care to understand for each case if the presentation is in per-unit or absolute values. In general, the pu subscripts have not been added to equation parameters. Here is an example of an absolute-value stator voltage space-vector equation in rotor coordinates (superscript r).

$$u_s^r = R_s i_s^r + \frac{d\psi_s^r}{dt} + j\omega_r\psi_s^r \tag{2.38}$$

When using per-unit values, the above can be divided by the base value of the voltage. If grouped appropriately, the expression becomes

$$\frac{u_s^r}{\hat{u}_N} = \frac{R_s \hat{i}_N}{\hat{u}_N} \frac{i_s^r}{\hat{i}_N} + \frac{d\psi_s^r}{d(\omega_N t)} \frac{\omega_N}{\hat{u}_N} + \frac{j\omega_r}{\omega_N} \frac{\omega_N \psi_s^r}{\hat{u}_N} \tag{2.39}$$

Therefore, the voltage equation given in per-unit values is as follows.

$$u_{s,pu}^r = R_{s,pu} i_{s,pu}^r + \frac{d\psi_{s,pu}^r}{d\tau} + j\omega_{r,pu}\psi_{s,pu}^r \tag{2.40}$$

This equation follows the same form as the absolute value Equation (2.38). However, the per-unit time may cause problems. If normal time is used in the per-unit value equations, the equation can be rewritten as

$$u_{s,pu}^r = R_{s,pu} i_{s,pu}^r + \frac{1}{\omega_N}\frac{d\psi_{s,pu}^r}{dt} + j\omega_{r,pu}\psi_{s,pu}^r \tag{2.41}$$

The absolute-value differential Equation (2.10) for rotating movement with electric torque T_e, load T_L, and J as inertia can be rewritten

$$T_e = \frac{3}{2}p\psi_s \times i_s = \frac{J}{p}\frac{d^2\theta}{dt^2} + T_L \tag{2.42}$$

This can be dividing by the base value of torque adapted according to mechanical angular velocity $\Omega = \omega_r/p$ with pole pair number p.

$$\frac{T_e}{\frac{3p}{2\omega_N}\hat{i}_N\hat{u}_N\cos\varphi_N} = \frac{\frac{3}{2}p\psi_s \times i_s}{\frac{3p}{2\omega_N}\hat{i}_N\hat{u}_N\cos\varphi_N} = \frac{\frac{J}{p}\frac{d^2\theta}{dt^2} + T_L}{\frac{3p}{2\omega_N}\hat{i}_N\hat{u}_N\cos\varphi_N} = \frac{p\frac{\frac{Jd^2\theta}{d\tau^2} + T_L}{\omega_N^2}}{\frac{3p}{2\omega_N}\hat{i}_N\hat{u}_N\cos\varphi_N} \tag{2.43}$$

The per-unit value equation for rotation becomes

$$T_{e,pu} = \psi_{s,pu} \times i_{s,pu} = \tau_J \frac{d^2\theta}{d\tau^2} + T_{L,pu} \tag{2.43}$$

This equation contains a mechanical dimensionless per-unit time constant $\tau_{J,pu}$.

$$\tau_{J,pu} = \frac{\frac{J\omega_N^2}{p}}{\frac{3p}{2\omega_N} \hat{i}_N \hat{u}_N \cos \varphi_N} = \frac{2J\omega_N^2}{\frac{3p^2}{\omega_N} \hat{i}_N \hat{u}_N \cos \varphi_N} = \omega_N \frac{\omega_N^2}{p^2} \frac{2J}{3\hat{i}_N \hat{u}_N \cos \varphi_N} \tag{2.43}$$

The absolute value of the mechanical time constant can also be written as follows.

$$T_J = \left(\frac{\omega_N}{p}\right)^2 \frac{J}{S_N \cos \varphi_N} = \left(\frac{\omega_N}{p}\right)^2 \frac{J}{\frac{3}{2}\hat{u}_N \hat{i}_N \cos \varphi_N} = \left(\frac{\omega_N}{p}\right)^2 \frac{J}{\sqrt{3}U_N I_N \cos \varphi_N}$$

$$= (\Omega_N)^2 \frac{J}{\sqrt{3}U_N I_N \cos \varphi_N} [s] \tag{2.44}$$

According to Equation (2.44), the mechanical time constant is the ratio of the kinetic energy of a rotor rotating at synchronous speed to the power of the machine. This constant is a measure of the time it takes for a motor to spin up from a standstill to synchronous speed given constant driving torque throughout the start-up period at the maximum torque magnitude made available by the motor's rated power. Half this value is known as relative kinetic energy (constant of inertia), and it is marked as $H = T_J/2$ (seconds).

3

The fundamentals of electric machines

Motors and generators are key elements in electrical drives. They are generally referred to as "electric machines". In principle, all electric machines can function in either of two rotating directions and in two torque directions, independently of rotation direction, making them four quadrant devices. Therefore, an electric machine can operate in either direction as a motor or a generator, which differentiates them from many other types of energy conversion machinery that limit rotational directionality. For example, an internal combustion engine can never convert mechanical work and exhaust gases back to fuel. However, an electric machine can be converting electrical power into mechanical power, then reverse function in a few milliseconds to begin converting mechanical power into electrical. For example, the kinetic energy of a hybrid car or a train can be converted into electricity and stored in a battery for later use or fed to a power grid for immediate use. Only the one-directional properties of some auxiliary systems inhibit the bidirectional functionality of an electric machine system. The electric machine itself, in principle, is not limited.

Three-phase alternating current (AC) machines represent the most important family of electric machines for industrial drives and other heavy-duty applications. In addition, traditional brushed direct current (DC) machines with separately controlled excitation windings still play an important role, but this importance is declining as AC drives need less maintenance and offer higher efficiencies. However, DC motor drives with low-cost PM-excited motors are extensively used in nonindustrial applications. These drives are used, for example, in automobile starters, windshield wipers, fans, pumps, etc. The universal motor, a DC series motor accommodated for operation with AC, also plays a significant role as the chosen power source for many powered hand tools such as drills, angle grinders, saws, etc. Switched reluctance (SR) machines have also gained some kind of market position, and one later chapter in this material is dedicated to this machine type. This textbook, however, concentrates mostly on rotating field machine drives. The most important electric machine types studied here are induction machines (IMs) and various synchronous machines (SMs).

Electrical Machine Drives Control: An Introduction, First Edition. Juha Pyrhönen, Valéria Hrabovcová and R. Scott Semken.
© 2016 John Wiley & Sons, Ltd. Published 2016 by John Wiley & Sons, Ltd.

Induction machines can be divided into two categories: squirrel-cage machines and wound rotor machines with slip rings. The squirrel-cage type is the real workhorse of the industry, and the wound rotor type, with three-phase rotor winding, is widely used for wind power applications in *doubly fed* systems where the IM stator is direct online (DOL) connected and the rotor is controlled via a four-quadrant frequency converter enabling subsynchronous and supersynchronous speed control. In addition, the rotor can be DC supplied when the wound-rotor IM operates as an SM. This can be explained by the fact that a three-phase winding creates a constant sinusoidal flux when supplied by DC, for example, so that one rotor slip ring is connected to positive and the two other ones to negative DC supply. The rotor then has to rotate synchronously with the air-gap flux.

Synchronous machines come in a wide variety of configurations. The traditional SM has DC slip rings to supply the rotor field windings. The rotor can have either salient (projecting) or nonsalient electromagnetic poles. It is possible to construct brushless excitation systems using machines in series. Permanent magnet excitation is popular in machines up to 10 MW in power. Lately, nonexcited synchronous electric machines— synchronous reluctance machines (SynRM)—using plain steel rotors have become popular. The rotors of these machines must offer high saliency. The SynRM rotor features an array of flux barriers (usually internal air pockets) to increase saliency and can be further enhanced by inserting PMs into the air pockets to produce a PM-assisted (PMa) SynRM. Both PMSMs and SynRMs come in a variety of configurations offering unique properties. A PMa SynRM is similar to an ordinary PM machine with PMs embedded within its rotor. What distinguishes one type from the other is which torque is bigger, the reluctance torque (PMa SynRM) or the PM torque.

Three-phase Brushless DC (BLDC) drives are also widely used, especially in low-power automation applications. In practice, there is little difference between the BLDC and a PMSM. In principle, however, the difference is clear: the back electromotive force (emf) of the BLDC is trapezoidal, while the back emf of the PMSM is sinusoidal. Moreover, BLDC commutation is accomplished using the actual timing of the DC current pulses, while the PMSM is normally vector controlled using sinusoidal currents.

All electric machines need a magnetic circuit and windings. A magnetic circuit consists typically of iron and air. Adding PM material significantly modifies the magnetic circuit, because the relative permeability of PM material $\mu_{r,PM}$ is approximately the same as that of air. The PM material increases the effective magnetic air gap. If the PMs are positioned on the gap surface of the rotor, this effective air gap is increased by the magnetic length $h_{PM}/\mu_{r,PM}$. As a consequence, a rotor surface magnet PM machine has significantly less magnetizing inductance than a similar electrically excited machine with a normal air gap.

Windings are normally set into slots or (less commonly) onto the gap surfaces of the stator or rotor. When PMs are used in the magnetic circuit, the windings and magnets together make up the portion of the magnetic circuit that strongly affects circuit permeance, as mentioned previously, and provides excitation. Using PM material brings an additional and predictable equivalent excitation current in the rotor.

Normally, insulated windings are set into stator slots that fall within the magnetic circuit. In principle, machines need at least two windings: a stationary winding and a rotational winding. Of these, the rotating winding can be replaced with PM material or, in case of reluctance machines, by high saliency. Electrical torque production results from the interaction between the stationary and rotating windings, or stationary windings and PMs on the rotor, or stationary windings and high rotor saliency.

There are multiple ways to mechanically configure an electric machine. It can be a radial flux machine with a cylindrical internal rotor or an annular external rotor. It can be an axial flux machine with a disk-shaped rotor or a transverse flux machine with a more complicated magnetic flux path. Or, it can also be a linear machine. However, here the focus will be on, and the text will refer to, the control properties of an internal rotor radial flux machine unless otherwise specified. It is, however, possible to claim that the machine user does not need to know accurately the actual construction of the machine. Even the converter does not need to know the machine internal construction as it only can see the machine via its terminals. Therefore, the machine construction is irrelevant from the control point of view. Only the machine type must be known.

Figure 3.1 depicts two magnetically connected windings that form the basis of an elementary electric machine. The coils with $k_{ws}N_s$ and $k_{wr}N_r$ effective turns represent, for example, the windings of an asynchronous machine. More accurately, they represent the coil system of a single-phase transformer, but for symmetrical multiphase machines, machine behaviour can be represented with single-phase equivalent circuits. Therefore, the equivalent circuit of Figure 3.1 represents a basic description of any electric machine having two windings (stator and rotor windings). In cases of more complicated machines, the equivalent circuit is modified to represent the behaviour of more-linked windings. Of course, the movement of the different windings with respect to each other and the number of phases add

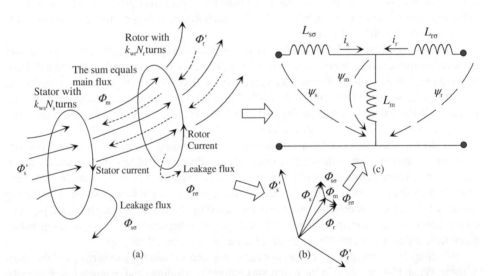

Figure 3.1 Image (a) represents a two-coil electric machine system with a main flux phasor Φ_m produced by the stator and rotor current flux components, that is, $\Phi_m = \Phi'_s + \Phi'_r$. Φ_m is the mutual sum flux phasor linking both electric circuits. Both windings have their own leakage fluxes $\Phi_{s\sigma}$ and $\Phi_{r\sigma}$. Image (b) is a phasor diagram showing the temporal dependence and phase shift of the fluxes, and image (c) is the equivalent circuit of the system ignoring resistances and magnetic circuit losses and referring rotor quantities to the stator. In (a) and (b) the flux components Φ'_s and Φ'_r are opposing. In this case, however, Φ'_s is stronger, and therefore the mutual flux Φ_m travels in the direction determined by the stator currents. Which current linkage will dominate depends on the machine type.

extra features to the equivalent circuit, but the machines can be described with the equivalent circuit of Figure 3.1, in principle.

The stator coil in Figure 3.1 carries a current i_s. This current alone produces the stator magnetic flux Φ'_s. When the flux travels from the stator to the rotor, part of it travels through the rotor coil. In addition, the rotor carries its own current that produces corresponding flux components. The common flux linking the stator and the rotor coils - Φ_m - is called the main flux. The part of the stator-produced flux that does not link to the rotor coil is called stator leakage flux $\Phi_{s\sigma}$. Correspondingly, as the rotor coil carries its own current, it produces its own flux including a leakage flux component $\Phi_{r\sigma}$. These fluxes are not temporally in phase or in the same spatial direction. Therefore, when calculating their sum, a phasor representation must be adopted for the currents, voltages, fluxes, and flux linkages.

Phasors describe temporally phase-shifted phenomena. Space vectors, which describe rotating fluxes, are covered in more detail later in Chapter 4. Space vectors will be used to represent rotating or travelling current linkages, fluxes, and flux linkages developed by multiwinding systems. The stator windings have $k_{ws}N_s$ effective turns; therefore, the space vector for stator flux linkage can be expressed as follows

$$\psi_s = k_{ws}N_s\Phi_s \tag{3.1}$$

The stator leakage flux similarly produces the following space vector for the leakage flux linkage.

$$\psi_{s\sigma} = k_{ws}N_s\Phi_{s\sigma} \tag{3.2}$$

The main flux linking the stator and rotor produces the flux-linkage space vector of the air gap with the stator windings

$$\psi_m = k_{ws}N_s\Phi_m \tag{3.3}$$

A basic principle in modelling electric machines for electrical drive control is to refer all flux linkages, inductances, and resistances to the stator voltage level. The transformation ratio between the stator and the rotor is, in practice, taken into account automatically. Especially in drives that cannot monitor the rotor windings during operation, the stator control electronics only *sees* the phenomena linked to the stator windings. Therefore, it is not even necessary to know how many effective turns there are in the rotor windings. In cases where the rotor windings can be accessed, for example, via slip rings in SMs or wound rotor IMs, it is important to know the transformation ratio between the rotor and stator windings.

Inductance is a measure of the windings' capability to produce flux linkage. Therefore, the air-gap flux-linkage (main flux linkage) space vector can be written

$$\psi_m = k_{ws}N_s\Phi_m = i_mL_m = (i_s + i_r)L_m \tag{3.4}$$

Correspondingly, the stator and rotor space vectors for leakage flux linkage can be written as follows.

$$\psi_{s\sigma} = i_sL_{s\sigma} \tag{3.5}$$

$$\psi_{r\sigma} = i_rL_{r\sigma} \tag{3.6}$$

When the components of the flux-linkage space vector are known, the stator and rotor flux linkage vectors can be written according to these equations.

$$\psi_s = \psi_m + \psi_{s\sigma} = L_{s\sigma}i_s + (i_s + i_r)L_m = L_si_s + L_mi_r \tag{3.7}$$

$$\psi_r = \psi_m + \psi_{r\sigma} = L_{r\sigma}i_r + (i_s + i_r)L_m = L_ri_r + L_mi_s \tag{3.8}$$

The stator and rotor inductances are L_s and L_r

$$L_s = L_m + L_{s\sigma} \tag{3.9}$$

$$L_r = L_m + L_{r\sigma} \tag{3.10}$$

For an electrical drive, flux linkage is also the integral of the voltage supplied to the windings. The voltage not consumed by resistive losses in the windings integrates into the flux linkage. Therefore, if the stator is supplied with a voltage defined as voltage space vector u_s, and the windings have resistance R_s, the space vector of stator flux linkage can be integrated thus

$$\psi_s = \int (u_s - i_sR_s)dt \tag{3.11}$$

There are, therefore, two equations that can be used to calculate stator flux linkage for supplied windings: (3.7) and (3.11). Further on in the discussion of motor control, the equations are referred to as the current model ($\psi_s = L_si_s + L_mi_r$ for the IM and similar equations for other types of machines) and the voltage model ($\psi_s = \int (u_s - i_sR_s)dt$).

When commissioning an electrical drive, the nameplate values and parameters, such as resistances, inductances, pole pair number, etc., must first be input into the embedded converter control software. The converter must know the equivalent circuit parameters to control the electric machine. Moreover, how the input parameters will vary as a function of motor operating condition must somehow be estimated. In principle, the converter is capable of performing no load test and other tests during drive initialization. However, it is important to have some basic knowledge of the machine's construction and its influence on the most important parameters. These areas are discussed next.

3.1   Energy conversion in electric machines

The fundamental idea of an electric machine is to convert electrical energy into mechanical and vice versa. To understand such a conversion process, one must become familiar with the different forms of power. Three-phase instantaneous power is the sum of the three phase components of electrical power.

$$P_{el}(t) = u_1i_1 + u_2i_2 + u_3i_3 \tag{3.12}$$

Mechanical power is the product of electromagnetic torque T_{em} and mechanical angular velocity Ω_r.

$$P_{mech}(t) = T_{em}\Omega_r \tag{3.13}$$

Neglecting losses, electrical power equals mechanical power.

$$P_{\text{mech}}(t) = P_{\text{el}}(t) \tag{3.14}$$

An electric machine has p magnetic pole pairs. Therefore, there is a steady state dependency between mechanical angular velocity and electrical angular velocity ω_r.

$$\omega_r = p\Omega_r \tag{3.15}$$

In electromechanical power conversion, the induction phenomenon is of principal importance. The following Figure 3.2 summarizes the basic principles of electromechanical power conversion in an electrical drive.

As Figure 3.2 indicates, power conversion in both directions is possible. Outgoing power is slightly less than incoming power in either direction, because of system losses that cannot be fully eliminated. One loss factor on the electrical power side is stator resistance R_s, which loses a fraction of the input electrical power to Joule losses (heat). These Joule losses manifest as reduced flux linkage and, in turn, reduced electromagnetic torque. In addition to Joule losses, there are losses in the magnetic circuit and the frictional losses of the rotating machinery (mechanical friction and windage losses). However,

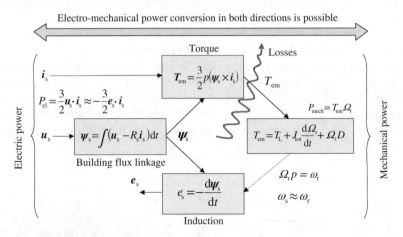

Figure 3.2 Basic phenomena in an electrical drive. The supplied stator voltage space vector u_s is integrated into the stator flux-linkage space vector ψ_s. The flux-linkage space vector and the current space vector i_s produce electromagnetic torque. The mechanics interact with the torque to produce speed Ω_r. This angular velocity corresponds to the stator angular velocity ω_s, which together with the stator flux linkage, produces the back emf of the system. The input and output powers are equal in an ideal lossless system, and in practice, electromechanical power conversion is highly efficient. 90% efficiencies and higher are common. Electro-mechanical power conversion is possible in both directions. Electrical power can be produced from mechanical power or mechanical power can be produced from electrical (generating or motoring). For a linear drive, the electromagnetic torque must be replaced by a corresponding force F_{em} and the angular velocity must be replaced with speed v.

electromechanical power conversion is typically one of the most efficient energy conversion methods. It is much more efficient, for example, than thermal conversion systems. In the power range of industrial drives, overall machine efficiencies are typically better than 90%. On the mechanical power side, viscous friction $D\Omega_r$ defines the mechanical losses of the system. The mechanical friction and windage losses of the electric machine could be considered elements of viscous friction and accounted for on the mechanical power conversion side.

3.2   Industrial machine windings

In AC machines, a travelling wave of magnetic flux is the basis for electromechanical energy conversion. The travelling magnetic field follows the air gap between the stator and rotor of the machine. In a rotating electric machine, the wave motion is rotary. In a linear machine, the motion is translation, following a rectilinear or curvilinear path.

Polyphase windings and a corresponding polyphase alternating current input are required to generate the moving wave. The magnetic axes of the windings are positioned in a spatial phase shift: 120° for three-phase windings and 90° for reduced two-phase windings. The currents have a corresponding temporal phase shift.

Figure 3.3 illustrates how a single stator coil forms a rectangular waveform of the current linkage and, because the air gap is smooth, essentially rectangular flux density in the air gap. A Fourier analysis for the unit amplitude of a trapezoid function reveals the amplitude of the fundamental, indicated by a dashed line, is 4/π.

If the winding of Figure 3.3 is fed with sinusoidal AC, a pulsating current linkage waveform is obtained that alternates according to Figure 3.4.

The curve form of the current linkage can be improved by adding slots; however, the current linkage of a single coil is still a pulse. Pulsating current linkage windings are currently popular in so-called tooth-coil or concentrated nonoverlapping windings machines (discussed in brief later in Section 3.3). The design principle for tooth coil or concentrated nonoverlapping windings is to interact with the PM rotor using an air-gap

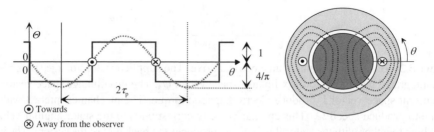

Towards
Away from the observer

Figure 3.3   Distribution of the current linkage Θ produced by a single full-pitch coil in the air gap of a rotating electric machine – If there are only two poles in the machine, its periphery will be encircled at the distance of two pole pitches $2\tau_p$. Since the circle is continuous, a single coil producing a continuous square wave can be assumed, and the Fourier analysis is straightforward. θ is an angle running on the air gap surface from 0 to $p2\pi$ that depends on the number of pole pairs p.

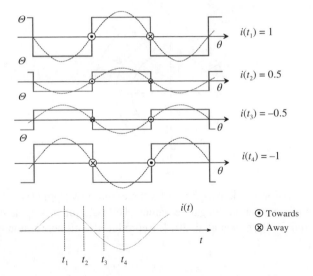

$i(t_1) = 1$

$i(t_2) = 0.5$

$i(t_3) = -0.5$

$i(t_4) = -1$

$i(t)$

⊙ Towards
⊗ Away

Figure 3.4 Pulsating current linkage waveform produced by a single loop at different times t and different current values – θ is an angle running on the air-gap surface from 0 to $p2\pi$ depending on the number of pole pairs p.

harmonic, usually not the fundamental. At the machine terminal connections, where the control electronics interfaces, this type of machine still looks like any other machine with rotating field windings. Therefore, the converter does not need to know how the windings are designed to control the machine. Feedback is not affected by the number of pole pairs p. The number of pole pairs is only needed to provide a speed feedback signal to the converter. Mechanical angular velocity must be multiplied by p to get accurate electrical angular speed information for motor control.

When the target is to produce a nicely rotating flux, at least two-phase windings are required. Two is the minimum number required to implement the necessary spatial phase difference of the windings and the temporal phase difference of currents. Two-phase windings are commonly used in various single-phase fed DOL machine systems. From the theoretical point of view, the two-phase windings are important, because the processor-based control systems of current electrical drives usually apply virtual two-phase constructions.

In principle, the phase number m can be selected freely, but in practice, due to the three-phase distribution network, most electric machines use three-phase constructions. Polyphase windings can be considered symmetrical when constructed as follows. The air gap periphery is divided evenly between all the poles so the pole arc, a pole pitch of 180 electrical degrees, is achieved. Pole pitch τ_p is given in metres.

$$\tau_p = \frac{\pi D}{2p} \tag{3.16}$$

Figure 3.5 shows how the periphery of the machine is divided into phase zones. In the figure, the number of pole pairs is $p = 1$, and the phase number is $m = 3$.

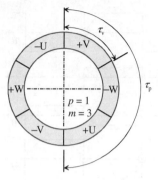

Figure 3.5 The division of the periphery of a three-phase two-pole ($m=3$, $p=1$) machine into phase zones of positive and negative values. The pole pitch is τ_p, and the phase zone distribution τ_v. *Source:* Pyrhönen et al. 2014. Reproduced with permission of John Wiley & Sons Ltd.

The phase zone distribution can be written

$$\tau_v = \frac{\tau_p}{m} \tag{3.17}$$

Therefore, the number of zones will be $2pm$. For a stator with Q slots, the number of slots per zone is expressed by q slots per pole and phase.

$$q = \frac{Q}{2pm} \tag{3.18}$$

The armature windings of a three-phase electric machine are usually constructed in the stator and spatially distributed in the stator slots so current linkage produced by the stator currents is distributed as sinusoidally as possible. Using the zone distribution described in Figure 3.5, the simplest possible three-phase rotating field windings result in three coils, which are inserted into slots according to Figure 3.6. Figure 3.7 illustrates the behaviour of the current linkage waveforms of the simplest windings.

Figure 3.7 illustrates that a current linkage waveform produced with simple three-phase winding deviates considerably from the sinusoidal waveform. Therefore, electric machines usually employ several additional slots per pole and phase. It is possible, however, to design a machine with satisfactory performance using only one slot per pole and phase ($q=1$). In a machine this simple, there are a large number of spatial harmonics in the magnetic flux traversing the air gap. These harmonics may result in unsmooth torque production, which is referred to as torque ripple, and potentially problematic machine vibration. The undesirable effects of torque ripple can be mitigated to some extent using special control techniques if nonsinusoidal currents are applied. It is more advisable, however, to design machines that minimize or eliminate torque ripple when applying sinusoidal currents. Smooth operation is extremely important when the mechanical power

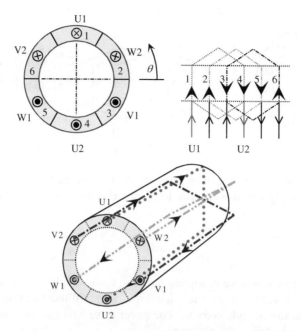

Figure 3.6 Simplest form of three-phase windings at the instant when $i_W = i_V = -\frac{1}{2} i_U$, $(i_U + i_V + i_W = 0)$. The lower part of the figure illustrates how the windings penetrate the machine. The coil end at the rear end of the machine is not illustrated realistically, but the coil comes directly from one slot to another without travelling along the rear end face of the stator. The ends of the phases U, V, and W at the terminals are denoted U1-U2, V1-V2, and W1-W2. *Source:* Pyrhönen et al. 2014. Reproduced with permission of John Wiley & Sons Ltd.

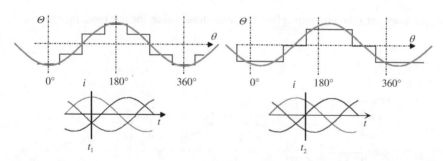

Figure 3.7 Current linkage waveforms of simple three-phase windings at the instants when $i_W = i_V = -\frac{1}{2} i_U$ and $i_W = -i_U$, $i_V = 0$. $i_U + i_V + i_W = 0$ always holds for a three-phase system without a neutral point conductor. The representations show the fundamental harmonics of a staircase current-linkage curve. The stepped curve is obtained by applying Ampère's law to the current-carrying tooth zone of the electric machine. *Source:* Pyrhönen et al. 2014. Reproduced with permission of John Wiley & Sons Ltd.

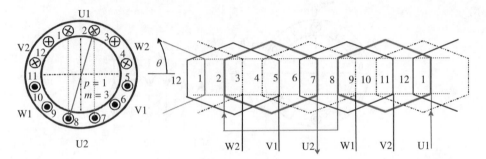

Figure 3.8   Three-phase windings with two coil sides per pole and phase. Slots per pole and per phase $q = 2$. *Source:* Pyrhönen et al. 2014. Reproduced with permission of John Wiley & Sons Ltd.

produced by an electrical drive is applied directly. For example, the torque ripple of a lift motor directly connected to rotate the cable drum of the lift must be less than 0.5% of the rated torque to ensure lift ride comfort. For direct drive systems, in general, the highest-quality torque motor drives are preferred.

Figure 3.8 and Figure 3.9 illustrate integral slot windings with $p = 1$, $q = 2$, and $m = 3$.

According to Figure 3.9, higher values of q (slots per pole and phase) result in a more sinusoidal current linkage of the stator windings. The harmonic amplitude of the current linkage for a polyphase ($m > 1$) rotating-field stator (or rotor) windings $\hat{\Theta}_{s\nu}$ can be expressed as follows.

$$\hat{\Theta}_{s\nu} = \frac{m}{2}\frac{4}{\pi}\frac{k_{ws\nu}\,N_s}{p\nu}\frac{1}{2}\sqrt{2}\,I_s = \frac{m k_{ws\nu}\,N_s}{\pi p\nu}\hat{i}_s \tag{3.19}$$

Inserting $\nu = 1$ into the equation gives the amplitude value for the fundamental.

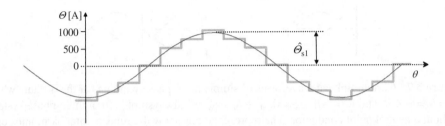

Figure 3.9   The current linkage produced by the windings on the surface of the stator bore of Figure 3.8 at time $i_W = i_V = -1/2i_U$. *Source:* Pyrhönen et al. 2014. Reproduced with permission of John Wiley & Sons Ltd.

EXAMPLE 3.1: Calculate the amplitude of the current linkage if $N_s = 100$, $k_{ws1} = 0.96$, $m = 3$, $p = 1$, and $i_{sU}(t) = \hat{i} = 10$ A. If the effective air gap is $\delta_{eff} = 0.001$ m, calculate also the flux density in the air gap caused by the current linkage.

SOLUTION: According to the given quantities, the effective value of the sinusoidal current is 7.07 A, and based on Equation (3.19), the amplitude of the current linkage is 916.7 A.

The field strength H is

$$H = \frac{\hat{\Theta}_{s\nu}}{\delta_{eff}} = \frac{916.7 \text{ A}}{0.001 \text{ m}} = 916700 \text{ A/m}$$

The corresponding flux density B is

$$B = \mu_0 H = 1.15 \text{ T}$$

3.3   Effective winding turns and spatial harmonics

Since the windings are spatially distributed in the slots on the stator surface, the flux penetrating them does not intersect all the windings conductors simultaneously. There is temporal phase shifting. Therefore, the emf of the windings is not calculated directly from the number of turns N_s. Electromotive force is calculated using the windings factor $k_{ws\nu}$, which corresponds to the spatial harmonics. The emf of a fundamental harmonic induced in a winding turn is calculated using the flux linkage ψ_s by applying Faraday's induction law.

$$e_\nu = -N_s k_{ws\nu} \frac{d\Phi}{dt} = -\frac{d\psi_{s\nu}}{dt} \tag{3.20}$$

Figure 3.10 illustrates the voltage phasor graph of the two-pole winding shown in Figure 3.8.

The coils of the phase U travel from slot 1 to slot 8 and from slot 2 to slot 7. Therefore, a voltage, the difference between phasors $\overline{U}_1$ and $\overline{U}_8$, is induced in coil 1. The total fundamental voltage of the phase can be expressed with the following equation.

$$\overline{U}_U = \overline{U}_1 - \overline{U}_8 + \overline{U}_2 - \overline{U}_7 \tag{3.21}$$

The fundamental winding factor k_{ws1} can be defined as the ratio of the geometric sum and the sum of absolute values.

$$k_{ws1} = \frac{|\overline{U}_1 - \overline{U}_8 + \overline{U}_2 - \overline{U}_7|}{|\overline{U}_1| + |\overline{U}_8| + |\overline{U}_2| + |\overline{U}_7|} = 0.966 \leq 1 \tag{3.22}$$

The winding factor is always ≤ 1. A value of $k_{ws1} = 1$ can be achieved when $q = 1$. In general, the winding factor for a spatial harmonic ν is calculated as the product of the pitch factor k_p,

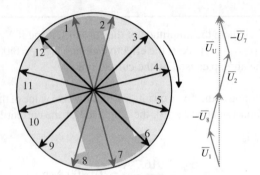

Figure 3.10 The voltage phasor graph for the windings of Figure 3.8 – $Q_s = 12$, $p = 1$, and $q_s = 2$. At the given time instant, the spinning SM salient pole rotor with nonvisible DC windings produces peak positive and negative voltage values in slots 1 and 7, respectively. Peak values are induced in slots 2 and 8 next, and so on. The phasor angles depict time difference, and the phasor lengths correspond to induced voltage magnitude. *Source:* Pyrhönen et al. 2014. Reproduced with permission of John Wiley & Sons Ltd.

the distribution factor k_d, and the skewing factor k_{sq}, which can be found, for example, in Pyrhönen et al. (2014).

$$k_{w\nu} = k_{p\nu}k_{d\nu}k_{sk\nu} = \sin\left(\nu\frac{W}{\tau_p}\frac{\pi}{2}\right) \cdot \frac{2\sin\left(\frac{\nu}{m}\frac{\pi}{2}\right)}{\frac{Q}{mp}\sin\left(\nu\pi\frac{p}{Q}\right)} \cdot \frac{\sin\nu\frac{\pi}{2}\frac{1}{mq}}{\nu\frac{\pi}{2}\frac{1}{mq}} \tag{3.23}$$

W is the coil width, ν is the order of the harmonic, Q_s is the number of stator slots, p is the number of pole pairs, and m is the phase number. Depending on the number of phases m, the slot windings develop spatial harmonics (k is a positive integer) given by

$$\nu = 1 \pm 2km \tag{3.24}$$

A symmetrical three-phase winding may develop the following spatial harmonics listed in Table 3.1.

The values for harmonic order ν shown in the Table 3.1 do not include −1 or any harmonics divisible by 3. A symmetrical polyphase winding does not produce a harmonic that propagates in the opposite direction at the frequency of the fundamental harmonic. In contrast, a single-phase winding, $m = 1$, does produce a harmonic with a $\nu = -1$ ordinal. This particularly harmful harmonic impedes the smooth efficient operation of single-phase machines. For instance, a single-phase induction motor with the field rotating in the negative direction will not start without assistance.

Table 3.1 Orders of the harmonics developed by a three-phase winding ($m = 3$)

k	0	1	2	3	4	5	6	7	8 . . .
ν	+1	+7	+13	+19	+25	+31	+37	+43	+49 . . .
		−5	−11	−17	−23	−29	−35	−41	−47 . . .

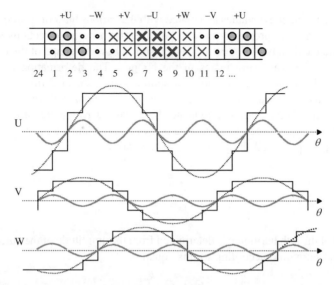

Figure 3.11 The cancellation of the third harmonic in a three-phase winding. The winding currents are $i_U = -2i_V = -2i_W$. When the third harmonic of phases V and W combines with the harmonic of phase U, the harmonics cancel. *Source:* Pyrhönen et al. 2014. Reproduced with permission of John Wiley & Sons Ltd.

In an electrical drive, harmonics in the windings can result in torque ripple. Especially for the lowest stator harmonics ($k = 1$), harmonic order values of -5 and $+7$ produce a significant sixth time harmonic in the machine torque if the rotor has the same order spatial harmonics.

From the winding factor, see Equation (3.23), the relative magnitude of the current linkage harmonic, (3.19), can be derived. Figure 3.11 illustrates how even a short pitch winding produces a third harmonic with significant amplitude, about 22% of the amplitude of the fundamental harmonic. This, however, is not very harmful, because the third harmonic is cancelled by the current linkage wave produced by the windings. The figure shows windings currents $i_U = -2i_V = -2i_W$ and $i_U = 1$.

An ideal polyphase winding, therefore, produces harmonics with orders that can be calculated by using Equation (3.24). However, the third harmonic may cause problems in real delta-connected windings if the machine has any anisotropy in the concentricity of its rotor and stator. Normally, it is recommended to use star-connected windings, if possible, to avoid a circulating current in delta. For example, rotor eccentricity in a delta-connected winding machine may cause significant extra losses caused by the third-harmonic current circulating in delta.

When the stator is fed at an angular frequency ω_s, the angular speed of the harmonic ν with respect to the stator is the following ratio.

$$\omega_{s\nu} = \frac{\omega_s}{\nu} \tag{3.25}$$

The movement direction for the spatial harmonic, denoted by the sign of the harmonic, must be taken into account in Equation (3.25). For example, the harmonic $\nu = -5$ produces a spatial wave that propagates slowly in the negative direction.

The positive and negative values associated with the various harmonic order values verify that flux wave harmonics propagate in both positive and negative directions in the air gap. This was illustrated already in Figure 3.7, which shows how the shape of the current linkage wave changes as the wave propagates in the air gap. The deformation of the wave, in particular, is an indication that harmonic amplitudes proceed at different speeds and in different directions. Each harmonic induces a voltage of fundamental frequency in the stator windings. The order of the harmonic indicates how many wavelengths of the harmonic fall within the distance $2\tau_p$ between a single pole pair of one fundamental harmonic. This yields the number of pole pairs and the pole pitch of the harmonic.

$$p_\nu = \nu p \tag{3.26}$$

$$\tau_\nu = \frac{\tau_p}{\nu} \tag{3.27}$$

The amplitude of the ν th harmonic is determined from the amplitude of the current linkage of the fundamental harmonic according to this equation

$$\hat{\Theta}_\nu = \hat{\Theta}_1 \frac{k_{w\nu}}{\nu k_{w1}} \tag{3.28}$$

Figure 3.12 shows how an increase in the number of slots per pole and phase (q) improves the shape of the current linkage curve, now $q_s = 3$. In the figure, the magnetic axis of phase U is in the direction of the arrow drawn in the middle of the curves on the left. For a three-phase machine, the directions of the phase V and W currents result in magnetic axes spaced 120° from the axis of phase U.

The current linkage wave propagating through the stator bore produces a rotating field that propagates along the periphery. This is a rotating-field machine winding. The main flux penetrating the winding varies almost sinusoidally as a function of time

$$\Phi_h(t) = \hat{\Phi}_h \sin \omega_s t \tag{3.29}$$

According to Faraday's induction law, the induced voltage fundamental is obtained by the flux linkage fundamental ψ_{s1}

$$e_{s1} = -\frac{d\psi_{s1}}{dt} = -N_s k_{ws1} \frac{d\Phi_h}{dt} = -N_s k_{ws1} \omega_s \hat{\Phi}_h \cos \omega_s t \tag{3.30}$$

If the induced voltage is assumed sinusoidal and neglecting the minus sign, the RMS value becomes

$$E_s = \frac{1}{\sqrt{2}} \hat{e}_{smv1} = \frac{1}{\sqrt{2}} \omega_s k_{ws1} N_s \hat{\Phi}_h \tag{3.31}$$

To be able to calculate this value, the peak value of the flux related to the fundamental must be known. This peak value must be calculated by integrating the flux density over the pole pitch and the effective machine length as illustrated by Figure 3.13.

$$\hat{\Phi}_h = \int_0^{\tau_p} \int_0^{l'} (B_\delta) d\tau \, dl \tag{3.32}$$

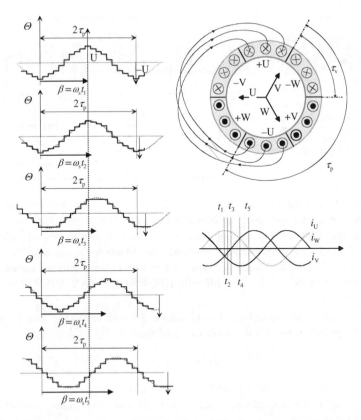

Figure 3.12 Current linkage waveforms produced by the three-phase winding at different values of three-phase current – The wave propagates to the right, and currents vary as a function of time. Flux peaks shift to the right with time as shown by angle β. U indicates the magnetic axis of phase U, which is stable. *Source:* Pyrhönen et al. 2014. Reproduced with permission of John Wiley & Sons Ltd.

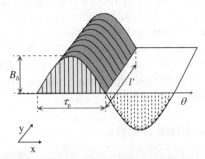

Figure 3.13 Air-gap fundamental flux density in a rotating field machine. *Source:* Pyrhönen et al. 2014. Reproduced with permission of John Wiley & Sons Ltd.

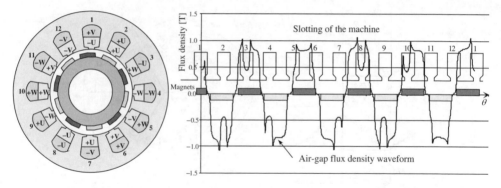

Figure 3.14 The air-gap flux density of a 12-slot stator by 10-pole rotor electric machine ($q = 0.4$). There is a small two-pole fundamental in the machine air gap, but torque production comes from the fifth air-gap spatial harmonic. *Source:* Modified from Jussila 2009. Concentrated Winding Multiphase Permanent Magnet Machine Design and Electromagnetic Properties—Case Axial Flux Machine. Doctoral thesis, Acta Universitatis Lappeenrantaensis, Finland (https://www.doria.fi/bitstream/handle/10024/50604/isbn 9789522148834.pdf).

Assuming the flux density distribution to be sinusoidal with respect to x and constant with respect to y, integration reveals the peak value of flux.

$$\hat{\Phi}_h = \alpha_i \hat{B}_\delta \tau_p l', \tag{3.33}$$

where l' is the equivalent core length, and α_i is a coefficient indicating the arithmetical average of the flux density in the x-direction, which with a sinusoidal flux density distribution yields the value $\alpha_i = 2/\pi$.

As discussed previously, the shape of the propagating waveform improves with increasing number of slots per pole and phase q. This type of stator winding is preferable for most AC machines. However, if a PM rotor is assumed nonconductive, it is possible to replace the traditional winding with a type of winding called a fractional-slot tooth-coil winding (concentrated nonoverlapping winding) with $q \leq 0.5$. Figure 3.14 illustrates a fractional slot winding machine with 12 stator slots and 10 rotor poles ($q = 0.4$).

The torque production for the 12–10 machine illustrated in Figure 3.14 comes from the fifth spatial air-gap harmonic. This suits the rotor configuration and results in relatively smooth torque, because this type of machine can have sinusoidal back emf at the machine terminals. Therefore, it can be controlled using the same vector control theory as machines with traditional windings. From the machine construction point of view, the rotor magnets should be nonconductive to minimize losses induced by the complex harmonic content of the air gap. This is, nonetheless, a machine design issue. From the electrical drive's viewpoint, the machine reacts like an ordinary PMSM.

3.4 Induction machine rotors

The most common industrial machine type is the squirrel-cage induction motor. In the rotor of such a machine, there is a short-circuited squirrel-cage arrangement of winding conductors. See Figure 3.15. The magnetic circuit of the rotor in low-speed industrial motors consists of electrical steel laminations and winding conductors. The circular array of conductor rails can

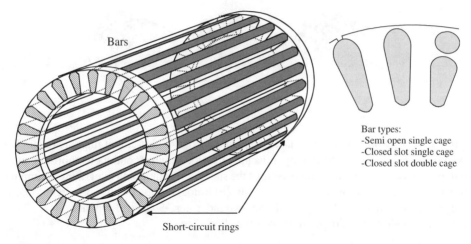

Figure 3.15 Single-cage winding conductors. Cooling fans are not shown, and $Q_r = 24$. The image on the right shows typical cross sections for the industrial IM bars of semi-closed or closed slot rotors with either a single or double cage. *Source:* Pyrhönen et al. 2014. Reproduced with permission of John Wiley & Sons Ltd.

be die cast into slots in the electrical steel, or a complete squirrel cage assembled from prefabricated copper bars and rings can be soldered or welded into the laminated steel core. The short-circuiting end rings often include cooling fins.

Double cages are frequently used in DOL motors to reduce starting current and increase starting torque. From the power electronic drives' efficiency point of view, the rotor construction has a clear adverse impact. Normally, industrial induction motors with die-cast rotors have closed rotor slots because of easy manufacturing. The slot closing iron bridges are prone to losses caused by the pulse width–modulated (PWM) voltage-caused harmonics. In addition, the double cage is far from ideal for a PWM supply. Actually, a single cage with semi-open slots should be a good choice for PWM-supplied motors. Double-cage motors with closed rotor slots, however, are frequently used also in PWM supply. A slightly lower efficiency results from this.

In high-speed IMs, solid steel rotors can be used. The performance characteristics of a solid rotor can be significantly improved by slitting the surface of the rotor as illustrated by Figure 3.16. Narrow axial slots are used to guide the flow of useful, torque-producing eddy currents in a direction favourable to torque production. Radial slots increase the path length for the harmful eddy currents produced by certain high-frequency phenomena. Adding a squirrel cage to the solid core makes the rotor perform similarly to a laminated squirrel-cage machine rotor. Figure 3.17 shows the behaviour of principal static torque for different rotor types.

The wound slip-ring rotor of an IM is designed following the same principles as for multiphase stator windings. A wound rotor must be equipped with the same number of pole pairs p as the stator. The rotor winding is connected to an external circuit via slip rings and brushes. Before power electronics became standard, the slip-ring induction motor was very popular in industrial applications, because its speed can be controlled to some extent by adjusting the rotor circuit resistance with an external three-phase resistor. Naturally, this type of control increases system losses, so it is no longer commonly used.

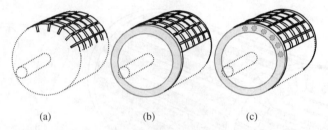

Figure 3.16   (a) Solid rotor with axial and tangential slots – Short-circuit rings are required in this rotor. They can be constructed by truncating the slots leaving continuous short-circuiting bands on the rotor ends that extend outside of the stator or by equipping integrating aluminium or copper short-circuit rings with the rotor ends. (b) Rotor equipped with short-circuit rings in addition to slots and (c) Slotted and cage-wound rotor. *Source:* Pyrhönen et al. 2014. Reproduced with permission of John Wiley & Sons Ltd.

Today, squirrel-cage induction motors are easily controlled with power electronic frequency converters. There are still, however, important applications for the slip-ring wound rotor IM. For example, modern wind turbines are an important new area of application making use of slip-ring doubly fed generators. The machine stator of these generators connects directly to the grid, and the rotor is supplied via a four-quadrant frequency converter driving the machine at either a subsynchronous or supersynchronous speed. This enables wind turbine speed control using a smaller-capacity frequency converter—typically only about 30% of rated drive power.

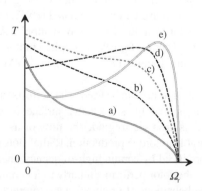

Figure 3.17 Schematic torque curves of different induction rotors as a function of mechanical angular speed Ω_r for (a) a smooth solid rotor without short-circuiting rings, (b) a smooth solid rotor equipped with copper short-circuiting rings, (c) an axially and radially slotted solid rotor equipped with copper short-circuiting rings, (d) a solid rotor with a copper cage, and (e) a rotor with a double-cage winding. *Source:* Pyrhönen et al. 2014. Reproduced with permission of John Wiley & Sons Ltd.

3.5   The damper winding

A damper winding resembles the squirrel cage of an induction motor. Typically, it is installed in a DOL SM to guarantee continued synchronous operation during a transient. DOL PMSMs with damper windings are intended for distributed power production, such as in small hydropower plants. In practice, some damping is present in all machine types, and even without the damper winding, induced oscillations will eventually attenuate. In nonsalient pole generators, by inducing eddy currents for attenuation into it, the solid frame of the rotor can be used as a virtual damper winding. Figure 3.18 illustrates the structure of the damper winding for a salient-pole SM.

Damper windings are typically configured for a particular application and purpose. For example, internal damping is essential for the proper operation of DOL machines. Although power-electronics–controlled position feedback drives can operate without the damping provided by a damper winding, the additional damping does have a remarkable effect on machine operation. Damped machine torque response is always faster than undamped response. The torque rise speed of a damped machine fed by power electronics is typically from 5 to 10 times that of a similar undamped machine. The faster torque rise time is based on the low subtransient time constant of damped machines that allows for fast stator current rise. In a frequency converter supply, however, the damper winding may suffer from high losses caused by the PWM supply of the stator. Coupled to a low switching frequency medium-voltage converter, the damper may even fail.

A damper winding makes it possible to start a DOL motor like an asynchronous machine. For a DOL generator, the damper winding damps the counterrotating fields produced by asymmetrical loads.

Damping parameters must be determined to model a damper-equipped PWM-supplied machine. This can be a challenge, because the damper is usually anisotropic and must be modelled on the direct and quadrature axes separately. It is possible, however, to tune the control using a suitable time constant approach.

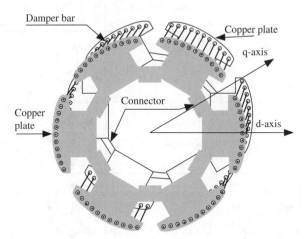

Figure 3.18   The configuration of a damper winding for a six-pole, salient-pole SM. To form a complete damper winding, all the bars must be shorted at both ends. In this figure, the final end plates of the pole shoes are made of copper and welded to the damper bars. The plates are electrically interconnected via copper connection paths, which are shown schematically. *Source:* Pyrhönen et al. 2014. Reproduced with permission of John Wiley & Sons Ltd.

3.6   AC winding systems

Most electric machines in use today have three-phase or two-phase windings. The two-phase system is common for low-power DOL drives. It is rare, however, to see a frequency converter operating a two-phase motor. The great majority of industrial AC electrical drives are three-phase machines, and most converters are designed for three-phase windings. From the power electronics' point of view, different winding system configurations offer some design freedom. Winding system configurations based on multiples of the three phases are interesting from the power increase point of view when using low-voltage converters to supply large machines. A pair of three-phase converters can supply, for example, a six-phase machine with a 30° temporal phase shift. This results in fewer spatial harmonics, increases the operating harmonic winding factor, and offers the possibility of increased drive power while still keeping voltage low.

Table 3.2 introduces winding systems commonly used in practice. For industrial motors, the most common winding type is the three-phase winding. In large SMs, a six-phase winding is used to some degree in the context of power electronic drives. Single- and double-phase

Table 3.2   Phase systems of the windings of electric machines. The fourth column introduces radially symmetric winding alternatives.

No. of Phases m	Normal System	Reduced System	Corresponding Nonreduced System
1			$m' = 2$
2			$m' = 4$
3			$m' = 6$
4			$m' = 8$
5			$m' = 10$
6			$m' = 12$

windings occur chiefly in small induction motors that are fed directly from a single-phase network.

There can be only one single-phase winding axis per magnetic axis of an electric machine. A polyphase system does not result if another phase winding is located on the same axis, because both windings produce collinear flux. Therefore, each nonreduced phase system that involves an even number of phases is reduced to involve only half of the original number of phases m' as illustrated in Table 3.2. If $m'/2$ is an odd number, a radially symmetric polyphase system, also called a *normal system,* is the result.

3.7   DC machine windings

DC machines with commutators have armature windings on the rotor and several other windings in the stator. Figure 3.19 shows all the important industrial DC machine windings including a lap-winding armature, compensating windings on the stator pole shoes, commutating windings wound on the commutating poles on the quadrature axis, and the excitation (field) winding on the stator poles.

A field winding produces current linkage on the direct axis of the DC machine. The direct axis (d-axis) is defined based on the minimum reluctance of the DC machine magnetic circuit. In practice, the d-axis is found in the middle of the exciting pole. An armature winding, however, produces current linkage on the quadrature axis (the q-axis found between poles) of the machine. The flux produced by the armature negatively affects machine performance. In a fully compensated machine, commutating poles or compensating windings mitigate this harmful influence. A fully compensated DC machine ensures, by its construction, the best possible torque production, because the flux and the armature current linkage are always perpendicular.

The air-gap flux produced by armature current ($i_A L_m$) in an electric machine is called the armature reaction. This armature reaction can be eliminated in some DC electrical machine architectures by adding compensating and commutating pole windings. Without armature reaction compensation, DC machine armature current commutation is more difficult. For PM-excited DC machines, the problem is less severe, because PM excitation results in inherently less armature reaction. Because of this, small PM-excited DC machines normally do not have commutating pole or compensating windings.

3.8   The brushless DC machine

In principle, the Brushless DC machine (BLDC) is built with the winding seen previously in Figure 3.8. The rotor of a BLDC differs from the rotor of a PM synchronous motor. The BLDC rotor produces, in principle, an air-gap flux density with a full square wave shape. Figure 3.20 illustrates the principal construction of a $p = 1$ BLDC machine. The windings are star connected, and the PMs span 180° electrical. The BLDC winding also can be connected in a delta configuration. For the delta configuration, the PMs span 120° electrical.

Figure 3.21 illustrates the principal behaviour of the BLDC machine when the machine operates with rectangular current pulses.

The three-phase currents of the BLDC build 120° DC pulses that follow each other. The converter of the BLDC has the same principal components as a three-phase two-level PWM converter for AC motors. The commutation is done, however, according to the rotor position

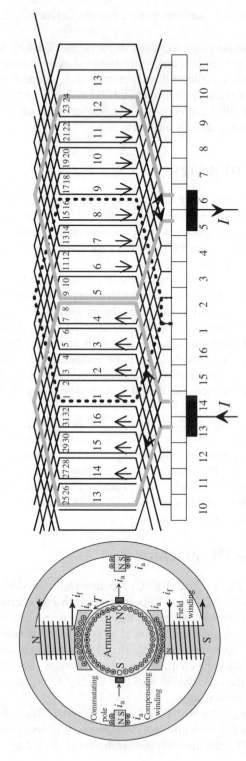

Figure 3.19 The image on the left illustrates the winding locations of a DC machine. On the right, the armature lap winding with connection to commutator and brushes is shown. *Source:* Pyrhönen et al. 2014. Reproduced with permission of John Wiley & Sons Ltd.

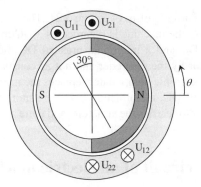

Figure 3.20 BLDC PM machine with an $m=3$, $q=2$, and $p=1$ winding and a two-pole rotor surface magnet to produce square wave flux density in the air gap. The winding is, in principle, a full-pitch three-phase winding. See Figure 3.8. The distributed winding has two coils for each phase. Shown here is the U phase (U_{11}–U_{12} and U_{21}–U_{22}).

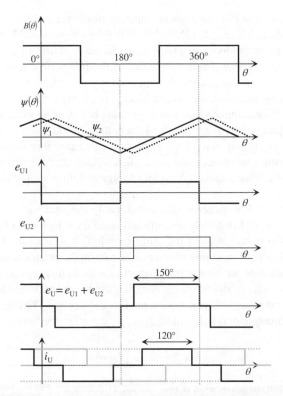

Figure 3.21 The flux density, flux linkages, induced voltages, and currents of a BLDC with $q=2$. The stator coils feature a 30° spatial phase shift. The phase of currents must match with the phase of induced voltages. For example, i_U in the figure produces constant power during 120 electrical degrees with the induced voltage e_U.

pulses resulting in the 120° current pulses of Figure 3.21. There is no need for vector control or a motor model. Therefore, the machine is regarded as a DC machine. The only principal difference between the BLDC and a traditional brushed DC motor is that DC pulses are supplied in the BLDC motor by the converter and not as constant DC via the brushes and the commutator. A power electronic commutator replaces the mechanical commutator of the brushed DC machine. In a BLDC, the armature is in the stator and PM excitation is in the rotor. In a DC machine, the rotor carries the armature, and there are either field coils or PMs in the stator.

3.9   The magnetic circuit of an electric machine

The magnetic circuit of an electric machine is governed by Ampère's law where the magneto motive force (MMF) is marked equal to the current linkage of the machine. In the following equation, the left hand line integral is the MMF, as recommended by IEC 27-1, and Θ on the right hand side is the current linkage.

$$\oint \boldsymbol{H} \cdot \mathrm{d}\boldsymbol{l} = \sum i = \Theta \tag{3.34}$$

All the electric currents and PM materials in a running machine combine to produce its current linkage Θ. The closed line integral (MMF, $\oint \boldsymbol{H} \cdot \mathrm{d}\boldsymbol{l}$) of the magnetic field strength $\boldsymbol{H}$ along the magnetizing route l corresponds, in practice, to the sum of the magnetic voltages from different parts of the machine. More reluctivity in the magnetic circuit results in a larger MMF, so more current linkage is needed to magnetize the machine to the desired flux level.

   With no PMs in the machine, the current linkage is mainly consumed in the air gap. This leads to a fairly stable inductance value when operating below iron saturation. However, if the machine saturates due to overvoltage, the field strength in various parts of the machine begins to increase, and more current is required to produce the same flux linkage, decreasing the machine's magnetizing inductance. Moreover, high torque saturates the magnetizing inductance considerably, because it causes the lines of force to follow longer paths than at the no-load condition.

   The permeability of PM material is approximately the same as the permeability of a vacuum. Therefore, the PM materials strongly influence the reluctance of a magnetic circuit; and consequently, the inductances of the armature winding of a rotating-field machine.

   Polyphase rotating-field machines play a central role on the electrical drive stage, so it will be constructive to calculate the magnetizing inductance for an unsaturated machine. As stated previously, the peak value of the flux in an electric machine depends on the pole pitch τ_p, the machine equivalent core length l', and the air-gap flux density B_m. The flux linkage of a single phase is obtained correspondingly by multiplying by the effective turns (of winding) $k_{ws1}N_s$.

$$\hat{\psi}_h = k_{ws1}N_s \frac{2}{\pi}\tau_p l' \hat{B}_m \tag{3.35}$$

For a single phase, the magnetic flux density in the effective air gap δ_{ef} (which takes also the effect of the iron into account) can be determined using the current linkage Θ of the phase.

$$\hat{B}_m = \frac{\mu_0 \hat{\Theta}_s}{\delta_{ef}} \tag{3.36}$$

The effective air gap δ_{ef} represents the reluctance of entire magnetic circuit half with one parameter. Depending on the saturation level of the iron, δ_{ef} can be close to the real air gap, but for heavily saturated iron parts, δ_{ef} can be significantly larger than the real air gap. The effect of saturated iron can be seen in the nonlinear behaviour of the magnetizing inductance as a function of the increasing flux density level.

The flux linkage of a single phase can be expressed as follows.

$$\hat{\psi}_h = k_{ws1} N_s \frac{2}{\pi} \frac{\mu_0 \hat{\Theta}_s}{\delta_{ef}} \tau_p l' \tag{3.37}$$

The current linkage of the phase winding is

$$\hat{\Theta}_s = \frac{4}{\pi} \frac{k_{ws1} N_s}{\nu 2p} \sqrt{2} I_s \tag{3.38}$$

Substitution yields the following for the flux linkage.

$$\hat{\psi}_h = \frac{2}{\pi} \mu_0 \frac{1}{2p} \frac{4}{\pi} \frac{\tau_p}{\delta_{ef}} l' (k_{ws1} N_s)^2 \sqrt{2} I_s \tag{3.39}$$

By dividing by the peak value for the current, the main inductance of the single phase becomes

$$L_p = \frac{2}{\pi} \mu_0 \frac{1}{2p} \frac{4}{\pi} \frac{\tau_p}{\delta_{ef}} l' (k_{ws1} N_s)^2 \tag{3.40}$$

The magnetizing inductance of the m-phase machine can be determined by multiplying the main inductance by $m/2$.

$$L_m = \frac{m}{2} \frac{2}{\pi} \mu_0 \frac{1}{2p} \frac{4}{\pi} \frac{\tau_p}{\delta_{ef}} l' (k_{ws1} N_s)^2 \tag{3.41}$$

The magnetizing inductance is the most important part of the machine synchronous inductance L_s.

$$L_s = L_m + L_{s\sigma} \tag{3.42}$$

The other part of the synchronous inductance is the leakage inductance of the winding $L_{s\sigma}$, which will not be formulated here. Formulation of the leakage inductance term can be found in the literature (Pyrhönen et al., 2014).

The inductance value expressed by Equation (3.41) gives some guidelines for the electrical drive designer. Magnetizing inductance is inversely proportional to the effective air gap δ_{ef}. When PM materials on the rotor surface are used in the magnetic circuit, the effective air gap will be significantly higher than in traditional machines. There will be an additional effective air gap caused by the PM materials of height h_{PM}.

$$\Delta \delta_{ef} = \frac{h_{PM}}{\mu_{rPM}} \tag{3.43}$$

As a result, the magnetizing inductance of a PM machine can be very low, resulting, for example, in needing a high switching frequency converter.

Examining (3.41) further reveals that magnetizing inductance will be inversely proportional to the square of the pole pair number p, because pole pitch is inversely proportional to the number of poles $2p$.

$$\tau_p = \frac{\pi D}{2p} \tag{3.44}$$

Inserting expression (3.44) into (3.41) yields the following expression for magnetizing inductance.

$$L_m = \mu_0 \frac{mDl'}{\pi p^2 \delta_{ef}} (k_{ws1} N_s)^2 \tag{3.45}$$

According to Equation (3.45), the level of magnetizing inductance L_m drops as the number of pole pairs p increases. Therefore, increasing the number of pole pairs in an induction motor, for example, makes it increasingly unsuitable, because the magnetizing inductance of an induction motor must be as high as possible. Because dropping of the magnetizing inductance with increasing pole pair number p does not actually harm SMs, most electric machines with a large number of pole pairs are SMs. Synchronous machines can even benefit from a lower value of magnetizing inductance, because their peak torque is inversely proportional to $L_m + L_{s\sigma}$.

Therefore, understanding the relationships expressed by Equation (3.45) is important when working in the field of electrical drives and when selecting a correct machine type for a specific purpose. How magnetizing inductance behaves as a function of voltage (flux level), torque, and pole pair number can be illustrated, in general, with the curves presented in Figure 3.22. The torque dependence is a bit more difficult to understand, but remembering the Faraday's principle of lines of forces, it is possible to understand that under stress (torque), the lines of force get longer and the MMF $= \oint H \cdot dl$ becomes larger, which results in less magnetizing inductance. For electrical drives, it is often important to model the behaviour of L_m appropriately to calculate the current model correctly.

Electric machines are usually designed so that the magnetic circuit somewhat affects effective air gap δ_{ef} already at the rated conditions. Therefore, at the rated flux level $\Phi_{m,pu} = 1$,

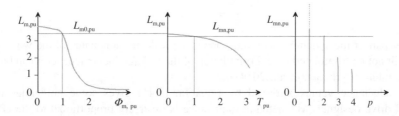

Figure 3.22 Principal behaviour of magnetizing inductance as a function of (left) level of air gap flux Φ_m, (middle) torque magnitude T, or (right) pole pair number p – The $L_{m0,pu}$ represents the magnetizing inductance at no load at the rated stator flux-linkage level. For example, for a 110 kW four-pole induction motor, $L_{m0,pu} > 3$. $L_{mn,pu}$ is the magnetizing inductance at the rated torque and rated stator flux-linkage level. In larger induction motors with $p = 2$, $L_{mn,pu} \approx 3$. And, as pole pair number increases, magnetizing inductance decreases as a function of $1/p^2$. In practice, multiple-pole IMs must be designed so magnetizing inductance is large enough to maintain an acceptable power factor.

the magnetizing inductance has already begun to saturate as Figure 3.20 indicated. Because iron saturates rapidly in the range of 2 T, the level of magnetizing inductance collapses if flux level increases substantially from its rated value. Finally, magnetizing inductance reaches a stable low level where the iron is fully saturated and the inductance corresponds approximately to the inductance of an air core winding. Saturation as a function of torque is more moderate but reaches significant values in the range of 3 times the rated torque. The inductance is inversely proportional to the square of the pole pair number p.

In PMSMs, the effective air gap is large and, therefore, magnetizing inductance is low. The magnetizing inductance per unit value falls typically in the range of $L_{m, pu} = 0.2$ to 0.5 to guarantee adequate peak torque.

3.10   Motor voltage, flux linkage, flux, field weakening, and voltage reserve

There are many drive types where a wide speed range is needed. Such drives are, for example, traction drives, different rolling mill drives, and winding drive in, for example, the paper and steel industries. In such cases, there is a need to provide high torque capability at low speeds and therefore select the rated operating point of the machine at a fairly low speed and then drive the motor also at speeds much higher than the rated speed. It is typical that at least 3 times the rated speed of the motor is used in such a drive, but much higher speed ratios may be used. In such cases, the motor is at low speeds operated with its rated or even increased flux level, and beyond the rated point the motor is driven into field weakening. Field weakening operation principles are common for all electric motor types. However, those machines that allow easy flux level controlling may more easily be used in the field weakening, and machines like PMSMs may have difficulties in going into the field weakening as the PMs cannot be controlled and the field weakening must be performed by demagnetizing stator current.

According to Faraday's law, the voltage induced in an electric motor—the back emf—is related to the electrical angular frequency ω_s and the stator flux linkage ψ_s of the drive.

$$\hat{e}_s = \hat{\psi}_s \omega_s \tag{3.46}$$

The stator voltage must be approximately equal to the induced voltage. In electrical motor drives, the stator voltage is slightly larger than the induced voltage, because there is a small voltage drop in the stator resistance R_s. Correspondingly, the induced voltage is a bit larger than the induced voltage in generator drives.

$$\hat{u}_s \approx \hat{e}_s; \quad U_s \approx E_s \tag{3.47}$$

According to Equations (3.46) and (3.47), the induced and terminal voltages of an electric machine increase with angular frequency if flux linkage remains constant. It is typical in electrical drives to keep flux linkage constant until maximum drive voltage is reached. Then, the field is weakened. Since ω_s increases with machine frequency, the value of stator flux linkage $\hat{\psi}_s$ must be decreased to keep the induced voltage at the maximum allowable level.

$$\hat{e}_s = \hat{\psi}_s \omega_s \approx \hat{u}_{s,max} \tag{3.48}$$

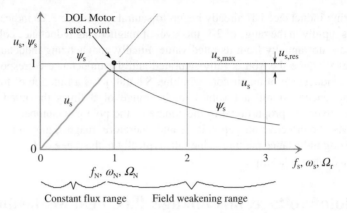

Figure 3.23 Voltage, back emf, flux behaviour, and voltage reserve in an electrical drive with a 50 Hz AC motor running at 3 times the rated per-unit speed. Field weakening begins at 45 Hz and not at 50 Hz, which is the rated frequency of the DOL motor used in the application.

DC machines exhibit this same behaviour, because rotor frequency depends on the rotor rotational speed and the number of poles.

In drives with high-performance dynamics, there is yet another significant feature that must be considered. The converter must provide some voltage reserve. Field weakening must begin at a lower frequency, below 50 Hz, for example, to maintain a suitable voltage reserve. Normally, a 5% to 10% voltage reserve is needed to respond to sudden changes in torque demand. Voltage reserve can be defined as follows.

$$\hat{u}_{s,res} = \hat{u}_{s,max} - \hat{e}_{s,max} = \hat{u}_{s,max} - (\hat{\psi}_s \omega_s)_{max} \tag{3.49}$$

Figure 3.23 illustrates voltage and flux linkage behaviours for an electrical drive in the constant flux and field-weakening regions. In the figure, a machine designed for DOL operation has been used in a speed-controlled drive. The DOL motor rated point corresponds to, for example, the 50 Hz rated torque point of the motor.

In Figure 3.23, the rotational speed and flux linkage per-unit values refer to the DOL drive values of the same machine. For example, field weakening begins at 360 V, 45 Hz for a 50 Hz, 400 V frequency converter-driven motor with about 10% voltage reserve ($U_{s,pu} = 0.9$, $f_{s,pu} = 0.9$). When motor torque must be increased quickly, the converter can deliver an extra 40 V to increase torque in the field-weakening region.

As said earlier, the field-weakening region is important, for example, for vehicle traction drives and for many electrical drives used in the paper and metal-processing industries. Examples include the motor drives in electric automobiles and trains and in the reelers and unwinders used by the pulp and paper industry. A common step in the paper-making process is to feed paper onto or off of a roll or reel at a constant linear paper speed. Maintaining this constant paper speed and maintaining suitable web tension require the roll to change its rotational speed and torque as the diameter of the paper on the roll varies. For example, as paper is fed onto it, a roll must continuously increase rotational speed and decrease torque.

For these applications, the electrical drive's voltage reserve can be small, because speed changes are gradual. In most cases, however, at least 5% voltage reserve is designed into the electrical drive.

> **EXAMPLE 3.2:** Calculate the rated flux linkage of a 400 V star-connected 50 Hz induction motor assuming a 1.5% voltage drop due to stator resistance. What is the value of the flux linkage at 75 Hz in the field-weakening region when 10% voltage reserve is used?
>
> **SOLUTION:** The motor peak phase voltage after the voltage drop in the stator winding is $0.985 \cdot \sqrt{2} \cdot 400 \, V/\sqrt{3} = \sqrt{2} \cdot 227.5 = 321.7$ V. The rated angular frequency is $\omega_s = 50 \cdot 2\pi$. Therefore, the peak value for stator flux linkage in the drive is $321.7/100\pi = 1.024$ Vs. This value can be used at low drive speeds.
>
> Because of the 10% voltage reserve, field weakening begins at 45 Hz. At 75 Hz, the flux linkage will be $(45 \, Hz/75 \, Hz) \cdot 1.024 \, Vs = 0.61$ Vs.

3.11   Motors in power-electronic electrical drives

AC and DC motors are subject to nonidealities-losses and torque quality problems. The presence of losses means that machines must be designed to manage the thermal condition of the materials. And because PWM-based control introduces additional losses, better cooling is needed for a power-electronic supply than that needed for a DOL drive. Moreover, the electrical insulations used must tolerate operating temperatures and endure for the design life of the machine. In general, however, electromechanical power conversion is very efficient, and losses need only be properly considered and effectively managed.

However, torque or force quality problems can be more problematic and can deteriorate the performance of the entire drive system. Torque quality can be especially critical in a direct-drive (DD) system. In a DD elevator, for example, the passenger experiences uncomfortable speed ripple or noise if the DD motor produces torque ripple that is more than 0.5% of the rated torque. Therefore, the lift DD motor must be very carefully designed to achieve a smooth torque level.

Motor control is one method of improving torque or force quality that is occasionally discussed, and it is possible to improve torque or force quality by profiling motor currents. In principle, if there is enough voltage reserve for current profiling, producing smooth output under all conditions is possible using this approach. However, current profiling control is demanding, and it requires a specially tuned drive system. No general models are available for motor current profiling. Complex controller systems with learning capability could possibly be used to tune motor current waveforms in real time during operation. In general, however, AC and DC motors are designed to produce torque and force quality that is good enough for the intended purpose using sinusoidal current input.

An example of a motor that inherently does not produce smooth torque is the switched reluctance motor. Its torque is always pulsed, and smooth torque can only be achieved via suitable current profiling. One example of such control is given in the chapter discussing SR-machine drive systems.

References

Further reading about rotating electric machine design and machine properties can be found, for example, in Pyrhönen, J., Jokinen, T., & Hrabovcova V. (2008, 2014). *Design of rotating electric machines* (1st and 2nd eds.) Chichester, UK: John Wiley & Sons.

T.J.E. Miller: Brushless Permanent–Magnet and Reluctance Motor Drives. Oxford Science Publications, ISBN 0-19-859369-4, 1989.

4

The fundamentals of space-vector theory

A fast calculation model is required to control an electric machine in real time. Modelling based on space-vector theory has been commonly adopted for this purpose. The theory was widely developed by Kovàcs and Ràcz in their books *Transiente Vorgänge in Wechsel-strommaschinen* (1954)

Despite its deficiencies, the space-vector model for rotating field machines is fairly well suited for control if feedback is used to compensate for modelling inaccuracies. For an electrical drive supplied with pulse-width-modulation (PWM) converter, the electric machine is always, in principle, in a transient state. The traditional single-phase equivalent circuit with time-harmonic RMS phasor quantities, applicable to the sinusoidal nontransient state, cannot be used. Space-vector theory was developed to describe the transient behaviours of electric machines.

Understanding the transient states of electric machines has become more and more important as electrical drive development has progressed. It is increasingly common to supply a motor using a frequency converter, which results in internal voltages that are far from sinusoidal. Even with a sinusoidal supply, electric machines experience transients. They occur, for example, at start-up and in the context of process control in response to changing machine loads.

Space-vector theory was developed by Kovács and Rácz in the 1950s. Their purpose was to develop a model for alternating current (AC) machines to model transients in direct online (DOL) drives. In the 1960s, Felix Blaschke of Siemens was working with the fundamentals of vector control for AC machines. Focusing on controlling magnetization and torque separately, his target was to produce fundamentals that would make it possible to apply DC machine control principles to AC machines. Space-vector theory provided a firm foundation on which to build.

In space-vector theory, the spatial distribution of current linkages, fluxes and flux linkages, and voltages are presented with "space vectors". In this book, these space vectors will normally be referred to as vectors and denoted with boldfaced italic symbols. For example, the space vector for stator current is i_s. It is common for experts to speak about the

Electrical Machine Drives Control: An Introduction, First Edition. Juha Pyrhönen, Valéria Hrabovcová and R. Scott Semken.
© 2016 John Wiley & Sons, Ltd. Published 2016 by John Wiley & Sons, Ltd.

vector control of drives and not space vector control. Here, in general, the terms *vector* and *space vector* are considered synonomous.

These space vectors are not vectors in the classic sense, but their calculus follows the mathematical forms of traditional vectors. In the space-vector presentation, the scalar product of a coefficient and the space vector is a vector, the dot (scalar) product two space vectors is a scalar, and the cross (vector) product two space vectors is a vector.

For example, the air-gap flux-linkage space vector ψ_m is the product of the magnetizing inductance L_m and the magnetizing space vector for current i_m.

$$\psi_m = i_m L_m \tag{4.1}$$

The instantaneous three-phase stator power P_s is the scalar product of the stator voltage space vector u_s and the stator space vector for current i_s.

$$P_s = \frac{3}{2} u_s \cdot i_s \tag{4.2}$$

The vector (cross) product is needed to calculate forces and, especially, to calculate the torque vector, which is expressed as in the following equation.

$$T = \frac{3}{2} p \psi_s \times i_s \tag{4.3}$$

Equation (4.3) indicates that torque will be a vector perpendicular to the plane defined by the stator-flux-linkage and stator-current space vectors. The torque vector will point outwards from or inwards to the machine shaft depending on operating mode. It will point outwards for a motor and inwards for a generator rotating in a mathematically positive (counterclockwise) direction. Normally, torque is expressed as a scalar value and assumed positive for motoring and negative for generating. Torque production will be examined in more detail in the next chapter.

Because space-vector theory was developed to evaluate transient behaviours in electric machines and was later adapted to electrical drive control, exact presentation is not necessary. Especially in feedback control, small modelling simplifications can be made. The following is a list of simplifying assumptions that are typically made.

1. Flux density distribution is sinusoidal in the air gap.

2. Magnetizing circuit saturation is constant.

3. There are no iron losses.

4. Resistances and inductances are independent of temperature and frequency.

5. The electric machine is treated as if it were a two-pole machine.

1. The assumption of sinusoidal flux density distribution usually brings good results, because AC machine structures are designed to yield sinusoidal terminal voltages regardless of the real air-gap flux density waveforms. The electrical drive converter works with the terminal voltages and not with the more complex air-gap flux density waveforms.
In most cases, an electrical drive will be most efficient if the air-gap flux density distribution produced is primarily sinusoidal. Fundamental space-vector theory assumes

that each single-phase winding is constructed to produce a purely sinusoidal flux density. While not true in practice, this assumption yields acceptable results. For a three-phase machine in particular, the three windings combine to produce a superposed flux density wave that is well formed if not sinusoidal.

2. The second assumption of constant saturation is a bit more complicated. Normally, good vector control necessitates saturation modelling of the inductances, especially the magnetizing inductance of the controlled machine. This modelling can be carried out using, for example, look-up tables to update the machine inductances over different operating conditions. In principle, the differential of the flux linkage vector, see Equation (4.1), should be written as follows.

$$\frac{d\psi_m}{dt} = \frac{d}{dt}(i_m L_m) = L_m \frac{di_m}{dt} + i_m \frac{dL_m}{dt} \tag{4.4}$$

The latter part of the differential should be taken into account in the calculation if the inductance changes. Experiments have shown, however, that numerical instability can easily result if the inductance is presented as a function of flux linkage and torque.

$$L_m = L_m(\psi_m, T) \tag{4.5}$$

Therefore, the previous expression for the flux linkage differential (4.4) can be reformulated as this equation.

$$\frac{d\psi_m}{dt} \approx L_m \frac{di_m}{dt} \tag{4.6}$$

Ignoring the latter term in Equation (4.4), the inductance Equation (4.5) is updated according to torque and flux linkage level. This approach seems adequate even for high performance AC drives.

3. Neglecting iron losses, the third simplification, is not significant from the control point of view; however, the energy efficiency of a drive cannot be accurately determined using space-vector theory without including the iron losses. Although not necessary for frequency converter control, in general, iron losses must be included to obtain accurate scientific modelling results. In particular, calculating the vector equivalent circuit becomes a great deal more complicated if any iron loss resistance parallel to the magnetizing inductance is not accounted for accurately. Further difficulties arise because the iron losses P_{Fe} are dependent on air-gap flux linkage level ψ_m and frequency f.

$$P_{Fe} \approx P_{Fe}\left(f^2, \psi_m^2\right) \tag{4.7}$$

Since a single-phase equivalent circuit describes iron loss as a function of air gap voltage u_m, loss should be written as a function of the rated iron loss resistance $R_{Fe,N}$, the rated frequency f_N, and the rated flux linkage level $\psi_{m,N}$.

$$P_{Fe} \approx \frac{3u_m}{R_{Fe,N}}\left(\frac{f}{f_N}\right)^2\left(\frac{\psi_m}{\psi_{m,N}}\right)^2 \tag{4.8}$$

The exponent of frequency can also vary, from 1.5 to 2.5, depending on the magnetic circuit material of the machine. For an accurate evaluation, the exponent should be determined, for example, via measurement.

4. The fourth simplification, assuming that resistances and inductances are independent of temperature and frequency, is valid only at certain operating points, and the modelled resistance and inductance values must be updated according to known behaviours.

The following expression gives an approximation for the DC resistance as a function of machine operating temperature T.

$$R_s(T) \approx R_{s(20\ °C)}(1 + \alpha \Delta T) \qquad (4.9)$$

ΔT is the temperature rise of the machine measured from room temperature (20 °C), and α is the temperature coefficient for the resistivity of the winding material. The temperature coefficient of resistivity for copper and aluminium is $\alpha_{Cu} = 3.81 \cdot 10^{-3}\ K^{-1}$ and $\alpha_{Al} = 3.7 \cdot 10^{-3}\ K^{-1}$, respectively.

The assumption of frequency independent inductances is reasonable and typically gives good results; however, winding resistance varies with AC frequency, and this dependence should be modelled if precise simulation results are required. In fact, at any given temperature, the operating frequency resistance in an electric machine winding can be 50% higher than its DC resistance. The analyst can find guidelines for calculating the AC resistances for an electric machine in machine design literature, such as in Pyrhönen et al. (2014). Alternatively, a simple model for AC resistance can be built that depends on the square of the operating frequency.

$$R_s(f) \approx R_s\left(1 + c \cdot f^2\right) \qquad (4.10)$$

In the equation, c is an experimentally determined coefficient that establishes the correct relationship between the known DC resistance value and the desired rated frequency resistance value.

5. The fifth simplification, assuming a two-pole machine, does not limit space-vector theory usability by any means. The current converter of the electrical drive *sees* only stator voltages and currents at the electric machine's input and output terminals. Pole pair number information is not available to the converter, nor does it need it to perform its function. Of course, several electric machine parameters depend, to some extent, on how many pole pairs it has; however, these dependencies do not manifest at the connection terminals. For example, the power factor of a multipole induction motor is lower than that of a two-pole machine. Furthermore, pole pair number is not the only determinant of power factor. Machine saturation or reduced per-unit magnetizing inductance also result in lower power factors for two-pole machines.

Only in the torque equation, where it is used as a multiplier, does number of poles have any practical effect on the control system of an electrical drive. Moreover, because the electrical degrees of rotation needed for space-vector theory modelling are p-multiples of a machine's cylindrical geometry, rotor rotation angle must be multiplied by pole pair number to determine rotor electrical angle. The same is valid for angular velocities: $\omega = p\Omega$, where ω is the electrical angular velocity, and Ω is the mechanical angular velocity.

These restrictions having been considered and possible straightforward remedies having been suggested, space vectors for three-phase quantities can now be constructed. These space vectors will comprise and be helpful in understanding the actual space-vector model. The mathematical treatment of space vectors almost certainly requires coordinate transformations. These transformations will be discussed in brief.

4.1   Introduction to the space vector for current linkage

Figure 4.1 illustrates a special winding arrangement and the magnetic axes U, V, and W of a two-pole three-phase machine. This winding configuration is designed to produce a sinusoidal current linkage for each one of the three phases individually that when superposed still results in a sinusoidal flux density distribution across the air gap. To accomplish this, each slot includes coil sides from all three phases with each composed of a different number of winding turns. This winding arrangement is mathematically interesting and demonstrates the feasibility of a sinusoidal-phase winding system; however, it is not practical, because it results in a relatively low winding factor and puts winding turns for all three phases into one slot, which presents insulation difficulties.

In actuality, this type of sinusoidal-phase winding arrangement is not really necessary. Because current is always being conducted in at least two phases at any given moment, the sum current linkages of the three individual phase windings always remain fairly sinusoidal even though each alone does not produce a sinusoidal current linkage distribution.

For a two-phase winding, however, there are moments when only one phase is actively conducting. To produce a sinusoidal flux density distribution across the air gap of a two-phase machine, a winding arrangement analogous to that shown in Figure 4.1 becomes a real advantage. Without it, a two-phase machine cannot match the performance of an equivalent three-phase machine.

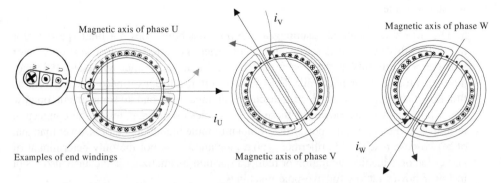

Figure 4.1   An ideal stator winding arrangement for a three-phase machine illustrating the three magnetic axes. The arrangement is designed so that each winding produces its own sinusoidal flux density distribution across the air gap. These superpose to produce a sinusoidal flux density distribution across the air gap. According to Ampère's law, the magnetic axis of a winding is along its flux path. The maximum current linkage lies 90° from the magnetic axis. The magnified slot on the left shows how each slot shares winding turns for each phase with this kind of a special winding arrangement.

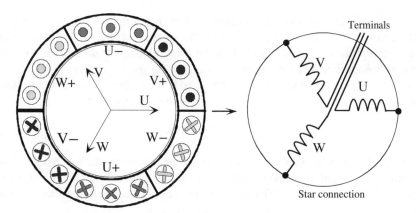

Figure 4.2   The stator windings for each phase are typically depicted symbolically as a coil aligned with the magnetic axis of each phase. The star physical connection of the phases has been moved to the outer terminals of the windings to make later studies of space vectors more acceptable. (See Chapter 7.)

According to the first simplifying assumption for space-vector theory modelling, each individual winding should produce a sinusoidal flux density distribution in the air gap. As just described, this can be accomplished using the "ideal" but somewhat impractical winding arrangement illustrated in Figure 4.1. Fortunately, more-workable three-phase winding configurations can still produce sinusoidal flux density distributions across the air gap, so they can be implemented without sacrificing overall machine performance.

In practice, three-phase windings produce total current linkages similar to those illustrated in the previous chapter. See Figures 3.9 through 3.12. Figure 3.11 shows the individual-phase current-linkage waveforms. As the figures indicate, the individual waveforms deviate substantially from the sinusoidal form, but the superposition of the three phases is more sinusoidal.

In the space-vector theory approach to electric machines, the stator windings for each phase are typically depicted symbolically as a vector aligned with the magnetic axis of each phase as illustrated by Figure 4.2.

Consider the magnetic field strengths caused by the phase U winding in Figure 4.3 at the point γ on the periphery of the machine. The field strength $H_U(\gamma)$ oriented radially in the air

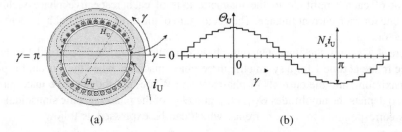

Figure 4.3   (a) Applying Ampère's law to a particular integration path. The circles arrayed circumferentially along the outside of the air gap represent conductors, and their diameters represent the amount of current (cross-sectional area) for each slot. (b) The waveform of the single-phase "sinusoidal" current linkage. The current linkage fundamental amplitude is $k_{ws1}N_s i_U/2$.

gap, of the phase U current, can be calculated by applying Ampère's law. In Figure 4.3, the integration path is from the angle γ to $\gamma + \pi$ on the machine periphery. The total number of turns in a stator phase is N_s. The amplitude of the current linkage in the system is $k_{ws1}N_s i_U/2$. The current linkage function can be written $(k_{ws1}N_s i_U/2) \cos\gamma$.

Assuming the phase current linkage is consumed in the air gap of the machine, phase field strength in the air gap can be expressed with the following equation.

$$H_U = \frac{k_{ws1}N_s i_U}{2\delta} \cos\gamma \qquad (4.11)$$

The subscript U refers to phase U, and H_U is the field strength for that phase alone. By convention, field strength from the rotor to the stator is considered positive. Field strength in the opposite direction is negative. The respective air-gap flux density for the phase at point γ is

$$B_U = \mu_0 H_U = \mu_0 \frac{k_{ws1}N_s i_U}{2\delta} \cos\gamma \qquad (4.12)$$

Consequently, the following expression for current linkage can be written

$$\Theta_U = \delta H_U = \frac{k_{ws1}N_s i_U}{2} \cos\gamma \qquad (4.13)$$

Without considering magnetic saturation of the electrical steel components, the values for current linkage, magnetic field strength, and air-gap magnetic flux density are linearly interdependent.

The two other phases can be treated similarly to obtain their current linkage equations.

$$\Theta_V = \frac{k_{ws1}N_s i_V}{2} \cos\left(\gamma - \frac{2}{3}\pi\right) \qquad (4.14)$$

$$\Theta_W = \frac{k_{ws1}N_s i_W}{2} \cos\left(\gamma - \frac{4}{3}\pi\right) \qquad (4.15)$$

The peak current linkage for each phase runs along the magnetic axis of the phase at any given instant in time. Furthermore, because conductor density is sinusoidal, the current linkage distributions are sinusoidal. This sinusoidal form remains unaltered irrespective of the instantaneous values of the currents.

The instantaneous amplitudes of each individual current linkage are Θ_U, Θ_V, and Θ_W. The alignment of each amplitude to the magnetic axis of each respective phase yields three vectors, one for each current linkage. The vector sum of the three current linkage vectors is the space vector.

Figure 4.4 shows the instantaneous phase currents i_U, i_V, and i_W fixed to the stator reference frame at time 1 for a symmetric three-phase machine, when the current for phase U is at a maximum, and the currents of phases V and W are half the negative maximum. The figure also depicts the amplitudes Θ_{U1}, Θ_{V1}, and Θ_{W1} of the phase-specific sinusoidal current linkages corresponding to these currents, which can be expressed as follows.

$$\begin{aligned} \Theta_{U1} &= k_{ws1}N_U i_U(t)/2 \\ \Theta_{V1} &= k_{ws1}N_V i_V(t)/2 \\ \Theta_{W1} &= k_{ws1}N_W i_W(t)/2 \end{aligned} \qquad (4.16)$$

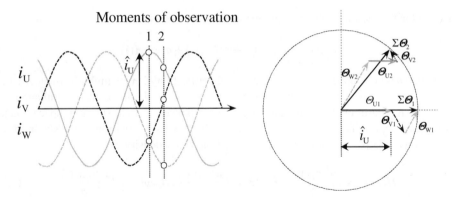

Figure 4.4 The formation of the vector sums of the phase-specific current linkage space vectors. The phase current-linkage vectors are proportional to the instantaneous phase currents. The vector sums of the phase current linkages are drawn at two time instants showing how the vector sum rotates as the phases are supplied with three-phase alternating currents. When supplied from a symmetric three-phase current source, the tip of the space vector follows a circle.

The individual amplitudes are at the peaks of each sinusoidal current linkage distribution. The peak values coincide with the magnetic axes of the phases, so the total three-phase winding current linkage is the sum of the three vectors produced when current linkage amplitude is given the direction of the magnetic axes. This sum is the space vector of the current linkage.

Using a space vector for current linkage makes it possible to represent the three-phase quantity with a single vector that rotates in the complex plane representing actually the cross-sectional plane of the AC machine. This facilitates the mathematical treatment of electric machinery. The rotation of the space vector also corresponds to the rotation of flux in the machine.

4.1.1   Mathematical representation of the space vector

In a three-phase machine, there is a local phase shift of 120 electrical degrees between the phases. Therefore, it is useful to formulate a phase-shift operator that is an exponential function based on circle thirds.

$$a = e^{j\frac{2\pi}{3}} \tag{4.17}$$

This function can be used to designate the directions of the magnetic axes for the three current phases. Shown in their positive directions, the three phase-shift operators are denoted thusly.

$$a^0, a^1, a^2 \tag{4.18}$$

Multiplying each of these operators by its respective phase current linkage (as a function of time) will result in three current linkage vectors, one for each phase. Taking the vector sum of the three vectors gives a vector for total current linkage. This vector $\boldsymbol{\Theta}'_s(t)$ is the current linkage space vector for the three phase windings and their currents.

$$\Theta_s'(t) = \left(a^0 \Theta_{sU}(t) + a^1 \Theta_{sV}(t) + a^2 \Theta_{sW}(t)\right) \tag{4.19}$$

If its air gap δ is constant and if its electrical steel permeability is very high, the magnetic field strength in an electric machine varies relative to the magnitude of the current linkage space vector. Each phase current linkage amplitude produces a field-strength space vector. However, a coupling factor k can be applied to indicate what proportion of the stator current linkage is used to produce air-gap field strength and flux density.

$$H_{mU}(t) = k\frac{a^0 \Theta_{sU}(t)}{\delta}, H_{mV}(t) = k\frac{a^1 \Theta_{sV}(t)}{\delta}, H_{mW}(t) = k\frac{a^2 \Theta_{sW}(t)}{\delta} \tag{4.20}$$

Analogously, the space vector for the field strength can be expressed with this equation.

$$H_m'(t) = \left(a^0 H_{mU}(t) + a^1 H_{mV}(t) + a^2 H_{mW}(t)\right) \tag{4.21}$$

The air-gap flux density space vector can be formulated by multiplying the field strengths by the permeability of free air μ_0 yielding the following expression, which shows that space-vector theory relies on the air-gap flux density of the electric machine.

$$B_m'(t) = \mu_0\left(a^0 H_{mU}(t) + a^1 H_{mV}(t) + a^2 H_{mW}(t)\right) = \mu_0 H_m'(t) \tag{4.22}$$

When the current linkages of Equation (4.19) are divided by the effective numbers of turns (e.g., by $k_{w1}N_U$), the phase currents can be combined to analogically formulate a space vector $i_s'(t)$. A time-based representation in the stator reference frame, this space vector for current is expressed in this manner.

$$i_s'(t) = \left(a^0 i_{sU}(t) + a^1 i_{sV}(t) + a^2 i_{sW}(t)\right) \tag{4.23}$$

This space vector for current illustrates the overall effect of the three-phase currents in the three-phase windings. The phase space vectors for current lie in the same directions as the phase current linkage space vectors.

Equation (4.23) suggests the magnitude of the space vector for three-phase stator current is 3/2 of the peak value of the phase sinusoidal current. This is shown clearly by Figure 4.4, where the first sum vector is built with the peak current of phase U. The vector resultant is 3/2 times the current linkage of phase U. Dividing by the effective number of winding turns, a figure similar to Figure 4.4, and looking much the same, could be prepared for the current space vectors.

When defining the magnetizing inductance L_m in the design of an m-phase electric machine, multiplying the single-phase main inductance L_p by $m/2$ describes the three electric current phases and their sum effects in a single-phase equivalent circuit for a three-phase machine.

$$L_m = \frac{3}{2}L_p \tag{4.24}$$

However, it is customary to use the single-phase phasor equivalent circuit to describe electric machine behaviours. The space-vector presentation offers a similar single-phase vector equivalent circuit, but uses space-vector variables in place of phasors. Multiplying i'_s by 2/3 reduces its magnitude, making it equal in amplitude to the single-phase sinusoidal currents supplying the electric machine. The result is the space vector for current i_s.

$$i_s(t) = \frac{2}{3}\left(a^0 i_{sU}(t) + a^1 i_{sV}(t) + a^2 i_{sW}(t)\right) = \frac{2}{3}i'_s(t) \tag{4.25}$$

Taking this approach to define the space vector for current makes it possible to write the space-vector equations directly using actual resistances and inductances, the parameters that describe the real equivalent circuit of the machine. Similarly, the vector equivalent circuit can easily be derived using the same parameters. For example, the voltage drop vector in response to stator resistance Δu_s can be derived using the same methodology as in the single-phase vector equivalent circuit description with no extra coefficients needed.

$$\Delta u_s = i_s R_s \tag{4.26}$$

The stator leakage flux-linkage vector also can be formulated by taking the product of the stator electric current vector and the stator leakage flux inductance $L_{s\sigma}$.

$$\psi_{s\sigma} = i_s L_{s\sigma} \tag{4.27}$$

The stator voltage drop and stator leakage flux-linkage space vectors lie in the same direction as the stator space vector for current.

In (4.26), the voltage drop vector was expressed as a function of the analogically defined space vector for current i_s. The voltage space vector $u_s(t)$ can be defined in the same way as detailed by

$$u_s(t) = \frac{2}{3}\left(a^0 u_{sU}(t) + a^1 u_{sV}(t) + a^2 u_{sW}(t)\right) = \frac{2}{3}U_{DC}\left(a^0 S_U(t) + a^1 S_V(t) + a^2 S_W(t)\right) \tag{4.28}$$

The latter part of Equation (4.28) expresses the voltage vector produced by a frequency converter with U_{DC} intermediate circuit voltage and switching commands $S_{U,V,W}$.

The physical nature of the space vector for current, which has the same vector direction as the phase current linkage space vector, is clear. The nature of the voltage space vector is not as clear.

For the voltage space vector, the same magnetic axes directions are used as in the definitions for current. Switching function variables S_U, S_V, and S_W describe the states for three converter switches. Ideally, each has a value of 0 or 1, corresponding to the OFF or ON condition. In reality, as they change state, switch voltages sweep from 0 to 1 or 1 to 0.

The air-gap flux-linkage space vector $\psi_m(t)$ can be formed analogically from phase quantities in the same way as were the field strength and air-gap flux density space vectors. Again, the reduction factor of 2/3 is used.

$$\psi_m(t) = \frac{2}{3}\left(a^0 \psi_{mU}(t) + a^1 \psi_{mV}(t) + a^2 \psi_{mW}(t)\right) \tag{4.29}$$

Multiplying Equation (4.22) by the effective numbers of turns and reducing the result by 2/3 yields another expression of the air-gap flux-linkage space vector.

$$\psi_m(t) = \frac{2}{3}k_{ws1}N_s B'_m(t) = \frac{2}{3}k_{ws1}N_s \mu_0 H'_m(t) \tag{4.30}$$

The stator flux-linkage space vector can now be found by adding Equation (4.27) to Equation (4.30).

$$\boldsymbol{\psi}_s = \boldsymbol{\psi}_{s\sigma} + \boldsymbol{\psi}_m = i_s L_{s\sigma} + \boldsymbol{\psi}_m \tag{4.31}$$

The current, voltage, and flux-linkage space vectors of the rotor can be determined accordingly. When changing over to a space-vector representation, all electric current, voltage, and flux-linkage components take on vector form, and all the ordinary relationships for the circuit analysis and three-phase system are expressed using the space vectors in a single-phase vector equivalent circuit.

It is also possible that the motor has a different number of phases (*m*). In such a case, the phase shift operator is defined as $\boldsymbol{a} = \mathrm{e}^{\mathrm{j}2\pi/m}$ and the equations defining the space vectors must be adjusted accordingly. In this material, we regularly assume that the motor has three phases unless otherwise specified.

4.1.2   Two-axis representation of the space vector

A two-axis electric machine presentation is commonly used, because it is often convenient to describe behaviours with respect to the two most relevant magnetic orientations, along the direct axis (d-axis) and along the quadrature axis (q-axis); d-axis usually aligns with the minimum magnetic reluctance of the machine magnetic circuit and q-axis with its maximum reluctance. In machines with no saliency d-axis is often selected to align with the rotor excitation direction. It is also possible to select the d- and q-axes freely in case of different coordinate systems.

In embedded computers, such as those used for electrical drive control, complex vector presentation is challenging. Instead, it is easier to use x- and y-coordinate scalar values. Furthermore, it is reasonable to represent three-phase windings using an equivalent two-phase winding mathematical model, making it straightforward and reasonable to move to the two-axis representation of the space vector.

Figure 4.5 illustrates symmetric three-phase and two-phase currents in the time plane. The three-phase currents i_U, i_V, and i_W are depicted on the left in Figure 4.5a. Figure 4.5b depicts the symmetric two-phase current components i'_{sx} and i'_{sy} corresponding to the space vector i'_s of the stator current. The xy-reference frame is fixed to the stator with the x-axis being collinear with the axis of phase U. Each depiction indicates with a dotted line the instant that the current of phase W has reached its negative peak value. At this moment, the phases U and V currents are half their positive peak values.

If there is an equal number of effective turns $k_{ws}N_s$ in the two-phase system, the peak values of the currents i'_{sx} and i'_{sy} of the symmetric two-phase system must be 1.5 times the symmetric three-phase currents to produce an equal current linkage. The space vector i'_s in a three-phase system is always formulated from at least two of the phase currents; however, in the case of a two-phase system, there are instants when the space vector of the current is produced as an effect of single-phase current (i'_{sx} or i'_{sy}) only. Such a case emphasized the need to produce sinusoidal flux density by a single phase. In a real motor, a two-phase system is inherently weaker than a three-phase system from the harmonic content point of view.

The instantaneous values for three-phase currents presented on the U, V, and W magnetic axes can be transformed mathematically into the space-vector components for stator current on the xy-axes to develop the two-axis electric machine presentation. Both reference frames are fixed, and there is a phase shift of the constant angle κ between the U-axis and the x-axis.

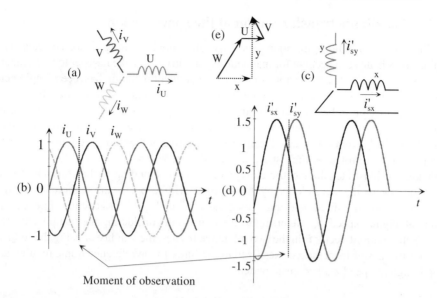

Figure 4.5 (a) Three-phase winding, (b) three-phase currents in the time plane, (c) two-phase winding with a neutral conductor, (d) two-phase currents causing respective current linkages when the number of winding turns in series is the same as in the three-phase system, and (e) vector presentation of the three-phase and two-phase currents.

Taking into account the current component i_{s0} for the possible zero-sequence network (currents flowing in all three phases in the same temporal phase toward the star point and further along neutral wire (normally not present) to ground), the following matrix notation can be introduced for the instantaneous values of the three-phase currents.

$$\begin{bmatrix} i_{sU} \\ i_{sV} \\ i_{sW} \end{bmatrix} = \begin{bmatrix} \cos\kappa & \sin\kappa & 1 \\ \cos(\kappa+120°) & \sin(\kappa+120°) & 1 \\ \cos(\kappa+240°) & \sin(\kappa+240°) & 1 \end{bmatrix} \begin{bmatrix} i_{sx}^{s} \\ i_{sy}^{s} \\ i_{s0}^{s} \end{bmatrix} \tag{4.32}$$

The three-phase current components can be transformed into two-phase components as follows.

$$\begin{bmatrix} i_{sx}^{s} \\ i_{sy}^{s} \\ i_{s0}^{s} \end{bmatrix} = \frac{2}{3} \begin{bmatrix} \cos\kappa & \cos(\kappa+120°) & \cos(\kappa+240°) \\ \sin\kappa & \sin(\kappa+120°) & \sin(\kappa+240°) \\ 1/2 & 1/2 & 1/2 \end{bmatrix} \begin{bmatrix} i_{sU} \\ i_{sV} \\ i_{sW} \end{bmatrix} \tag{4.33}$$

In symmetric cases, the three-phase system has no zero sequence component. It can come into play only for star-connected three-phase windings with a neutral line connection, a configuration only found in a generator with a star point ground connection. In practice, the angle κ is usually set to zero as a simplification and the zero sequence component of the phase currents is nonexistent.

4.1.3   Coordinate transformation of the space vector

Coordinate transformations are often necessary when implementing vector controls. Transforming coordinates can account for magnetic asymmetries or make it easier to implement the control logic. For instance, vector control of an induction motor has been implemented traditionally in the reference frame (coordinate system) oriented to rotor flux linkage. For synchronous machines, because of their anisotropic rotor, it is natural to use a rotor reference frame. Figuratively speaking, this transformation to the rotor reference frame corresponds to a situation in which events are viewed from the perspective of the rotor. From this perspective, the stator seems to rotate at high speed around the viewpoint.

Figure 4.6 illustrates the electric current vector corresponding to the constant state of a symmetric system, the locus plotted by the vector tip with respect to the stator reference frame (the xy-axes) and the direct and quadrature axes (d^g and q^g) in a general reference frame rotating at angular speed ω_g. Referring to the figure and considering the components of the stator electric current vector i_s in the two reference frames, the validity of using the following equations to transfer from the xy-axes fixed to the stator to the reference frame rotating at the general angular speed ω_g becomes evident

$$i_d^g = i_{sx} \cos \omega_g t + i_{sy} \sin \omega_g t \tag{4.34}$$

$$i_q^g = -i_{sx} \sin \omega_g t + i_{sy} \cos \omega_g t \tag{4.35}$$

Using a polar complex representation for the space vectors offers the simplest representation for the coordinate transformation. Now the transformation of the stator current from the xy-axes frame of reference (i_s^s) to the $d^g q^g$-axes frame of reference is accomplished via this equation.

$$i_s^g = i_d^g + j i_q^g = i_s^s e^{-j\omega_g t} = \left(i_{sx} + j i_{sy} \right) e^{-j\omega_g t} \tag{4.36}$$

The complex vector needs only be rotated to point in the required direction. The angular relationships involved in the coordinate transformation are illustrated in Figure 4.7. In the

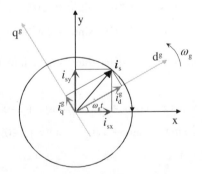

Figure 4.6   Stator current can be represented with two components in the stator reference frame (the xy-axes) or with direct and quadrature axes (d^g and q^g) in a general reference frame if a transfer is made from the reference frame fixed to the stator to the general reference frame rotating at angular speed ω_g. The figure illustrates the electric current vector corresponding to the constant state of a symmetric system and the locus plotted by the vector tip.

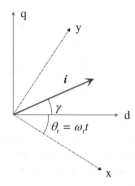

Figure 4.7 The vector i in different frames of reference. The dq-axes reference frame corresponds to the rotor reference frame, and the xy-axes reference frame corresponds to the reference frame of the stator.

figure, the dq-axes reference frame corresponds to the rotor reference frame, and the xy-axes reference frame corresponds to the reference frame of the stator.

The electric current vector can be written in the rotor reference frame.

$$i^r = |i|e^{j\gamma} \tag{4.37}$$

The same current can also be expressed in the stator reference frame as follows.

$$i^s = |i|e^{j(\gamma+\theta_r)} = i^r e^{j\theta_r} \tag{4.38}$$

With this simple vector presentation, it is easy to see the coordinate transformation of the vector i^s from the rotor reference frame to the stator reference frame.

$$i^s = i^r e^{j\theta_r} \tag{4.39}$$

From the stator reference frame to the rotor reference frame, it is

$$i^s e^{-j\theta_r} = i^r e^{j\theta_r} e^{-j\theta_r} = i^r \tag{4.40}$$

Therefore, the following equation is the expression mapping the stator electric current vector to the rotor electric current vector.

$$i^r = i^s e^{-j\theta_r} \tag{4.41}$$

When expressed as a complex vector, coordinate transformation is accomplished by merely adding or subtracting the angle between the two different coordinate systems.

4.2   Space-vector equivalent circuits and the voltage-vector equations

In the previous chapter, Figure 3.1 illustrated the magnetic connection between two stationary winding coils. This equivalent circuit can be regarded, for example, as describing an induction machine having two windings: the three-phase stator winding and the rotating multiphase

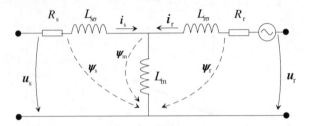

Figure 4.8 Vector equivalent circuit of an induction motor in the stator reference frame. On the rotor side, there is an extra voltage source, because the spinning rotor has been described in a different frame of reference, the stator reference frame.

rotor winding. Both of these multiphase windings are regarded as single-phase vector windings rotating with respect to one another. This rotation induces its own effect, which is manifest as an additional voltage source in the equivalent circuit. Figure 4.8 is the circuit diagram for the vector equivalent circuit of an induction machine in the stator reference frame. In this text, this type of parameter-based equivalent circuit is referred to as the current model. Its circuit currents result in flux linkages across the inductances. Stator flux linkage can also be estimated by integrating stator voltage. This approach is referred to here as the voltage model.

The stator voltage vector for an AC machine consists of the resistive and inductive voltage components. Its derivation begins by defining the stator voltage equation in the stator frame of reference, the natural reference frame of the winding. Adding the remaining components results in an absolute-value space-vector equation of the following form.

$$u_s^s = R_s i_s^s + \frac{d\psi_s^s}{dt} \tag{4.42}$$

However, it is often necessary to transform this equation to another reference frame, for example, the rotor reference frame, which for the stator winding is a foreign frame of reference. To develop the equation for the rotor reference frame, the principles introduced in Figure 4.7 are applied.

$$u_s^s = u_s^r e^{j\theta_r} = R_s i_s^r e^{j\theta_r} + \frac{d}{dt}\left(\psi_s^r e^{j\theta_r}\right) \tag{4.43}$$

Deriving the latter term of the equation results in the following expression.

$$u_s^s = u_s^r e^{j\theta_r} = R_s i_s^r e^{j\theta_r} + \frac{d\psi_s^r}{dt} e^{j\theta_r} + j\omega_r \psi_s^r e^{j\theta_r} \tag{4.44}$$

Dividing Equation (4.44) by $e^{j\theta_r}$ yields the equation for stator voltage in the rotor reference frame.

$$u_s^s e^{-j\theta_r} = u_s^r = R_s i_s^r + \frac{d\psi_s^r}{dt} + j\omega_r \psi_s^r \tag{4.45}$$

When the winding voltage equation is derived in a foreign reference frame, for example, the voltage equation of the stator winding in the rotor reference frame, it picks up a

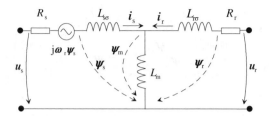

Figure 4.9 Vector equivalent circuit of an induction motor in the rotor reference frame. There is an extra voltage source on the stator side, because the stator has been moved to a foreign reference frame. The rotor, however, is described in its own reference frame, and therefore no additional voltage source is included.

rotation-induced voltage term $j\omega_r\psi_s^r$, caused by the speed difference of the reference frames. While this additional term can be justified both mathematically and physically, it presents, in fact, the most demanding challenge when carrying out the coordinate transformations. Deriving the product $\psi_s^r e^{j\theta_r}$ yields, according to the product derivation rule, the term $d\theta_r/dt = \omega_r$ for the multiplier. As a result, the vector equivalent circuit diagram presented previously in Figure 4.8 can now be drawn as a vector equivalent circuit in the rotor reference frame: Figure 4.9.

4.3   Space-vector model in the general reference frame

Space-vector modelling of the asynchronous machine can also be accomplished in the general reference frame. The voltage equations for the stator and rotor should first be represented in a foreign frame of reference that rotates at an angular speed ω_g. When the observation frame of reference rotates, the two voltage equations pick up an additional term for motion voltage $j\omega_g\psi_s^g$.

$$u_s^g = R_s i_s^g + \frac{d\psi_s^g}{dt} + j\omega_g\psi_s^g \tag{4.46}$$

$$u_r^g = R_r i_r^g + \frac{d\psi_r^g}{dt} + j(\omega_g - \omega_r)\psi_r^g \tag{4.47}$$

In Equation (4.47), ω_r is the rotor electrical angular speed ($\omega_r = p\Omega$). The stator and rotor flux linkages in Equations (4.46) and (4.47) can be written as follows.

$$\psi_s^g = L_s i_s^g + L_m i_r^g \tag{4.48}$$

$$\psi_r^g = L_m i_s^g + L_r i_r^g \tag{4.49}$$

In Equations (4.48) and (4.49), L_m is the magnetizing inductance, $L_s = L_m + L_{s\sigma}$ is the total inductance of the stator, and $L_r = L_m + L_{r\sigma}$ is the total inductance of the rotor. $L_{s\sigma}$ and $L_{r\sigma}$ are the leakage inductances of the stator and rotor (rotor quantities being referred to the stator). Using Equations (4.46) through (4.49), a vector equivalent circuit in the general reference frame can be constructed. See Figure 4.10.

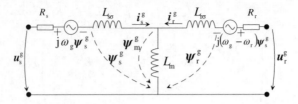

Figure 4.10 The T equivalent circuit of an asynchronous machine in general reference frame.

In a reference frame fixed to the stator ($\omega_g = 0$), Equations (4.46) through (4.49) take on a now familiar absolute-value form.

$$u_s^s = R_s i_s^s + \frac{d\psi_s^s}{dt} \tag{4.50}$$

$$u_r^s = R_r i_r^s + \frac{d\psi_r^s}{dt} - j\omega_r \psi_r^s \tag{4.51}$$

$$\psi_s^s = L_s i_s^s + L_m i_r^s \tag{4.52}$$

$$\psi_r^s = L_m i_s^s + L_r i_r^s \tag{4.53}$$

For a squirrel-cage motor, the rotor voltage vector-see Equation (4.51)-is set to zero.

In Figure 4.10, the dynamic T equivalent circuit for the induction motor was introduced. In some texts, researchers like to use the so-called dynamic Γ equivalent circuit instead. Using the Γ-model may slightly simplify the mathematics, because it collects all the leakage flux phenomena in the stator and needs, thereafter, only four parameters to describe the total machine model (DeDoncker and Novotny, 1994). The parameters needed are magnetizing inductance L_M, total leakage inductance L_σ, stator resistance R_s, and rotor resistance R_R. The differences between the T and Γ approaches are summarized in Table 4.1.

Table 4.1 Comparing T and Γ equivalent circuits

Quantity	T Model	Γ Model
Rotor current	i_r^s	$i_R^s = \dfrac{L_m + L_{r\sigma}}{L_m} i_r^s$
Rotor flux linkage	ψ_r^s	$\psi_R^s = \dfrac{L_m}{L_m + L_{r\sigma}} \psi_r^s$
Magnetizing current	i_m^s	$i_M^s = i_s^s + i_R^s$
Magnetizing inductance	L_m	$L_M = \dfrac{L_m^2}{L_m + L_{r\sigma}}$
Leakage inductance	$L_{s\sigma}, L_{r\sigma}$	$L_\sigma = L_{s\sigma} + \dfrac{L_m L_{r\sigma}}{L_m + L_{r\sigma}}$
Rotor resistance	R_r	$R_R = \left(\dfrac{L_m}{L_m + L_{r\sigma}}\right)^2 R_r$

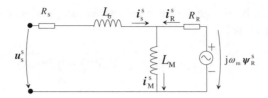

Figure 4.11 Γ equivalent circuit of induction machine in the stator reference frame.

Figure 4.11 illustrates the Γ equivalent circuit.The equations governing the Γ equivalent circuit of the motor are

$$u_s^s = R_s i_s^s + L_\sigma \frac{di_s^s}{dt} + L_M \frac{di_M^s}{dt} \tag{4.54}$$

$$u_R^s = 0 = R_R i_R^s + L_M \frac{di_M^s}{dt} - j\omega_m \psi_R^s \tag{4.55}$$

$$\psi_s^s = L_\sigma i_s^s + L_M i_M^s \tag{4.56}$$

$$\psi_R^s = L_M i_M^s \tag{4.57}$$

Using this equation system may in some cases simplify the presentation of the induction motor. In this book, however, we mainly use dynamic T equivalent circuits.

4.4   The two-axis model

A two-axis model fixed to the stators can be developed using the methodology shown earlier. For the direct x-axis and quadrature y-axis, the equivalent circuits shown by Figure 4.12 are valid. They are the result of separating out the real and imaginary parts of Equations (4.50) through (4.53).

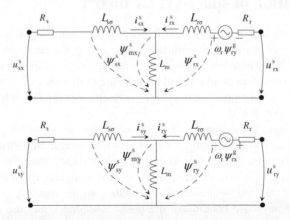

Figure 4.12   Equivalent circuits corresponding to the two-axis model fixed to the stator. L_m is the magnetizing inductance, $L_{r\sigma} = L_r - L_m$ is the leakage inductance of the rotor referred to the stator, and $L_{s\sigma} = L_s - L_m$ is the leakage inductance of the stator.

The following family of equations can be derived using the relationships $\boldsymbol{u} = u_x + ju_y$, $\boldsymbol{i} = i_x + ji_y$, and $\boldsymbol{\psi} = \psi_x + j\psi_y$.

$$u_{sx} + ju_{sy} = R_s\left(i_{sx} + ji_{sy}\right) + \frac{d\left(\psi_{sx} + j\psi_{sy}\right)}{dt} \tag{4.58}$$

$$u_{rx} + ju_{ry} = R_r\left(i_{rx} + ji_{ry}\right) + \frac{d\left(\psi_{rx} + j\psi_{ry}\right)}{dt} - j\omega_r\left(\psi_{rx} + j\psi_{ry}\right) \tag{4.59}$$

$$u_{sx} = R_s i_{sx} + \frac{d\psi_{sx}}{dt} \tag{4.60}$$

$$u_{sy} = R_s i_{sy} + \frac{d\psi_{sy}}{dt} \tag{4.61}$$

$$u_{rx} = R_r i_{rx} + \frac{d\psi_{rx}}{dt} + \omega_r\psi_{ry} \tag{4.62}$$

$$u_{ry} = R_r i_{ry} + \frac{d\psi_{ry}}{dt} - \omega_r\psi_{rx} \tag{4.63}$$

The four flux linkages shown in Figure 4.12 can be formulated with these equations.

$$\psi_{sx}^s = L_{s\sigma}i_{sx}^s + L_m\left(i_{sx}^s + i_{rx}^s\right) = L_s i_{sx}^s + L_m i_{rx}^s \tag{4.64}$$

$$\psi_{rx}^s = L_{r\sigma}i_{rx}^s + L_m\left(i_{rx}^s + i_{sx}^s\right) = L_r i_{rx}^s + L_m i_{sx}^s \tag{4.65}$$

$$\psi_{sy}^s = L_{s\sigma}i_{sy}^s + L_m\left(i_{sy}^s + i_{ry}^s\right) = L_s i_{sy}^s + L_m i_{ry}^s \tag{4.66}$$

$$\psi_{ry}^s = L_{r\sigma}i_{ry}^s + L_m\left(i_{ry}^s + i_{sy}^s\right) = L_r i_{ry}^s + L_m i_{sy}^s \tag{4.67}$$

4.5   Application of space-vector theory

The space-vector approach simplifies electric machine representation. It is considerably more straightforward, for example, than formulating and solving the equations for each winding needed to define all the transient behaviours of an electric machine. Space vectors can be applied to the representation of all types of asymmetric or distorted three-phase currents. For example, the space-vector representation is applicable to a motor fed by a frequency converter based on, for example, PWM.

Representing asymmetric sinusoidal states serves as an interesting example. For an asymmetric three-phase system, the tip of the rotating electric current vector plots an ellipse. If there is no zero component, the electric current vector comes from the positive-sequence phasor $\boldsymbol{i}_1$ rotating in the positive direction at the electrical angular speed ω_e and the complex conjugate $\boldsymbol{i}_2^*$ of the negative-sequence phasor rotating in the negative direction at angular speed $-\omega_e$. For steady state conditions, the positive-sequence and the negative-sequence component lengths are constant ($\hat{i}_1$ and $\hat{i}_2$). Figure 4.13 shows the locus of the tip of the electric current space vector for an asymmetric three-phase system in steady state when the phase currents have no zero component.

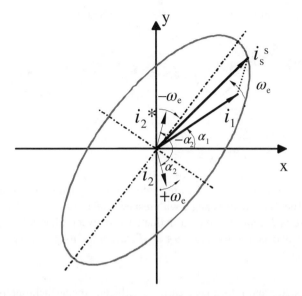

Figure 4.13 The locus plotted by the tip of the electric current space vector for an asymmetric three-phase system in steady state when the phase currents have no zero component. Reproduced by permission of Dr. M. Niemelä.

According to the figure, the major (transverse) axis of the produced ellipse is $\hat{i}_1 + \hat{i}_2$, and the length of the minor (conjugate) axis is $\left|\hat{i} - \hat{i}_2\right|$. The phasors of the positive and negative sequence components can be written as follows.

$$i_1 = \hat{i}_1 e^{j(\omega_e\, t + \alpha_1)} \tag{4.68}$$

$$i_2 = \hat{i}_2 e^{j(\omega_e\, t + \alpha_2)} \tag{4.69}$$

The angle of the major axis is determined from this equation.

$$\frac{\alpha_1 - \alpha_2}{2} \tag{4.70}$$

In practice, this kind of an asymmetric system can be represented by adding suitable temporal components to the definition of the electric current vector, which will automatically result in the above elliptical orbit. If, for instance, one of the phase currents is relatively low, the electric current circle starts to become elliptical. If one phase current is completely missing, the result is a single-phase system, which plots as a line in the xy-plane.

The transformation from a three-phase to a two-phase system and the resulting representation in the two-axis reference frame were presented previously. Figure 4.14 illustrates the stator and the rotor of an asynchronous machine with three-phase windings. Formulating and solving the winding equations to define an electric machine's transient behaviours is a challenging undertaking. It will be useful to demonstrate the relative simplicity of the space-vector approach by first developing the winding equations for an asynchronous machine then the space vectors.

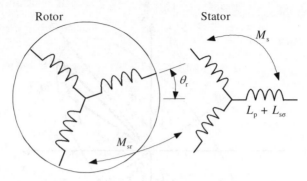

Figure 4.14 The stator and short-circuited rotor of an asynchronous machine with three-phase windings. The rotor rotation angle with respect to the stator is θ_r. The main inductance of the stator phase is L_p, and the mutual inductance between phases is M_s. The maximum mutual inductance between stator and rotor is M_{sr} when the magnetic axes of the stator and rotor phases coincide.

All six of the stator and rotor phase currents influence the asynchronous machine's total flux linkage. To begin the space vector ψ_s^s for the stator flux linkage in a reference frame fixed to the stator will be constructed. The instantaneous flux linkages for the different stator phases can be represented by the following equation set.

$$\psi_{sU} = \left(L_{s\sigma} + L_p\right)i_{sU} + M_s i_{sV} + M_s i_{sW} + M_{sr}\cos(\theta_r)i_{rU} + M_{sr}\cos\left(\theta_r + \frac{2\pi}{3}\right)i_{rV}$$
$$+ M_{sr}\cos\left(\theta_r + \frac{4\pi}{3}\right)i_{rW} \tag{4.71}$$

$$\psi_{sV} = \left(L_{s\sigma} + L_p\right)i_{sV} + M_s i_{sU} + M_s i_{sW} + M_{sr}\cos\left(\theta_r + \frac{4\pi}{3}\right)i_{rU} + M_{sr}\cos(\theta_r)i_{rV}$$
$$+ M_{sr}\cos\left(\theta_r + \frac{2\pi}{3}\right)i_{rW} \tag{4.72}$$

$$\psi_{sW} = \left(L_{s\sigma} + L_p\right)i_{sW} + M_s i_{sU} + M_s i_{sV} + M_{sr}\cos\left(\theta_r + \frac{2\pi}{3}\right)i_{rU} + M_{sr}\cos\left(\theta_r + \frac{4\pi}{3}\right)i_{rV}$$
$$+ M_{sr}\cos(\theta_r)i_{rW} \tag{4.73}$$

$L_{s\sigma}$ is the leakage inductance of the stator, L_p is the main inductance (single-phase inductance), M_s is the mutual inductance between the stator windings, M_{sr} is the maximum value of the mutual inductance between the stator and rotor circuits, θ_r is the angle between the magnetic axes of the rotor and stator, and the instantaneous phase currents of the rotor are i_{rU}, i_{rV}, and i_{rW}.

Because there is a phase difference of $2\pi/3$ between the magnetic axes of the stator windings, when the flux density is sinusoidally distributed in the air gap, the mutual inductance between stator windings phases can be expressed by this next equation. Again, L_p is the main (single-phase) inductance.

$$M_s = L_p\cos\left(\frac{2\pi}{3}\right) = -\frac{L_p}{2} \tag{4.74}$$

Because of this phase difference, the current in one phase, for example, phase U, produces a flux linkage in another, such as phase V. The magnitude of this induced flux linkage is half that of the flux linkage produced in phase U by the U-phase current flow. If the stator currents have no zero components, $i_U + i_V + i_W = 0$, and the expression of the first three components in Equation (4.71) can be simplified as the following equation illustrates.

$$(L_{s\sigma} + L_p)i_{sU} + M_s i_{sV} + M_s i_{sW} = (L_{s\sigma} + L_p)i_{sU} - \frac{L_p}{2}(i_{sV} + i_{sW}) = \left(L_{s\sigma} + \frac{3}{2}L_p\right)i_{sU}$$

$$= (L_{s\sigma} + L_m)i_{sU} = L_s i_{sU} \qquad (4.75)$$

The maximum value M_{sr} for the mutual inductance between the stator and rotor circuits is reached when the magnetic axes of the stator and rotor coincide. When addressing primary flux, the coupling factor between the stator and rotor circuits is $k = 1$. Referring the rotor currents to the stator side results in this equivalency.

$$M_{sr} = L_p \qquad (4.76)$$

Instantaneous values for the flux linkages of the different phases can be substituted into the equation for the stator flux-linkage space vector ψ_s^s.

$$\psi_s^s = \frac{2}{3}\left(a^0 \psi_{sU} + a^1 \psi_{sV} + a^2 \psi_{sW}\right) \qquad (4.77)$$

Simplification yields

$$\psi_s^s = L_s i_s^s + L_m i_r^r e^{-j\omega_e t} = L_s i_s^s + L_m i_r^s \qquad (4.78)$$

Since it was developed in the complex plane, the currents and flux linkages of the equation are complex vectors.

So, instead of the more complex Equations (4.71) through (4.73), a simple space-vector representation can be applied as presented by Equation (4.78). The result is the same as the previously presented, Equation (4.52). The space-vector approach offers a simple and efficient method to achieve a control model solution for an asynchronous electric machine.

Figure 4.15 relates the three space-vector representation methods to the electric current vector components for the windings of an electric machine. The instantaneous flux linkage components ψ_{sU}, ψ_{sV}, and ψ_{sW} caused by the instantaneous phase currents are depicted in Figure 4.15a. The space vector ψ_s' describing the flux-linkage components is illustrated by Figure 4.15b. In the figure, the windings are located in the stator in the direction of the direct and quadrature axes.

The sum flux-linkage ψ_s^s can be obtained from the following equation if the windings are fed with the appropriate instantaneous current components i_{xs} and i_{ys}.

$$\psi_s^s = \psi_{sx}^s + j\psi_{sy}^s \qquad (4.79)$$

When simulating or designing the digital control of electric machines, they are often modelled according to the above two-phase system. For instance, when using microprocessor-based control, the equations of the motor must be represented in a two-axis system, since the

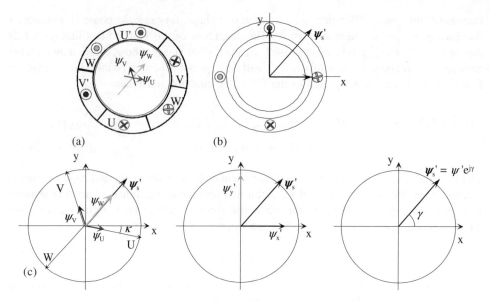

Figure 4.15 (a) The flux-linkage components ψ_U, ψ_V, and ψ_W of the windings for phases U – U', V – V', and W – W', (b) The corresponding space vector ψ'_s of the stator flux linkage represented by the two-phase components ψ'_x and ψ'_y, (c) Three different means to represent the space vector. The example is given at the time instant when $i_W = -1$, $i_U = i_V = \frac{1}{2}$. 1) three-phase components, 2) two axis components, 3) space-vector given in single-phase complex form with its amplitude ψ' and its position angle γ. The phase shift κ is usually set to zero. In this presentation the reduction by 2/3 is omitted, and therefore the primed symbols, e.g. ψ'_s are used.

processors are incapable of handling the vectors in a polar coordinate system, the best method mathematically.

The two-axis system is particularly useful when modelling magnetically nonisotropic electric machines, such as salient-pole synchronous machines. When applying the two-axis model, the rotor reference frame is often used, since synchronous machines are genuinely nonisotropic either with respect to the magnetic circuit *and* the electric circuit or with respect to the electric circuit *alone*, which leads naturally to selection of the rotor reference frame.

So far, three representation methods have been developed to simulate the behaviours of rotating electric machinery: three-phase, two-phase, and single-phase complex. In the three-phase method, the resulting equations include numerous terms, and their solution is challenging. The two-phase method results in considerably simpler equations, since the orthogonal windings have, in principle, no mutual inductance, which makes the representation easier. In practice, there exists cross-saturation (the flux on d-axis affects the properties of q-axis and vice versa) between the d- and q-axes that must be taken into account in the modelling in some cases. Another benefit of the two-axis model is that quantities can be treated in the complex plane.

Figure 4.15c illustrates the generation of the space vector and its three representation methods. The space vector is a single-phase complex quantity, which can naturally be used as such; however, it is usually decomposed into its components. Therefore, the representation

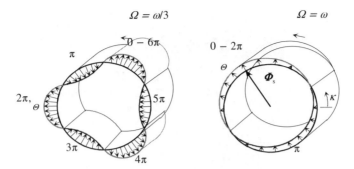

Figure 4.16 The flux of a six- and a two-pole machine as three-dimensional illustrations along the machine length. Flux aligns with the three pole pairs ($p = 3$) of a six-pole machine as shown on the left in the figure. In a two-pole machine ($p = 1$), flux vector $\boldsymbol{\Phi}_s$ aligns with the single pole pair. The flux vector outline for the six-pole machine is difficult to visualize. The six-pole current-linkage fundamental propagates at a physical angular speed of $\Omega = \omega/3$, and a two-pole ($p = 1$) fundamental propagates at $\Omega = \omega$. If the angular frequency ω of the input current is equal for both windings, the current linkage and air-gap flux in a six-pole winding propagates at one-third the speed of the two-pole winding flux. Both distributions propagate one wavelength for each supply frequency period resulting in the propagation speed difference. This difference forms the basis for the influence of pole pair number on machine rotation speed. *Source:* Pyrhönen et al. 2014. Reproduced with permission of John Wiley & Sons Ltd.

method ultimately returns to the two-axis model. The treatment of complex numbers, for instance in signal processors, is still best carried out in component form rather than as a polar representation, and therefore, returning to the two-phase representation makes computation easier.

Figure 4.16 illustrates the connection between the physical flux distribution and the flux space vector. The distribution-to-vector connection is clear for a two-pole machine; however, it is less clear for a multipole polyphase machine, where the definition of total flux depends both on machine structure and on how the windings are connected. Nevertheless, the physical flux distribution for the multipole machine is easier to visualize.

When applying space-vector theory, the mathematical treatment of the electric machine reverts, in practice, to the two-pole representation. A simple connection to the input and output terminals of an electric machine does not give the motor controller any information about its pole pair quantity, unless the operator of the machine inputs this information to the frequency converter. Only the angular data supplied by integrated angle sensors must be handled in connection with the number of pole pairs, since a motor control based on the space-vector theory operates based on rotor electrical angle. The number of pole pairs also affects machine torque.

Figure 4.17 illustrates the relationship of the real currents flowing in the slots of an induction machine. It shows the electric currents, current space vectors (i_s, i_r, i_m), the flux lines, and the air-gap flux-linkage space vector ($\boldsymbol{\psi}_m$). The magnetizing current vector and the air-gap flux linkage are both in the direction of maximum air-gap flux linkage.

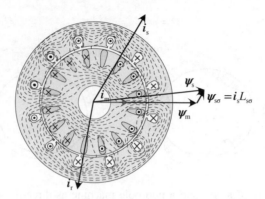

Figure 4.17 Induction machine currents, electric current vectors, flux lines, air-gap flux linkage, stator-leakage flux linkage, and stator flux linkage. The air-gap flux-linkage vector is clearly bound to the air-gap flux of the machine, and therefore has a clear physical interpretation. According to Equation (4.31), the stator flux-linkage vector is $\psi_s = \psi_m + \psi_{s\sigma}$. The leakage flux-linkage space vector $\psi_{s\sigma}$ is calculated with the stator current vector i_s and the stator leakage inductance $L_{s\sigma}$. Adding a vector such as this to the air-gap flux-linkage vector is clear mathematically, but the physical interpretation of the stator flux-linkage vector and especially the stator leakage flux-linkage space vector is not clear, because the leakage of the machine is distributed among several of the machine's components. However, stator flux linkage is clearly the integral of the stator voltage.

References

DeDoncker, R. W., & Novotny, D. W. (1994). The universal field oriented controller. *IEEE Transactions on Industry Applications*, 30(1), 92–100.

Kovàcs, K. P., & Ràcz, I. (1954). *Transiente Vorgänge in Wechselstrommaschinen* (1st and 2nd eds.) Budapest, Hungary: Verlag der Ungarischen Akademie.

Pyrhönen, J., Jokinen, T., & Hrabovcova V. (2014). *Design of rotating electric machines* (2nd ed.). Chichester, UK: John Wiley & Sons.

5

Torque and force production and power

A core task for a rotating electric machine is to produce the torque needed to achieve the required rotation speed under load. In linear machines, correspondingly, force production is the key element. Torque production is based on forces affecting the stator and the rotor. There are several ways to study force and torque production. The most important ways from the electrical drive's point of view are explored in this chapter.

5.1 The Lorentz force

Torque production in electric machines is based on two phenomena. In round-rotor machines, torque production is most easily explained in terms of Lorentz forces. Lorentz force also explains the force production in linear machines. The following study is dedicated for rotating machines, but a similar approach can be used for linear machines.

For salient-pole rotor machines, torque production also includes the magnetic force effects that result when the reluctance of the magnetic circuit is different in different directions. Naturally, torque production for salient-pole rotor machines can also be explained by the Lorentz forces acting on the stator windings.

Because electric machines must always be magnetized to produce torque, one can assume a flux density B [Vs/m^2] in the air gap of the machine. According to the equation for Lorentz force, the charge element dQ [As] moving at speed v [m/s] experiences the following force element.

$$dF = dQ(E + v \times B) \tag{5.1}$$

Here, E is the vector expressing the electric field strength [V/m]. In Equation (5.1), the vectors describe electromagnetic field quantities; they are not space vectors. The electric field strength

Electrical Machine Drives Control: An Introduction, First Edition. Juha Pyrhönen, Valéria Hrabovcová and R. Scott Semken.
© 2016 John Wiley & Sons, Ltd. Published 2016 by John Wiley & Sons, Ltd.

proportion of the force is in the direction of the field strength E (the units of the multiplication give $AsV/m = Ws/m = J/m = Nm/m = N$). It does not depend on the motion of the charge element. The force produced by the magnetic field is perpendicular to the plane determined by both the speed v and the flux density B, and it depends on these two according to the cross product of the vectors (the units of the multiplication give $As \times m/s \times Vs/m^2 = N$). The absolute value of the force effect caused by the magnetic field is

$$dF = dQvB \sin \beta \qquad (5.2)$$

Here, β is the angle between the velocity vector of the charge element and the flux-density vector as shown in Figure 5.1.

In ordinary electric machines, torque is produced by the interaction of the air-gap magnetic field and currents running through the windings. Therefore, depending on electric field strength, the force term can be ignored. In the case of a current-carrying conductor, the following expression can be written for a charge element dQ moving at speed v.

$$dQv = dQ \frac{dl}{dt} = \frac{dQ}{dt} dl = idl \qquad (5.3)$$

By substituting this expression of the current-carrying element into the equation of the Lorentz force, dF can be rewritten. See Figure 5.2.

$$dF = idl \times B \qquad (5.4)$$

The force acting on the conductor element is perpendicular to the plane determined by the element dl and the flux density B. This force reaches its maximum when dl and B are perpendicular ($\sin \beta = 1$). Otherwise, the force drops in the ratio of $\sin \beta$ and reaches zero for parallel current and flux density.

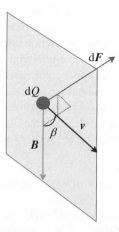

Figure 5.1 Lorentz force produced by a magnetic field acting on a moving charge. A charge dQ moving at speed v in the magnetic field B experiences a force dF. The vectors v and B determine a plane, and the force is normal to that plane. The dotted lines indicate perpendicularity between v and dF or B and dF.

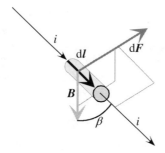

Figure 5.2   Application of the magnetic component of the Lorentz force to a current-carrying conductor with length $\mathrm{d}l$. The dotted lines indicate perpendicularity between $\mathrm{d}l$ and $\mathrm{d}F$ or B and $\mathrm{d}F$.

Good electric machine design, as much as possible, seeks to establish and maintain perpendicularity between current i and flux density B. If there is no significant saturation of the magnetic circuit, air-gap flux penetration into the electrical steel is perpendicular. However, this is not enough, because in squirrel-cage machines, for example, the stator must both magnetize and carry torque-producing currents. As a result, the spatial distribution of stator currents (the linear current density distribution) is out of phase with the spatial distribution of air-gap flux (and this cannot be fixed with the correct rotor excitation). For synchronous machines, the spatial distributions of stator linear current density and air-gap flux can be kept in phase by applying the correct rotor excitation. Operating in this way makes it possible to achieve complete air-gap flux density vector and stator current density vector perpendicularity.

The flux Φ penetrating the rotor from the air gap of electric machines intersects the current-carrying rotor elements, thus generating Lorentz forces that manifest on the rotor surface. The current depicted in Figure 5.2 flows through the current-carrying rotor elements, such as the copper bars of a squirrel-cage rotor. The total force exerted on a single bar is obtained by integrating over the length of the bar. If flux density and current remain constant over the length of the bar, this integration becomes a simple multiplication operation.

In an electric machine, any conductor in the presence of a magnetic field that is carrying current will experience a force. As a result, there will be opposite forces produced in the stator and rotor, resulting in opposite torques.

5.2   The general equation for torque

The control of electric machines requires the expression of torque in terms of space vectors, which were the focus of the previous chapter. Figure 5.3 shows the sinusoidal flux-density distribution $B_\delta(\alpha)$ acting in the air gap of a machine and the stator linear current density fundamental $A_{s1}(\alpha)$ on the air-gap surface. The linear current density A is an imaginary quantity that represents the currents as an infinitesimally thin layer on the stator or rotor surface.

Linear current density is the sum of currents $z_Q i(t)$ (with z_Q being the number of conductors in a slot and $i(t)$ their instantaneous current) divided by the slot opening width b_1. The peak value of the linear current density fundamental occurs at the location of actual peak current. In Figure 5.3, the linear current density spikes indicate that one of the stator phases is at its peak

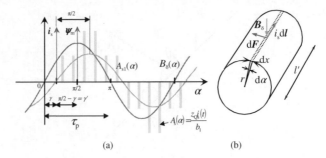

(a)   (b)

Figure 5.3 (a) The fundamental waves of flux density and linear current density in the air-gap region of a $q = 2$ machine and (b) the inner diameter dimensions of the machine. τ_p is the magnetic pole pitch, r is the radius of the air-gap inner diameter, and l' is the effective length of the machine. l' is not simply the machine's physical length, and its accurate determination can be complicated. However, for induction motors without cooling ducts, $l' \approx l + 2\delta$, with δ being the air-gap length. The air-gap flux-linkage ψ_m is directly proportional to B_δ. See Equation (5.5) and Figure 5.4.

positive current and the other two are at half of their negative currents, and that there are two slots per pole and phase in the winding. The figure also shows where the stator current space vector and the air-gap flux-linkage space vector (i_s and ψ_m) are located. There is a 90° phase shift between the peak of A_{s1} and the locations of the current space vectors. The right-hand side of Figure 5.3 depicts the location of a current element of width dx on the air-gap surface, the air-gap flux-density B_δ acting on it, the force dF, and the dimensions of the air-gap inner diameter.

Air-gap surface flux and linear current density distribution can be observed on either the stator or the rotor surface. Because stator and rotor torque are equal but opposite, the torque determined with either approach will be equal in magnitude but opposite in direction. The stator surface linear current density approach is taken here, and the air-gap distributions are assumed sinusoidal.

$$B_{\delta 1}(\alpha) = \hat{B}_{\delta 1} \sin \alpha,$$
$$A_{s1}(\alpha) = \hat{A}_{s1} \sin (\alpha - \gamma). \tag{5.5}$$

Here, γ is the spatial phase shift between the fundamental waves of the stator linear current density A_{s1} and the air-gap flux density $B_{\delta 1}$. In machines with separate rotor excitation, γ values can range, in theory, from −90° to +90° and include 0°, which is most efficient for torque production. In addition, γ has a relationship with the power factor angle φ, but it is not the same. And φ includes the effect of stator leakage, whereas γ observes only the air-gap phenomena.

The magnetic flux is nearly perpendicular to the air-gap surface in electric machines, so the Lorentz force equation can be applied to a conductor of width dx located on the air-gap surface. Assuming there is a current di_s flowing in the conductor,

$$d\boldsymbol{F}_{\tan} = \int_0^{l'} di_s d\boldsymbol{l} \times \boldsymbol{B}_\delta \tag{5.6}$$

For the sake of generality, the general air-gap flux-density vector $\boldsymbol{B}_\delta$ is taken here without finding the fundamental. Because the flux-density distribution can be assumed with sufficient accuracy to be independent of the coordinate of the stator surface in the axial direction, the absolute value for the force acting on an imaginary conductor bar of width dx can be obtained from this equation.

$$dF_{\text{tan}} = di_s B_\delta l' \tag{5.7}$$

Calculated anywhere on the stator surface, this force is tangential.

Considering sinusoidal quantities, the magnitude of the current di_s in the region of width dx can be obtained by applying the expression for the sinusoidal linear current density distribution.

$$di_s = \hat{A}_{s1} \sin(\alpha - \gamma)dx \tag{5.8}$$

Regardless of the number of pole pairs, the air-gap torque can be calculated by integrating over the full perimeter using a two-pole approach. The sum of the absolute values of the tangential forces acting on the surface of a two-pole stator is obtained by integrating over the pole pair, that is, over two pole pitches $2\tau_p$.

$$\sum |F_{\text{tan}}| = l' \int_0^{2\tau_p} \hat{A}_{s1} \sin(\alpha - \gamma)\hat{B}_{\delta 1} \sin \alpha \, dx \tag{5.9}$$

The variable can be changed into $dx = rd\alpha$ to simplify the integration.

$$\sum |F_{\text{tan}}| = \hat{A}_{s1}\hat{B}_{\delta 1} l' r \int_0^{2\pi} \sin(\alpha - \gamma)\sin \alpha \, d\alpha$$

$$\sum |F_{\text{tan}}| = \hat{A}_{s1}\hat{B}_{\delta 1} l' r \int_0^{2\pi} (\sin^2 \alpha \cos \gamma - \cos \alpha \sin \gamma \sin \alpha)d\alpha \tag{5.10}$$

$$= \hat{A}_{s1}\hat{B}_{\delta 1} l' r \pi \cos \gamma$$

For rotating machines, the produced electromagnetic torque can be determined by multiplying the result by the radius r of the air-gap inner diameter.

$$T_e = r \sum |F_{\text{tan}}| = \hat{A}_{s1}\hat{B}_{\delta 1}\pi r^2 l' \cos \gamma \tag{5.11}$$

Equation (5.11) clearly shows how the torque is proportional to the air-gap inner diameter volume $V_\delta = \pi r^2 l'$ and the linear current and air-gap flux-density peak values $\hat{A}_{s1}$ $\hat{B}_{\delta 1}$.

In principle, the peak value of the linear current density is obtained by dividing the peak value of the slot current sum by the slot pitch taking into account the winding factor k_{ws1} of the fundamental harmonic. If the number of slots per pole and phase is q_s, the number of coil turns in series per p pole pairs is N_s, the number of phases is $m_s = 3$, and the number of slots is Q_s. The effective peak current sum of the slot can be expressed as follows.

$$\hat{i}_{us} = \frac{k_{w1s}N_s\hat{i}_s}{q_s p} \tag{5.12}$$

The slot pitch is

$$\tau_{us} = \frac{2\tau_p}{Q_s/p} \tag{5.13}$$

In a three-phase machine, there is a connection between the number of slots per pole and phase q_s and the number of slots Q_s in the $2p$-pole case based on the dimensions of the winding. This connection is revealed by

$$Q_s = m_s 2pq_s = 6pq_s \tag{5.14}$$

In the previous Figure 5.3, the maximum slot sum currents are divided by the slot opening to get the linear current density peaks. The value is averaged over the slot pitch, and by taking the fundamental winding factor k_{w1s} into account, the fundamental peak value of the linear current density can be expressed.

$$\hat{A}_{s1} = \frac{\dfrac{k_{w1s}N_s\hat{i}_s}{q_s p}}{\dfrac{2\tau_p p}{Q_s}} = \frac{k_{w1s}N_s\hat{i}_s Q_s}{2\tau_p p^2 q_s} = \frac{3k_{w1s}N_s\hat{i}_s}{\tau_p p} \tag{5.15}$$

By substituting this result to the expression of the torque produced by the air-gap inner diameter, torque formula can be rewritten as follows.

$$T_e = \hat{A}_s\hat{B}_\delta l' r^2 \pi \cos\gamma = \frac{3\hat{i}_s k_{w1s}N_s}{\pi r}\hat{B}_\delta l' \pi r^2 \cos\gamma = 3\hat{i}_s k_{w1s}N_s\hat{B}_\delta l' r \cos\gamma$$
$$= 3\hat{i}_s k_{w1s}N_s\frac{2}{\pi}\hat{B}_\delta l'\frac{\pi}{2}r \cos\gamma \tag{5.16}$$

Because the pole pitch in a multpole case is $2\pi r/2p = \pi r/p$, the expression can be further developed.

$$T_e = \frac{3}{2}p\hat{i}_s k_{w1s}N_s\frac{2}{\pi}\hat{B}_\delta l'\frac{\pi r}{p}\cos\gamma$$
$$= \frac{3}{2}p\hat{i}_s k_{w1s}N_s\frac{2}{\pi}\hat{B}_\delta l'\tau_p \cos\gamma. \tag{5.17}$$

Equation (5.17) includes the peak flux term $\hat{\Phi}_m = \frac{2}{\pi}\hat{B}_\delta l'\tau_p$ of the sinusoidal air-gap flux-density distribution, which is the surface integral of the flux density over the pole pitch. When the peak value of the air-gap flux is multiplied by the effective number of turns $k_{w1s}N_s$, the

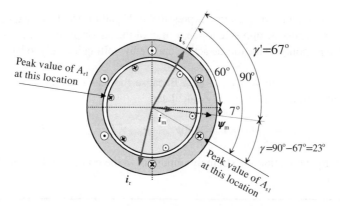

Figure 5.4 The stator and rotor currents of a two-pole asynchronous machine at any given instant and their respective vectors. The asynchronous machine is well suited for torque production, since there is a large phase difference between the vector of the air-gap flux linkage $\boldsymbol{\psi}_\mathrm{m}$ and the stator $\boldsymbol{i}_\mathrm{s}$ or the rotor $\boldsymbol{i}_\mathrm{r}$ current vectors. The figure also illustrates the 90° phase shift between the peak value of the linear current density and the stator current vector as well as how the stator and rotor currents are approximately opposing.

peak value $\hat{\psi}_\mathrm{m}$ of the air-gap flux linkage and thus the expression for the torque can be simplified into this formula.

$$T_\mathrm{e} = \frac{3}{2}p\hat{i}_\mathrm{s}\hat{\psi}_\mathrm{m}\cos\gamma.\tag{5.18}$$

Figure 5.4 illustrates the currents of an electric machine and the respective current vectors according to space-vector theory. The direction of the current vector is the same as the direction of the current linkage of the winding in question, which deviates $\pi/2$ from the direction of the peak value of the linear current density used in the definition of the torque.

When adopting the space vectors, the lengths of which are scaled to be equal to the peak values of the respective temporal phase quantities, a torque vector in accordance with space-vector theory is obtained from Equation (5.18). Taking the $\pi/2$ phase shift between the linear current density and the current space vector into account (see Figure 5.3) and replacing the peak value of the real $\hat{i}_\mathrm{s}$ current in Equation (5.18) with the current space vector $\boldsymbol{i}_\mathrm{s}$, Equation (5.18) can be rewritten accordingly.

$$T_\mathrm{e} = \frac{3}{2}p|\boldsymbol{\psi}_\mathrm{m}||\boldsymbol{i}_\mathrm{s}|\sin\gamma'\tag{5.19}$$

In Equation (5.19), the angle $\gamma' = \pi/2 - \gamma$ according to the earlier Figure 5.3; γ' represents the angle between the air-gap flux-linkage vector and the stator current vector. The latter part of (5.19) can be expressed as a vector product presentation.

$$T_\mathrm{e} = \frac{3}{2}p\boldsymbol{\psi}_\mathrm{m}\times\boldsymbol{i}_\mathrm{s} = -\frac{3}{2}p\boldsymbol{\psi}_\mathrm{m}\times\boldsymbol{i}_\mathrm{r}.\tag{5.20}$$

There is a negative sign on the latter expression in Equation (5.20) as expected, since the torque of the rotor must be opposite to the torque of the stator. In practice, the stator is kept from rotating in response to the torque acting upon it by bolting it firmly to base of the electric machine.

The stator flux linkage $\psi_s = \psi_m + \psi_{s\sigma}$. Because the leakage flux linkage $\psi_{s\sigma}$ does not produce torque, torque can be equally calculated.

$$T_e = \frac{3}{2}p\psi_s \times i_s = \frac{3}{2}p(\psi_m + \psi_{s\sigma}) \times i_s = \frac{3}{2}p\psi_m \times i_s \qquad (5.21)$$

This general torque equation is extremely important in the context of the controlled drive of electric machines.

EXAMPLE 5.1: The 400 V, star-connected, 50 Hz induction machine from Figure 5.4 is running at its rated operating point. The machine has a peak air-gap flux-linkage value of 0.95 Vs, which at the instant of observation happens to be at an angle of -7°. The stator current space vector is at an angle of 60°. The stator current RMS value is 52.3 A. Calculate the motor torque at this instant. The rotational speed is 2955 min^{-1}. Calculate the power and the approximate efficiency if the power factor of the two-pole machine is given by $\cos \varphi = 0.9$.

SOLUTION: Because the RMS value is 52.3 A, the magnitude of the stator current space vector will be $|i_s| = 52.3\,A\sqrt{2} = 74\,A$. The torque is the cross product of the stator flux linkage and the stator current vector $T_e = \frac{3}{2}p\psi_s \times i_s$. Applying Equation (5.19), the torque $T_e = \frac{3}{2}1 \cdot 0.95\,Vs \cdot 74\,A \sin(60° + 7°) = 97\,Nm$. Because the induction machine is a 400 V motor with 52.3 A RMS, the motor apparent power is 36.1 kVA. The specified values for rotation speed and calculated torque result in $P = \Omega_r T = 2955 \cdot 2\pi/60/s \cdot 97\,Nm = 30\,kW$. The product of power factor and efficiency of the machine is therefore $30/36.1 = 0.832$. Because the power factor is $\cos \varphi = 0.9$, the efficiency will be $0.832/0.9 = 0.924$. Because the iron loss current of the motor is neglected, this is an approximate result.

Linear motor drives utilize use the same torque presentation even though there is no torque present but electromagnetic force instead. Dividing Equation (5.2) by the rotor radius r, used in the derivation of the rotor torque, results in force F_e affecting the rotor surface

$$\frac{T_e}{r} = F_e = \frac{3}{2r}p\psi_s \times i_s = \frac{3}{2r}p(\psi_m + \psi_{s\sigma}) \times i_s = \frac{3}{2r}p\psi_m \times i_s = \frac{3\pi}{2\tau_p}\psi_m \times i_s \qquad (5.22)$$

The force in linear motor drives is proportional to the sentence of Equation (5.21), $F_e \sim T_e$. Therefore, in principle the same controller that is dedicated to the use of torque controlling in rotating machines can be used in the force control of linear machines.

5.3   Power

Ignoring losses, the torque equation determined previously correlates with power. Taking into account that $j\omega_s\psi_s = u_s$, an expression for power P_e results when Equation (5.21) is multiplied by the mechanical angular speed $\Omega_r = \omega_s/p$ (assuming no slip and losses), as shown by:

$$P_e = \frac{\omega_s}{p}|T_e| = \frac{3}{2}\omega_s|\boldsymbol{\psi}_s \times \boldsymbol{i}_s| = \frac{3}{2}\left|\frac{\boldsymbol{u}_s}{\mathrm{j}} \times \boldsymbol{i}_s\right| \Rightarrow P_e = \frac{3}{2}\boldsymbol{u}_s \cdot \boldsymbol{i}_s \qquad (5.23)$$

The latter part of the equation contains a scalar product of space vectors, so the expression for power can be revised as follows.

$$P_e = \frac{3|\boldsymbol{u}_s||\boldsymbol{i}_s|}{2}\cos\varphi \qquad (5.24)$$

This formula resembles the well-known expression for power in terms of RMS values. The lengths of the voltage and current space vectors were selected to be equivalent to the peak values of the sinusoidal phase quantities, and power calculated using Equation (5.24) corresponds to power calculated in terms of the RMS values in case of no losses.

$$P_e = \Omega_r T_e = 3U_{\mathrm{ph}}I_{\mathrm{ph}}\cos\varphi \qquad (5.25)$$

If there is no zero component, the instantaneous value $P_e(t)$ for the power of a three-phase asynchronous machine may be determined according to

$$P_e(t) = \frac{3}{2}\mathrm{Re}\{\boldsymbol{u}_s\,\boldsymbol{i}_s^*\} = \frac{3}{2}\mathrm{Re}\{\boldsymbol{u}_s^*\,\boldsymbol{i}_s\} = \frac{3}{2}\boldsymbol{u}_s \cdot \boldsymbol{i}_s \qquad (5.26)$$

The validity of Equation (5.26) can be demonstrated by reviewing the definition of the space vector. Because it holds for the phase-shift operator that $a^* = a^2$ and $a^{*2} = a$, in the stator reference frame, power can be expressed:

$$\begin{aligned} P_e(t) &= \frac{3}{2}\mathrm{Re}\{\boldsymbol{u}_s^s\,\boldsymbol{i}_s^{*s}\} \\ &= \frac{3}{2}\mathrm{Re}\left\{\frac{2}{3}\left(u_{sU} + au_{sV} + a^2u_{sW}\right)\frac{2}{3}\left(i_{sU} + a^2i_{sV} + ai_{sW}\right)\right\} \end{aligned} \qquad (5.27)$$

Taking into account that $\mathrm{Re}\{a\} = \mathrm{Re}\{a^2\} = -\frac{1}{2}$, the equation can be rewritten as follows.

$$\begin{aligned} P_e(t) = \frac{2}{3}\Big\{ &u_{sU}i_{sU} + u_{sV}i_{sV} + u_{sW}i_{sW} - \frac{1}{2}[u_{sU}(i_{sV} + i_{sW}) \\ &+ u_{sV}(i_{sU} + i_{sW}) + u_{sW}(i_{sU} + i_{sV})]\Big\}. \end{aligned} \qquad (5.28)$$

For the instantaneous values of the phase currents, $i_{sU} + i_{sV} + i_{sW} = 0$. Therefore, Equation (5.28) can be further simplified to produce the following equation.

$$P_e(t) = u_{sU}(t)i_{sU}(t) + u_{sV}(t)i_{sV}(t) + u_{sW}(t)i_{sW}(t) \qquad (5.29)$$

So, $P_e(t)$ is the instantaneous sum of the phase powers, and in steady state, the value of $P_e(t)$ corresponds to the value of P_e determined using Equation (5.25).

5.4   Reluctance torque and co-energy

The operation of a doubly salient pole reluctance machine, which is illustrated in Figure 5.5, is based on the saliency of both the stator and the rotor. This machine cannot use the same space-vector approach as rotating field machines, and therefore torque production must be evaluated differently.

Torque production can be examined by analysing the energy stored in the magnetic circuit of the machine. The voltage of a double-salient pole reluctance machine can be expressed by applying Faraday's induction law and Ohm's law.

$$u = Ri + \frac{\partial \psi(i, \theta_r)}{\partial t} \tag{5.30}$$

Flux linkage ψ depends on both the current i and the rotor angle θ_r. The power fed to a phase is obtained by multiplying the voltage by the current.

$$ui = Ri^2 + i\frac{\partial \psi}{\partial i}\frac{di}{dt} + i\frac{\partial \psi}{\partial \theta_r}\frac{d\theta_r}{dt} \tag{5.31}$$

The energy consumed in the phase represents the mechanical work dW_{mec} and the change of the energy dW_e stored in the magnetic field. Furthermore, some of the energy is wasted in resistive losses.

$$uidt = Ri^2dt + i\frac{\partial \psi}{\partial i}di + i\frac{\partial \psi}{\partial \theta_r}d\theta_r = Ri^2dt + dW_e + dW_{mec} \tag{5.32}$$

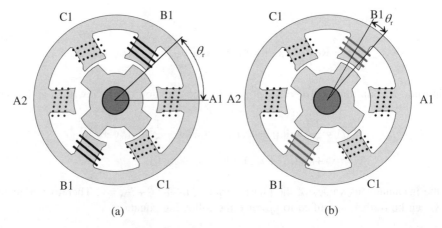

(a)   (b)

Figure 5.5 Doubly salient reluctance motor with six stator and four rotor poles. (a) The rotor is in unaligned position with respect to phase A. The rotor angle θ_r is measured from the aligned position with phase A. (b) Phase B is energized, and the rotor approaches the aligned position. The rotor angle θ_r is measured toward the aligned position of phase B.

The change of magnetic energy can be expressed with the current i and the rotation angle θ_r as follows.

$$dW_e = \frac{\partial W_e}{\partial i} di + \frac{\partial W_e}{\partial \theta_r} d\theta_r \tag{5.33}$$

Consequently, based on (5.32) and (5.33) the derivative of the mechanical energy can be written:

$$dW_{mech} = \left(i\frac{\partial \psi}{\partial \theta_r} - \frac{\partial W_e}{\partial \theta_r} \right) d\theta_r + \left(i\frac{\partial \psi}{\partial i} - \frac{\partial W_e}{\partial i} \right) di \tag{5.34}$$

The energy stored in the phase at any given instant can be calculated using this equation.

$$W_e = \int_0^\psi i d\psi = i\psi - \int_0^i \psi di \tag{5.35}$$

Therefore, the derivative of magnetic energy with respect to current is as follows.

$$\frac{\partial W_e}{\partial i} = i\frac{\partial \psi}{\partial i} + \psi - \int_0^i \frac{\partial \psi}{\partial i} di = i\frac{\partial \psi}{\partial i} \tag{5.36}$$

Inputting the solution for Equation (5.36) into Equation (5.34) leads to this expression for the derivative of mechanical energy:

$$dW_{mec} = \left(i\frac{\partial \psi}{\partial \theta_r} - \frac{\partial W_e}{\partial \theta_r} \right) d\theta_r \tag{5.37}$$

Torque T is the change of the mechanical energy with respect to the angle of rotation as expressed by the following equation.

$$T = \frac{dW_{mec}}{d\theta_r} = i\frac{\partial \psi}{\partial \theta_r} - \frac{\partial W_e}{\partial \theta_r} \tag{5.38}$$

This formula for torque can be simplified by replacing the change of magnetic energy W_e with the change of magnetic co-energy W^*. Magnetic co-energy is defined as:

$$W^* = \int_0^i \psi di \tag{5.39}$$

The geometric interpretation for the magnetic co-energy is the area between the magnetizing curve represented in the (i,ψ) plane and the i-axis. See Figure 5.6.

Based on the illustration, the sum of the magnetic energy and the co-energy can be written:

$$W_e + W^* = i\psi \tag{5.40}$$

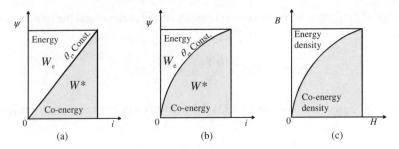

Figure 5.6 Determination of energy and co-energy with current and flux linkage for (a) a linear case with L constant and (b, c) a nonlinear case where L saturates as a function of current. If the figure is used to illustrate the behaviour of an SR machine, the rotor position remains constant.

By deriving W^* with respect to the angle θ_r, the expression for the derivative of the co-energy becomes:

$$\frac{\partial W^*}{\partial \theta_r} = i\frac{\partial \psi}{\partial \theta_r} - \frac{\partial W_e}{\partial \theta_r} \qquad (5.41)$$

Comparing this result with the torque Equation (5.38) shows that torque for a reluctance machine is equal to the change of magnetic co-energy per differential angular change.

$$T = \frac{\partial W^*}{\partial \theta_r} \qquad (5.42)$$

Calculation of this requires knowledge of the field solution, and therefore it is limited to numerical calculations. It is essential for the modelling of an SR machine to know its ψi mappings for different operation situations.

5.5 Reluctance torque and the cross-field principle in a rotating field machine

The saliency of an electric machine produces torque if rotor movement results in a reduction in the reluctance of the main flux path. Electric machinery tends to settle at a reluctance minimum, which corresponds to the minimum value for energy stored in the magnetic circuit at any particular voltage. This principle holds for the operation of both the doubly salient pole reluctance machine and the synchronous reluctance machine.

To determine how a synchronous reluctance machine produces torque when applying the cross-field principle, the direct and quadrature-axis inductances of the machine must be known. Consider the machine of Figure 5.7. The stator current vector and the d-axis of the rotor form an angle κ. It is most convenient to investigate the machine in the dq reference frame of the salient pole. The flux linkage is expressed by the direct and quadrature inductances as follows.

$$\psi = L_d i_d + j L_q i_q = \psi_d + j\psi_q \qquad (5.43)$$

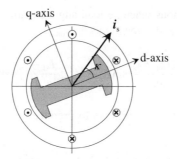

Figure 5.7 Current vector of a synchronous reluctance machine with respect to the rotor of the machine. The current angle κ is a controlled value in the vector control of a synchronous reluctance machine.

According to Figure 5.7, the space-vector current and flux-linkage components can be expressed as listed next in the following equations.

$$\begin{aligned}
i_d &= |i_s|\cos\kappa \\
i_q &= |i_s|\sin\kappa \\
\psi_d &= L_d|i_s|\cos\kappa \\
\psi_q &= L_q|i_s|\sin\kappa.
\end{aligned} \tag{5.44}$$

Substituting the expressions of Equation (5.44) in the torque equation results in the following.

$$|T_e| = \left|\frac{3}{2}\boldsymbol{\psi}_s \times \boldsymbol{i}_s\right| = \frac{3}{2}\left(\psi_d i_q - \psi_q i_d\right) \tag{5.45}$$

Using Equation (5.44), this can be rewritten using the amplitudes i of the vector $\boldsymbol{i}$.

$$\Rightarrow T_e = \frac{3}{2}\left(L_d i_s \cos\kappa \cdot i_s \sin\kappa - L_q i_s \sin\kappa \cdot i_s \cos\kappa\right) \tag{5.46}$$

$$\Rightarrow T_e = \frac{3}{2}\left((L_d - L_q)i_s^2 \cos\kappa \cdot \sin\kappa\right) = \frac{3}{2}\left((L_d - L_q)i_s^2\frac{1}{2}\sin(2\kappa)\right) \tag{5.47}$$

Equation (5.45) can, therefore, be rewritten to arrive at

$$T_e = \frac{3}{4}i_s^2\left(L_d - L_q\right)\sin 2\kappa \tag{5.48}$$

The maximum torque produced by a synchronous reluctance machine at a given current is achieved when the angle between the current vector and the d-axis of the rotor is $\pi/4$. The fraction 3/4 appears in the equation, because the space-vector amplitudes are measured with the peak value of the stator sinusoidal current. Using the RMS values, the fraction becomes the more familiar 3/2. Equation (5.48) must not be confused with the synchronous reluctance machine load angle equation, which looks similar. The load angle equation is a function of the synchronous machine load angle δ_s. This torque equation (5.48) is a function of the electric current angle κ. The synchronous machine load angle equation is normally written in RMS values of phase voltage U_{sph} and rotor-induced E_{fph}.

$$P = 3\left(\frac{U_{\mathrm{sph}} E_{\mathrm{fph}}}{\omega_s L_d} \sin \delta_s + U_{\mathrm{sph}}^2 \frac{L_d - L_q}{2\omega_s L_d L_q} \sin 2\delta_s\right) \tag{5.49}$$

The equation can be reformulated as follows to produce an expression for torque.

$$T = \frac{P}{\Omega_s} = 3\left(\frac{U_{\mathrm{sph}} E_{\mathrm{fph}}}{\Omega_s \omega_s L_d} \sin \delta_s + U_{\mathrm{sph}}^2 \frac{L_d - L_q}{\Omega_s 2\omega_s L_d L_q} \sin 2\delta_s\right)$$

$$T = 3\left(\frac{U_{\mathrm{sph}} E_{\mathrm{fph}}}{\Omega_s p \Omega_s L_d} \sin \delta_s + U_{\mathrm{sph}}^2 \frac{L_d - L_q}{\Omega_s 2p\Omega_s L_d L_q} \sin 2\delta_s\right) \tag{5.50}$$

$$T = 3\left(\frac{U_{\mathrm{sph}} E_{\mathrm{fph}}}{p\Omega_s^2 L_d} \sin \delta_s + U_{\mathrm{sph}}^2 \frac{L_d - L_q}{2p\Omega_s^2 L_d L_q} \sin 2\delta_s\right)$$

In terms of space-vector amplitudes $\hat{u}_s$ and $\hat{e}_f$, Equation (5.50) becomes

$$T = \frac{3}{2p\Omega_s^2}\left(\frac{\hat{u}_s \hat{e}_f}{L_d} \sin \delta_s + \hat{u}_s^2 \frac{L_d - L_q}{2L_d L_q} \sin 2\delta_s\right) \tag{5.51}$$

The latter part of Equation (5.51) expresses reluctance torque as a function of load angle. Its similarity with Equation (5.48) is obvious, and that may be confusing. A complete derivation of the load angle equation may be found, for example, in Pyrhönen et al. (2014).

5.6   Maxwell's stress tensor in the definition of torque

When applying numerical methods, Maxwell's stress tensor is often used for the calculation of torque. The idea is based on Faraday's statement, which relates flux-line tension to rubber bands trying to contract. In electric machines, there are often configurations without discrete current-carrying conductor paths on the rotor surface. The previously examined synchronous reluctance machine is a typical example of such a structure. Furthermore, a separately magnetized salient-pole synchronous machine operating in steady state has no rotor surface currents, while the exciting current conductors are located on the pole cores, not the pole surfaces. Figure 5.8 depicts the flux solution for the air gap of an asynchronous machine as it operates under heavy load.

The figure illustrates that most of the flux lines are crossing the air gap diagonally, so if the flux lines act like tension bands, they produce notable torque forcing the rotor to rotate in the counterclockwise direction. According to Maxwell's tension theory, the magnetic field strength between objects in a vacuum creates a tension force σ_F on the object surfaces. The magnitude of this force can be expressed as follows.

$$\sigma_F = \frac{1}{2}\mu_0 H^2, \; [\mathrm{N/m^2}] \tag{5.52}$$

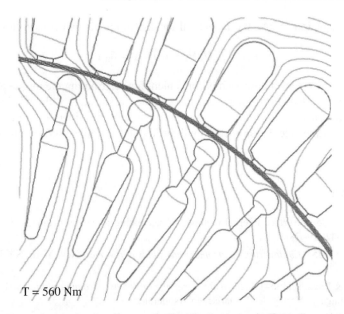

$T = 560$ Nm

Figure 5.8 Flux solution for the air gap of a heavily loaded asynchronous machine. The rotor torque is in the counterclockwise direction. *Source:* Reproduced with the permission of John Wiley & Sons.

These lines of force can be considered a pressure perpendicular to the flux lines. The following equations result when the tension term is divided into its normal and tangential components with respect to the object in question.

$$\sigma_{Fn} = \frac{1}{2}\mu_0\left(H_n^2 - H_t^2\right) \tag{5.53}$$

$$\sigma_{Ft} = \mu_0 H_n H_t. \tag{5.54}$$

Considering torque production, the tangential component σ_{Ft} is most interesting. The total torque exerted on the rotor can be obtained by integrating the stress tensor, for example, over a cylinder Γ that encloses the rotor

$$T = \frac{l}{\mu_0}\int_\Gamma r \times \left((\boldsymbol{B}\cdot\boldsymbol{n})\boldsymbol{B} - \frac{B^2\boldsymbol{n}}{2}\right)d\Gamma \tag{5.55}$$

The hypothetical cylinder can be sized and positioned so it encloses the rotor precisely. Torque is obtained by multiplying the resulting force by the radius of the rotor. For a synchronous motor, force can be obtained by integrating along the rotor contour. However, calculating torque in this way would require the rotor radius and the normal component of the force to be defined everywhere along the contour of the rotor. Note that no steel may be left inside the surface to be integrated-it must be totally located in air to get the correct result.

The linear current density A creates a tangential field strength in the electric machine.

$$H_t = A \Leftrightarrow B_t = \mu_0 A \tag{5.56}$$

According to Equation (5.54), the tangential tension in the air gap is as follows.

$$\sigma_{Ft} = \mu_0 H_n H_t = \mu_0 H_n A = B_n A. \tag{5.57}$$

The Maxwell's stress tensor approach is convenient for determining the torque production of electric machinery. However, there are inaccuracies associated with existing numerical solution methods, and the torque values predicted must be considered with caution. Often, more than one torque calculation method will be applied to gain confidence in the predicted torque values.

References

Carpenter, C. J. (1959). Surface integral methods of calculating forces on magnetised iron parts. *IEE Monograph* 342, 19–28.

Johnk, C. T. A. (1975). *Engineering electromagnetic fields and waves*. New York, NY John Wiley & Sons.

Kovàcs, K. P., & Ràcz, I. (1954). *Transiente Vorgänge in Wechselstrommaschinen* (1st and 2nd eds.) Budapest, Hungary: Verlag der Ungarischen Akademie.

Pyrhönen, J., Jokinen, T., & Hrabovcova V. (2014). *Design of rotating electric machines* (2nd ed.). Chichester, UK: John Wiley & Sons.

Ulaby, F. T. (2001). *Fundamentals of applied electromagnetics*. Upper Saddle River, NJ: Prentice Hall.

6

Basic control principles for electric machines

This chapter will explore the control basics for direct current (DC) machines and the basic alternating current (AC) machine control principles of field-oriented control (FOC), direct flux-linkage control (DFLC), and direct torque control (DTC). The principles are applicable both to linear and rotating machine control despite the fact that big majority of the technologies have first been developed to rotating machines. Only a few pages have been dedicated to the relatively straightforward topic of DC machine control. In a separately excited and compensated DC machine, it is possible to control air gap flux by controlling field winding current. Since torque (or force for a linear machine) is directly proportional to electric current, its control is accomplished by controlling armature current. The clever construction of a fully compensated DC machine keeps the air-gap flux-linkage *vector* perpendicular to the *armature current vector*. These vectors are not as precisely defined as in AC machine drives, but the concepts are similar.

FOC adapts the basic principles of DC machine control to AC machinery. The parts of the current that affect flux and that produce force and torque are determined so excitation and torque (or force) can be separately controlled. The FOC approach requires coordinate transformation and motor equivalent circuit analysis; however, their principles are clear. The biggest problem with the FOC control method is its total reliance on machine inductance and resistance parameters, which can vary significantly as functions of torque (or force) and air-gap flux level. A model of inductance behaviour is required to make FOC work well. In addition, rotor position information is typically needed.

Direct flux-linkage control (DFLC) is simple, in principle, requiring only two space vector equations for control – the voltage integral, which gives an estimate of stator flux linkage, and the torque (or force) equation. However, stability issues with DFLC implementations make the method less desirable. Direct torque control (DTC) takes advantage of the simplicity of DFLC, but avoids stability problems by introducing stabilization techniques. The electric current model, using the same inductance parameters as FOC, is a good stabilizer, but other

Electrical Machine Drives Control: An Introduction, First Edition. Juha Pyrhönen, Valéria Hrabovcová and R. Scott Semken.
© 2016 John Wiley & Sons, Ltd. Published 2016 by John Wiley & Sons, Ltd.

stabilization techniques based on clever signal processing, can also be used to develop a satisfactory DTC implementation based on the DFLC approach. The DTC method can be considered a kind of synthesis method that tries to combine the good properties of alternative control methodologies.

The control system for a rotary electrical drive must be able to produce a desired torque at a specific rotational speed. For a servomotor, rotor position must also be precisely controlled. Linear drives must produce a desired force at a specific linear speed, while controlling the position of the linear rotor, which is called the *forcer*. High-performance control of electric machines, therefore, focuses first on the control of torque or force. Chapter 5 described how the force production of an AC machine is a product of the Lorentz force, which can be expressed in terms of space vectors. Furthermore, this torque or force can be controlled by regulating the AC machine's flux linkages and electric currents. The best present-day approaches apply the space-vector method to model flux-linkage and current space-vector behaviours, and the primary control system tool is vector control.

Electric machine control methods have improved with advances in information technology. The analog electronics of the past offered only limited control. Modern digital electronics and today's greater computational power make it possible to realize control systems based on more accurate digital machine models. Fundamentally, electric machine control comprises torque control (force control for linear drives), speed control, and position control. However, it is not possible, currently, to control an electric machine in real time using a finite-element analysis methodology. Calculation speeds are still far too low. It can easily take tens of minutes for a powerful personal computer with silicon-based processors to resolve a single operating point in a voltage-supply calculation for an induction motor (IM). For true real-time control, flux and current solutions must be carried out 40,000 times a second, which is substantially larger (more than 6 orders of magnitude) than is now possible. Therefore, space-vector modelling and control of electric machines represent today's best available option.

Torque can be described as a vector cross product of a flux-linkage space vector and a current space vector ($T \sim \psi \times i$). To control torque, flux linkage and electric current must be regulated. In DC machines, flux-linkage control is independent of current control. But in AC machines, the flux linkage and current space vectors are interdependent and must be controlled simultaneously using the model of the machine to determine their interactions. Speed is regulated by controlling torque magnitude. Applying more torque increases speed and applying less torque reduces speed. Position control, important for servo drives, uses angle of rotation (of the rotor) as the feedback control input. The basis of all the control tasks is accurate torque control. Therefore, torque control loops are always the innermost for any control application. For a torque-only electrical drive, the torque control loop is the only one needed.

In an electric car (basically also in an internal combustion engine car), the accelerator pedal is the input reference used to control the torque output of the car's traction motor. Acting as the speed controller, the driver speeds up or slows down the car by depressing or releasing the accelerator pedal to increase or decrease the torque output of the motor. Position control is also important for the electric car. Along with controlling vehicle speed, the driver also controls vehicle position, stopping the car when it arrives at its destination. A car is also an example of linear drives in which rotating machines are used to produce linear movement.

To automatically control the speed of an electrical drive, a speed control system that produces a reference for the torque controller similar to that of the human driver must be included outside the torque control loop. If vehicle speed is too low, the input reference should call for an increase in motor torque output. If speed is too high, the reference should call for a torque output reduction.

This discussion of electrical drives control principles begins with a look at DC machine control. A good understanding of DC machine control is important in itself, and since the main principles of vector control are based on the ideal control properties of the fully compensated DC machine, it takes on added importance. Figure 6.1 illustrates a separately excited DC machine and its flux-linkage diagram.

Ideally, a fully compensated DC machine operates according to the cross-field principle, because armature reaction ψ_A is cancelled by the effects of compensating winding flux-linkage ψ_C and commutating pole winding flux-linkage ψ_B. Without these compensating effects, total machine flux linkage would be a distorted ψ'_{tot}, which would cause problems with commutation. (In DC machines, the electric current has to change instantaneously its direction in the winding loop whose terminals are connected in the very commutator segments touching the brushes while the rotor is running. This delicate process is called commutation. It can be studied in more detail in *Design of Rotating Electrical Machines* by Pyrhönen et al. [2014]). With full compensation, the actual total flux linkage of the machine ψ_{tot} remains equivalent to the flux linkage ψ_F produced by the field winding.

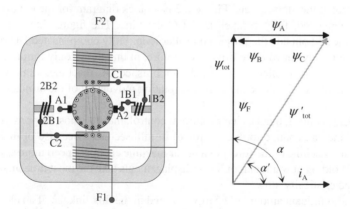

Figure 6.1 The image on the left depicts a separately excited DC machine with field windings F, armature windings A, compensating windings C, and commutating pole windings B, which consist of two parts connected to both sides of the armature winding. The image on the right is the "vector" diagram of the fully compensated DC machine. In the vector diagram, directions correspond to the DC currents fed in via A1 and F1. The corresponding output terminals are B2 and F2, respectively. The armature current "vector" is perpendicular to the flux-linkage "vector", i.e., $\alpha = 90°$. If the machine is not compensated, the angle between the flux-linkage and current "vectors" will be smaller, i.e., $\alpha' < 90°$, which complicates control and weakens the commutation of the armature current. This also leads to other means of improving commutation. For example, the position of the brushes must be altered to enable better commutation of a machine without armature reaction compensation. Moving the brushes to a new position also changes the direction of the armature reaction. However, saturation of the pole edges finally stops the air-gap flux rotation, and good commutation can be reached by moving the brushes. However, one brush position is valid for only one torque value.

Even though space vectors are not defined for DC machine control as they are for AC machine control, the basis of the control approach is analogous. One can imagine that a hypothetical armature "current space vector" I_A lies in the same direction as the armature-produced flux component ψ_A (Figure 6.1). This hypothetical current space vector then can be said to react with the air-gap flux linkage to produce torque. The cross-field principle is well realized in this machine type.

6.1   The control of a DC machine

Control systems for electric machinery always include a single torque (or force) control loop or cascaded control loops. If a cascaded-loop approach is taken, the outer loop always produces a reference value for the next control stage. According to the differential equations, angular or linear speed can be expressed as an integral of torque or force. Position can be expressed as an integral of speed. These dependencies suggest that torque (or force) dynamics is most responsive, speed dynamics is less responsive, and position dynamics is least responsive. That is, changes to torque or force are fastest and changes to position are slowest. Modern controllers for electric machines are designed based on this principle with force production control operating at the fastest rate, speed control at the next fastest rate, and position control at the slowest rate. Figure 6.2 is a block diagram for the control system of a separately magnetized, fully compensated DC machine. The figure shows how the speed controller is located outside the torque controller loop. Fully compensated DC machines are, in practice, built only for rotating drives. Therefore, in this figure only torque and rotational speed are shown as controlled variables. With a brushed DC machine, a linear drive can be built using suitable mechanics converting the rotating movement of the armature into a linear one.

Flux-linkage control, that is, electric current control, must be most responsive, because an electric machine can change torque (or force) in milliseconds. Because of their inertia, the rotor of a rotary machine or the forcer of a linear machine changes speed more slowly. Finally, reaching and holding new angular or linear displacements take the most time, making position control characteristically slow.

Because a significant amount of energy is stored in the flux linkage of an electric machine ($W = \frac{1}{2}\psi I = \frac{1}{2}Li^2$) and because changing flux-linkage magnitude takes time, holding the flux-linkage vector length constant is a general rule for the different control methods. Machine control normally starts by establishing a suitable constant flux linkage. Actual torque or force control then begins once this constant flux linkage has been established.

Expensive and complex sensing technologies are required to directly measure the flux and force production of an electric machine. Consequently, all electrical drive control methods are based on a virtual motor model that predicts flux linkage and torque (or force). The motor model is either a motor parameter–based "electric current model", where flux linkages are calculated with the help of inductances and currents ($\psi = iL$), or a voltage integration–based "voltage model", where voltages are directly integrated into flux linkages ($\psi \approx \int u \, dt$). Using both models simultaneously makes it possible to correct for errors in either of the models and achieve better control.

DC machine control is straightforward and relatively easy to implement. The development of thyristors in the late 1950s and in the 1960s introduced new hardware to DC motor control applications. Thyristor-based hardware was quickly adapted for particularly demanding

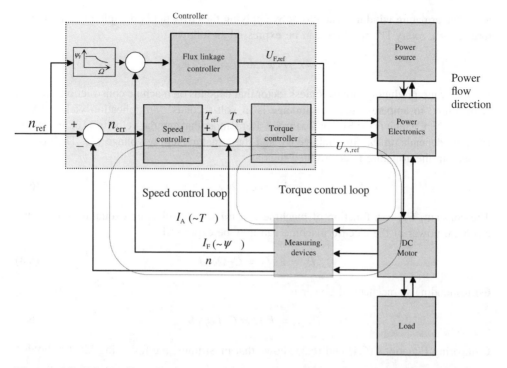

Figure 6.2 Block diagram for the control system of a separately magnetized, fully compensated DC machine. The figure clearly illustrates that armature current I_A and magnetizing current i_F are controlled separately. The controller receives the speed reference n_{ref} as input. Moreover, field-weakening control at high speeds is easy to implement for a DC machine by controlling the magnetizing current. The torque, speed, and flux controllers are included to eliminate potential torque, speed, or flux errors. With these errors eliminated, the drive follows the references.

electrical drive applications. Today, thyristors are still used as power switches in larger industrial DC drives.

The performance characteristics of a fully compensated DC machine are excellent in terms of control engineering. However, DC machine efficiency is lower than that of a corresponding AC machine, because the DC machine has so many extra windings to overcome armature reaction. Moreover, the brush contacts used in DC machines introduce additional losses.

In a separately excited, fully compensated DC machine, the electrical torque and the flux-linkage ψ_F can be controlled independently. The flux linkage of a fully compensated DC machine depends only on the excitation current I_F, because its design compensates for the effect of armature current I_A on the air gap flux, that is, the armature reaction.

$$\psi_F = f(I_F) \tag{6.1}$$

For a DC machine, electrical torque depends on the controlled flux linkage and armature current, because the structure of the machine guarantees the perpendicularity of the flux linkage and the armature current "vector" ($\alpha = 90°$). The field winding produces flux-linkage

ψ_F in the armature winding. Therefore, by applying the cross-field principle in scalar mode, torque for a rotary DC machine can be expressed as follows.

$$T_e = C_T I_A \psi_F \sin \alpha = C_T I_A \psi_F \tag{6.2}$$

C_T is a torque-producing dimensionless factor that depends on machine construction. Current is measured in amperes, and flux linkage is in volt-seconds. Multiplication yields VAs = Ws = J = Nm, which is the SI unit for torque. Remember that angular velocity Ω of an electric machine depends on the flux-linkage ψ_F and the armature supply voltage U_A. The voltage induced in the armature is determined thus as

$$E_A = C_E \Omega \psi_F \tag{6.3}$$

The constant, C_E, is a function of machine construction. Neglecting excitation power, the electrical power of the electric machine can now be expressed

$$P_e = E_A I_A = C_E \Omega \psi_F I_A \tag{6.4}$$

Correspondingly, mechanical power is

$$P_{mech} = T_e \Omega = C_T I_A \psi_F \Omega \tag{6.5}$$

Comparing Equations (6.4) and (6.5) shows that in SI units, the following identity holds.

$$C_E \left[\frac{V}{Vs \cdot \frac{1}{s}} \right] = C_T \left[\frac{Nm}{A \cdot Vs} \right] \tag{6.6}$$

The armature voltage equation is as follows.

$$U_A = I_A R_A + L_A \frac{dI_A}{dt} + E_A \tag{6.7}$$

Therefore, for a rotating electric machine in steady state, $dI_A/dt = 0$, and constant flux linkage and rotational speed are mainly determined by stator supply voltage.

$$\Omega = \frac{U_A - I_A R_A}{C_E \psi_F} \tag{6.8}$$

Below rated speed, control systems commonly maintain the rated value for flux linkage ($\psi_F = \psi_{F,N}$). At higher speeds, flux linkage is reduced in magnitude by an amount that is inversely proportional to speed ($\psi_F \sim 1/\Omega$) to make high-speed operation possible as was explained in Chapter 2. This is commonly referred to as field weakening. Since armature reaction is compensated for by design and flux linkage is controlled as a function of speed, armature current is the single parameter that influences torque, which makes rapid torque control possible.

The electrical torque output of an electrical drive satisfies the differential equation for rotation speed. For a rotary DC machine, the equation for electrical torque T_e can be written in terms of load torque T_L, inertia J, mechanical angular speed Ω, and the rotation friction coefficient D.

$$T_e = T_L + J\frac{d\Omega}{dt} + D\Omega \tag{6.9}$$

As with all electric machinery, a DC machine must first be excited to produce torque. Since a field winding with resistance R_F and inductance L_F acts as its own separate LR circuit, its voltage equation in the Laplace domain can be written as follows.

$$U_F(s) = I_F(s)(R_F + L_F s) \tag{6.10}$$

Magnetizing winding inductance consists of magnetizing inductance and leakage inductance.

$$L_F = L_{Fm} + L_{F\sigma} \tag{6.11}$$

Referred to the armature winding, the magnetizing flux linkage of the machine is

$$\psi_F = L'_{Fm} i_F \tag{6.12}$$

The upper left corner of Figure 6.2 shows how a flux-linkage reference generator produces first a constant reference value for the flux linkage and then a weakened reference after the rated speed. As mentioned previously, flux linkage is normally kept constant from zero to the rated speed Ω_N. Beyond rated speed; flux linkage is gradually weakened to enable high-speed operation. However, if an extremely high starting torque is desired, such as in a traction drive, flux linkage can be increased for its normal value at start-up.

The armature voltage equation and the torque equation (see next equation) written in the Laplace domain are

$$U_A(s) = I_A(s)(R_A + L_A s) + E_A(s) \tag{6.13}$$

Armature inductance consists of magnetizing inductance and armature leakage inductance as follows.

$$L_A = L_{Am} + L_{A\sigma} \tag{6.14}$$

For a rotating DC machine, the torque equation can be written with the load working machine torque $T_L(s)$, friction D, and the sum of the load and motor inertias J.

$$T_e(s) = T_L(s) + (D + Js)\Omega(s) \tag{6.15}$$

With these equations, it is easy to draw a simplified DC machine drive model in the Laplace domain. See Figure 6.3.

Examining Figure 6.3 reveals that the armature circuit of the machine responds to armature voltage changes with the armature time constant.

$$\tau_A = \frac{L_A}{R_A} \tag{6.16}$$

Similarly, field winding behaviour is described by an LR circuit with a time constant analogous to the armature. The magnetizing inductance L_{Am} describes the production of

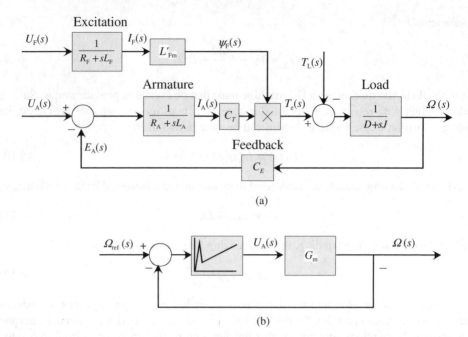

(a)

(b)

Figure 6.3 Image (a) is a Laplace domain model of a loaded DC machine with fixed separate excitation of voltage U_F and armature voltage U_A. The motor rotation will produce a back electromotive force E_A. Image (b) shows a motor speed block diagram for closed-loop control with a PID controller with G_m representing the transfer function of the motor including mechanical and electrical time constants. See upcoming Example 6.1.

armature flux linkage by the excitation current. L_F is the total inductance including leakage of the field winding. See Equation (6.11). A transformation ratio relates the field winding to armature, but in this case, the system has been simplified by assuming this ratio is equal to 1.

In simple cases, speed and torque (or force) control can both be accomplished using proportional-integral (PI) controllers. Controller time constants are chosen based on the torque (or force) controller being fast and the speed controller having a time constant that is an order of magnitude greater.

As Figure 6.2 showed, both controllers react to an error. The speed controller amplifies the speed error and produces a torque reference. The torque controller reacts to the difference between actual torque and reference torque, producing a voltage reference to the power electronics. The power electronics to control an industrial DC machine simply consists of a thyristor bridge or a transistor-based DC chopper in cases of lower power drives. As previously mentioned, control is straightforward since only armature current and voltage must be managed to achieve torque and speed control. The controllers are normally the PI type, which amplify and integrate error using suitable controller parameters. Depending on power electronics type, the voltage reference given by the controller is converted into a firing angle command for a thyristor bridge or a duty cycle reference for a DC chopper.

EXAMPLE 6.1: Design a PID controller for a DC machine electrical drive having its armature and mechanical time constants $\tau_{el} = L_A/R_A = 1$ ms and $\tau_{mech} = 100$ ms.

SOLUTION: The DC motor drive's simple transfer function is constructed using the armature and mechanical time constants and can be expressed as

$$G_m = \frac{1}{(1 + s\tau_{el})(1 + s\tau_{mech})} = \frac{1}{(1 + s0.001)(1 + s0.1)}$$

The traditional way to select the controller parameters is to compensate for the process time constants. Therefore, the selected transfer function for the PID controller is

$$G_{PID} = K_P \frac{(1 + s0.1)(1 + s0.001)}{s0.001}$$

The transfer function of the closed-loop system becomes

$$G_{cl} = \frac{G_{PID} \cdot G_m}{1 + G_{PID} \cdot G_m} = \frac{\frac{K_P}{s0.001}}{1 + \frac{K_P}{s0.001}} = \frac{K_P}{s0.001 + K_P} = \frac{1}{s\frac{0.001}{K_P} + 1}$$

If $\tau = 3$ ms is set as a suitable time constant for the closed-loop system, amplification must be selected as follows.

$$\frac{0.001}{K_P} = 0.003 \Leftrightarrow K_P = \frac{0.01}{0.003} = 3.33$$

Modern controllers are digital, and the same PID algorithm can be used for digital control. Instead of the Laplace transform, the Z-transformation applies for digital controllers.

This is the basic design for the DC drive speed controller, taking both the motor armature behaviour and the motor mechanical behaviour into account. Modern controllers are digital, but the same PID algorithm can be used for digital control. Instead of the Laplace transform, Z-transformation applies for digital controllers.

An industrial brushed DC machine is expensive and has lower efficiency than a corresponding AC machine, and its brushes and commutator require regular maintenance. A long-time goal and expectation has been to replace brushed DC drives with brushless AC or DC machines. Vector control, the core of which is based on simple DC machine control, has been developed to accommodate AC machinery.

6.2 AC machine control basics

Some AC motor types can be activated with a direct online (DOL) starter, which connects the motor terminals directly to the supply power. DOL AC motors are not speed controlled. Induction motors, in particular, are often connected to a three-phase network and allowed to run according to their load. They are just switched ON and OFF based on demand. The speed of a DOL induction drive depends on the frequency of the supply, the motor pole-pair number p, and the load, because the slip of an induction machine varies slightly as a function of load. Synchronous motors and generators also can run DOL, if they are equipped with adequate damper windings to guarantee synchronous operation.

Controlling the voltage amplitude and frequency of the incoming AC power is the only means of controlling an AC machine. In practice, a frequency converter can provide this control, delivering all the needed supply frequencies. With this method of control, the machine operates in the artificial supply network according to its functional properties. The ratio of voltage to frequency U/f is held constant as "network" frequency varies. The control of an AC machine using a frequency converter to regulate supply frequency is called scalar control. More sophisticated versions of scalar control have been introduced that manage more than just frequency and voltage, but since the electromagnetic state of an AC machine is not accurately known, machines controlled via the more-sophisticated approaches do not perform well in dynamic conditions.

Scalar control systems can have a speed control loop with a PI controller for accurate steady state control of speed, but there is no accurate torque (or force) control. The assumption is made that increasing input supply frequency results in a momentary increase in force production and a subsequent increase in speed or vice versa. Speed transients are not precisely controlled. Instead, they take place according to the functional properties of the electric machine system, settling into a new steady state little by little according the dynamic properties of the system.

Different scalar control systems have been developed especially for IM drives. IMs respond well to the changes in supply frequency coming from the converter. Many practical tasks can be carried out using scalar-controlled IMs. For example, they are ideal for pump and fan applications that do not call for accurate and highly dynamic torque control.

For more demanding applications, a more-sophisticated electrical drive control system is needed that monitors the electromagnetic state of the motor or generator to more accurately manage torque. DFLC vector control or the more-stable DTC use analytical space-vector models to accurately predict and then control motor or generator status. A DFLC or DTC controller measures motor currents and calculates voltages and can accurately estimate machine state and control torque in real time.

The control systems based on space-vector theory are suitable for demanding torque, speed, and position control applications. Transient control using the DFLC or DTC space-vector approaches is superior to what is available using scalar control approaches. A status change in torque (or force) that would take a scalar-controlled electric machine 200–500 ms can typically be accomplished by a DFLC in less than 20 ms.

A DTC can increase the force production of an electric machine from zero to its rated value in a few milliseconds. This is especially important for servo drives. Because DTC also improves drive properties in general, it is widely implemented. DTC is seen in both the most-demanding and the least-demanding applications.

6.3   Vector control of AC motors

In general, AC machines require less maintenance, have better endurance at high speeds, and offer notably higher powers than DC machines. At present, an AC drive is generally a better choice for most applications than a DC drive. However, the complicated dependence between flux linkage and torque (or force) control makes the control of AC machinery more difficult.

In DC machines, where flux linkage and torque can be treated independently, drive control is more straightforward.

A very common AC induction machine is the squirrel-cage motor, which uses a laminated steel rotor embedded with a squirrel-cage configuration of copper or aluminium conductors. Squirrel-cage machines can be driven DOL and are often controlled using the simplest and oldest scalar control methods. Other enhanced scalar control methods use steady state phasor equations to predict motor behaviours. With these approaches, control is precise at steady state but imprecise during transients, becoming less and less precise for faster and faster changes in status.

Pulse-width-modulation (PWM) is a voltage control method of drive control that supplies voltage to a rotary or linear motor as a regular series of pulses. Voltage is regulated by modulating the width of these pulses. PWM-based control was first applied to squirrel-cage motors.

Torque (or force) control methods based on the space-vector differential equations are more efficient, and they make it possible to control electric machinery during transients. A three-phase electric motor can be controlled using the complex space vector that represents all three of its phase quantities instead of building three individual phase current controllers. The present-day AC machine vector control approaches are either field-oriented control or direct torque control (FOC or DTC). In FOC, AC machine currents are controlled to produce a desired flux linkage and torque or force. DFLC is the core of the DTC approach.

German engineer Felix Blaschke developed the principles for IM FOC working with Siemens early in the 1970s. This control approach gave induction AC motor drives properties similar to those of DC drives. Siemens called the method transvector control, obviously referring to the cross-field principle discussed earlier in this text.

Blaschke's original squirrel-cage motor vector approach controlled stator space-vector current components in a rotor flux-linkage–oriented coordinate system. Separating the stator current space vector into one component parallel to the rotor flux-linkage space vector component (the flux-producing component) and one perpendicular component (the torque-producing component) makes it possible, in principle, to control rotor flux linkage and electrical torque separately, as in DC machine control.

Carrying out the FOC calculations in a coordinate system rotating at a synchronous speed makes control easier, because the normal AC drive components can be treated as DC components. At an asynchronous speed, motor speed information must be available for flux-linkage coordinate system control, and until recently, this speed information could only be provided by a speed sensor. Performing motor model calculations in different coordinate systems is laborious and time consuming. As a result, these calculations are carried out at a relatively low rate. The controller calculates new reference values for frequency and voltage, and a separate PWM supplies actual voltage.

As was presented in Chapter 5, electrical torque in rotary AC machinery can be expressed as the cross product of the stator current space vector and the space vector for stator flux linkage as follows.

$$T_e = \frac{3}{2} p \boldsymbol{\psi}_s \times \boldsymbol{i}_s = \frac{3}{2} p \frac{L_m}{L_r} \boldsymbol{\psi}_r \times \boldsymbol{i}_s \qquad (6.17)$$

The right-hand side of the expression is valid for an induction machine and used especially in FOC. Each torque equation is the product of the absolute value and the sine of the angles between the two vectors. Torque is a vector parallel to the shaft of the machine. The first part of the equation can be expressed as the following scalar quantity.

$$T_e = \frac{3}{2}p|\psi_s||i_s|\sin\gamma' \qquad (6.18)$$

T_e is the electromagnetic torque vector,
T_e is the absolute value of the electrical torque,
p is the number of pole pairs,
i_s is the stator current space vector,
ψ_s is the space vector for stator flux linkage, and
γ' is the angle between the above two vectors

     This expression results in a positive value when the current rotates in the positive direction ahead of the flux linkage ($\gamma' > 0$). This would be a motor drive. If ($\gamma' < 0$) or the flux linkage is ahead of the current, it would be a generator drive.

EXAMPLE 6.2: Derive the latter part of Equation (6.17).

SOLUTION: The equivalent circuit (see Figure 4.8) of an induction machine shows that rotor flux linkage consists of the air gap portion, produced by the sum of the stator and rotor current vectors, and the rotor leakage

$$\psi_r = (i_s + i_r)L_m + L_{r\sigma}i_r = i_rL_r + L_mi_s$$

Therefore, the rotor current vector is as follows.

$$i_r = \frac{\psi_r - L_mi_s}{L_r}$$

The stator flux linkage is formulated similarly. It consists of the air-gap flux linkage, which is produced by the sum of the stator and rotor current vectors, and the stator leakage.

$$\psi_s = (i_s + i_r)L_m + i_sL_{s\sigma} = L_si_s + L_mi_r$$

Inserting this solution for rotor current vector yields

$$\psi_s = L_si_s + L_m\frac{\psi_r - L_mi_s}{L_r}$$

Inserting in the torque equation gives the desired result.

$$T_e = \frac{3}{2}p\psi_s \times i_s = \frac{3}{2}p\left(L_si_s + \frac{L_m}{L_r}(\psi_r - L_mi_s)\right) \times i_s = \frac{3}{2}p\frac{L_m}{L_r}\psi_r \times i_s$$

Therefore, IM torque can be expressed as the cross product of the rotor flux linkage and the stator current.

According to the cross-field principle, force production reaches a maximum, when the current and flux-linkage vectors are perpendicular. In other words, the angle between these vectors is 90 electrical degrees. These vectors are always perpendicular in fully compensated DC machines, because of their compensating and commutating-pole windings. However, the angle between these vectors in AC machines varies depending on the situation, on machine type, and on rotor control, if the rotor is controllable. The basic idea behind a vector-based controller is to bring the current and flux-linkage space vectors as close to perpendicular as possible. As a result, the approach is commonly referred to as *vector control*.

In traditional vector control, the goal is to establish current references so magnetization and electrical torque (or force) can be controlled independently and during transients. To achieve this goal, references must be calculated separately for the current component that produces torque or force and for the magnetizing current component. For the calculations, a general reference frame can be chosen that has one axis running along the air-gap flux linkage (or stator or rotor/forcer flux linkage). The other will be the torque (or force) production axis. In the strictest sense, of course, only air-gap flux linkage is needed for the so-called ψT reference frame, because it also comprises both components. A similar evaluation is valid for the rotor flux linkage of a rotary induction machine. Depending on case, it is also possible to fix the reference frame with the stator flux-linkage or rotor flux-linkage vector.

The electric currents of a synchronous rotary machine can be represented as components in the ψT reference frame fixed to the air-gap flux linkage or in a dq reference frame fixed to the rotor as shown in Figure 6.4.

According to Figure 6.4, increasing current in the ψ-axis direction increases machine flux linkage. Accordingly, increasing current in the T-axis direction increases torque. Unfortunately, increasing current in the ψ-axis also slightly increases torque if T-axis current remains constant, because of the increased flux linkage. Moreover, increasing current in the T-axis direction also changes the position of the flux linkage. Therefore, both components must

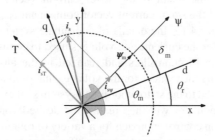

Figure 6.4 Examples of different reference frames used in the vector control of a synchronous machine. The xy reference frame is fixed to the stator (axis of phase U), the dq frame is fixed to the rotor (d is identical to the magnetic axis of the field winding), and the ψT reference frame is fixed to air-gap flux linkage. ψT indicates the controlled flux linkage (i.e., the air-gap flux linkage) and is in the ψ-axis direction. The T-axis is perpendicular to the ψ-axis. Therefore, the ψ-axis direction current component only affects flux linkage, and in principle, the T-axis direction current only produces torque. The angle between the dq and ψT reference frames is the load angle δ_m of the air-gap flux linkage. The rotation angle between the dq and xy reference frames is the rotor position angle θ_r. The rotation angle θ_m, between the ψT and xy reference frames, is the position angle of the air-gap flux-linkage vector in the xy reference frame.

be solved simultaneously and their interaction taken into account. Saturation of the air-gap flux linkage must also be kept in mind when trying to increase the flux-linkage level.

In vector control, the machine model must be used to position the ψT reference frame. The location of the dq reference frame is found by measuring or estimating rotor position. Of course, the position of the stator is known, as it is normally bolted to the motor base. The x-axis can be fixed in the direction of the U-phase winding magnetic axis.

Figure 6.5 is a block diagram of the vector control for an AC motor. Comparing the block diagrams for a DC machine and an AC machine (Figures 6.2 and 6.5) shows that both control approaches are based on the same principle. As mentioned previously, the goal is to control flux linkage and force production separately. However, there are two basic differences between the DC and AC machine control methods. First, electric machine currents are handled differently. In the DC machine, the flux linkage and torque currents are directly measured to provide controller feedback. For vector control of the AC machine, coordinate transformations and calculations are required to determine the flux linkage and torque (or force) current components. Correspondingly, inverse coordinate transformations must be carried out to determine the phase voltages supplied to the motor. Furthermore, there is a direct link to the actual flux-linkage and torque values in a DC machine. In the AC control approach, these are calculated based on the analytical model of the machine.

Figure 6.6 illustrates, in more detail, the signal-processing chain of Figure 6.5 from the motor phase current sensor outputs to the $i_{\psi,\text{act}}$ and $i_{T,\text{act}}$ components used in the control of the flux linkage and torque.

As shown by Figure 6.6, the motor phase currents $i_U(t)$, $i_V(t)$, and $i_W(t)$ are measured first. The $3 \rightarrow 2$ block converts the instantaneous phase current values to corresponding instantaneous two-phase values. Rotor position measurement information θ_r is used to transform the currents to the rotor coordinate system (dq). In the motor model calculation, the i_d and i_q values are used to determine the movement of the flux linkage in the rotor coordinate system. After further calculation, the result is the stator current vector component $i_{\psi,\text{act}}$, which produces flux linkage, and the stator current vector component $i_{T,\text{act}}$, which produces torque (or force). Refer to Chapter 4 for the details of coordinate system transformation.

The reference values for the currents and voltages used to control flux linkage and torque or force are calculated via PI controllers and realized in the air-gap flux-linkage aligned coordinate system. The reference values for the voltages that produce flux linkage and torque (or force) must be transferred to the stator coordinate system so they can be used by converter. For example, a vector modulator will then realize the desired voltage.

At present, the common control approach for a variety of rotating-field machines is vector control. Most frequency converter manufacturers offer some sort of vector control in their products. One challenge for vector control implementations is the determination of accurate machine model parameters. For example, if inductance parameter values are incorrect, then vector control will be imperfect. The following paragraphs will examine the principle of direct flux-linkage control and the so-called voltage model to define stator flux linkage by integrating the stator voltage. Stator flux linkage can be defined in two ways: using the current model (see Chapter 4) and relying on the machine parameters or using the voltage model and relying on the voltage integral. Combining both methods improves accuracy since each can be used to validate the other. With vector control, errors that result from inaccurate machine parameter information can be corrected using the voltage integral. For example, the current model is used to stabilize the voltage integral in the DTC discussed in the following text.

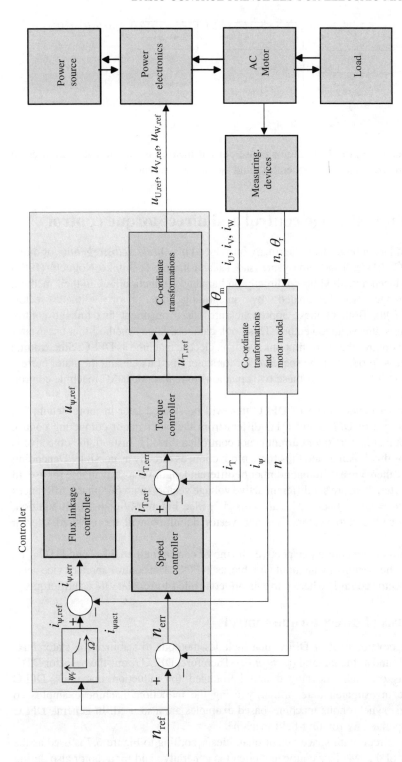

Figure 6.5 Block diagram of traditional vector control for an AC machine. The goal is to control flux linkage and torque (or force) separately to attain a desired speed n_{ref}. Coordinate transformations are required so control can be carried out in the ψT coordinate system. A machine model is necessary to determine the position of the ψT-coordinate system. The three-phase currents $i_{U,V,W}$, rotor speed n, and position θ_r are measured to gather sufficient feedback information. In practice, DC-link voltage is also measured.

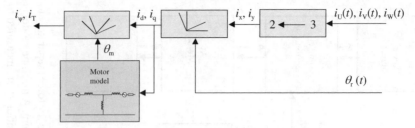

Figure 6.6 Block diagram of the motor phase current measurement and calculation chain resulting in the space vector components i_ψ and i_T for vector control.

6.4   Direct flux-linkage control and direct torque control

In 1984, Manfred Depenbrock from Germany introduced his *Direkt Selbstregelung* or direct self-control (DSC, 1985). That same year, Isao Takahashi and Toshihiko Noguchi (1986) from Japan introduced a method based directly on Faraday's induction law. In their method, stator flux linkage was estimated by integrating the stator voltage vector: $\psi_{s,est} = \int (u_s - R_s i_s) dt$. Both of these approaches are based on direct flux-linkage control of the stator using stator voltage vectors. This book categorizes any method that uses stator voltage vectors to control the stator flux linkage of an electric machine as DFLC. Since torque (or force) is the cross product of the space vector for stator flux linkage and the stator current space vector, the DFLC only needs these two equations to realize ideal AC machine control. See Equation (6.17).

DTC is a further development of DFLC that will be studied later in more detail. The fundamental principle of DFLC and DTC differs from the other major competing control platform, FOC, in that motor stator current is not controlled directly. Instead, the objective is to influence motor flux linkage, and thus force production, as directly as possible. Depending on converter type, there are a different number of different length vectors available for use. In a two-level converter, there are six different active voltage vectors (with 60° phase differences beginning from $u_1 = \hat{u}e^{j0}, u_2 = \hat{u}e^{j\pi/3}, ...u_5 = \hat{u}e^{j5\pi/3}$). See Figure 6.9 or 6.10. In addition, there are two so-called zero vectors. Voltage vector definition will be revisited later in Chapter 7.

From the control engineering perspective, the motor current in the DFLC and DTC is an output variable of the system, not an input variable, as in FOC. The control operates vice versa compared with vector control, where currents are controlled to regulate flux and torque.

6.4.1   The basis of direct torque control

DTC is a further development of DFLC that includes methods to stabilize the stator flux-linkage estimate found as the integral ($\psi_s \approx \int u_s dt$). Therefore, DFLC forms the basis for DTC. Since DFLC was originally developed and introduced for induction machines, DFLC fundamentals are investigated here primarily using the induction machine example. To make some points, synchronous machine–based examples are also used. In general, DFLC can be used to operate any rotating-field machine.

An equivalent circuit with space vector quantities according to Figure 6.7 is used as the equivalent circuit of the IM. This equivalent circuit is generally valid for a motor also during

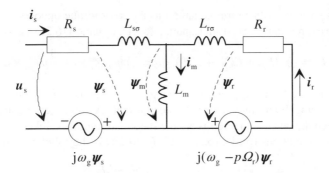

Figure 6.7 Equivalent circuit of an IM according to space-vector theory in a reference frame rotating at a general speed ω_g. The currents i, the voltages u, and the flux linkages ψ are space vectors. In the stator reference frame, the angular frequency $\omega_g = 0$ and, for example, in the rotor flux-linkage reference frame $\omega_g = \omega_r$.

transients. The equivalent circuit does not consider iron losses, harmonics, or saturation. However, it is sufficiently accurate from the control-engineering point of view. The parameters of the equivalent circuit are usually updated for each of the machine's operating modes, but the time derivatives of the inductances are not usually taken into account.

The asynchronous machine equations are used in deriving the laws of direct flux-linkage control. Based on the equivalent circuit (the current model), the following expressions represent the voltages and flux linkages in the general reference frame.

$$u_s = R_s i_s + \frac{d\psi_s}{dt} + j\omega_g \psi_s \qquad (6.19)$$

$$u_r = R_r i_r + \frac{d\psi_r}{dt} + j(\omega_g - p\Omega_r)\psi_r \qquad (6.20)$$

$$\psi_s = L_s i_s + L_m i_r \qquad (6.21)$$

$$\psi_r = L_r i_r + L_m i_s \qquad (6.22)$$

$$\psi_m = L_m(i_s + i_r) = L_m i_m \qquad (6.23)$$

According to Equation (6.23), the motor magnetizing current vector i_m is the sum of the stator and rotor currents. Here, the so-called stator inductance is the sum of the leakage inductance of the stator and the magnetizing inductance.

$$L_s = L_{s\sigma} + L_m \qquad (6.24)$$

Rotor inductance is determined correspondingly (rotor quantities are referred to the stator).

$$L_r = L_{r\sigma} + L_m \qquad (6.25)$$

The instantaneous torque of the machine can be expressed by applying the familiar cross-field principle.

$$T_e = +\frac{3}{2}p\psi_s \times i_s = +\frac{3}{2}p\psi_m \times i_s = -\frac{3}{2}p\psi_r \times i_r \qquad (6.26)$$

The negative sign in the latter form is natural as the rotor torque opposes the stator torque. Bringing in the expression for ψ_s from Equation (6.21), Equation (6.26) can be reformulated.

$$T_e = \frac{3}{2}p((L_{s\sigma} + L_m)i_s + L_m i_r) \times i_s = \frac{3}{2}p\psi_m \times i_s = -\frac{3}{2}p\psi_m \times i_r \tag{6.27}$$

Applying the formula for air-gap flux linkage ($\psi_m = i_m L_m = (i_s + i_r)L_m$) makes possible an expression including both stator and the rotor current. A torque expression can also be derived, in which the determining terms are the stator and rotor flux linkages and the leakage factor. Applying Equations (6.21), (6.22) and (6.26) yields the following equations.

$$T_e = \frac{3}{2}p\psi_s \times i_s = \frac{3}{2}p\psi_s \times \frac{\psi_r - i_r L_r}{L_m}$$

$$T_e = \frac{3}{2}p\left[\psi_s \times \frac{\psi_r}{L_m} - \psi_s \times \frac{\psi_s - i_s L_s}{L_m}\frac{L_r}{L_m}\right] \tag{6.28}$$

$$T_e = \frac{3}{2}p\psi_s \times i_s = \frac{3}{2}p\left[\psi_s \times \frac{\psi_r}{L_m} + \psi_s \times \frac{i_s L_s L_r}{L_m^2}\right]$$

The torque equation (6.28) can be rewritten as follows using the factor common to the left and right side of the expression, $\frac{3}{2}p\psi_s \times i_s$.

$$T_e = \frac{3}{2}p\frac{1}{\left(1 - \frac{L_s L_r}{L_m^2}\right)L_m}\psi_s \times \psi_r \tag{6.29}$$

The flux leakage factors for the stator leakage flux, the rotor leakage flux, and the total leakage flux are as follows.

$$\sigma_s = \frac{L_{s\sigma}}{L_m}, \quad \sigma_r = \frac{L_{r\sigma}}{L_m}, \quad \sigma = 1 - \frac{1}{(1 + \sigma_s)(1 + \sigma_r)} \tag{6.30}$$

The denominator in Equation (6.29) can be reformulated.

$$\frac{1}{\left(1 - \frac{L_s L_r}{L_m^2}\right)L_m} = \frac{1}{(L_m - (1 + \sigma_s)(1 + \sigma_r)L_m)} = \frac{1}{\left(L_m - \frac{L_m}{(1 - \sigma)}\right)} = \frac{(\sigma - 1)}{\sigma L_m} \tag{6.31}$$

With this expression introduced, torque equation can be expressed in simpler form.

$$T_e = -\frac{3}{2}p\frac{\sigma - 1}{\sigma L_m}\psi_s \times \psi_r \tag{6.32}$$

This equation justifies the fundamental premise of DFLC and DTC, which is controlling stator flux linkage directly. Rotor flux linkage ψ_r is stable, and it rotates steadily. Moreover, stator flux linkage ψ_s responds quickly, because it includes the stator leakage flux-linkage component $\psi_{s\sigma} = i_s L_{s\sigma}$. The time constant of the rotor τ_r is defined according to this equation.

$$\tau_r = L_r/R_r = (L_m + L_{r\sigma})/R_r \tag{6.33}$$

In induction machines, the rotor time constant is relatively large, 100–1500 ms, and it is proportional to the machine power. Therefore, the torque of the IM can be adjusted efficiently by controlling the angle between the stator and rotor flux linkages. Since the flux linkages also include the leakage components, only the machine's leakage flux constrains the rate of change. Consequently, direct flux-linkage control results in the fastest possible state changes in electrical motor drives. From this point of view, DFLC is superior to and quicker to respond than other AC machine control methods based on controller dynamics.

Instead of rotor flux linkage, air-gap flux linkage could be used in Equation (6.32). The resulting equation looks like this.

$$T_e = -\frac{3}{2}p\frac{1}{L_{s\sigma}}\boldsymbol{\psi}_s \times \boldsymbol{\psi}_m \qquad (6.34)$$

In an induction machine, air-gap flux linkage is also quite stable. In general, Equation (6.34) applies also to DFLC for synchronous machines in which similar rotor flux linkage as in IMs cannot be found.

Modern industrial electronic control systems make use of insulated-gate bipolar transistors (IGBTs) or metal-oxide semiconductor field-effect transistors (MOSFETs) as electronic switches. An IGBT switches in less than 100 ns, and MOSFETs are even faster. If processor capacity is sufficient, switching frequencies up to several kilohertz and even a couple tens of kilohertz can be used. Therefore, the time constants of rotor and air-gap flux linkages are easily a few decades longer than the time required for the computation of the motor model and for making and implementing the switching decisions. Figure 6.8 depicts the steady behaviour

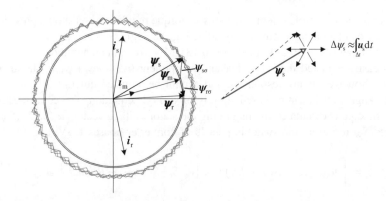

Figure 6.8 Principles of the DFLC for an induction machine. In steady state, the stator flux linkage rotates on a circular path at a nearly constant angular velocity and amplitude. The possible changes in the stator flux linkage based on different voltage vectors are illustrated on the right. As a result, there is ripple in the stator flux-linkage path, because of the discrete voltage vectors supplied by the pulse width modulation of the converter. The different machine time constants almost completely filter out the consequences of this ripple effect when observing the loci plotted both by the point of the air-gap flux-linkage vector and by the point of the rotor flux-linkage vector. Unlike the rippled locus plotted by the point of the stator flux-linkage vector, the locus drawn by the point of the air-gap flux-linkage vector, and particularly, the locus plotted by the point of the rotor flux-linkage vector, is almost a perfect circle. In steady state, the angular speed of the rotor flux linkage $\omega_{\psi r}$ is smooth, whereas the angular speed $\omega_{\psi s}$ of the stator flux-linkage vector is uneven. However, the average is equal to the angular speed of the rotor flux-linkage vector or the air-gap flux-linkage vector.

of air-gap flux linkage and rotor flux linkage regardless of the somewhat unstable stator flux linkage.

Equations (6.29) and (6.34), as well as Figure 6.8, indicate clearly that the torque of a rotating-field machine can be controlled rapidly by making quick adjustments to stator flux linkage. This principle is fundamental to DFLC.

DFLC also has weaknesses. One relates directly to the main control principle. Since currents are output variables, they cannot be directly controlled. It can be somewhat difficult to understand the current producing mechanism in a DFLC. Figure 6.8 illustrates that the stator current always flows in the direction of the difference $(\boldsymbol{\psi}_s - \boldsymbol{\psi}_m)$, which means the stator leakage flux linkage is as follows.

$$\boldsymbol{\psi}_{s\sigma} = i_s L_{s\sigma} \tag{6.35}$$

In addition, $\boldsymbol{\psi}_{s\sigma}$, $\boldsymbol{\psi}_s$, and $\boldsymbol{\psi}_m$ always form a triangle. Therefore,

$$i_s = (\boldsymbol{\psi}_s - \boldsymbol{\psi}_m)/L_{s\sigma} \tag{6.36}$$

Equation (6.36) also indicates that current responds quickly to changes in either of the flux linkages, because stator leakage inductance is usually small.

6.4.2 DFLC implementation

The basis of a good control system is responsive torque regulation. As shown previously, AC motor torque can be changed quickly by adjusting the angle between the stator and rotor or air-gap flux linkages. See (6.29), (6.34), or (6.18). Moreover, the magnitude of the stator flux linkage is easily adjusted due to its leakage component $(\boldsymbol{\psi}_s = \boldsymbol{\psi}_{s\sigma} + \boldsymbol{\psi}_m)$. The air-gap flux linkage $\boldsymbol{\psi}_m$, however, is not responsive and cannot be adjusted quickly.

Flux-linkage control can be based entirely on a single equation. The control estimate for stator flux linkage is calculated by integrating the stator voltage vectors $\boldsymbol{u}_s(S_U, S_V, S_W)$, where S_U, S_V, and S_W represent the switching states of converter phases U, V, and W.

$$\boldsymbol{\psi}_{s,\text{est}} = \int (\boldsymbol{u}_s(S_U, S_V, S_W) - R_s i_s) dt \approx [\boldsymbol{u}_s(S_U, S_V, S_W) - R_s i_s] \Delta t + \boldsymbol{\psi}_{s,0} \tag{6.37}$$

Equation (6.37) is often referred to as the *voltage model* for an AC machine, since it is mainly based on the voltage integral. The equation also illustrates the principle of space vector integration. The change in flux linkage is in the direction of the voltage vector $\boldsymbol{u}_s$. The value of change in volt-seconds (Vs) is calculated by multiplying the expression by Δt, which is the incremental time that voltage affects the winding. The effect of resistive voltage drop is integrated similarly, and the result is a flux-linkage change in the direction of the current vector i_s.

S_U, S_V, and S_W refer to the three converter switching functions, which receive values between 0 and 1. Usually only 0 or 1 is taken into account, but as they switch, voltages swing continuously from 0 to 1 and vice versa. The tip of the space vector for stator flux linkage travels approximately in the direction of the stator voltage vector, because the voltage drop caused by the current in the stator resistance is relatively small. According to the definition of the voltage vector, $\boldsymbol{u}_s(t) = (2/3)U_{DC}(a^0 S_U(t) + a^1 S_V(t) + a^2 S_W(t))$, the speed of stator flux-linkage change is proportional to the DC-link voltage. By selecting one of the six active vectors and two zero vectors, it is possible to rotate the stator flux-linkage vector almost along a circular path. The path become more and more circular with increasing switching frequency. The most appropriate voltage vector is the one that keeps the absolute value of the flux linkage close to constant but maximizes the stator flux-linkage vector tip travel speed.

Only stator resistance R_s is required in the voltage model. Estimated electrical torque is a function of the estimated stator flux linkage and the measured stator current vector as expressed by the next equation.

$$T_{e,est} = \frac{3}{2} p \psi_{s,est} \times i_s \tag{6.38}$$

Therefore, for a rotating-field machine, DFL control provides the simplest principal equations. To implement DFLC, position information on the flux-linkage vector with respect to the flux circle is also needed. The flux-linkage circle is divided into six sectors so that sector borders bisect the angles between the voltage vectors of the two-level inverter. See Figure 6.9.

In DFLC and DTC implementations, stator flux linkage and torque are held within the set hysteresis limits (hysteresis band). See Figure 6.10, which illustrates flux-linkage vector movement in a DFLC. Power switches are adjusted only when torque or the absolute value of the flux linkage deviate too much from their set point values. When the hysteresis limit is reached, the next suitable voltage vector is selected that will bring the stator flux-linkage vector to the right orientation.

Stator flux linkage can be regulated directly by controlling the switches of the inverter. In each sector, two voltage vectors can be used in both rotation directions. One vector increases and the other decreases stator flux linkage. Normal two-point control regulates the magnitude. In other words, flux is raised until it reaches the upper hysteresis limit, and then it is lowered until the lower limit is reached. The hysteresis limits are determined based on the allowable switching frequency. If switching frequency begins to exceed the allowed upper limit, the

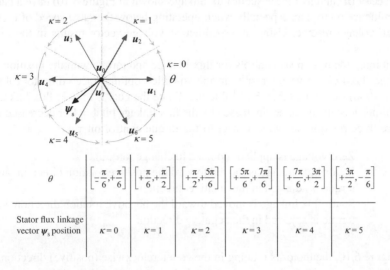

Figure 6.9 Flux-linkage circle division into sectors κ determined by the voltage vectors. Stator flux linkage is moving in sector $\kappa = 4$.

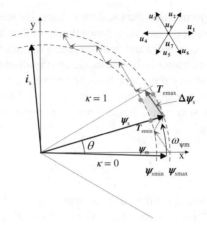

Figure 6.10 Simplified flux-linkage vector movement in a DFLC showing how voltage vector changes are integrated with stator flux-linkage changes. The voltage vectors available for use in a two-level converter are shown in the upper right corner. The shaded area illustrates the hysteresis control limits showing the maximum and minimum values for both flux linkage and torque. The stator flux-linkage vector tip moves in the direction of voltage vector u_3 in sector $\kappa = 0$. The torque controller only requests more torque, and therefore the u_3 voltage vector is selected. This changes stator flux linkage. $\Delta\psi_s = u_3\Delta t$. If the magnitude of the stator flux linkage is driven to the upper hysteresis limit, which maximizes its angle θ with respect to air-gap flux linkage ψ_m, then torque T_{emax} also reaches its maximum.

hysteresis limits must be extended. On the other hand, correct hysteresis settings can optimize energy efficiency, etc. The voltage vectors that are about perpendicular to the stator flux-linkage vector (u_3 and u_6 for the stator flux linkage shown in Figure 6.10) have a particularly strong influence on torque, especially when operating below the rated speed of the machine with high voltage reserve. Using a perpendicular voltage vector results in the most rapid torque response.

Combining the hysteresis controls for flux linkage and torque with the position information of the flux-linkage vector results in the so-called optimal (according to Takahashi & Noguchi, 1986) switching table. See Table 6.1. Since the goal is to hold flux linkage within certain limits, it is always either increased by the flux linkage bit $\phi = 1$ or decreased by $\phi = 0$. There are three possibilities with respect to the torque control bit τ.

$\tau = 0$ Zero voltage is applied (no need to change torque).

$\tau = +1$ Stator flux linkage is moved toward the positive rotation direction, and motor torque is adjusted in the positive direction.

$\tau = -1$ Stator flux linkage is moved toward the negative rotation direction, and motor torque is adjusted in the negative direction.

In Figure 6.10, the motor is rotating in the counterclockwise (positive) direction and flux linkage happens to be located in sector $\kappa = 0$. More torque is needed ($\tau = 1$), but no change in stator flux linkage ($\phi = 0$) is required. According to Table 6.1, u_3 is the appropriate voltage

Table 6.1 Optimal switching table showing selection of voltage vector based on location and desired change in flux linkage

Torque Bit τ	Flux Linkage Bit ϕ	Instantaneous Flux-Linkage Vector Location Sector $\kappa=0$	$\kappa=1$	$\kappa=2$	$\kappa=3$	$\kappa=4$	$\kappa=5$
−1	0	$\boldsymbol{u}_5$	$\boldsymbol{u}_6$	$\boldsymbol{u}_1$	$\boldsymbol{u}_2$	$\boldsymbol{u}_3$	$\boldsymbol{u}_4$
	1	$\boldsymbol{u}_6$	$\boldsymbol{u}_1$	$\boldsymbol{u}_2$	$\boldsymbol{u}_3$	$\boldsymbol{u}_4$	$\boldsymbol{u}_5$
1	0	$\boldsymbol{u}_3$	$\boldsymbol{u}_4$	$\boldsymbol{u}_5$	$\boldsymbol{u}_6$	$\boldsymbol{u}_1$	$\boldsymbol{u}_2$
	1	$\boldsymbol{u}_2$	$\boldsymbol{u}_3$	$\boldsymbol{u}_4$	$\boldsymbol{u}_5$	$\boldsymbol{u}_6$	$\boldsymbol{u}_1$
0	0	$\boldsymbol{u}_0$ or $\boldsymbol{u}_7$*	$\boldsymbol{u}_0$ or $\boldsymbol{u}_7$	$\boldsymbol{u}_0$ or $\boldsymbol{u}_7$	$\boldsymbol{u}_0$ or $\boldsymbol{u}_7$	$\boldsymbol{u}_0$ or $\boldsymbol{u}_7$	$\boldsymbol{u}_0$ or $\boldsymbol{u}_7$

The selected zero vector is one that can be achieved, following an active vector, by switching a single switch. See Chapter 7.

vector to select to raise stator flux linkage speed in the positive direction. As shown by Figure 6.10, the stator flux linkage meets the hysteresis limits of the flux linkage amplitude only in cases when the torque controller lets the flux linkage travel to the boundaries.

This core functionality of a DFLC system is illustrated by Figure 6.11. Torque and flux-linkage control and flux-linkage position information are combined in the figure.

The behaviour of hysteresis-controlled torque as a function of time is depicted by Figure 6.12. In the figure, *T*− indicates torque is decreasing after reaching a peak, and *T*+ indicates increasing torque. A zero voltage vector is used so stator flux linkage remains stationary, and it is slightly shortened by the influence of the resistances. The instantaneous value of torque T_e drops until its value is below the lowest torque determined by the hysteresis band. Next, an active voltage vector is selected that keeps the length of the stator flux linkage appropriate and increases the angle between the flux-linkage vectors, thus increasing torque.

There is no inherent switching frequency in a DFLC. Frequency is defined by the hysteresis limits and load conditions, making it difficult to evaluate the thermal losses in a DFLC converter. The switching frequency control must be based on the hysteresis band control. If the switching frequency gets too high, the hysteresis band widens and vice versa.

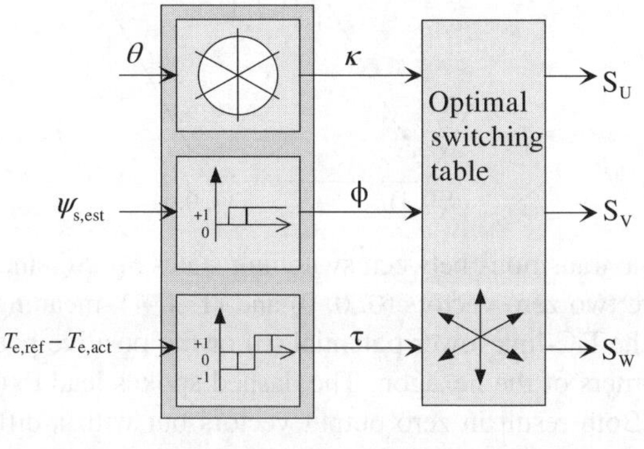

Figure 6.11 A block diagram of combined torque and flux-linkage control.

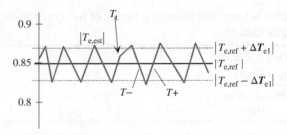

Figure 6.12 Hysteresis control of air gap torque. The magnitude of the electrical torque oscillates around the reference value. Resulting torque ripple is mild and not harmful. Since the switching frequency is typically in the kilohertz range, the mechanical system filters out the ripple. The asymmetry of the torque pattern at position T_t is because flux linkage is reduced by selecting a new voltage vector, and the torque rate of increase changes. The hysteresis ΔT_{e1} is regulated by a PI controller, which thus also determines the average switching frequency.

DFLC modulation at low speeds, additional hysteresis, and double switching

Optimally, inverter bridge switches are actuated individually. All transitions are carried out by switching one of the converter branches at a time. Drawing a hexagon to indicate possible transitions from one switching state to another makes it is easy to see which are allowable when simultaneous switching is prohibited. The voltage vector hexagon of Figure 6.13 results when switching functions S_U, S_V and S_W are set to values from (0, 0, 0) to (1, 1, 1). Note: A more detailed derivation of the voltage vectors is offered in Chapter 7.

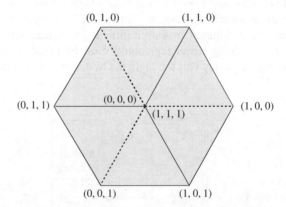

Figure 6.13 Possible transitions between switching states S_U, S_V, and S_W. In the middle of the hexagon there are two zero vectors (0, 0, 0) and (1, 1, 1), meaning that all switches are connected either to the DC-link lower potential (0) or the positive potential (1). The active vectors are in the corners of the hexagon. The dashed spokes lead to (0, 0, 0) and the solid spokes to (1, 1, 1). Both result in zero output vectors but with a different common mode voltage.

Following a route that changes two switches simultaneously is prohibited in principle. This means that, for example, following (0, 0, 1), the zero voltage vector must be (0, 0, 0). And, from any corner, it is possible to move to one of three positions: either of the two neighbouring active vectors or the nearest zero vector. Two adjacent voltage vectors or a zero vector should be used in each sector. This is normal DFLC functionality. The selection of a zero vector or one of the active vectors is determined from the three torque controller hysteresis levels, $\tau \in (-1, 0, +1)$. The flux-linkage controller defines which one of the two active vectors is selected.

In normal DFLC modulation, stator flux-linkage control is subordinate to torque control. With this type of control, if the reference value of the torque becomes zero, stator flux-linkage control is not regulated, and a zero vector is regularly used for a long period. This situation may occur often, especially for an induction machine, where the resistive voltage drop $-i_s R_s$ begins dominating when operating at low torque and low speed. Switch conduction losses should also be added here, and the relative influence of such parasitic elements can become dominant at the lowest speeds.

In induction machines or synchronous reluctance machines, stator flux linkage must be maintained by the stator current magnetizing component. The magnetizing current component of an IM or a synchronous reluctance motor is large. If the per-unit value for stator inductance $L_s = 2$, which is the case in low-power motors or multipole motors, the resulting no load current of the motor is 0.5 p.u. Resistive voltage drop in the stator windings, therefore, is high even at no load, and the $-R_s i_s$ component in $\psi_s = \int (u_s - R_s i_s)\mathrm{d}t$ rapidly reduces stator flux linkage. For synchronous machines equipped with field windings and for permanent magnet synchronous machines, the danger of losing stator flux linkage is not obvious. Therefore, an additional means of controlling flux linkage amplitude is not needed for these machines. However, synchronous reluctance machines are extremely sensitive to flux linkage level, so they require an additional correction method, as do IMs.

Figure 6.14 examines additional hysteresis in the DFLC of an induction machine. If the torque control does not require an active voltage vector, the magnitude of stator flux linkage is less due to resistive loss, and it can be driven outside the original hysteresis limits. Subsequently, if stator flux linkage is not immediately strengthened; the motor may have insufficient torque capacity for the next phase. This situation can be avoided by applying the separate and larger hysteresis limits $\psi_{\min 2}$ and $\psi_{\max 2}$ for the stator flux linkage.

The hysteresis limits $\psi_{\min 1}$ and $\psi_{\max 1}$ are used when the torque reference is other than zero. As the torque reference approaches zero, the stator flux-linkage control does not function, and therefore the stator flux-linkage control comprising the hysteresis limits $\psi_{\min 2}$ and $\psi_{\max 2}$ must be applied. When flux linkage deviates too much from its reference value, an active voltage vector is selected independent of the torque controller. In this situation, the magnitude of the flux linkage is adjusted using the voltage vectors most strongly affecting the flux-linkage vector. Voltage vectors located in the middle of the stator flux linkage sector κ are chosen, because with them, the effects of stator flux-linkage control on torque are less pronounced than when using the vectors suggested by the optimal switching table. Once torque increases, the basis of voltage vector selection switches back to the optimal switching table.

In Figure 6.14, low induction machine torque results in the small angle between the stator flux linkage and the stator current vector. Therefore, the resistive voltage drop vector $-R_s i_s$ works almost fully against the stator flux-linkage vector. More current is needed to maintain the machine's flux linkage, and additional hysteresis is needed to maintain the magnetizing

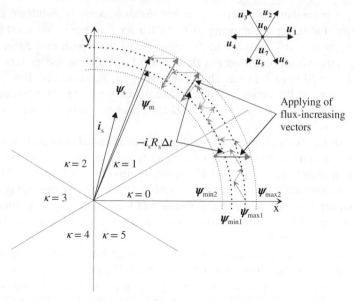

Figure 6.14 Additional hysteresis in the DFLC of an induction machine illustrating stator flux-linkage control at low values of motor torque. In the figure, points of influence are indicated by arrows. For torque-oriented control, stator flux linkage can drop because of resistive losses. This behaviour must be taken into account by setting hysteresis limits for stator flux linkage that are independent of torque control. Voltage vectors are used that best correspond to the direction of the stator flux linkage, and therefore the magnitude of stator flux linkage can be increased so the angle between the stator and air-gap flux linkage remains practically unchanged. The problem related to this correction is that it may result in double switching. For example, if u_3 (0, 1, 0) has been used as the previous active vector (upper example) and u_0 (0, 0, 0) and if the flux linkage restoring vector u_2 (1, 1, 0) is then needed, two change-over switches must be switched simultaneously. This increases burden to the motor insulation. This scenario rarely occur, however, and may therefore be acceptable. Another possibility is to use the minimum pulse active vector as an intermediate phase.

current if the optimal switching table is not reacting because no additional torque is needed. The additional hysteresis guarantees that stator flux linkage is held at the desired level and that the drive remains fully operable, even when it is driven at no load or long periods of low load. When producing higher levels of torque, the additional hysteresis is usually not needed, because normal (optimal switching table) operation results in adequate flux linkage levels.

Using extra hysteresis is problematic, because flux-correcting vectors can only be realized by double switching. Double switching adversely affects the average switching frequency, common mode voltage, and motor cable transients. Each phase voltage is a projection of the total voltage vector on the three magnetic axes. If there is double switching between active voltage vectors, the phase voltages produce higher than normal stresses in the phases whose switches are changed directly from 1 to 0 or vice versa, for example, from (1, 1, 0) to (1, 0, 1).

The term "double switching" is used only when switching between active vectors. When using zero vectors, two switches must often be turned, but these situations do not result in exaggerated voltage stresses. However, common mode voltage changes increase, which might result in premature bearing failure. Bearing currents will be studied later in Chapter 13. There are, however, later DFLC developments that modulate differently to avoid the usage of these flux-correcting vectors.

Finally, the DFLC is vulnerable to several other conditions that can result in control malfunctioning. In the following discussion, these conditions are reviewed, and possible corrective methods are examined that when applied are able to transform the simple DFLC into the well-known DTC, a rugged and good-performing machine control method.

6.4.3   Shortcomings of direct flux-linkage control

The observant reader may have noticed that DFLC does not include the feedback elements typical of ordinary control systems. For this reason, the method has been subject to criticism: it has been referred to as merely an advanced open-loop (nonfeedback) scalar controller. The only feedback information used by the DFLC are the motor currents. If stator voltage is properly integrated to estimate stator flux linkage, current measurement is all that is needed, because torque can then be exactly calculated.

Unfortunately, however, voltage integration according to Equation (6.37) alone is not always useful as such; particularly when operating at lower frequencies. Flux linkage estimation by integration can be problematic, since even small errors can have a significant cumulative effect. Motor voltage is not usually measured, because making the motor voltage measurement is not economical in commercial applications. And, voltage drops in an electric machine (e.g., the nonlinear voltage drops in power switches and resistances) may account for as much as 50−70% of the total voltage when operating at low or zero speed. In addition, switching delays affect estimation accuracy.

So far, estimating these voltage drops computationally has proved to be impossible given the present performance level of available commercial processors. In particular, the voltage drop across power switches, which can vary substantially with production lot, is difficult to model. If the integral of the stator voltage is not accurate, the error accumulates as time progresses in the flux linkage of the motor. Because of errors in the stator flux linkage estimate, the progression of the flux linkage space vector of the motor does not remain origin centred; the greatest weakness of the DFLC. Figure 6.15 illustrates how the trajectory plotted by the point of the real stator flux linkage of the motor drifts gradually out of an origin-centric trajectory.

Stator current, the only DFLC feedback measurement, includes information about stator flux linkage drift, because drifting produces a DC current component. Furthermore, the integration of stator flux linkage per Equation (6.37) produces negative feedback. In principle, the $-R_s i_s$ term stabilizes the situation. But since stator resistance is small, particularly in large machines, motor flux linkage will become seriously eccentric before the $R_s i_s$ term grows large enough to begin opposing the drift. Fortunately, this drift phenomenon is relatively slow, and various methods can be used to counter it. The effect of different errors has been investigated, for instance, by Kaukonen (1999). In small machines with larger R_s, the effect of resitive voltage drop is large, so DFLC is more stable.

Integrating the stator voltage vectors in the stator reference frame provides the stator flux linkage estimate for DFLC. The stator reference frame is a natural choice, and unlike the FOC case, no extra coordinate transformations are required for DFLC. The integration is carried out digitally using an efficient signal processor. In addition to R_s, only the stator current and voltage vectors u_s and i_s are required.

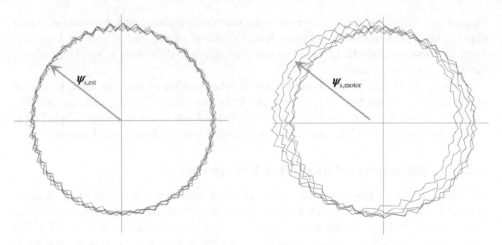

Figure 6.15 The trajectory vector moves away from the origin as stator flux linkage drifts. The DFLC assumes an origin-centric estimate. However, voltage integration errors move the actual stator flux linkage away from the origin. *Source:* Adapted from Pyrhönen (1998). Reproduced with permission of O. Pyrhönen.

Voltage vectors are formed as the voltage source inverter supplies voltage pulses into the windings of the motor. The phase currents of the stator are measured and used to construct the stator current vector. Additionally, the intermediate circuit voltage U_{DC} is typically measured, as are the inverter switching states S_U, S_V, and S_W. Furthermore, a switches model of the voltage losses is developed. Based on the above, values of 1 or 0 are entered to represent switching state. The stator voltage vector, therefore, can be written as follows

$$\boldsymbol{u}_s(S_U, S_V, S_W) = \frac{2}{3} U_{DC}\left(S_U e^{j0} + S_V e^{j\frac{2\pi}{3}} + S_W e^{j\frac{4\pi}{3}}\right) - \frac{2}{3}\left(u_{U,drop} e^{j0} + u_{V,drop} e^{j\frac{2\pi}{3}} + u_{W,drop} e^{j\frac{4\pi}{3}}\right)$$

$$(6.39)$$

Here, $u_{U,drop}$, $u_{V,drop}$, and $u_{W,drop}$ model the voltage drops in the power switches for both the switched and unswitched states. The prediction of stator flux linkage can be written

$$\boldsymbol{\psi}_{sest,1} = \int (\boldsymbol{u}_s(S_U, S_V, S_W) - i_s R_s)dt + \boldsymbol{\psi}_{s,est0}$$

$$\approx \boldsymbol{u}_s(S_U, S_V, S_W)\Delta t - \int i_s R_s dt + \boldsymbol{\psi}_{s,est0}$$

$$(6.40)$$

Equation (6.40) shows that when stator resistance is low, the tip of the stator flux-linkage vector travels mainly in the direction of the instantaneous voltage vector. When applying a nonzero vector, the velocity can be high depending on the voltage reserve. When applying zero voltage vectors, the velocity of the movement of the stator flux-linkage vector tip is small and proportional to the drop $-i_s R_s$. In principle, and ignoring the problems caused by error accumulation, predicting flux linkage is relatively easy.

Construction of the stator flux-linkage vector, therefore, is independent of machine inductances. This characteristic makes DFLC and DTC superior to traditional current control methods for fast transients. In contrast, FOC handles transients poorly. The FOC current model is based on machine inductance values that can change dramatically during a transient. These rapid inductance changes introduce significant error to the FOC flux linkage estimates. In DTC, the only critical machine parameter is stator resistance R_s, which is easily measured in practice.

Equations (6.39) and (6.40) reveal the four different factors that influence integration error.

- DC-link voltage measuring error

- stator current measuring error

- power switch voltage drop estimation error

- stator resistance estimation error

Measuring errors can be divided into two categories. They are either gain or offset errors. Gain errors introduce a constant error to the integration of stator flux linkage. Torque error is directly proportional to the intermediate voltage measuring error and to the square of the current measurement. It is directly proportional to stator resistance estimation error. Torque estimate error is inversely proportional to frequency, and therefore insignificant at high frequencies. At zero speed, torque estimation error is extremely large. Offset errors produce DC components to phase quantities and cause the stator flux-linkage vector to rotate eccentrically about its origin. A pulsating error proportional to the rotation speed is introduced to the torque.

Therefore, the two error types for voltage determination result in a constant stable error or an unstable drifting error that drives the stator flux linkage out of an origin-centric trajectory, a serious problem.

DC-link voltage measurement error

With respect to the intermediate circuit voltage, the effects of gain error and offset error are similar, since the measurement signal is a DC quantity. Two different types of errors can be detected, both of which result from the erroneous measuring of the intermediate voltage. One is a stable DC error and the other the unstable drifting of the flux linkage. Figure 6.16 illustrates these errors.

Measured intermediate voltage $U_{DC,meas}$ can be expressed according to the following equation, where k_{gain} is a gain coefficient, U_{DC} is the actual intermediate voltage, and $\Delta U_{DC,offs}$ is the offset voltage.

$$U_{DC,meas} = \left(1 - k_{gain}\right)U_{DC} + \Delta U_{DC,offs} = U_{DC} + \Delta U_{DC} \tag{6.41}$$

ΔU_{DC} is the total measuring error at a certain instant of time. If $\Delta U_{DC} \geq 0$ ($U_{DC\ meas} \geq U_{DC}$), a stable flux-linkage error occurs. If $\Delta U_{DC} < 0$ ($U_{DC,meas} < U_{DC}$), unstable drifting of the stator flux linkage occurs.

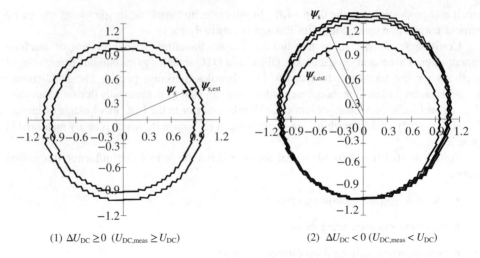

(1) $\Delta U_{\mathrm{DC}} \geq 0$ $(U_{\mathrm{DC,meas}} \geq U_{\mathrm{DC}})$ (2) $\Delta U_{\mathrm{DC}} < 0$ $(U_{\mathrm{DC,meas}} < U_{\mathrm{DC}})$

Figure 6.16 Examples of the behaviour of the flux linkage integral as a result of the DC-link voltage measurement error. *Source:* Adapted from Kaukonen (1999). Reproduced with permission of J. Kaukonen.

Phase current error

The problems associated with gain and offset errors in measured phase currents are similar to those of the gain and offset errors for DC-link voltage measurements. There are two different behaviour types: the constant stable error in the flux linkage and the alternating unstable drift of the flux linkage. The conditions for these two errors are the same as for the intermediate voltage error. The gain error for current measurement can be considered a voltage drop calculation error and, eventually, a voltage error. Measured stator voltage $u_{\mathrm{s,meas}}$ can be expressed

$$u_{\mathrm{s,meas}} = u_{\mathrm{s,calc}}(S_{\mathrm{U}}, S_{\mathrm{V}}, S_{\mathrm{W}}) - \left(1 - k_{\mathrm{gain}}\right)i_{\mathrm{s}}R_{\mathrm{s}} = u_{\mathrm{s}} + \Delta u_{\mathrm{s}} \qquad (6.42)$$

Where u_{s} is the actual stator voltage, $u_{\mathrm{s\ calc}}$ is the voltage calculated from the intermediate voltage and the switch states, and k_{gain} is the gain coefficient. Δu_{s} is the total measuring error at a given time instant. If $\Delta u_{\mathrm{s}} \geq 0$ $(|u_{\mathrm{s\ meas}}| \geq |u_{\mathrm{s}}|)$, a stable stator flux linkage error occurs. If $\Delta u_{\mathrm{s}} < 0$ $(|u_{\mathrm{s\ meas}}| < |u_{\mathrm{s}}|)$, unstable drifting of the stator flux linkage occurs. See Figure 6.17. The offset error in the measured phase currents produces a DC component, which leads inevitably to the drifting of the integral.

Voltage drop estimation error in power switches

Accurate power switch modelling is not practical with current computational tools; therefore, power switches are usually modelled as a resistance (R_{d}) and a threshold voltage u_{th}.

$$u_{\mathrm{drop}} = u_{\mathrm{th}} + i_{\mathrm{s}}R_{\mathrm{d}} \qquad (6.43)$$

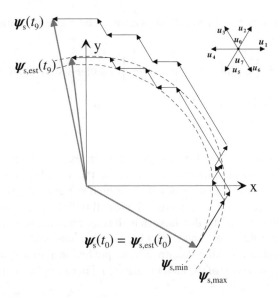

Figure 6.17 Stator flux-linkage drift resulting from incorrect voltage drop estimation
Source: Adapted from Kaukonen (1999). Reproduced with permission of J. Kaukonen.

The voltage drop estimation error in switch components is similar to stator resistance error. Uncertainty regarding exactly when a switch turns ON or OFF is another switch-related problem. Accurate time intervals are required for successful integration. This problem can be mitigated with instrumentation to detect the threshold voltages, so integration can begin precisely when the switch has really turned on or off.

Stator resistance error

Stator resistance, the most significant variable in voltage integration, is a measure of DFLC stability. Moreover, flux linkage integration is sensitive to errors in stator resistance. For an IM, both analytical and empirical evidence shows that the stator flux linkage error term increases in the direction of the flux linkage resulting in unstable DFLC (positive feedback), if stator resistance is over estimated ($R_{s,est} > R_{s,act}$). If the stator resistance is underestimated ($R_{s,est} \leq R_{s,act}$), the flux linkage error tends to decrease, and this negative feedback results in a stable DFLC. The same holds also for other machine types. The integral of the stator voltage can be written as follows.

$$\psi_s = \int u_s dt - R_s \int i_s dt \tag{6.44}$$

The stator resistance estimate can be expressed

$$R_{s,est} = R_{s,act} + \Delta R_s \tag{6.45}$$

$R_{s,est}$ is the estimated resistance and ΔR_s is its error. The error term of the stator flux linkage may be expressed as

$$
\begin{aligned}
\psi_{s,err} &= \psi_s - \psi_{s,est} = \int u_{s,act}dt - R_{s,act}\int i_{s\ act}dt - \int u_{s,act}dt + (R_{s,act} + \Delta R_s)\int i_{s,act}dt \\
&= \Delta R_s\int i_{s,act}dt = \Delta R_s\left[\int i_{sx,act}dt + j\left(\Delta R_s\int i_{sy,act}dt\right)\right] = \psi_{sx,err} + j\psi_{sy,err} = (\psi_{sd,err} + j\psi_{sq,err})e^{j\theta_r}
\end{aligned}
$$

(6.46)

Figure 6.17 illustrates stator flux drift resulting from errors in voltage drop prediction for a synchronous machine having different inductances in the d- and q-axes. The voltage drops have been overestimated, so actual stator voltage is higher than predicted. The DFLC applies the selected voltage vector until the predicted stator flux linkage exceeds its set hysteresis limit. The actual magnitude of the flux linkage in the electric machine moves way beyond the hysteresis limit during this time, and the DFLC becomes unstable.

If the instantaneous error in the stator resistance prediction at time t_1 results in stator flux-linkage drift, the current vector of the motor is affected. For example, for a synchronous motor

$$
\begin{aligned}
i_s &= i_{sd} + ji_{sq} = \frac{1}{L_{sd}}\left[\psi_{sd} - L_{md}(i_f + i_D)\right] + j\left[\frac{1}{L_{sq}}\left(\psi_{sq} - L_{mq}i_Q\right)\right] \\
&= \frac{1}{L_{sd}}\left[\psi_{sd,est} + \psi_{sd,err} - L_{md}(i_f + i_D)\right] + j\left[\frac{1}{L_{sq}}\left(\psi_{sq,est} + \psi_{sq,err} - L_{mq}i_Q\right)\right] \\
&= \frac{1}{L_{sd}}\left[\psi_{sd,est} - L_{md}(i_f + i_D)\right] + \frac{\psi_{sd,err}}{L_{sd}} + j\left[\frac{1}{L_{sq}}\left(\psi_{sq,est} - L_{mq}i_Q\right) + \frac{\psi_{sq,err}}{L_{sq}}\right]
\end{aligned}
$$

(6.47)

As expected, the stator current vector moves in the same direction as the flux linkage error term $\psi_{s,err}$. The error accumulating during one electric cycle can be calculated by assuming that the angular speed of the stator current vector is constant.

$$
\Delta\psi_{s,err} = (\Delta\psi_{sx,err} + j\Delta\psi_{sy,err})e^{j\theta_r} = \Delta R_s\Delta t(\Delta i_{sd} + j\Delta i_{sq})e^{j\theta_r}
$$

(6.48)

Based on Equation (6.47), the current components Δi_{sd} and Δi_{sq} caused by the estimation error can be written as follows.

$$
\Delta i_{sd} = \psi_{sd,err}/L_{sd} \text{ and } \Delta i_{sq} = \psi_{sq,err}/L_{sq}
$$

(6.49)

The estimation error difference is expressed by

$$
\Delta\psi_{s,err} = \Delta\psi_{sx,err} + j\Delta\psi_{sy,err} = \Delta R_s\Delta t\left(\frac{\psi_{sd,err}}{L_{sd}} + j\frac{\psi_{sq,err}}{L_{sq}}\right)e^{j\theta_r}
$$

(6.50)

This equation indicates that unstable drifting occurs when $R_{s,est} > R_{s,act}$, because the flux linkage increases into the direction of the error. Stable drift behaviour results if stator resistance is estimated to be smaller than actual resistance ($R_{s,est} \leq R_{s,act}$). This is easy to understand also based on the energy principle. A system with larger losses than expected tends to be more stable than a system with smaller than expected losses. Figure 6.18 depicts situations in which estimated resistance is larger than actual resistance ($R_{s,est} > R_{s,act}$) and vice versa ($R_{s,est} < R_{s,act}$).

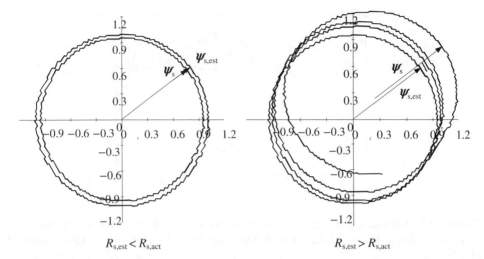

Figure 6.18 Estimated and actual flux linkages when there is an estimation error in the stator resistance. When $R_{s,est} < R_{s,act}$, actual Joule losses are greater than estimated losses. According to the energy principle, this leads to a stable error. When $R_{s,est} > R_{s,act}$, the Joule losses of the actual system are smaller than predicted, leading to unstable operation. However, the drift is slow. There is, in principle, sufficient time to make necessary corrections: four full periods in this case. *Source:* Adapted from Kaukonen (1999). Reproduced with permission of J. Kaukonen.

Further, Figure 6.19 illustrates a typical nonsinusoidal stator current waveform caused by stator flux-linkage drift. The currents are heavily distorted but, for example, the peak values are about the same. Moreover, the waveform periods are roughly equivalent. The half-wave integrals differ, however.

Because of these inherent weaknesses, DFLC, in itself, is not a usable control method in most cases. It must be supplemented by introducing a method to correct the instability problems resulting from nonidealities. The methods primary task is to prevent the machine's stator flux-linkage vector from drifting away from the origin. A natural and efficient method is to apply the electric current model based on motor parameters. The method of control referred to as direct torque control, or DTC, is, in fact, just the DFLC methodology with a supplemental electric current model to mitigate instabilities.

Because DFLC offers the best possible dynamics for the drive, using it instead of the more stable FOC to develop good and stable electric machine control is a worthwhile endeavour. Although it does not drift, FOC is subject to the same errors as the DFLC. It suffers severely from incorrect machine inductance parameters, which the DFLC does not need at all.

6.5   Improving DFLC to achieve DTC

The previous section introduced the basics of the DFLC and discussed it main weakness, drifting of the stator flux-linkage vector. For highly reliable control, a more stable method is

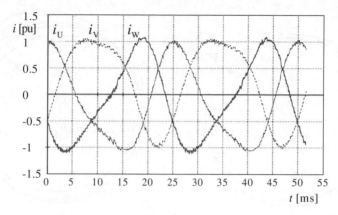

Figure 6.19 Nonsinusoidal waveforms of the stator currents caused by the eccentricity of the stator flux linkage. *Source:* Adapted from Niemelä (1999). Reproduced with permission of M. Niemelä.

needed. DTC is a well-known method applied by ABB, such as, in its high performance drives. At ABB, DTC refers to a fully developed method that combines the above-mentioned DFLC with a reliable method of stabilizing the integration of stator flux linkage. The primary function of the stabilization method is to keep the actual flux linkage of the motor origin-centric. With this improvement, DFLC become a highly reliable and dynamic controller for all AC electrical drive types.

There are at least two obvious methods to control DFLC in a way that the machine's stator flux linkage will rotate centred about the origin. In the following discussion, the most obvious method, using electric current model correction, is introduced. In many cases, this method requires rotor position feedback, so a method based on signal processing to correct stator flux linkage eccentricity (drift correction) has also been introduced. Both methods will be discussed here in brief.

6.5.1 Current model correction

In DTC, the voltage model $\boldsymbol{\psi}_{\mathrm{s,est}} = \int (\boldsymbol{u}_\mathrm{s} - R_\mathrm{s}\boldsymbol{i}_\mathrm{s})\mathrm{d}t$ is calculated frequently. For instance, in ABB's classic DTC converters, the computation of the voltage model is repeated every 25 μs. Drifting of the motor flux-linkage vector away from the origin because of errors in the voltage model occurs in tens of milliseconds. Therefore, the eccentricity of the flux linkage calculated by the voltage model must be corrected about every millisecond.

In electric current model correction, stator flux linkage integrated from the voltage $\boldsymbol{\psi}_{\mathrm{su}}$ is corrected by an error vector $\Delta\boldsymbol{\psi}_\mathrm{s}$. The current model is the vector equivalent circuit of the electric machine including resistances and inductances. Since motor current is measured and the equivalent circuit parameters are known, the stator flux linkage $\boldsymbol{\psi}_{\mathrm{si}}$ for an induction machine, based on the electric current model, can be calculated thus.

$$\boldsymbol{\psi}_{\mathrm{si}} = \boldsymbol{i}_\mathrm{s}L_{\mathrm{s\sigma}} + \boldsymbol{i}_\mathrm{m}L_\mathrm{m} \tag{6.51}$$

The error vector is constructed as a difference of the stator flux-linkage vector integrated from the voltage (subscript u) and the stator flux-linkage vector calculated by the current model (subscript i).

$$\Delta \boldsymbol{\psi}_s = \boldsymbol{\psi}_{si} - \boldsymbol{\psi}_{su} \tag{6.52}$$

The flux linkage difference of Equation (6.52), suitably weighted, is used to correct the stator flux linkage for instance at every 100th microsecond.

$$\boldsymbol{\psi}_{su(n+1)} = \boldsymbol{\psi}_{su(n)} + k_{cm} e^{j\omega_s \Delta t} \Delta \boldsymbol{\psi}_s \tag{6.53}$$

In discrete system n, the time-dependent index k_{cm} is the weighting coefficient for the electric current model. The correction's ($\Delta \boldsymbol{\psi}_s$) direction must be turned appropriately by the angular correction term $e^{j\omega_s \Delta t}$, which depends on the electric angular frequency and describes the progression of the flux linkage.

$$\Delta \boldsymbol{\psi}_{s,n+1} = e^{j\omega_s \Delta t} \Delta \boldsymbol{\psi}_{s,n} \tag{6.54}$$

This is because the flux linkage correction term ($\Delta \boldsymbol{\psi}_{s,n}$) contains information that can be no older than 1 ms. The time Δt of the correction term is calculated from the instant of the determination of the error term. The magnitude of the error is clearly dependent on the frequency, and therefore the definition of the error vector can be modified to be frequency dependent, if necessary, by suitably adjusting the weighting coefficient. In practice, however, a fixed weighting coefficient is usually used.

In particular, the superiority of the DTC to the traditional FOC is based on combining the advantages of the voltage and the electric current models. When applying a large weighting coefficient typical of the voltage model during fast transient phenomena, outstanding dynamic drive characteristics can be achieved.

The electric current model tends to fail, especially in dynamic states, because motor inductances vary with torque- and flux-level–induced saturation (Chapter 3) and with eddy currents. The calculation error becomes especially big during dynamic transients, because the terms related to the derivatives of the inductances are ignored. This can be clarified by observing the differential of the flux linkage.

$$-e_s = \frac{d\boldsymbol{\psi}_s}{dt} = \frac{d(i_s L_s + i_r L_m)}{dt} = i_s \frac{dL_s}{dt} + L_s \frac{di_s}{dt} + i_r \frac{dL_m}{dt} + L_m \frac{di_r}{dt} \tag{6.55}$$

Equation (6.55) includes the differentials of stator inductance and magnetizing inductance. Normally, these differentials are ignored, because the inductances are held constant. The change in inductance is difficult to model, however, and taking the differentials in the mathematical solution easily leads to mathematical instability. The error associated with leaving the differentials out of the observation is smaller.

The voltage model in the DTC does not suffer similar difficulties, because it does not use inductance parameters. The voltage integration accounts for all known phenomena, and in principle, the voltage integral should give better and more reliable information about flux linkage behaviour than does the electric current model. Previously, however, it was explained why the integration fails. Nonetheless, the voltage model remains more correct than the electric current model for fast transients. Therefore, the DTC relies on the voltage model during transients. For more stable operating conditions, the DTC applies the electric current model to mitigate integration errors.

During a fast transient, the DTC gives more weight to corrections coming from the voltage model and less weight to corrections suggested by the electric current model. Therefore, fast transients are chiefly handled by the DFLC core of the DTC. As the transient stabilizes, more and more weight is transferred to the electric current model to keep the motor stator flux-linkage vector from drifting into an eccentric orbit.

To control magnetically anisotropic machines, knowledge of the rotor position is required to enable coordinate transformations and make corrections using the electric current model. This is clearly a weakness, and it is tempting to search for alternative stator flux linkage estimate stabilization methods. For an IM, however, the current model does not need rotor position information. It is possible to implement high-performance DTC for an induction machine without measuring angular position. Stator flux linkage drift correction works well without rotor position feedback information.

6.5.2   Stator flux-linkage eccentricity correction

Based on the work of Niemelä (1999) and Luukko (2000), when operating at moderate frequencies or about 2–5% of rated frequency, stator flux linkage eccentricity can be corrected by applying the scalar product of the stator flux linkage estimate $\psi_{s,est}$ and the stator current i_s. To accomplish this, the estimate components $\psi_{sx,est}$ and $\psi_{sy,est}$ are represented in the x and y directions of the stator reference frame. Figure 6.20 illustrates a situation in which the actual stator flux linkage of the motor has drifted and become eccentric by $\Delta\psi_s$. This eccentricity results in an extra magnetizing current component Δi_s in isotropic machines in approximately the same direction, especially, as the machine starts to saturate in the direction of the eccentricity.

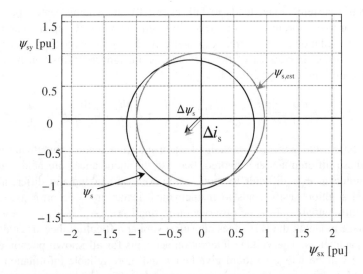

Figure 6.20 Eccentric actual stator flux linkage ψ_s, the estimated noneccentric stator flux linkage $\psi_{s,est}$, and a current component in approximately the same direction as $\Delta\psi_s$ in case of an isotropic machine.

In salient pole machines, the reluctance differences modulate the direction of the extra eccentricity-induced current. It is, however, still possible to apply the correction (based on estimating the scalar product of the stator flux linkage and the stator current) to salient pole machines. To accomplish this, extra analysis must be done. For normal salient-pole synchronous machines with damper windings, the correction can be applied without incurring significant problems, because the damper windings compensate for the anisotropy phenomena.

Based on Figure 6.20, there will be a positive nonzero vector scalar product component $\psi_{s,est}\cdot\Delta i_s$ at the instant when the estimated flux linkage $\psi_{s,est}$ and the extra stator current Δi_s are at the same position.

The vector scalar product (the dot product) of the stator flux linkage estimate $\psi_{s,est}$ and the stator current i_s is constructed as follows.

$$\psi_{s,est}\cdot i_s = \psi_{sx,est}i_{sx} + \psi_{sy,est}i_{sy} \tag{6.56}$$

Low-pass filtering, with a filtering time constant sufficiently longer than the periodic time of the supply frequency, makes it possible to find the average value of the vector scalar product. Once found, the original signal is subtracted from the average to formulate a negative expression for the AC value of the vector scalar product. The correction terms $\psi_{sx,corr}$, $\psi_{sy,corr}$ of the stator flux-linkage estimate are formed as a product of the difference between the calculated and filtered vector scalar product and the components of the stator flux-linkage estimate.

$$\psi_{sx,corr} = K_{\psi corr}\left[\psi_{s,est}\cdot i_s - \left(\psi_{s,est}\cdot i_s\right)_{filt}\right]\cdot\psi_{sx,est} \tag{6.57}$$

$$\psi_{sy,corr} = K_{\psi corr}\left[\psi_{s,est}\cdot i_s - \left(\psi_{s,est}\cdot i_s\right)_{filt}\right]\cdot\psi_{sy,est} \tag{6.58}$$

Figure 6.21 illustrates the formation of these stator flux-linkage correction terms. Applying them within the DTC enables full utilization without the need for feedback information on rotor position. Furthermore, the DTC becomes capable of controlling a rotating-field machine over a wide range of rotation speeds even though the motor parameter information is incomplete. Only the value of the stator resistance, which can be easily measured, must be known. In principle, the method enables the application of the DTC to the power transmission of all rotating-field machines and in the DC-to-AC inversion to the network. Therefore, the method is universal. Because of the filtering time required to determine the average of the scalar product, the method is not applicable for zero speed. Consequently, the

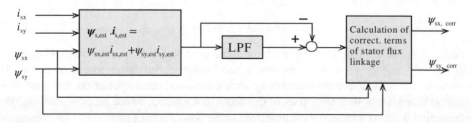

Figure 6.21 Formation of the correction terms for the stator flux linkage (LPF $\hat{=}$ low pass filter).

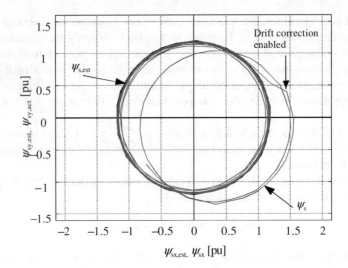

Figure 6.22 Drift correction returns flux linkage rapidly back to the origin. Subscript act refers to the electric current model calculated value. The method is applied to a salient-pole synchronous machine with damper windings. *Source:* Adapted from Niemelä (1999). Reproduced with permission of M. Niemelä.

correction method based on the electric current model must be available and applied in demanding drives at close to zero speeds.

Figure 6.22 depicts the influence of drift correction application on a test drive when operating a 50 Hz machine at a supply frequency of 1 Hz. Drift correction works efficiently in this region.

Luukko (2000) improved the dynamic performance of the original implementation by replacing the conventional low pass filter by a phase lag compensator.

$$G_{\text{LPF1}}(s) = \frac{sk_T T_1 + 1}{sT_1 + 1} \tag{6.59}$$

In the expression, k_T is made adaptive $0 \leq k_T \leq 1$. The correction gain k_ψ is also made adaptive and dependent on the parameter k_T.

$$k_\psi = (1 - k_T)k_{\psi 0} \tag{6.60}$$

The $k_{\psi 0}$ is the base value. Drifting resulting in incorrect flux-linkage estimation takes placer fairly slowly, and therefore, no correction is necessarily needed during torque transients. To be able to correctly detect the eccentricity after a transient, the output of the low-pass filter must be adapted to a new DC offset of the vector dot product, which takes some time. The filter output is allowed to follow the unfiltered value faster during a transient by increasing the coefficient k_T. Figure 6.23 shows the block diagram of the improved version of the flux linkage correction method.

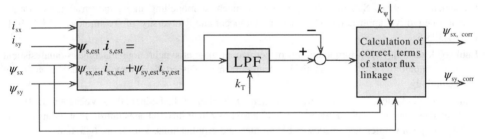

Figure 6.23 Block diagram of improved correction method.

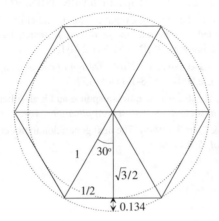

Figure 6.24 DSC modulation using a hysteresis bandwidth of 0.134 resulting in the use of only one active vector in each segment.

6.6   Other control principles

The direct self control (DSC) suggested by Depenbrock can be regarded as a version of DFLC or DTC. In DSC, the flux path is kept hexagonal, and torque is limited by applying zero vectors. In theory, DSC can be derived from DFLC by applying a sufficiently large hysteresis band. Figure 6.24 shows that if a hysteresis bandwidth of 0.134 per unit is selected, only one active vector and zero vectors are used in the modulation, and the flux linkage travels a hexagonal route. A similar hexagon results when full overmodulation is used with any two-level converter.

As a function of time, DSC modulation results in square wave stator voltages. This modulation, therefore, produces low-frequency torque ripple.

There are several variations of DTC and FOC, but these dual methods currently form the basis for most motor control solutions. How these basic methods are applied to the primary machine types will be discussed in upcoming chapters.

References

Depenbrock, M. (1985). Direkte Selbstregelung (DSR) für hochdynamische Drehfeldantriebe mit Stromrichterspeisung. *Etz Archiv Bd*, 7, H.7, 211–218.

Kaukonen, J. (1999). Salient pole synchronous machine modelling in an industrial direct torque controlled drive application. Dissertation. Lappeenranta University of Technology. ISBN 951-764-305-5.

Luukko, J. (2000). Direct torque control of permanent magnet synchronous machines – analysis and implementation. Acta Universitatis Lappeenrantaensis. Dissertation, Lappeenranta University of Technology.

Luukko, J., Kaukonen, J., Niemelä, M., Pyrhönen, O., Pyrhönen, J., Tiitinen, P., & Väänänen, J. (1997). Permanent magnet synchronous motor drive based on direct flux-linkage control. Proceedings of the 7th European Conference on Power Electronics and Applications, vol. 3. pp. 683–688.

Niemelä, M. (1999). Position sensorless electrically excited synchronous motor drive for industrial use, based on direct flux linkage and torque control. Acta Universitatis Lappeenrantaensis. Dissertation, Lappeenranta University of Technology. http://urn.fi/URN: ISBN: 978-952-265-042-9.

Pyrhönen, O. (1998). Analysis and control of excitation, field weakening and stability in direct torque controlled electrically excited synchronous motor drives. Research Papers 74. Dissertation, Lappeenranta University of Technology. ISBN 951-764-274-1.

Pyrhönen, J., Jokinen, T., & Hrabovcova, V. (2014). *Design of rotating electric machines*. Chichester, UK: John Wiley & Sons, Ltd. ISBN: 978-1-118-58157-5.

Takahashi, I., & Noguchi, T. (1986). A new quick-response and high-efficiency control strategy of an induction motor. *IEEE Transactions on Industry Applications*, IA-22, 820–827.

Tiitinen, P., Pohjalainen, P., & Lalu, J. (1995). The next generation motor control method: direct torque control DTC. *EPE Journal*, 5, 14–18.

7

DC and AC power electronic topologies – modulation for the control of rotating-field motors

In principle, alternating current (AC) motors can be controlled using a direct converter or a converter with an intermediate direct current (DC) link (DC-link converters). A direct converter transforms incoming AC voltage to produce a different outgoing AC voltage. A DC-link converter changes incoming AC voltage into either DC using a current source inverter (CSI) or direct voltage using a voltage source inverter (VSI). Then, the DC-link current or voltage is inverted to produce the new outgoing AC voltage.

A DC motor converter is simpler than an AC motor converter, because in industrial cases, only the controlled interaction between the network and the DC machine is needed. A DC motor converter may also have a DC link if a PWM chopper is used to supply the motor.

Early on, using a current source inverter was the most common approach to frequency conversion. However, the popularity of AC-drive current source inverters has declined continually; today, frequency converters used in commercial applications are almost without exception voltage source inverters. The large majority of commercial inverters are two-level DC-link inverters. Three-level, low-voltage, DC-link inverters are now becoming available, and they may become increasingly popular. Currently, three-level, low-voltage, DC-link inverters are used mostly for large power applications at medium voltages. A five-level, medium-voltage converter is also available. Some manufacturers are also offering matrix converters, which can be regarded as direct converters.

DC-link converters have shortcomings. The necessary energy-storing DC-link coil in the CSI-based converter is relatively large and needs a lot of space. In the VSI-based converter, the relatively large DC-link electrolytic capacitor can become unreliable after 6 to 10 years of operation, and its failure can destroy the converter. To reduce costs, more reliable capacitor types are being introduced and rated capacitance values are being reduced.

Electrical Machine Drives Control: An Introduction, First Edition. Juha Pyrhönen, Valéria Hrabovcová and R. Scott Semken.
© 2016 John Wiley & Sons, Ltd. Published 2016 by John Wiley & Sons, Ltd.

If capacitance is reduced to the minimum, the VSI begins to resemble a direct converter. In that case, the control technology must focus on controlling DC-link voltage. With minimized capacitance, DC-link voltage is relatively unsteady, tending to follow the converter input voltage. With large capacitance, the DC-link voltage can be considered nearly constant.

In a direct converter, the DC link and related components can be avoided. However, the direct converter also has disadvantages. Thyristor-based cycloconverters (CCVs) used to be popular for converting frequencies in low-speed high-power motor drives. Recently, however, high-voltage VSIs are increasingly being used instead of these CCVs. In high-power ship propulsion, because they are compact and reliable, CCVs are still competitive.

The load-commutated inverter (LCI), a current source inverter, is rarely used today. The LCI, however, is still used as the frequency converter in large, high-speed, synchronous machine drives. It also holds the present-day 101 MW power record for electric motor drives.

The matrix converter is an emerging frequency converter technology that promises to replace DC-link converters. This converter type requires bidirectional switches. In a matrix converter, each input phase can be connected to each output phase. It is possible, in principle, to always select the desired input voltage for each output. A bidirectional switch must be able to change the positive and negative polarities of current or voltage. Because there are no individual power electronic components with this capability, it must be accomplished using a bidirectional switching circuit made up of several switching components.

Figure 7.1 illustrates the basic topologies of the main converter types.

The following text examines, in more detail, the operating principles of the most popular converter technologies.

7.1   The thyristor bridge as a power-electronic drive component

Industrial DC motor drives normally rely on thyristor bridges for DC-motor-armature-current and field-winding-current control. The six-pulse thyristor bridge shown in Figure 7.2a comprises a positive commutation group of three thyristors P and a corresponding negative N.

The line-commutated six-pulse thyristor bridge can operate in both inversion and rectification. The bridge can be used to operate with a DC voltage U_{DC}, the mean value of which is either positive or negative. The current I_{DC} of the bridge can flow only in one direction, as determined by the thyristors. Therefore, the bridge may operate in two quadrants of the current-voltage plane. See Figure 7.2b.

Rectification into DC of the three-phase AC voltage is straightforward and easy to understand. If the thyristors are triggered with zero firing angle ($\alpha = 0$), the positive commutating group P always selects the most positive value of the input, while the negative commutating group N selects the most negative voltage of the input. The output of the bridge is the difference between these two voltages and is similar to the output of a corresponding diode bridge.

As firing angle α increases, the output voltage of the thyristor bridge decreases. If the bridge is connected to an active load capable of producing DC power, it is possible to drive the firing angle to over 90° when negative output voltages with positive current are available (the IV quadrant). A thyristor bridge can also be used, therefore, as a DC-to-AC inverter. This builds on the principle of a DC-link between two AC networks. By using a rectifier at the first AC network and an inverter at the second AC network, it becomes possible to transmit power between two

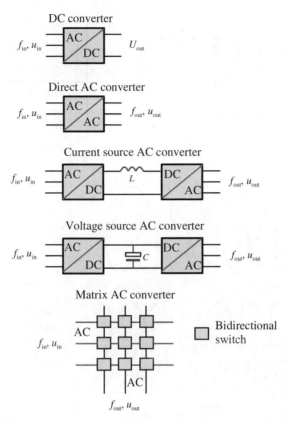

Figure 7.1 Principal topologies of converters, As shown in the figure, each AC converter type produces a three-phase output system from the three-phase input system. The numbers of phases may be other than shown. For instance, the input current for small voltage source inverters is usually single phase. The voltage source converter can be connected directly to a DC source when it, for example, converts battery DC to AC or vice versa. The DC converter may either rectify the AC voltage from the network or invert DC to AC.

AC networks having either the same or the different frequencies. This principle is used in high-voltage DC system for long-distance power transmission.

As the introduction of the bridge illustrates, it is possible to operate in two quadrants (I and IV) using one thyristor bridge. Adding another bridge antiparallel with the first makes it possible to cover quadrants II and III, as shown in Figure 7.3. Current I_{DC} can then flow in the opposite direction from that shown in the figure resulting in both polarities of voltage U_{DC}.

Thus, the converter can operate in a four-quadrant current-voltage plane, as shown in Figure 7.4a. Figure 7.4b depicts the schematic diagram of the antiparallel six-pulse thyristor bridges. By using the antiparallel connection, the direction and magnitude of current and voltage may freely change independent of each other. Since power is a product of current and voltage, its direction may also change freely. Thus, the converter with the antiparallel connection can be used to supply power to the load, or an active load may supply power

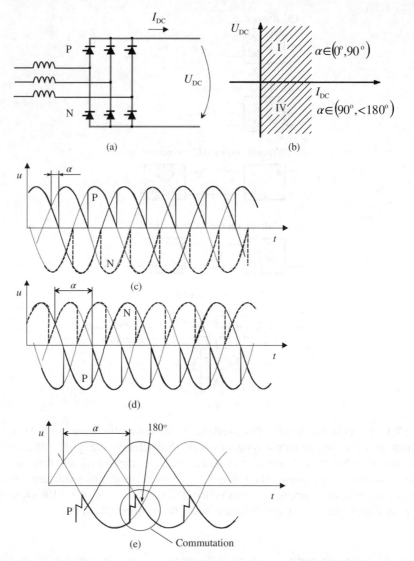

Figure 7.2 Image (a) represents a six-pulse thyristor bridge. I_{DC} is the DC conducted by the bridge, and U_{DC} is the generated DC voltage. Image (b) shows the current-voltage plane and the operating quadrants for the six-pulse thyristor bridge. In (c), the bridge produces positive voltage with firing angle $\alpha \approx 30°$, and in (d), it produces negative voltage with $\alpha \approx 150°$ (positive group P produces negative voltage and negative group N positive voltage). This represents the maximum possible practical firing angle for a thyristor bridge, because there must be a commutation voltage reserve to enable bridge commutation. At $\alpha = 180°$, there is no remaining commutation driving voltage, and therefore commutation must take place before the angle is reached. Image (e) reveals that during commutation, the output voltage of the positive commutating group follows the average of the two supply voltages. Commutation must end before reaching $180°$.

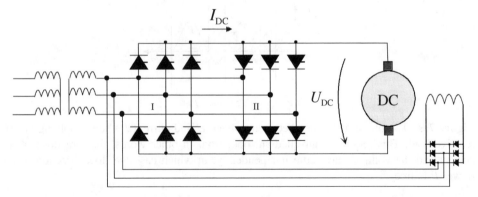

Figure 7.3 Antiparallel thyristor bridges (I and II) in the supply of a DC machine four-quadrant drive where the machine may act as both a motor and a generator. I_{DC} is the DC conducted by the antiparallel connection, and U_{DC} is the generated DC voltage. The field winding has its own six-pulse thyristor bridge to control excitation.

through the converter to the grid. This converter is used in DC drives to implement four-quadrant operation.

With the three-phase AC line-to-line U_{LL} input voltage, the maximum possible output voltage of a thyristor bridge corresponds to $U_{DC} = 3\sqrt{2}U_{LL}/\pi$, which is also the output of the corresponding diode bridge often used in power electronics. In a 400 V AC supply, the theoretical output voltage of a thyristor bridge with zero firing angle will be 540 V DC. When the output is $U_{DC} = 1.35U_{LL} \cos \alpha$, it is possible with a thyristor bridge to control it using a suitable firing angle α. In DC drives, the rated DC voltage, however, is limited to ensure commutation during inversion. That is why in 400 V AC networks rated DC machine voltage is typically set to 440 V, which corresponds to a firing angle of $\alpha = 35°$ (positive voltage) or 145° (negative voltage). Operation of the inversion commutation for a corresponding antiparallel bridge is normally safe between these extremes. For example, if a DC motor is running at its maximum load with a bridge I, $\alpha = 35°$, power can be reversed to generation and braking by first applying $\alpha = 145°$ to bridge II, and then lowering the firing angle gradually based on controlling DC machine current and torque.

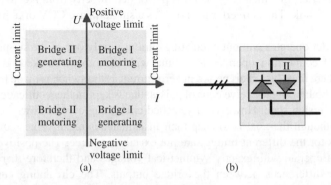

Figure 7.4 (a) The four operating quadrants of the current–voltage plane for the antiparallel six-pulse bridge and (b) the block diagram symbol for the antiparallel converter.

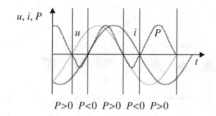

Figure 7.5 The phase difference between voltage U and current I when supplying an AC reactive load. This requires four-quadrant operation. In other words, the directions of the current and the voltage may settle independently, in which case the flow direction of the power P is free.

There is a network-side problem with the displacement power factor in DC drives. With constant DC current being supplied by a thyristor bridge, the displacement power factor is dependent on firing angle. It remains high if the firing angle $\alpha = 0°$, but collapses with larger firing angles. At zero speed and high torque, the reactive power can reach values that correspond to the rated power of the DC drive, because near $\alpha = 90°$, the firing angle displaces voltage and current by $90°$ ($\cos \varphi \approx \cos \alpha$). Reactive power caused by commutation can further deteriorate the power factor and induce unwanted current harmonics. The total power factor of a thyristor bridge is about $\Gamma \approx (I_1/I) \cos \alpha$ with I_1 being the RMS value of the fundamental.

For DC drives, converter output voltage is naturally DC. However, to a DC motor that is frequently reversing direction, the voltage supplied by the converter looks much like very low frequency AC voltage. It follows that nothing prohibits a DC converter from outputting a low frequency AC supply. The converter then becomes an AC voltage source, in which the current and voltage directions may settle independently with respect to each other, as shown in Figure 7.5. In AC systems, the phase angle between current and voltage is determined by the load impedance. Also, the flow direction of active and reactive power may settle according to the load. In other words, the four-quadrant DC thyristor drive operates completely in four quadrants with AC.

The frequency and amplitude of the output voltage are altered by changing the frequency and amplitude of the voltage reference. These switching properties enable a frequency converter, in which conversion is performed directly from AC to AC without an intermediate DC link. The converter is known as a single-phase CCV or a direct frequency converter.

The converter operates without circulating current if only one of the six-pulse bridges of the antiparallel connection operates at a time with the others blocked. Between bridge switchovers, there must be a short period of zero current as the switch is made from one bridge to the other. If both bridges receive control pulses, one is kept in the rectification state and the other in the inversion state. However, a so-called circulating current flows in the converter, because even though the goal is to keep their instantaneous mean values equal, the output voltage values for the different bridges are not exactly the same. The positive and negative bridges do not operate with exactly symmetrical voltages and therefore there are instantaneous voltage differences between the bridge outputs. The circulating current must be restricted by a choke so the current does not increase excessively. Manufacturers usually configure CCVs to operate without circulating current.

7.2   The cycloconverter

A CCV is a direct frequency converter that converts an incoming AC waveform of constant frequency and voltage into an outgoing AC waveform of varying frequency and varying voltage. CCVs are suitable for applications ranging from megawatts to tens of megawatts and have been used as to control mine hoists, rolling mill motors, ball mills, cement kilns, and ship propulsion systems. Their variable-frequency output can be reduced essentially to zero so they can be started very slowly on full load, and they offer variable speed and reversing. CCV-based drives require minimal maintenance are quite reliable and very efficient. The topology of a CCV is shown in Figure 7.6.

Combining three CCVs, one for each phase, enables three-phase frequency conversion. The converters for each phase are usually star connected, or they may supply the phases of the load separately. The star connection and the separate-phase connection are illustrated in Figure 7.7. Transformers or isolated phase windings must be used, since forming the star point by interconnecting the negative buses of the CCVs would result in a short circuit between phases if a different phase of the grid is connected to the negative bus in different CCVs. This problem can be avoided using transformers, since they galvanically separate the different phases of the grid. Unfortunately, adding transformers increases cost, weight, and bulk. Nonetheless, using CCVs with transformers seems to be a more popular approach than the separately wound machine method.

The separately wound, galvanically isolated method usually requires a separate choke to manage the zero-sequence current, through which all the phase currents are transferred. The choke maintains the zero-sequence flux linkage to simultaneously guarantee an instantaneous zero sum of the phase currents, which is important for the operation of rotating-field machinery.

Usually supplied via transformers, the topology of an ordinary three-phase CCV is illustrated in Figure 7.8.

Simulations reveal the curve form of the output voltage of each phase to be constructed of the cycles of the different supply line-to-line voltages, as shown in Figures 7.9 and 7.10. In the two figures, the supply frequency is 50 Hz. Sine control is applied in Figure 7.9. In other words, the controller works to vary the instantaneous mean value of the output voltage of the CCV sinusoidally. The control principle for sine modulation that produces a

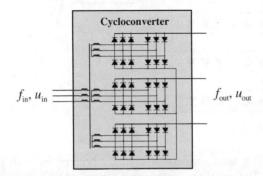

Figure 7.6 The topology of a CCV showing direct AC-AC conversion; f_{in} is the line frequency, and u_{in} is the line voltage. f_{out} is the output frequency, and u_{out} is the output voltage of the CCV.

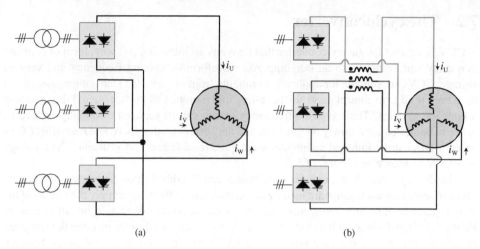

(a)   (b)

Figure 7.7 A three-phase cycloconverter (a) with the phases connected in a star configuration and (b) with the phases galvanically separated and equipped with a zero sequence current choke that guarantees $i_U + i_V + i_W \approx 0$.

motor phase voltage u_{ph} with a motor electrical angular frequency ω_m can be simply expressed as follows.

$$U_{d0} \cos \alpha = \hat{u}_{ph} \cos \omega_m t \qquad (7.1)$$

Here, $U_{d0} = 3\sqrt{2}U_{LL}/\pi$ is the DC voltage of the six-pulse bridge corresponding to the firing angle $\alpha = 0$. By adjusting the firing angle α, the "instantaneous" DC voltage (left-hand side of

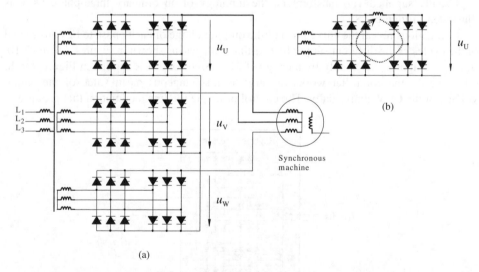

(a)

Figure 7.8 (a) A CCV with antiparallel thyristor bridges for each phase and with non-circulating currents. The galvanic separation between phases is provided by the supply transformer. u_U, u_V, and u_W are the phase voltages. (b) One phase of a converter operating both bridges simultaneously. Circulating currents between converters are allowed. A choke is needed to limit the circulating current.

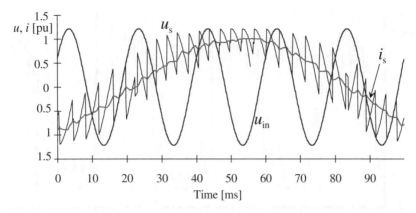

Figure 7.9 Curve form of the output voltage u_s and current i_s of a CCV when applying sine control. The curve form of one of the line-to-line voltages u_{in} of the grid supply voltage is also presented for a supply frequency of $f_1 = 50\,\text{Hz}$ and a motor frequency of $f_s = 10\,\text{Hz}$ (one period is 100 ms). And 1 pu corresponds to the peak value of the motor voltage or current.

the equation) is made equal to the low frequency AC voltage (right-hand side). Therefore, the modulation reference becomes:

$$\alpha = \arccos = \frac{\hat{u}_{ph} \cos \omega_m t}{U_{d0}} \qquad (7.2)$$

The trapezoidal method is another voltage modulation method. In the trapezoidal method, the goal is to make the instantaneous mean value of the output voltage change according to a trapezoidal curve. See Figure 7.10.

In practice, the output frequency of a CCV is limited to half the frequency of the supply grid at maximum. At higher frequencies, the output voltage, and therefore the current, increasingly deviates from the sine form. In addition to the usual harmonics, subharmonic frequency components occur in both the output voltage and the input current of the CCV. Therefore, the target is an output voltage that is sinusoidal or trapezoidal on average. With trapezoidal voltage, the control reactive power can be made lower than with sinusoidal voltage, because the thyristors in that case conduct longer at the peak voltage value.

The power factor of a CCV is always inductive due to its control reactive power, irrespective of the power factor of the load. For a modulation index of 1.0 (full output voltage), the power factor of a CCV drive using sine control is 0.707. As output voltage decreases and the modulation index becomes smaller, the power factor drops rapidly. The 0.707 power factor for a 1.0 modulation index value holds for single-phase and three-phase drives supplied with a six-pulse or 12-pulse CCV. It is also valid for a six-pulse CCV and a separate winding. In that case, the machine is synchronous with the power factor set to 1.0. The power factor of a CCV drive using trapezoidal control can exceed 0.707.

Figure 7.11 illustrates the basic configuration of a six-pulse, CCV-fed, three-phase synchronous machine. To increase power, voltage, or current must be increased. However, to withstand the greater values of voltage or current, thyristors must be added to the circuit, either in series or in parallel. When connecting the thyristors in series, the even distribution of

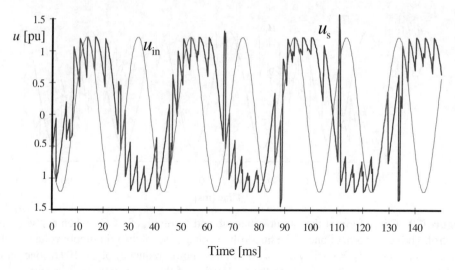

Figure 7.10 Curve form of the motor phase voltage u_s of the CCV when applying trapezoidal control and the curve form of the line-to-line voltage u_{in} of the supply grid voltage when the supply frequency is $f_1 = 50\,\text{Hz}$ and the output frequency is $f_m = 24\,\text{Hz}$.

voltages is problematic. When connecting them in parallel, the even division of current is uncertain. As a result, various new connection alternatives have been developed for the CCVs.

Output voltage can be approximately doubled by connecting two CCVs in series. Since CCVs connected in series to a single phase are supplied from different output windings of the three-winding transformer, the voltages are evenly distributed. Furthermore, the difference in the vector groups of the secondary windings results in a phase difference between the voltages, thus producing a 12-pulse series connection. This yields an improved curve form for both the output voltage and the phase current of the power line, since there are less harmonics in both the voltage and the current. The 12-pulse connection also results in current control that

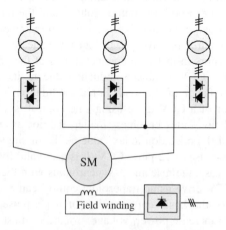

Figure 7.11 Six-pulse, CCV-fed, three-phase motor. A smaller six-pulse thyristor bridge is supplying the field winding.

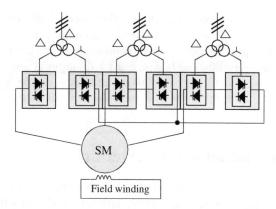

Figure 7.12 A three-phase motor fed by a 12-pulse CCV. Transformers (Ddy) supplying three-phase voltages in 30°-phase shift (y/d) are needed to supply the converter.

is better than with a six-pulse connection, because there are twice as many voltage control alternatives in the 12-pulse connection. See Figure 7.12.

The power of a CCV drive can be increased by configuring the windings as a parallel-connected pair with each winding being fed by a single six-pulse or a 12-pulse CCV. This motor windings connection, which is called a double star, results in increased total machine current. Figure 7.13 illustrates this CCV connection. In the figure, there are two four-winding transformers with connections Yyyy and Yddd. The motor has two star-connected three-phase windings.

The output frequency of the CCV is relatively low, being a half of the supply frequency (at maximum). Since the rotation speed of a synchronous machine is directly proportional to the supply frequency, CCV-fed synchronous drives run at relatively low speeds. Their power range is usually 1.5 to 30 MW. Both synchronous and asynchronous motors can be used with CCV control. Since the power factor of a synchronous machine can be set to 1.0, the power factor of the synchronous motor CCV drive is usually higher than the power factor of an asynchronous motor drive.

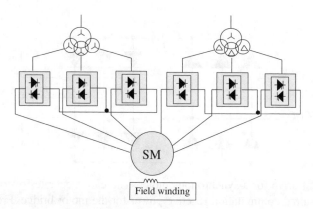

Figure 7.13 A 2×3–phase motor fed by two six-pulse CCVs. This is one arrangement that makes it possible to increase drive power.

Cycloconverters are commonly applied to slow-speed high-power drives, such as propeller motor drives in ships (e.g., icebreakers, multipurpose vessels, and cruise ships). CCVs are also applied to provide rotation speed control for ore and cement mills, pumps, and blowers; for mining industry lifts and hoists; and for rolling-mill drives used in the metal industry. The CCV drive provides accurate motor control. It can be applied to the implementation of the vector control; however, it is not as versatile as the VSI, which will be discussed later.

7.3   The load commutated inverter drive

In terms of power electronics, the simplest power stage is the LCI drive. The LCI drive is completely based on thyristor bridge technology; however, the six-pulse bridge in the drive is used as an inverter. The synchronous machine establishes with its back electromotive forces the "electrical power grid", which can produce the commutation reactive power for the thyristor bridge acting as the inverter.

Figure 7.14 illustrates a drive of this type. In its design, the inductance L_d included in the bridge connection must be large enough to appear as a current source. The coil ensures the magnitude of the DC, and the bridge acting as the inverter guides this DC to the over-magnetized synchronous machine, which is then able to take care of the commutation power required by the bridge. Therefore, an LCI drive is a CSI with indirectly determined machine terminal voltage. DC link current is regulated by controlling the input bridge.

An LCI drive is difficult to start, because there is no electromotive force in the stopped motor, and therefore the drive cannot generate the required commutation reactive power. Starting is facilitated by controlling the bridge on the line side, cutting the DC I_d of the DC link at the appropriate instants. The thyristors of the inverter bridge are left free to switch off naturally. When the motor has reached an appropriate speed (10% of rated), DC link current can be held continuous, and the overmagnetized synchronous machine enables inverter bridge commutation. Figure 7.15 represents the principal curve forms of the current and the voltage in an LCI drive.

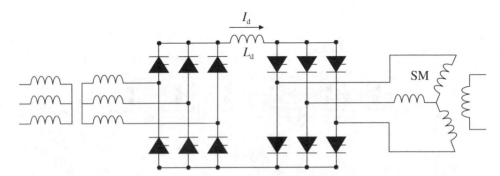

Figure 7.14 LCI drive for a synchronous motor. The motor operates overmagnetized and produces the required commutation reactive power for the motor bridge. From the point of view of the thyristor bridge inverter, the motor forms an "electrical power grid" that provides the commutation reactive power to the motor-side converter.

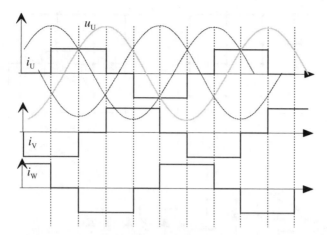

Figure 7.15 Idealized voltage and current waveforms for an LCI drive. Current must be leading voltage to provide commutation power to the inverting bridge. Therefore, there must be a leading power factor, and the most suitable load is an overexcited synchronous motor.

Because of these properties, an LCI drive is most appropriate in cases of simple dynamics, such as for pump or blower drives. However, some rolling mill drives have been developed using the control technology. In 1997, ABB supplied NASA with a 101 MW LCI drive for a wind tunnel installation. ABB claims this to be the largest and most powerful electrical variable speed drive in the world. LCI control technology can also be applied to generators. For generator use, the motor bridge operates as a rectifier and the input bridge as an inverter.

7.4 Voltage source inverter power stages

Commercially, VSI drives represent, at present, the most significant control topology. The two VSI drive types that are commonly available are referred to as two-level and three-level converters. In the two-level configuration, the outputs are connected to the + or − level of the DC link. In the three-level converter, the outputs connect to +, 0, or − terminals. At present, it is possible to reach such high power levels with three-level VSI converters that both CCV- and LCI-based drives are gradually losing market share in favour of VSI-based drives. Because it does not require the DC link, a CCV is more compact than a VSI, so it is preferred in restricted-space environments, such as in ships, where physical size is a critical factor. Two-level VSI technology is generally applied to industrial and marine drives ranging in power from 100 W to 5 MW and above (e.g., ABB 0.12–5,600 kW).

Three-level VSI technology has advantages also at lower powers. Lower-voltage transistors can be used with three-level drives. For example, 600 V transistors can be used in a 400 V three-level AC drive, but 1200 V transistors are needed for a 400 to 500 V two-level AC drive. Converter losses are smaller with the three-level converter's 600 V transistors than they are with the two-level converter's 1200 V transistors. The same holds true for 690 V converters using either 1200 or 1700 V transistors in three- or two-level converter topologies. In three-level converters, the common mode voltages behave more moderately than in

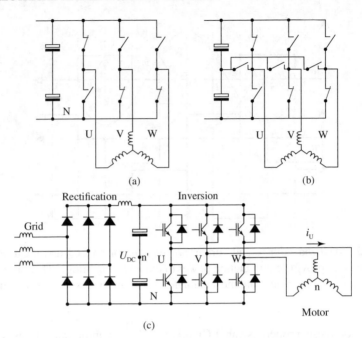

(a)   (b)

Rectification   Inversion

Grid

U_{DC} •n' U   V   W

N

i_{U}

Motor

(c)

Figure 7.16   Topologies of the inverter sections of (a) two-level and (b) three-level VSIs plus (c) the most common topology of a two-level frequency converter with line connectors, a diode bridge for rectification, a DC voltage link, an IGBT inverter, and motor connectors. A single symbolic switch in (a) or (b) includes in (c) an IGBT transistor and an antiparallel diode. The (n') potential of the capacitor-link middle point is close to, but varies slightly around, system neutral.

two-level converters, and therefore bearing current problems, for example, are somewhat smaller in three-level converter drives.

VSIs are applied to control of all kinds of rotating-field machines. Using a three-level IGCT (integrated-gate-controlled thyristor) or GTO (gate-turn-off thyristor) thyristor technology, a medium-voltage range and powers up to 30 MW can be reached. Figure 7.16 illustrates the basic topologies of the three-level and two-level voltage source inverters.

By applying pulse-width-modulation (PWM), a two-level VSI produces a three-phase output voltage, as shown in Figure 7.17. The figure illustrates the application of the so-called sine-triangle comparison technique to construct output voltage. There is a reference value curve for each phase and a triangular wave for all phases. The reference value curves have been synchronized to the triangular wave. This modulation method was developed in the days of analog technology, and it is still best understood by looking at the analog curves. Naturally, the same principles can be realized digitally.

Digital versions of modulation based on sine-triangle comparison have been introduced that replace the analogue triangle wave with a fast up-and-down counter. The reference signals are also digitally produced, and only a comparison of counter and reference numeric values is needed to produce the PWM signals. The comparator forms the switch references for the change-over switches of the two-level inverter so that output is connected to the upper voltage U_{DC} of the DC link or to the negative bus N. Together with the modulation method,

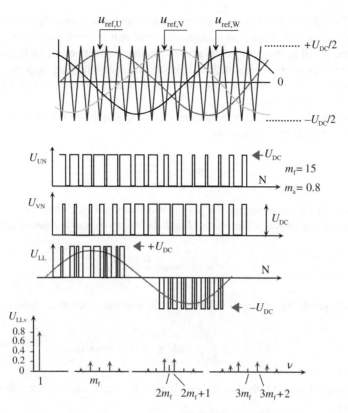

Figure 7.17 Modulation based on a sine-triangle comparison, the produced voltages, and the harmonic content of the line-to-line voltage U_{LL}. The amplitude (m_a) and frequency (m_f) modulation ratios are 0.8 and 15. The harmonic content of the line-to-line voltage U_{LL} and the fundamental harmonic appear near the switching frequency and its multiples. Voltages U_{UN} and U_{VN} are the voltages between converter output terminals and the negative DC-link terminal N. In the figure, U_{DC} and N illustrate the potentials of the DC link. *Source:* Adapted from Mohan et al. (1995).

the frequency modulation ratio is determined as the ratio of the switching frequency and the fundamental (output) frequency.

$$m_f = \frac{f_{sw}}{f_1} \tag{7.3}$$

If $m_f < 21$, synchronous modulation is recommended to avoid subharmonic components. In synchronous modulation, the frequency of the triangular wave varies along with the reference value. Furthermore, in the so-called synchronous PWM, all derivatives (slopes) must be of opposite polarity at the common zero positions of the curves. The switching frequency is determined by the frequency of the triangular wave, and therefore it varies constantly with this frequency. To achieve the best overall result, the frequency modulation ratio is usually lowered as frequency increases (to maintain low enough switching frequencies).

Today, the possible frequency converter switching frequencies are so high that synchronous modulation is normally not required. However, the switching frequency can be kept constant resulting in a modulation ratio that is no longer an integer and changes constantly with increasing output frequency.

In addition to the frequency modulation ratio, the amplitude modulation ratio is determined by the sine-triangle comparison.

$$m_a = \frac{\hat{u}_{ref}}{\hat{u}_{triangle}} \tag{7.4}$$

For the linear modulation range, the modulation ratio $m_a \leq 1$. The peak value for the fundamental voltage between the phase U and the negative bus is determined as

$$\hat{u}_{U,N,1} = m_a \frac{U_{DC}}{2} \tag{7.5}$$

Equation (7.5) yields the effective value of the line-to-line voltage as follows.

$$U_{LL,inv} = \frac{\sqrt{3}}{2\sqrt{2}} m_a U_{DC} \approx 0.612 m_a U_{DC} \tag{7.7}$$

If the DC voltage link is supplied by a diode bridge, the mean value of the voltage of the DC link is written:

$$U_{DC} = \frac{1}{\frac{\pi}{3}} \int_{-\frac{\pi}{6}}^{\frac{\pi}{6}} \sqrt{2} U_{LL} \cos \omega t d(\omega t) = \frac{3\sqrt{2}}{\pi} U_{LL} \approx 1.35 U_{LL} \tag{7.8}$$

In a 400 V AC supply, the output will be 540 V DC and, correspondingly, in a 690 V AC supply it will be 931.5 V DC. The maximum output voltage in the linear modulation based on the sine-triangle comparison is therefore

$$U_{LL,inv} \approx 1.35 \times 0.612 U_{LL} = 0.83 U_{LL} \tag{7.9}$$

For a frequency converter operating within a 400 V grid, the output voltage in linear modulation mode is 330 V. To reach a voltage corresponding to the line voltage in sine-triangle modulation, overmodulation ($m_a > 1$) is required. This overmodulation is realized by introducing a uniform block in the middle region of the pulse pattern that forms the phase voltage, which results in a number of odd low-frequency harmonics in the output voltage. This overmodulation situation is illustrated in Figure 7.18.

A modulator based on sine-triangle comparison produces an output voltage that is directly proportional to the amplitude modulation ratio as long as m_a remains ≤ 1. However, inverter bridge output voltage is still low compared to line-to-line input voltage, and overmodulation is required for, for example, a frequency converter connected to the 400 V grid to produce a 400 V

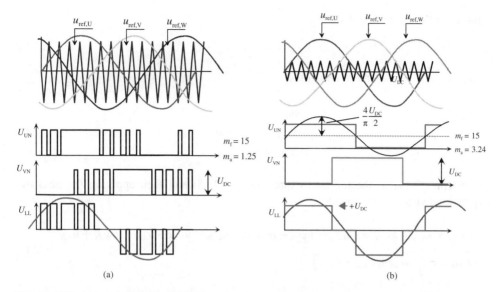

Figure 7.18 Overmodulation of a two-level inverter for two cases when applying a sine-triangle comparison. With a full square wave, the peak value of the phase voltage is $2/\pi \times U_{DC}$. A full square wave is realized at the amplitude modulation ratio $m_a = 3.24$ when the frequency modulation ratio is $m_f = 15$.

output voltage. With sufficient overmodulation, a two-level inverter produces a square wave output with a fundamental harmonic that exceeds that of the supply voltage.

For the full square wave of Figure 7.18, the maximum value $2/\pi \times U_{DC}$ is obtained (by Fourier analysis) for the fundamental harmonic of the phase voltage. Consequently, the maximum value of the effective converter RMS line-to-line voltage U_{LLconv} can be expressed as a function of the network voltage U_{LLnet}.

$$U_{LLconv} = \frac{\sqrt{3}}{\sqrt{2}} \frac{4}{\pi} \frac{U_{DC}}{2} = \frac{\sqrt{6}}{\pi} U_{DC} = \frac{3\sqrt{2}}{\pi} \frac{\sqrt{6}}{\pi} U_{LL} \approx 0.78 U_{DC} \approx 1.053 U_{LLnet} \qquad (7.10)$$

This square voltage contains harmonics of order:

$$\nu = 6n \pm 1, \quad n = 1, 2, 3, \ldots \qquad (7.11)$$

The amplitudes of the harmonics are inversely proportional to the order of the harmonic.

$$U_{LL,\nu} = \frac{0.78}{\nu} U_{DC} \qquad (7.12)$$

Referring to Figure 7.16, the voltage of a star-connected motor in these drives can be determined by first writing expressions for the phase voltages.

$$u_{U,n} = u_{U,N} - u_{n,N}, u_{V,n} = u_{V,N} - u_{n,N}, u_{W,n} = u_{W,N} - u_{n,N} \qquad (7.13)$$

And, for a three-phase system:

$$u_{U,n} + u_{V,n} + u_{W,n} = 0 \tag{7.14}$$

Equation (7.13) yields these equations

$$-u_{n,N} = u_{U,n} - u_{U,N}$$
$$-u_{n,N} = u_{V,n} - u_{V,N} \tag{7.15}$$
$$-u_{n,N} = u_{W,n} - u_{W,N}$$

Introducing (7.14) to the topmost expression of (7.15) and then applying the next two expressions yields the following.

$$-u_{n,N} = u_{U,n} - u_{U,N} = -u_{V,n} - u_{W,n} - u_{U,N} = -u_{U,N} - u_{V,N} + u_{n,N} - u_{W,N} + u_{n,N}$$
$$\Rightarrow u_{n,N} = \frac{1}{3}\left(u_{U,n} + u_{V,n} + u_{W,n}\right) \tag{7.16}$$

Plugging (7.16) into (7.13) produces (7.17).

$$u_{U,n} = \frac{2}{3}u_{U,N} - \frac{1}{3}\left(u_{V,N} + u_{W,N}\right) \tag{7.17}$$

Respective equations can also be derived for the V and phase W voltages. Figure 7.19 illustrates two different cases for the curve forms of the phase voltage of a star-connected motor. The lower illustration depicts square wave modulation according to Equation (7.12).

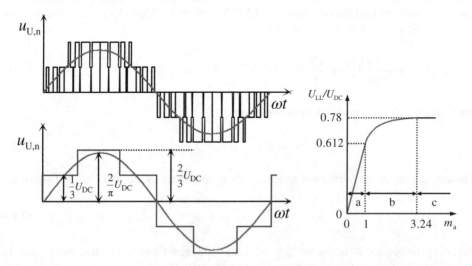

Figure 7.19   Voltage waveforms produced by the modulation in a two-level PWM frequency converter. The right-hand illustration shows the ratio of the output voltage fundamental to the DC-link voltage as a function of modulation index.

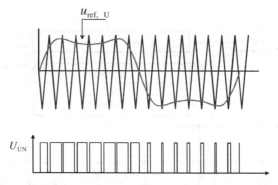

$u_{\text{ref, U}}$

U_{UN}

Figure 7.20 Modulation based on a sine-triangle comparison. The third harmonic has been included in the reference harmonic wave making it possible to reach the voltage level of the supply grid without inducing low-frequency harmonics.

Overmodulation results in a pure square wave with an effective fundamental harmonic value according to $0.78 \times 1.35\, U_{\text{LL}} = 1.053\, U_{\text{LL}}$. For a 400 V grid, the effective value of the fundamental harmonic produced by the square wave is $U_{\text{LL}} = 421$ V. The phase voltage is 243 V. Naturally, any voltage drops inherent to the drive reduce the magnitude of its output.

Since in the sine-triangle comparison the upper limit of the linear range is already at 330 V, different variations of the system have been developed to avoid the low-frequency harmonics in the overmodulation. By adding enough third harmonic to the reference value curves, it is possible to "extend" the range of linear modulation to approach square wave modulation.

Figure 7.20 shows sine-triangle modulation modified by the third harmonic. The harmonic can be applied without affecting supply, because the phases match and cannot produce currents in the motor windings when the star point of the motor is disconnected. Since this makes it possible to avoid the long uniform voltage block in the voltage waveform caused by the sine-triangle comparison-based overmodulation of the system, the low-frequency harmonics harmful to the operation of the motor can also be avoided, and 400 V output voltage can be achieved.

Currently, three-level VSIs are typically implemented using GTO or IGTC switches. An IGB transistor is usually used as the switch of a two-level inverter. However, this is changing. High-voltage IGBTs are now also being used in medium-voltage converters, and SiC metal oxide semiconductor field-effect transistors (MOSFETs) are being used in low-voltage applications. Both can potentially outperform the existing market-dominant approaches.

The circuit diagram of a three-level NPC (neutral point clamped) inverter is shown in Figure 7.21. There are three input terminals in an NPC inverter, and the connection resembles the series connection of a pair of two-level inverters. The difference is the clamping diodes used in the NPC inverter. DC-link voltage is divided into two parts using series-connected capacitors. Successful voltage division must be controlled in these inverters. GTO thyristors or IGCT switches are usually used as the switching components. Low-voltage three-level inverters equipped with IGBT switches also can be found.

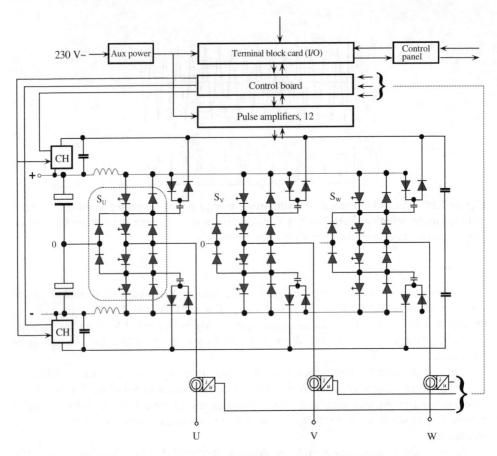

Figure 7.21 Schematic circuit diagram of an NPC inverter (SAMI Megastar) according to ABB. The switching component is a GTO thyristor. The two choppers (CH) restore the energy of the protection circuits to the DC link. *Source:* Reproduced with permission from ABB.

The rated voltages for the NPC inverters are typically 2400 V and 3300 V, and the power range is 1 to 30 MW.

Figure 7.22 illustrates the implementation of a single phase for a three-level inverter using GTO thyristors. It consists of four legs similar to the ones in the two-level inverter, as well as two clamping diodes V9 and V10.

The operation of a three-level, three-phase inverter can be demonstrated by the change-over switch illustrated in Figure 7.22. By appropriately controlling the switches (S_U, S_V, S_W in Figure 7.21), three-phase voltage is applied to the motor terminals. Figure 7.23 shows the voltage patterns produced by the NPC inverter given pulse number 1 (7.23a) and pulse number 3 (7.23b).

In Figure 7.22, the load for the inverter can come from a three-phase motor with a single phase, such as phase U, connected to the positive (+) terminal of the DC voltage source. When thyristors V1 and V2 are switched on, motor current flows away from the inverter through thyristors V1 and V2 and in the direction of the inverter through diodes V5 and V6 (Figure 7.24a).

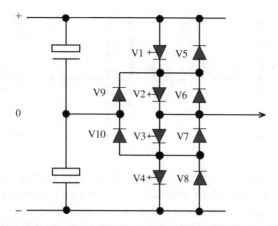

Figure 7.22 Implementation of a three-level NPC inverter by GTO thyristors or IGC thyristors, V1–V4 and fast switch diodes, V5–V10.

Then, when thyristor V1 is switched off and thyristor V3 switched on, phase U becomes connected to the centre point of the DC voltage source, and current flows to the motor through diode V9 and thyristor V2. Alternatively, if the current is flowing to the opposite direction, the current flows through thyristor V3 and diode V10 to the centre point (Figure 7.24b). Phase U is connected to the negative terminal (−) of the DC voltage source, when thyristor V2 is switched off and thyristor V4 is switched on, in which case the current flows to the direction of the inverter through thyristors V3 and V4. For opposite current flow, it passes through diodes V7 and V8 (Figure 7.24c).

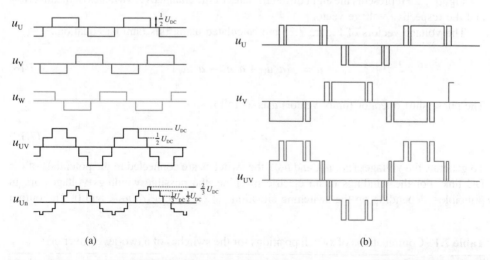

(a)   (b)

Figure 7.23 (a) The potentials U_U, U_V, and U_W applied to the motor phases by the change-over switches plus the line-to-line voltage U_{UV} and motor phase voltage U_{Un} shown with pulse number 1 and (b) the potentials applied to the motor phases by the change-over switches plus the line-to-line voltage with pulse number 3.

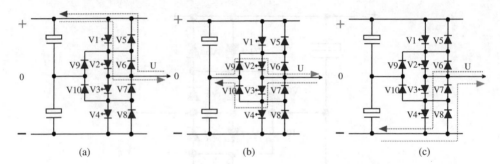

(a)   (b)   (c)

Figure 7.24 (a) The current directions of phase U of the NPC-PWM inverter when thyristors V1 and V2 or diodes V5 and V6 are switched on, (b) the current directions of phase U of the NPC-PWM inverter when diode V9 and thyristor V2 or diode V10 and thyristor V3 are switched on, and (c) the current directions of phase U of the NPC-PWM inverter when thyristors V3 and V4 or diodes V7 and V8 are switched on.

7.4.1 Voltage vectors as the combined effect of the inverter and the winding

Voltage vectors in the two-level inverter

Thus far, the inverter bridge has been considered a PWM voltage source; however, considered together with the winding, the inverter can also be regarded as the voltage vector generator. Voltage vectors generated by the inverter and the motor can be defined using a state vector representation. In a two-level, three-phase inverter, there are $2^3 = 8$ combinations of switch positions (S_U, S_V, S_W), which are listed in Table 7.1. There are six actual voltage vectors and two zero-sequence vectors.

Figure 7.25 represents the eight different states of the changeover switches of the inverters and the respective voltage vectors.

The voltage vectors of Figure 7.25 are calculated using this familiar equation.

$$u = \frac{2}{3}\left(a^0 u_U + a^1 u_V + a^2 u_W\right) \tag{7.18}$$

The phase shift operator (unity vector) a is as follows.

$$a = e^{j\frac{2\pi}{3}} \tag{7.19}$$

To generate the voltages (u_U, u_V, and u_W), the switches are connected to the potentials of the DC link. For the windings of an electric machine, the possible windings voltages are in principle—depending on the switching situation—$\pm 2/3\ u_{DC}$, $\pm 1/3\ u_{DC}$, and 0. The output

Table 7.1 Combinations of switch positions for the switches of a two-level inverter

Switch	Combinations of Switch Positions							
S_U	+	+	−	−	−	+	+	−
S_V	+	+	+	+	−	−	−	−
S_W	+	−	−	+	+	+	−	−

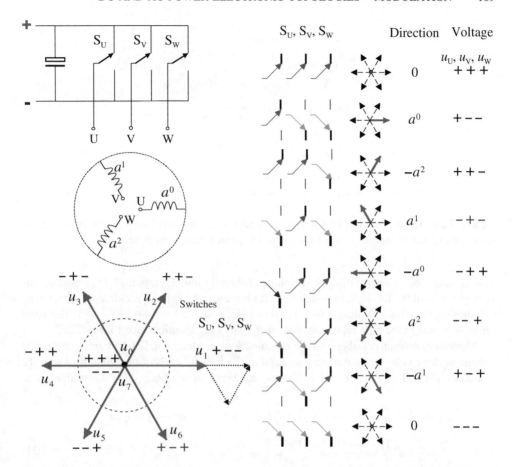

Figure 7.25 The switching alternatives of a three-phase inverter with a VSI and the directions of the possible output voltage vectors, which are the positive and negative directions of the magnetic axes of the phase windings. The zero values of the output voltage have no direction.

voltage vector values become:

$$u_0 = 0,$$

$$u_1 = \frac{2}{3}u_{DC}a^0,$$

$$u_2 = -\frac{2}{3}u_{DC}a^2,$$

$$u_3 = \frac{2}{3}u_{DC}a^1,$$

$$u_4 = -\frac{2}{3}u_{DC}a^0,$$

$$u_5 = \frac{2}{3}u_{DC}a^2,$$

$$u_6 = -\frac{2}{3}u_{DC}a^1,$$

$$u_7 = 0.$$

(7.20)

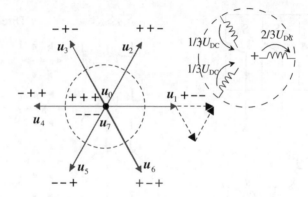

Figure 7.26 Representation of the voltage vector of a two-level inverter by applying the directions of the magnetic axes of the stator of a three-phase machine.

The instantaneous switch voltages are plugged directly into Equation (7.18) based on the voltage levels of the DC link, in which case, for instance, the voltage vector u_1 is obtained by substituting the relative voltages $(+1, 0, 0)$ or $(+\frac{1}{2}, -\frac{1}{2}, -\frac{1}{2})$ or $(+\frac{2}{3}, -\frac{1}{3}, -\frac{1}{3})$. The latter series is probably the best representation of the physical situation. See Figure 7.26.

Therefore, several possible voltage combinations produce parallel voltage vectors. The voltage vector can be thought to be generated as the forward ends of the windings touch the potentials $+\frac{1}{2}$ and $-\frac{1}{2}$ or $+1$ and 0. This assumption can be checked by determining u_1.

$$
\begin{aligned}
u_1 &= \frac{2}{3}\left(a^0 u_A + a^1 u_B + a^2 u_C\right) = \frac{2}{3}\left(a^0\frac{2}{3} + a^1\frac{-1}{3} + a^2\frac{-1}{3}\right) = \frac{2}{3}\left(a^0\right) \\
&= \frac{2}{3}\left(a^0\frac{1}{2} + a^1\frac{-1}{2} + a^2\frac{-1}{2}\right) = \frac{2}{3}\left(a^0\right) \\
&= \frac{2}{3}\left(a^0 1 + a^1 0 + a^2 0\right) = \frac{2}{3}\left(a^0\right).
\end{aligned}
\tag{7.21}
$$

Voltage vectors in a three-level NPC inverter

Figure 7.27 illustrates a model for a changeover switch of a three-level, three-phase inverter and the possible voltage levels $-$, 0, $+$ of each switch plus the line-to-line voltage U_{UV}.

Since the output voltage in the NPC inverter can be one of three different values, its switches have $3^3 = 27$ combinations of switch positions (Tables 7.2 and 7.3). These combinations generate 18 different vector directions and zero (Figure 7.28). Vector definition is carried out as it was for the two-level inverter.

NPC inverters are best adapted to high-power electrical drives. Their power range reaches 30 MVA, which covers most needs. The most typical applications include high-power pumps, blowers, compressors, propeller drives, locomotive drives, rolling mill drives, high-power hoists and cranes, and induction heaters. NPC inverters are also used for high-speed induction motors in, for instance, high-speed trains and metal machining tools. A frequency converter with an NPC-PWM inverter section is a real alternative for high-power electrical drives that require good control. An NPC inverter is more versatile than a traditional two-level inverter,

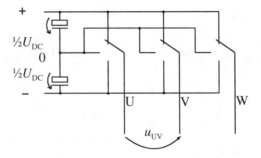

Figure 7.27 Topology of a three-phase three-level inverter and the voltage division between the DC capacitors in an ideal situation.

which has reduced harmonic content. Furthermore, the voltage stresses on the main components in the NPC inverter is one-half that of the voltage stresses on the two-level inverter components. Therefore, the NPC inverter can accommodate twice the voltage of the two-level inverter without connecting switches in series.

If three-level technology is applied to 400 V inverters, the inverter bridge can be constructed using 600 V IGBT modules. In a common 400 V two-level inverter, to ensure sufficient voltage headroom, 1200 V IGBT switches must be used. For a 690 V two-level inverter, 1700 V switches are required. The 600 V IGBT switches have considerably lower losses and are faster than higher-voltage switches.

Therefore, in addition to producing a superior curve form, three-level inverters can achieve efficiencies equal to those of two-level inverters even though three-level inverter circuits comprise more components. For motor function, the curve forms produced by the three-level inverter are much better than the corresponding two-level inverter curve forms.

Table 7.2 Combinations of switch positions for a three-level NPC inverter

	Combinations of Switch Positions																										
U	+	+	+	+	+	+	+	+	+	0	0	0	0	0	0	0	0	0	−	−	−	−	−	−	−	−	−
V	+	+	+	0	0	0	−	−	−	+	+	+	0	0	0	−	−	−	+	+	+	0	0	0	−	−	−
W	+	0	−	+	0	−	+	0	−	+	0	−	+	0	−	+	0	−	+	0	−	+	0	−	+	0	−

Table 7.3 Alternatives for the generation of the line-to-line voltage U_{UV} of the three-level NPC inverter: five different U_{UV} voltage levels are possible

U_{UV}	Combinations of Switch Positions		
$+U_{DC}$	+−		
$+\frac{1}{2}U_{DC}$	+0	0−	
0	++	00	−−
$-\frac{1}{2}U_{DC}$	0+	−0	
$-U_{DC}$	−+		

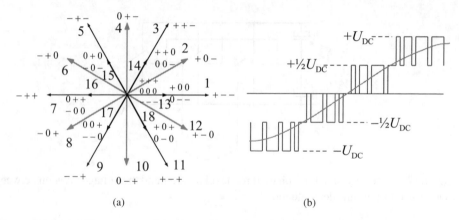

(a)   (b)

Figure 7.28   (a) The vectors of the three-phase, three-level inverter and (b) an example of the time function form of the line-to-line voltage. The desired vectors are obtained by plugging individual value sets (e.g. "+00") into the voltage vector equation.

7.4.2   Space vector modulation for two-level voltage source converters

Previously, the inverter bridge and motor windings in combination were shown to generate voltage vectors. These vectors can be applied directly to implement space vector modulation (SVM), which is another commonly used voltage control method. The reference vector of the modulator for the voltage is written as follows.

$$\boldsymbol{u}_{\mathrm{ref}} = u_{\mathrm{ref}} e^{j\theta} = u_{\alpha,\mathrm{ref}} + j u_{\beta,\mathrm{ref}} \tag{7.22}$$

This voltage can be constructed using four switching states in a sequence. In the case of Figure 7.29, to produce $\boldsymbol{u}_{\mathrm{ref}}$ as a time-weighted average, the vectors $\boldsymbol{u}_0$, $\boldsymbol{u}_1$, $\boldsymbol{u}_2$, and $\boldsymbol{u}_7$ must be applied in sequence. First $\boldsymbol{u}_0$ (000) is selected, then $\boldsymbol{u}_1$ (001) then $\boldsymbol{u}_2$ (011), and finally $\boldsymbol{u}_7$ (111). See Figure 7.29. The three-phase changeover-switches change state one at a time.

In Figure 7.29, the length of the active voltage vector is $2/3\,U_{\mathrm{DC}}$. The maximum length for the amplitude of the sinusoidal voltage is $(\sqrt{3}/2) \times |\boldsymbol{u}_{1\ldots6}| = (1/\sqrt{3})U_{\mathrm{DC}} \approx 0.866 \times |\boldsymbol{u}_{1\ldots6}| \approx 0.577 U_{\mathrm{DC}}$. This yields, for instance in a 400 V system, $\hat{u} = \frac{\sqrt{3}}{2} \times \frac{2}{3} \times \frac{3\sqrt{2}}{\pi} U_{\mathrm{LL}} = \frac{\sqrt{6}}{\pi} U_{\mathrm{LL}} \approx 0.78 U_{\mathrm{LL}} = 312$ V. The value corresponds to an $\approx$220 V RMS voltage, which is thus 95.4% of the line voltage (230 V). With diode rectification, it is not possible to reach the peak line-to-line voltage of $\sqrt{2} U_{\mathrm{LL}}$. The value reached is instead $\frac{3\sqrt{2}}{\pi} U_{\mathrm{LL}}$, which explains the drop from 100%. The ratio of these values is $3/\pi = 0.95$.

The main operating principle of the SVM in the linear region is that the time weighted average voltage of four voltage vectors (two active vectors and two zero vectors $\boldsymbol{u}_0$ and $\boldsymbol{u}_7$) is used to create the desired output during a switching period T_{sw}. In the case of Figure 7.29, to produce the reference voltage $\boldsymbol{u}_{\mathrm{ref}}$, $\boldsymbol{u}_0$ at t_0 is used, then $\boldsymbol{u}_1$ at t_1, then $\boldsymbol{u}_2$ at t_2, and finally, $\boldsymbol{u}_7$ at t_7. Then, the period is repeated in reverse order to realize the reference value $\boldsymbol{u}_{\mathrm{ref}}$. The modulator average time-weighted output during T_{sw} is $\boldsymbol{u}_{\mathrm{out}} = \boldsymbol{u}_{\mathrm{ref}} = 2(t_0 \boldsymbol{u}_0 + t_1 \boldsymbol{u}_1 + t_2 \boldsymbol{u}_2 + t_7 \boldsymbol{u}_7)/T_{\mathrm{sw}}$.

The operating range for SVM is divided into the following three sections.

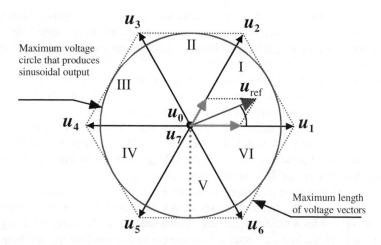

Figure 7.29 Voltage vectors u_0–u_7 of the two-level VSI. The figure illustrates the active and zero vectors for a static frame of reference. The reference vector u_{ref} is shown. Also shown is how the reference vector is the sum of the active vectors for a given time instant. The outer hexagon indicates the maximum length of the time-weighted average voltage vector at each point. The circle inside the hexagon represents the locus plotted by the point of the voltage vector when producing maximum sinusoidal voltage.

Linear modulation - In this range, the phase angle θ of the voltage vector in steady state travels at a constant speed ω. $u_{ref} = u_{ref}e^{j\theta} = u_{ref}e^{j\omega t}$, and the length of the voltage vector is $(1/\sqrt{3})U_{DC}$ at maximum *according to Figure* 7.29. When operating in the linear modulation range, zero vectors are applied as described earlier. The produced motor current is almost

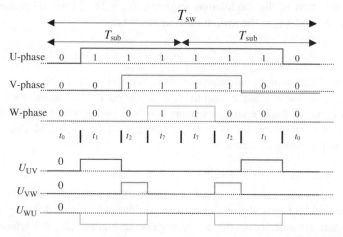

Figure 7.30 Vector representation for the switching sequence of the SVM and the respective switching durations in sector I. Voltage vectors u_0, u_1, u_2, and u_7 are applied so that no two change-over switches are ever switched simultaneously. The figure shows that phases U, V, and W produce switching states, i.e., voltage vectors according to Figure 7.29. The figure also illustrates how the pulses of the line-to-line voltage U_{LL} are produced from the switching references (zero level is indicated with a dotted line).

sinusoidal, because there are no low order harmonics in the voltage output. Only switching frequency-induced harmonics can be seen in the motor currents. In converters targeting high dynamic performance, operation is often kept within the linear modulation area. This ensures at least some voltage reserve for the drive.

Overmodulation range I - In this range, the phase angle θ of the voltage vector in steady state also travels at a constant speed ω. $u_{ref} = u_{ref}e^{j\theta} = u_{ref}e^{j\omega t}$. However, the amplitude varies, because the adjacent active voltage vectors are too short to create a long enough voltage vector in the centre portion of the hexagonal side. Zero vectors are no longer used in the area where the converter runs out of voltage reserve (the centre portion of the hexagonal side). More low-order harmonics are seen in the voltage output, which distorts motor current.

Overmodulation range II - In this range, the state corresponding to the full over-modulation of the sine-triangle comparison, the square wave, is gradually reached. The process progresses until the full voltage vector is held locked (the vector remains fixed) at the hexagon corners for set time durations. As the lock-in angle increases along with the modulation index M, the voltage vector is kept nonmodulated until the angular frequency has produced an angle corresponding to half the arc (hexagon side). Then, it is changed over to the next voltage vector. No zero vectors are used. Each voltage vector jumps nonmodulated from one sector edge to another. This corresponds to full overmodulation of the sine-triangle comparison, and the phase voltage of the motor appears as a square wave. Square wave output guarantees the highest possible voltage output for the inverter. Since it was derived above the maximum output, the line-to-line fundamental voltage of the square wave inverter is 1.053 U_{LL}, which results in a 421 V voltage for a 400 V AC supply.

In the linear modulation range, the length of the reference voltage vector $\hat{u}_{ref}$ depends on the induction law in steady operation and takes a value $\omega\hat{\psi}_s$. In dynamic states, however, rapid changes may occur. The angle θ is set to establish the desired rotation speed of the voltage vector. In steady state, $\theta = \omega t + \theta_0$. Here, θ_0 represents the initial angle of modulation.

The time duration of the modulation sequence $T_{sw} = 2T_{sub}$ (two subsequences) can be determined as dependent on the switching frequency f_{sw}.

$$T_{sw} = 2T_{sub} = \frac{1}{f_{sw}} \tag{7.23}$$

Here, T_{sub} is the duration of the subsequence of the modulation. The voltage vector, according to the sampling of the reference vector; is constructed by the space vectors illustrated previously in Figure 7.29. The complex plane is subdivided according to the figure into six sectors of equal size, the active vectors acting as the sides of the sectors. Modulation in sector I is based on the following equation.

$$u_{ref}T_{sub} = t_1 u_1 + t_2 u_2 + t_0 u_0 + t_7 u_7 \quad T_{sub} = t_1 + t_2 + t_0 + t_7 \tag{7.24}$$

The construction of the reference vector by modulation requires two active vectors and possibly corresponding zero vectors. In each sector, the two vectors defining the sector and both zero vectors are selected as the active vectors. In Figure 7.29, the reference vector is generated in sector I by selecting the active vectors u_1 and u_2 and both the zero vectors u_0 and u_7. Switching durations t_1 and t_2 are calculated for the selected active vectors, the on-duration t_1 being the switching duration of the vector on the leading edge and t_2 being the on-duration on the trailing edge.

The formulas for the switching durations for active vectors are given in Table 7.4. For instance, in sector I, the on-duration t_1 can be calculated by the sine rule from the triangle defined by the voltage vectors u_1 and u_2.

Table 7.4 Switching durations of the active voltage vectors in the SVM

Sector	Location Angle θ of the Reference Vector	Active Voltage Vectors Used in Modulation	Switching Durations of the Active Voltage Vectors
I	$0 \leq \theta < \dfrac{\pi}{3}$	u_1 u_2	$t_1 = \dfrac{\sqrt{3}}{2} MT_{\text{sub}} \sin\left(\dfrac{\pi}{3} - \theta\right)$ $t_2 = \dfrac{\sqrt{3}}{2} MT_{\text{sub}} \sin(\theta)$
II	$\dfrac{\pi}{3} \leq \theta < \dfrac{2\pi}{3}$	u_2 u_3	$t_1 = \dfrac{\sqrt{3}}{2} MT_{\text{sub}} \sin\left(\dfrac{2\pi}{3} - \theta\right)$ $t_2 = \dfrac{\sqrt{3}}{2} MT_{\text{sub}} \sin\left(\theta - \dfrac{\pi}{3}\right)$
III	$\dfrac{2\pi}{3} \leq \theta < \pi$	u_3 u_4	$t_1 = \dfrac{\sqrt{3}}{2} MT_{\text{sub}} \sin(\pi - \theta)$ $t_2 = \dfrac{\sqrt{3}}{2} MT_{\text{sub}} \sin\left(\theta - \dfrac{2\pi}{3}\right)$
IV	$\pi \leq \theta < \dfrac{4\pi}{3}$	u_4 u_5	$t_1 = \dfrac{\sqrt{3}}{2} MT_{\text{sub}} \sin\left(\dfrac{4\pi}{3} - \theta\right)$ $t_2 = \dfrac{\sqrt{3}}{2} MT_{\text{sub}} \sin(\theta - \pi)$
V	$\dfrac{4\pi}{3} \leq \theta < \dfrac{5\pi}{3}$	u_5 u_6	$t_1 = \dfrac{\sqrt{3}}{2} MT_{\text{sub}} \sin\left(\dfrac{5\pi}{3} - \theta\right)$ $t_2 = \dfrac{\sqrt{3}}{2} MT_{\text{sub}} \sin\left(\theta - \dfrac{4\pi}{3}\right)$
VI	$\dfrac{5\pi}{3} \leq \theta < 2\pi$	u_6 u_1	$t_1 = \dfrac{\sqrt{3}}{2} MT_{\text{sub}} \sin(2\pi - \theta)$ $t_2 = \dfrac{\sqrt{3}}{2} MT_{\text{sub}} \sin\left(\theta - \dfrac{5\pi}{3}\right)$

The symbol M in the definitions denotes the modulation index determined, for instance, by Holtz (1994), which is different from the modulation index based on the sine-triangle comparison.

$$M = \frac{\hat{u}_{\text{ref}}}{\hat{u}_{6p,1}} = \frac{\hat{u}_{\text{ref}}}{\dfrac{2}{\pi} U_{\text{DC}}}; \quad \hat{u}_{6p,1} = \frac{2}{\pi} U_{\text{DC}} \tag{7.25}$$

$\hat{u}_{\text{ref}}$ is the length of the reference vector (i.e., the peak value of the respective phase voltage curve) and U_{DC} is the voltage of the intermediate DC link. M is, thus, the ratio of the peak voltage to the peak value of the fundamental harmonic of the phase voltage obtained by the six-pulse modulation.

When the modulation index reaches the value $M = 1$, only active vectors are used, and the time functions of the voltage are square waves. This is the operating point that produces the maximum voltage of overmodulation range II.

In the linear modulation range, the switching periods of the zero vectors are determined by the on-durations of the active vectors.

$$t_0 + t_1 = T_{\text{sub}} - t_2 - t_7 \Leftrightarrow t_0 = t_7 = \frac{1}{2}(T_{\text{sub}} - t_1 - t_2) \tag{7.26}$$

Aggregating all the information gathered on the SVM thus far, Figure 7.29 can be represented as a diagram that connects the switching durations to the voltage vectors. The diagram illustrates the principle of SVM. Figure 7.30 illustrates with a single sequence of the SVM for the reference vector of Figure 7.29.

Figure 7.31 depicts how the on-durations change as the reference vector u_{ref} travels in sector I toward the voltage vector u_2. The on-durations change respectively in the other sectors as well.

In the linear modulation region, the maximum output voltage of a two-level converter is $\hat{u}_{\text{ref}} = (1/\sqrt{3})U_{\text{DC}} \approx 0.577U_{\text{DC}} \approx 0.866 \times |u_{1...6}|$. Consequently, the modulation index value is:

$$M = \frac{\hat{u}_{\text{ref}}}{\frac{2}{\pi}U_{\text{DC}}} = \frac{\frac{1}{\sqrt{3}}U_{\text{DC}}}{\frac{2}{\pi}U_{\text{DC}}} = \frac{\pi}{2\sqrt{3}} = 0.907 \tag{7.27}$$

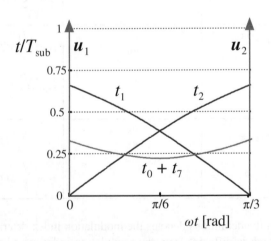

Figure 7.31 Behaviours during switching. The figure abscissa depicts the angle of travel for the reference vector u_{ref} and the active voltage vectors u_1 and u_2 of sector I at angles 0 and $\pi/3$, respectively. The times t_1, t_2, and $t_0 + t_7$ are valid for modulation index $M = 0.65$. The figure also shows the relative lengths of the voltage vectors u_1 and u_2. The voltage vector length is $2/3 \times U_{\text{DC}}$. Correspondingly, the full square wave voltage has amplitude $2/\pi \times U_{\text{DC}}$ for the fundamental harmonic of the voltage in the output. Their ratio is 1.047. *Source:* Adapted from Saren (2005). Reproduced with permission from H. Saren.

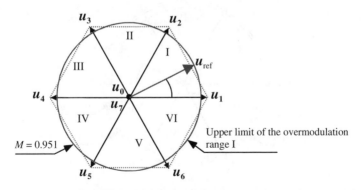

$$0.907 < M < 0.951$$

Figure 7.32 The upper limit of the overmodulation range I where $M = 0.951$.

Overmodulation range I corresponds to values of the modulation index M exceeding 0.907. In this overmodulation range, the angular speed of the reference value of the voltage vector is kept constant, but the amplitude crosses the edges of the hexagon as shown in Figure 7.32. With respect to modulation index values, overmodulation range I is defined as:

$$0.907 < M < 0.951 \tag{7.28}$$

In overmodulation range I, the locus plotted by the point of the reference vector is partly inside and partly outside the hexagon. The upper limit of overmodulation range I is reached when the areas inside and outside the hexagon are equal. According to Holtz (1994), the equations of Table 7.4 are used for the switching durations of the voltage vectors when the point of the reference vector is inside the hexagon. When the point of the reference vector is outside the hexagon, the following equations are used.

$$t_1 = \frac{T_{sub}}{3} \frac{\sqrt{3}\cos\theta - \sin\theta}{\sqrt{3}\cos\theta + \sin\theta}$$

$$t_2 = \frac{T_{sub}}{3} - t_1 \tag{7.29}$$

$$t_0 = 0$$

Bolognani and Zigliotto (1999) derived, analytically, the following equation for the maximum voltage of the fundamental harmonic in overmodulation range I.

$$\hat{u}_1 = \frac{\sqrt{3}\ln 3}{\pi} U_{DC} = 0.606 U_{DC} \tag{7.30}$$

Consequently, the modulation index becomes:

$$M = \frac{\sqrt{3}\ln 3}{2} \approx 0.951 \tag{7.31}$$

Overmodulation range I is followed by overmodulation range II ($M > 0.951$). In this range, the voltage vectors are locked at the hexagon corners for a gradually increasing time duration so

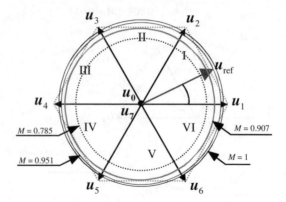

Figure 7.33 Limits of different modulation ranges. $M=0.785$ shows the limit at which the modulation index based on the sine-triangle comparison reaches the value $m=1$, and $M=0.907$ shows the maximum limit for the sinusoidal output voltage of the SVM. $M=0.951$ shows the highest modulation index for overmodulation range I of the SVM. $M=1$ produces the maximum voltage for overmodulation range II. In the modulation based on a sine-triangle comparison, this corresponds to full overmodulation. For example, if $m_f=15$, the value $m_a=3.24$ results in a full square wave.

the voltage vector is not turned at all. In overmodulation range I, the voltage reference is kept continuous, however, in overmodulation range II, the voltage reference gets discrete values. In overmodulation range II, the voltage vectors are kept on for the time duration of the lock-in angle. As the lock-in angle increases to a half of the sector width (hexagon side), that is, to a value $\pi/6$, hexagonal modulation is reached, and $M=1$. Figure 7.33 illustrates the limits of different modulation ranges. With $M=1$, the switching frequency is only determined by the output frequency of the converter and is, therefore, much lower than in the linear range and overmodulation range I, where PWM is used.

7.4.3   Dimensioning of a DC link capacitor

The size of the DC link capacitor is important to the operation of a voltage source inverter. Usually, the capacitance of the intermediate DC link of the VSI is dimensioned to be approximately $20\,\mu F$ per RMS ampere of frequency converter–rated current. Figure 7.34 illustrates the typical curve form of the line current of a frequency converter equipped with

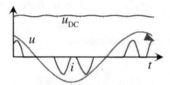

Figure 7.34 The line current and the voltage of the DC link of a frequency converter equipped with a three-phase diode bridge. The current is typical of the charging current for a large loaded capacitor.

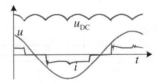

Figure 7.35 The line current of a frequency converter with a three-phase diode bridge and the DC link voltage when the DC link capacitor is reduced in scale to 1% of the original size. In practice, the DC voltage follows the curve form of the diode bridge. *Source:* Adapted from Saren (2005). Reproduced with permission from H. Saren.

such a capacitor. The voltage of the intermediate DC link remains relatively constant irrespective of the pulsating charging current.

Progressively, efforts have been made to reduce DC link dimensioning. In one extreme case, Hannu Sarén (2005) introduced in his dissertation a technique in which the capacitance of the intermediate DC link is reduced in scale by two orders of magnitude. In the reported test setup, the DC link of an inverter with a 10 A rated current uses a 2 μF capacitor made of metallized polypropylene film (MPPF) instead of the usual 200 μF electrolytic capacitor. Figure 7.35 illustrates line current for a frequency converter using a 1% capacitor.

As shown in the figure, reducing the size of the DC capacitor results in some switching-frequency interferences propagating to the grid. There is some ringing in the network current of Figure 7.35 caused by the switching of the inverter. Furthermore, special attention must be paid to the magnitude of the DC link voltage in the modulation. A 1% DC link capacitance compared to the traditional dimensioning cannot withstand the backfeed into the DC link of energy stored in the magnetic field.

7.5   The matrix converter

Matrix converter technology provides direct AC-to-AC power conversion. The operating principle of the matrix converter resembles that of the CCV. Constant-frequency and constant-voltage AC is converted directly without a DC link into AC of varying frequency and voltage to the output side of the transformer. However, the CCV requires twice as many components (36 thyristors) as the matrix converter. The CCV also requires a supply transformer for each phase, and therefore it is notably larger and heavier than the matrix converter. Furthermore, the frequency of the CCV is, in practice, limited to the half of the frequency of the supply grid at maximum, whereas the matrix converter does not have such a constraint. The power factor of the CCV on the grid side is always inductive; the power factor of the matrix converter instead can always to be adjusted.

The notable advantage of the matrix converter is that it avoids the constraints of the CCV technology without adding the DC link. The matrix converter consists of a number of bidirectional switches, arranged so that any converter input phase can be connected to any output phase. The curve forms for the output voltage of the matrix converter are constructed by switching one supply phase at a time to the load for a certain duration. Usually this duration is equal for all outgoing phases. Figure 7.36 illustrates a possible low-voltage matrix converter topology and a typical curve form for the output voltage when applying a relatively low switching frequency. The output voltage is generated by appropriately

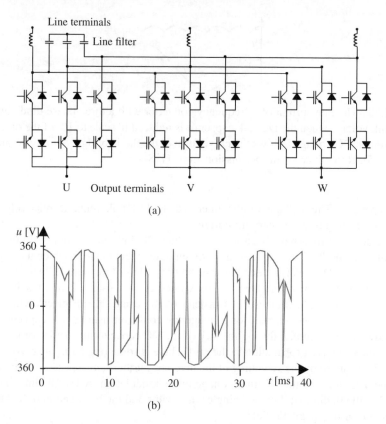

(a)

(b)

Figure 7.36   (a) Example of matrix converter topology and (b) example of the output voltage of a matrix converter.

controlling the three supplying phases at the switching frequency to the output. Often, a sufficiently large *LC* filter is connected to the input side to filter the high-frequency disturbances in the input current.

The most common matrix converters have three-phase topologies on both the input and output sides, that is, there are nine (3×3) bidirectional switches. A circuit of this kind can be used to directly replace an inverter that supplies a standard three-phase induction motor. One advantage, when compared with a traditional inverter, is the capability of supplying a load directly from the supply grid without any adjustment to voltage amplitude or frequency. Naturally, this is possible only when the load has as many phases as the supply network, that is, one incoming phase is connected to one outgoing phase and the other switches are kept off.

Matrix converter operation requires a controllable bidirectional switch that allows bidirectional current. At present, the commutation cell must be constructed of discrete components. The most common controllable switch used in the construction of the commutation cell is an IGBT. Other types are also used.

A commutation cell composed of a diode bridge comprises a single-phase diode bridge with an IGBT in the middle. See Figure 7.37. An advantage of this technology is that current flows in both directions through the same switch (IGTB), and therefore only one gate driver is required for each commutation cell. A disadvantage is that current must flow in both directions through

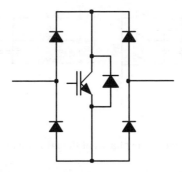

Figure 7.37 A bidirectional commutation cell constructed with a diode bridge. In principle, the freewheeling diode in parallel with the IGBT is not necessary.

three different components, which results in extra losses. Total losses can become quite large. The direction of current moving through the commutation cell cannot be controlled.

A commutation cell implemented by a common-emitter configuration consists of two antiparallel IGBTs and two diodes. See Figure 7.38. The IGBTs control current direction, and the diodes prevent current flow in the wrong direction. There are certain advantages to this kind of commutation cell over the previous example, the most important being that current direction can be freely controlled. Another advantage is that current flows only through two components, so losses are smaller than in the previous case. A disadvantage is that each commutation cell requires a power supply of its own to control the gate drivers, although both IGBT gates of each single cell can be supplied from a single supply. For a 3p → 3p (three-phase–to–three-phase) converter, nine different power supplies are required. This configuration of the matrix converter is applied chiefly in high-power drives.

A common-collector back-to-back commutation cell is similar to the previous example but puts the collectors at equal potential instead of the emitters. See Figure 7.39. Losses are equivalent to the common-emitter arrangement; however, gate control can be implemented so that only six separate power supplies are required in a 3p → 3p converter. One gate controller is required for each emitter potential, and therefore in a common-collector configuration, there

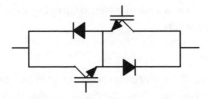

Figure 7.38 Common-emitter back-to-back commutation cell.

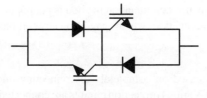

Figure 7.39 Common-collector back-to-back commutation cell.

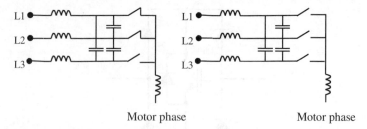

Motor phase   Motor phase

Figure 7.40   Short circuit between the phases of the grid or a circuit open to the motor phase.

are only six emitter potentials (three input and three output phases). Correspondingly, in a common-emitter configuration, the number of emitter potentials is nine (nodes). However, the mutual inductance of the commutation cells becomes so high in this configuration that it may hamper the operation of the converter. Therefore, to implement a bidirectional commutation cell, the common-emitter configuration is usually preferred.

Reliable current commutation is far more difficult to achieve with a matrix converter than with, for example, a traditional inverter, because there is no natural path between the source and the load. Each path includes a commutation cell, and current must flow through its controllable switch. The commutation must be controlled constantly and done so according to the following two principles. First, two incoming phases may not be connected simultaneously to a single outgoing phase, since this would result in a short circuit between the phases. This is illustrated by Figure 7.40 using a simplified (3p → 1p) converter. Second, none of the outgoing phases of the converter can be disconnected from the grid by opening the commutation cells. This would cut the current flow from that output phase and induce high instantaneous overvoltage due to the inductive load current that could destroy the converter. This situation also is illustrated in Figure 7.40.

These two conditions are controversial, since the semiconductor switches have a finite response time due to propagation delays and finite switching duration. Each has been a significant obstacle in matrix converter development.

The two simplest commutation methods are referred to as overlapping commutation and dead-time commutation. These methods break the basic rules presented here. Therefore, additional components must be added to avoid destroying overlapping commutation and dead-time commutation converters.

Overlapping commutation

In overlapping commutation, the connecting phase is switched to the load just before the disconnecting phase is switched out. This introduces a momentary short circuit between the two phases. Extra inductance is added to the circuit to inhibit the short circuit current transient and keep it from becoming too large before the disconnecting phase switches out. Because the induction coils are bulky and expensive, this commutation method is not popular. Moreover, longer time is required for overlapping commutation, which may also cause problems for the control.

Dead-time commutation

Dead-time commutation is based on completely disconnecting one of the outgoing (output) phases from the supply grid. Various protection circuits are connected between the commutation cells to ensure continuous current flow to the load. Because of the additional losses incurred in

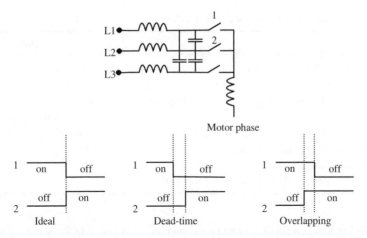

Figure 7.41   3p → 1p matrix converter and the scheduling diagrams for different commutation methods.

the protection circuits, dead-time commutation is less efficient. Bidirectional power flow further complicates the design of the protection circuits, and therefore this method is not very popular. Figure 7.41 illustrates a 3p → 1p converter and the scheduling diagram for the switches of the supply phase for different commutation methods.

Figure 7.42 depicts a two-phase to one-phase matrix converter, implemented by the so-called semi-soft (semi-natural) commutation method.

The semi-soft method can be explained by stepping through the commutation sequence. To begin, assume that the load-current direction is as depicted in the figure. Further, the phase 1 switches a and b are receiving the ON control signal, and the phase 2 switches a and b are not. According to the circuit, bidirectional power flow through the upper commutation cell is enabled, but phase 1 current is passing only through switch 1a.

To switch out phase 1 and switch in phase 2, the controller first terminates the ON signal to switch 1b (which is not flowing current), and then sends an ON control signal to switch 2a. At that moment, phase current 1 and phase current 2 are flowing simultaneously through switches 1a and 2a respectively. The controller immediately terminates the ON signal to switch 1a, which stops the phase 1 current. After sending the OFF signal to switch 1b and the ON signal to switch 2a, load current may transfer immediately to phase 2, or it may transfer after switch 1a disconnects the phase 1 current. Switch activation delays depend on switch properties. Finally, the controller sends an ON signal to switch 2b reenabling bidirectional power flow.

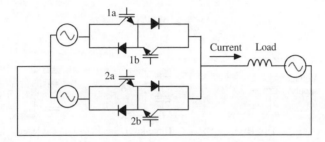

Figure 7.42   A 2p → 1p matrix converter based on the semi-soft commutation method.

Table 7.5 Pros and cons of the matrix converter

Cons	Pros
So far, there are no bidirectional switches on the market	Small physical size
Requires plenty of components → complicated structure	No space-consuming DC link required.
Ratio of output and input voltage max. 87%, may need special machine	Four-quadrant operation
Harder to control than traditional converters	High efficiency ($\approx$0.98 at full load)
Requires separate protection circuits	
Large LC filter in the input side	
The converter is completely dependent on the grid	

By applying this semi-soft commutation method, it is possible to switch phase currents (from one commutation cell to another) reliably without short-circuiting between phases or disconnecting the out-going (output) phases from the supply grid. Moreover, switching losses are reduced, because half the commutation process is carried out via *soft switching*. In soft switching, there is no current or voltage acting on the switch when it is turned on or off.

The pros and cons of the matrix converter are summarized in Table 7.5.

7.6   Multilevel inverters

Multilevel inverters have recently been the subject of intensive research activity. By adding frequency converter levels, a more ideal output curve form can be achieved even without modulation. In addition, frequency converter voltage levels can be increased using common low-voltage components. Figure 7.43 illustrates the principle of multilevel inverters. Output A can be connected to different potentials according to the circuit legs in the figure.

In two-level and three-level inverters, which were previously discussed, contact A can be connected to two or three potentials, respectively. This number of the potentials can be increased indefinitely, at least in principle. The right-hand image in the figure depicts an n-level inverter circuit leg. If m is the number of possible voltage levels of point A with respect to negative bus N, the number of steps k of the voltage between two phases is as follows.

$$k = 2m + 1 \tag{7.32}$$

In a two-level inverter, $k = 2 \times 1 + 1 = 3$. Consequently, if the load for this inverter is a three-phase star-connected motor, there are p levels for its phase voltage.

$$p = 2k - 1 \tag{7.33}$$

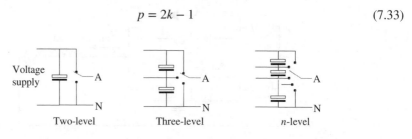

Figure 7.43 Two-level, three-level, and n-level inverter circuits.

Multilevel inverters are particularly interesting, because they offer several advantages.

- It is possible to produce nondistorted voltage.

- du/dt values remain relatively low with respect to voltage level.

- When used as rectifiers, multilevel inverter currents are nondistorted.

- The common-mode voltage of a multilevel inverter is steadier (than two-level converters).

- Switching frequency can be kept low.

The three common implementation methods for multilevel inverters are:

1. clamping to a certain DC voltage by diodes (diode-clamped),

2. clamping to a certain DC voltage by capacitors (capacitor-clamped), and

3. cascade connecting (galvanically separated DC voltage sources produce voltage levels)

Diode clamping

Figure 7.44 illustrates the diode clamping of three-level and five-level inverters. The output voltages are clamped to the centre point of the DC link.

In Figure 7.44, the number of diodes indicates the voltage stress on the diodes. In principle, therefore, different clamping diodes are required. To get full positive voltage $+U_{DC}/2$ from the five-level inverter, switches S1 through S4 are turned on. Correspondingly, when half the positive voltage $+U_{DC}/4$ is needed from the five-level inverter, switches S2 through S5 are turned on. When zero voltage is required, switches S3 through S6 are turned on. To get half-negative voltage $-U_{DC}/4$ from the five-level inverter, switches S4 through S7 are activated, and to get full negative voltage $-U_{DC}/2$, switches S5 through S8 are turned on.

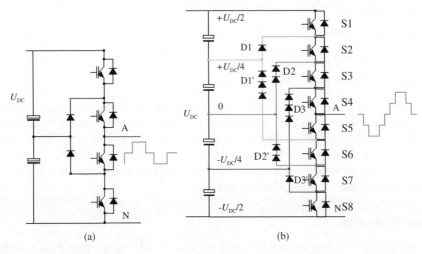

(a)   (b)

Figure 7.44 Diode clamping for one leg of (a) a three-level and (b) a five-level inverter.

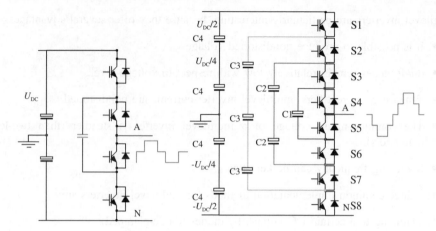

Figure 7.45 A capacitor-clamped three-level and five-level inverter.

Capacitor clamping

The voltage levels can also be clamped with capacitors, as shown in Figure 7.45.

The capacitor-clamped multilevel inverter provides various alternatives to produce different voltage levels. When a full positive voltage $+U_{DC}/2$ is needed from a capacitor-clamped five-level inverter, switches S1 through S4 are activated; the principle being the same as for the diode-clamped inverter.

When half positive voltage $+U_{DC}/4$ is required for the five-level inverter, switches S1, S2, S3, and S5 can be switched on ($+U_{DC}/2$ of C4 is less than $U_{DC}/4$ of C1). However, the same voltage is produced by switching on S2, S3, S4, and S8 or S1, S3, S4, and S7.

There are six possible combinations for the zero voltage level:

- S1, S2, S5, and S6; or
- S3, S4, S7, and S8; or
- S1, S3, S5, and S7; or
- S1, S4, S6, and S7; or
- S2, S4, S6, and S8; or
- S2, S3, S5, and S8.

When half negative voltage $-U_{DC}/4$ is required of the five-level inverter, switches S1, S5, S6, and S7 can be switched on ($+U_{DC}/2$ of the upper C4 is less than $3U_{DC}/4$ of C3); however, the same voltage can also be obtained by switching on S4, S6, S7, and S8 or S3, S5, S7, and S8.

Full negative voltage $-U_{DC}/2$ is produced by switching on S5, S6, S7, and S8.

By appropriately selecting various switch combinations to produce each voltage level, capacitor charges can be appropriately managed.

Cascade connecting

A cascade converter produces each voltage level individually using a galvanically separated voltage source. A single-phase inverter of this kind, comprised of DC sources and IGBT cells,

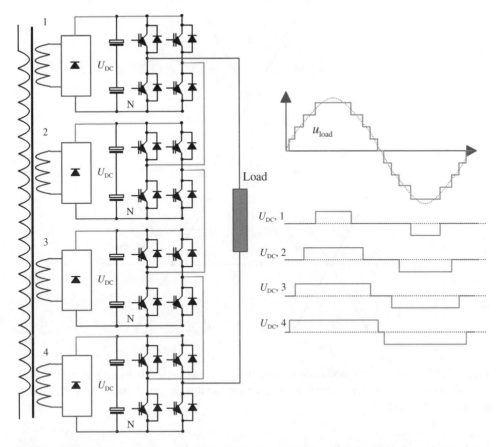

Figure 7.46 Basic topology of a cascade inverter. The connection presents nine voltage levels to the load. The number of voltage levels is twice the number of H-bridges plus 1. In this case, $2 \cdot 4 + 1 = 9$.

is illustrated in Figure 7.46. The cascade connection method enables implementing frequency converters at any voltage level. However, cascade converters are heavy, because they include a transformer that galvanically isolates all voltage levels. On the other hand, in a cascade converter, the voltage levels do not have to be secured in the same way as in the diode or capacitor-clamped systems.

The problem with single-phase converter systems is that reactive energy must be taken back into the DC-links of the multilevel converter. This is different compared to two- and three-level converters where the DC-link is common for all the phases. In these converters, the reactive power coming back from one phase is directly fed to the adjacent phase and the DC-link capacitor does not need to store the machine reactive energy as in multilevel H-bridge converters. In principle, H-bridge converters need bigger DC-link capacitors than two- or three-level converters or the DC-link voltage fluctuation must be taken into account in the modulation.

According to Figure 7.46, the output voltage of the cascade inverter is produced by first connecting the lowest cell to the positive voltage (the left upper transistor and the right lower

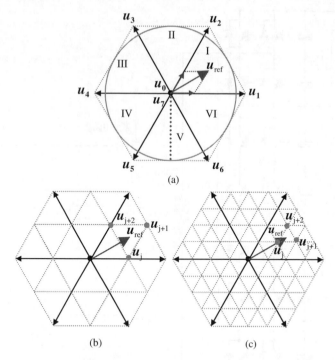

Figure 7.47 Space vector modulation alternatives to (a) a two-level, (b) a three-level, and (c) a five-level inverter. Active vectors can be generated so that the tip of the active vector points to each node of the grid.

transistor given the ON signal) and by connecting all the other cells to the upper zero voltage (both upper transistors ON). Next, cell number three is switched to positive voltage by turning off the control of its right upper transistor and by simultaneously turning on the lower transistor, *etc*. Correspondingly, negative voltages are produced by switching on the right upper transistor and the left lower transistor of cell number four, and simultaneously switching on the lower transistors of the other cells.

SVM of multilevel inverters can be carried out similarly as for two-level and three-level inverters. Figure 7.47 illustrates the division of the hexagon into a grid, the density of which increases with the number of voltage levels in the inverter.

A two-level inverter produces six active vectors, and correspondingly, a three-level inverter produces eighteen different active vectors. A five-level inverter yields 60 different active vectors and the zero vectors. Space vector modulation is implemented with vectors located at the corners of the location triangle of the point of the reference vector – similarly as for the two-level inverter. In a multilevel inverter, the reference voltage vector is implemented by active vectors, unless operating inside the innermost hexagon, in which case zero vectors are applied (as in a two-level inverter). The voltage vector is implemented as follows.

$$u_{\text{ref}} = \frac{1}{T} \left(t_j u_j + t_{j+1} u_{j+1} + t_{j+2} u_{j+2} \right)$$
$$T = t_j + t_{j+1} + t_{j+2}$$

(7.34)

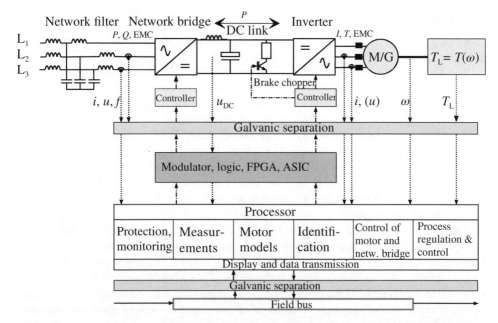

Figure 7.48 Some basic components of a modern motor control system with a VSI. Usually, the input bridge for a small drive is a diode bridge. However, there are also active input bridges available to apply PWM. With an active input bridge, the power factor can be regulated on the line side, and harmonic content can be reduced. The issues of electromagnetic compatibility (EMC) are becoming increasingly relevant. If an active input bridge is used, the DC chopper shown above in the intermediate DC link is not usually required.

7.7   The structure and interfaces of a frequency converter

Thus far, inverters for rotating-field machines have been discussed in detail, and inverter bridges are always linked to frequency converter configuration. In addition to the inverter power stage, there are other structures related to frequency converters. Figure 7.48 illustrates a modern motor controller based on a combination of information technology and power electronics.

As shown in this diagram, the main circuit consists of the following parts:

1. Line interface

2. Line filter

3. Line bridge (active or passive)

4. DC link

5. Brake chopper (if there is a passive line bridge and efficient braking is needed)

6. Inverter

7. Motor filter

8. Motor cabling

9. Motor

In addition to the main circuit, a regulation and control system is also required. The regulation and control system communicates with the main circuit via various galvanically separated control and measurement signals. Galvanic separation constitutes an important part of the frequency converter configuration. All measurement signals must be separated galvanically from the main circuit. There are various methods for separation, such as optical separation methods, galvanic separation methods based on induction, etc. Currents and voltages are often measured by galvanically separating measuring devices based on the Hall phenomenon; these devices can also be used to measure DC. In addition, resistance measurements are common; however, galvanic separation must be taken care of individually with resistance measurement.

The control of the switch components for both the line bridge and the inverter bridge must be implemented via galvanic separation. Each transistor requires a control power supply connected to the emitter potential, and each control signal of the transistor must be brought to the gate galvanically separated. This can be easier if the control power for the transistors connected to the potential of the negative bus is supplied from a single supply. In some cases, the entire motor control system is at the negative bus potential, and the upper leg transistors require galvanically controlled separation.

The above-mentioned motor control, the core task of the control system, usually demands most of the calculation capacity of the processor. In addition to motor control, the processor must control the user interface. There is also a link to higher-level logic field bus systems, through which the motor drive communicates with the industrial automation network. Finally, measurement capabilities will probably be implemented in the future to monitor the condition of the drive system. These measurements will also be managed by the processor.

References

Bolognani, S., & Zigliotto, M. (1996). Space vector Fourier analysis of SVM inverters in the overmodulation range. Proceedings of the 1996 International Conference on Power Electronics, Drives and Energy Systems for Industrial Growth, 1996, Vol. 1, pp. 319–324.

Holtz, J. (1993). On continuous control of PWM inverters in the overmodulation range including the six-step mode. *IEEE Transactions on Power Electronics*, 8(4), 546–553.

Holtz, J. (1994). Pulsewidth modulation for electronic power conversion. *Proceedings of the IEEE*, 82 (8), 1194–1214.

Mohan, N., Undeland, T. M., & Robbins, P. (1995). *Power electronics: Converters, applications, and design* (2nd ed.). New York, NY: John Wiley & Sons.

Peltoniemi, P. (2005). Vektorimodulointimenetelmien ja verkkosuotimien vertailu jännitevälipiiriverk-kovaihtosuuntaajassa. Master's thesis, Lappeenranta University of Technology.

Rodriguez, J., Lai, J-S., & Peng, F. Z. (2002). Multilevel inverters: A survey of topologies, controls and applications. *IEEE Transactions on Industrial Electronics*, 49(4), 724–738.

Sarén, H. (2005). Analysis of the voltage source inverter with small DC-link capacitor. Acta Universitatis Lappeenrantaensis 223. Dissertation, Lappeenranta University of Technology. ISBN 952-214-118-6.

8

Synchronous electrical machine drives

An examination of rotating-field machine drives can begin with a look at synchronous machine drives. They are the most versatile alternating-current (AC) machine type, and they essentially exhibit the characteristics of asynchronous machines. For instance, the equivalent circuits developed for synchronous machines need only be simplified to undertake an asynchronous machine analysis. Currently, several types of synchronous machines are available, and they are playing an increasing role in industry for drive motor applications. New applications are emerging for both permanent magnet synchronous machines and synchronous reluctance (SynRM) machines. Machines are typically available with three-phase windings, but at higher powers, six-phase machines are also common.

A polyphase synchronous machine is a rotating-field machine, in which the rotor, that is, the pole wheel, rotates synchronously with the rotating magnetic field generated by the armature winding of the machine when the machine is in steady state. The stator of the synchronous machine is composed of a laminated stack composed of sheets of electrical steel mounted to a steel frame. In the stator stack, there are usually slots for the normally three-phase stator winding. The rotor of the machine can be implemented in various ways: as a cylindrical nonsalient-pole rotor, as a salient-pole rotor with separate magnetic poles on the rotor axis, as a synchronous reluctance machine rotor, or as a permanent magnet rotor with magnetic poles generated by permanent magnets.

Different combinations of the aforementioned rotor types are also often used. The rotors can be solid or implemented as laminated constructions. In both nonsalient- and salient-pole machines, the field-winding current (rotor excitation) is conducted to the rotor either through slip rings and brushes or via a separate excitation generator mounted on the shaft of the machine to produce field-winding current. The latter case represents the so-called brushless synchronous machine. A permanent magnet machine employs permanent magnet excitation. Unlike the previous cases, permanent magnet excitation cannot be controlled. The final excitation state of

Electrical Machine Drives Control: An Introduction, First Edition. Juha Pyrhönen, Valéria Hrabovcová
and R. Scott Semken.
© 2016 John Wiley & Sons, Ltd. Published 2016 by John Wiley & Sons, Ltd.

an AC machine is determined by the magnitude of the stator supply voltage as it dictates the stator flux linkage $(\boldsymbol{\psi}_s \approx \int \boldsymbol{u}_s dt)$. Therefore, the air-gap flux of the machine will also be influenced by armature reaction, that is, by magnetizing the machine with stator current.

In the nonsalient-pole machine, there are slots on two-thirds of the rotor periphery for the field winding as there are in the stator for the stator winding. The field winding of the salient-pole machine is wound around the iron core of the magnetic poles. These assemblies are mounted on the rotor shaft. On the outer surface of the magnetic poles, there may be a damper winding constructed as a cage winding. It comprises bars on the surface of the pole shoe that are connected together at both ends by short-circuit rings.

Damper windings are used in synchronous machines to improve the stability of the drive. Direct-on-line machines in particular should be damped to ensure synchronous on line operation. However, damper windings may be useful also in controlled synchronous machine drives. In vector-controlled drives, stability can be achieved without damping. The effect of the damper windings is to retard the rate of change of air-gap flux linkage and speed up the rate of change of stator current. Because of the improved stator current response made available by the added damping, the torque of the machine can be changed rapidly improving machine dynamics.

The structures and space-vector diagrams (subsequently referred to as "vector diagrams") of a two-pole nonsalient- and salient-pole motor are illustrated in Figure 8.1.

Synchronous machines constitute a significant portion of all AC electric machines. Different machine types can be categorized, for instance, according to the schematic diagram illustrated in Figure 8.2.

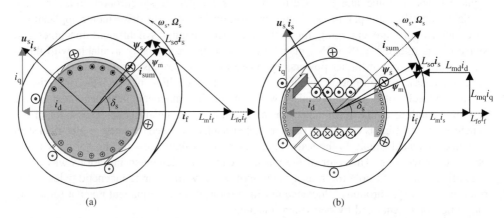

(a)   (b)

Figure 8.1 (a) Nonsalient-pole and (b) salient-pole synchronous machines with vector diagrams for steady-state counterclockwise motoring: stator voltage $\boldsymbol{u}_s$, stator current $\boldsymbol{i}_s$, field-winding current i_f, vector sum of currents $\boldsymbol{i}_{sum}$, stator flux linkage $\boldsymbol{\psi}_s$, and air-gap flux linkage $\boldsymbol{\psi}_m$. In salient-pole machines, unlike nonsalient-pole machines, the vector sum of currents and air-gap flux-linkage directions typically differ so the flux linkage turns towards the d-axis. Two-pole salient-pole machines are rare, but from the theoretical point of view, they illustrate how the vector diagram is always drawn for a two-pole machine. The electrical and mechanical angular velocities ω_s and Ω_s are the same. The machine per-unit parameters are $L_{md} = 1$, $L_{mq} = 0.5$, $L_{s\sigma} = 0.1$, $u_s = i_s = 1$, $i_{fnonsal} = 1.5$, and $i_{fsal} = 1.45$. At equal power $P = 1$, the load angle δ_s for the nonsalient-pole machine is significantly larger than the load angle for the salient-pole machine.

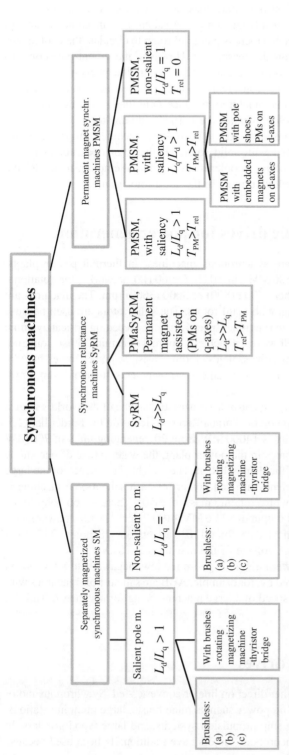

Figure 8.2 The family of synchronous machines. There are three types of synchronous machines: traditional separately excited machines, synchronous reluctance machines, and permanent magnet machines. The machine types fall into different categories based on the ratio of d- and q-axis inductances. Field-winding machines use various methods to supply excitation current to the windings including brushed or brushless commutation, for example. Brushless, that is, electronically commutated systems include (a) outer pole synchronous generators that typically come with a three-phase armature on the main machine shaft extension plus a rotating diode bridge to supply the main machine field winding, (b) axial transformers with a rotating diode bridge, and (c) three-phase, wound-rotor, wound-stator machines with an extension of the main machine shaft to supply the main machine field winding via a rotating diode bridge. The case (c) architecture enables standstill excitation. This is accomplished by rotating the excitation machine flux. Excitation systems are discussed in more detail at the end of this chapter. The reluctance and PM torque of a machine distinguish it as either a PMSM or a PM-assisted SynRM (PMaSynRM).

Low-speed drives often use salient-pole machines, and therefore the machine cross section (the xy- or dq-planes) is magnetically anisotropic. A salient-pole machine is easier to cool than a nonsalient-pole machine, since there is plenty of room to circulate the cooling air. Salient-pole construction is also technically reasonable. However, the salient-pole configuration leads to magnetic anisotropy and more complicated saturation phenomena. Consequently, numerical modelling becomes considerably more problematic. The modelling problems are one reason why some manufacturers produce nonsalient-pole machines also for low-speed drives where salient-pole machines are normally dominating.

However, in high-speed drives nonsalient-pole synchronous machine configurations are used. Generally, they are implemented with solid rotors. If the rotor does not include a separate, easily modelled damper winding, problems arise when attempting to model the complicated damping effects of the eddy currents in the solid rotor frame.

8.1   Synchronous machine drives for power generation

The majority of power generators are synchronous machines. In thermal power plants, machine powers may reach up to 1500 MW. In 50 Hz (or 60 Hz) networks, the rotational speeds of big turbo generators are either 1500 (1800) or 3000 (3600) rpm. The machines are typically nonsalient-pole constructions with solid rotors. The solid rotors are used because they are better able to withstand the high rotational speeds without exceeding critical dynamic limits. Excitation is often brushless. However, to achieve optimal dynamics, brush constructions are often used. In the biggest machines, field-winding resistive power losses may be several megawatts, which naturally favours brushless arrangements to avoid slip rings, brushes, and their maintenance.

Hydropower machines are typically salient-pole constructions. One of the world's largest hydropower stations is located in Itaipu on the Paraná River, which is on the border between Brazil and Paraguay. The total capacity is 14000 MW from 20 generating units of 700 MW each. There are 715 MW Francis turbines in the power plant, the water intake of one single turbine being 700 m^3/s. The height of the dam is 196 m. The weighted efficiency of the large turbine is 93.8%. The generator efficiency at rated power for each salient-pole machine is 98.6%. The rotor is 16 m in diameter with a rotating mass of 2650 t. There are 66 poles in the 50 Hz generators, and each has a rated output of 823.6 MVA (90.9 min^{-1}) with a power factor of 0.85_{ind}. The corresponding technical data for 60 Hz variants are 78 poles, 737 MVA (92.3 min^{-1}) and 0.95_{ind}. The terminal voltage of the machines is 18 kV.

Diesel or gas motor generator power ratings range from a few megawatts to a few dozen megawatts. The machines manufactured by, for example, ABB are salient-pole machines with 4 . . . 8 . . . 12 poles (the maximum speed of a large machine being for instance 750 min^{-1}, 8 poles, 50 Hz).

8.2   Synchronous motor drives

Synchronous motor drives may be either direct on line or converter-fed. Synchronous motor drives are used in systems requiring big power such as mine hoists, large pumping stations, ventilation systems, rolling mills, big ship propulsion systems, and large wood grinders. In direct online (DOL) applications, synchronous motors have traditionally been used because they offer *stepless* compensation of the reactive power of other loads connected to the same common coupling point. Low-speed machines are often salient-pole constructions; however, some manufacturers also produce nonsalient-pole machines. For example, a nonsalient-pole machine might be used for a rolling mill drive. High-speed machines are typically nonsalient-pole machines. For example, natural gas is pumped from Norway into the European markets by high-speed LCI-fed ca. 40 MW, 4000 min^{-1} nonsalient-pole synchronous machines. The highest power electronics-fed synchronous motor drive is probably the 101 MW LCI drive used in NASA's wind tunnel, which was supplied by ABB.

A damper winding is critical to establishing good DOL motor performance. Without it, DOL operation is impossible, because any torque disturbance would result in undesirable undamped oscillation in motor speed. A motor without a damping winding is comparable to a car without shock absorbers. However, a damper winding is not required in many vector-controlled converter drives, because vector control makes it possible to control the state of even an undamped machine. Nevertheless, the machine's ability to respond to fast torque steps is improved by a damper winding.

Using a higher resistance damper winding facilitates DOL starting; however, the increased resistivity degrades machine performance in synchronous operation. Therefore, a DOL synchronous motor is often started up with an auxiliary drive.

As shown previously in Figure 8.2, there are a large number of synchronous machine types. However, the same fundamental theory applies to all, and therefore this material addresses, for the most part, only the most versatile synchronous machine example, which is the salient-pole machine equipped with a damper winding. The equations presented for salient-pole synchronous machines hold for other machine types as well. When analysing different machine types, appropriate terms can be omitted from the equations to achieve machine-specific results. In the case of a permanent magnet synchronous machine, the behaviour of the magnets can, if desired, be expressed as a virtual field-winding current source.

8.3   Synchronous machine models

A two-axis model derived by applying the space vector theory is employed for the synchronous machine. Firstly, the frames of reference required for the analysis of the model must be established. See Figure 8.3. In the figure, the windings are shown as concentrated. In other words, the actual winding geometry has been replaced by an equivalent winding depicted on its magnetic axis. The magnetic axes of the stator phase windings U, V, and W are fixed to the respective phase windings. The *stator reference frame* can be fixed in the direction of phase winding U. The axes of this two-phase stator reference frame are denoted x and y. The fixed *rotor reference frame* is aligned with the magnetic pole of the rotor. The axes of this two-phase rotor reference frame are denoted d and q. The rotation angle between the rotor and stator reference frame is equal to the rotor electrical angle θ_r.

Next, a flux-linkage reference frame fixed to the air-gap flux-linkage vector is introduced. See Figure 8.4. The axes of the reference frame are the flux-linkage axis ψ and the torque axis T. The air-gap flux-linkage reference frame is most relevant to torque production. However, in the air-gap flux-linkage reference frame, calculating machine voltage requires more effort, because voltage is dependent on stator flux linkage ψ_s and not air-gap flux linkage ψ_m. (Their relationship is seen e.g. in Figure 8.1). Therefore, it may be appropriate to use the stator flux-linkage reference frame (see Figure 8.4b).

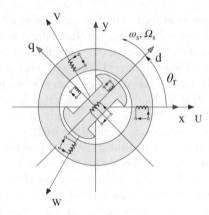

Figure 8.3 Frames of reference related to a synchronous machine. U, V, and W indicate the directions of the magnetic axes of the phase windings for a three-phase stator. The xy reference frame is two-phase and stationary with axes fixed in the direction of and perpendicular to the stator phase winding U. The dq reference frame is a two-phase reference frame fixed on the rotor with axes in the direction of and perpendicular to the magnetic pole. The angle between the xy and dq frames of reference is equal to the rotor position angle θ_r. The rotor spins at mechanical angular velocity Ω_r, and the dq reference frame spins at electrical angular velocity ω_r. For a two-pole machine, the angular velocities Ω_r and ω_r are equal.

In the case of a synchronous machine, it is important to operate in a reference frame that has the same electrical angular velocity as the rotor. This is illustrated by Figure 8.5, which shows the behaviour of the measured stator magnetizing inductance depending on rotor position. Sticking with a fixed stator reference frame, the stator magnetizing inductance varies, which complicates the determination of the equations considerably.

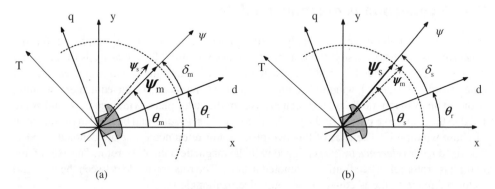

Figure 8.4 (a) The dq reference frame fixed to the rotor, the ψT reference frame fixed to the air-gap flux linkage, and the xy reference frame fixed to the stator. The angle between the dq and ψT reference frames is the load angle δ_m of the air-gap flux linkage. The angle between the dq and xy reference frames is the rotor position angle θ_r. The angle θ_m between the ψT and xy reference frames is the position angle of the air-gap flux-linkage vector in the xy reference frame. (b) The corresponding stator flux-linkage oriented ψT reference frame and angles such as the load angle δ_s and θ_s.

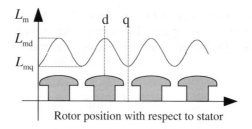

Rotor position with respect to stator

Figure 8.5 The behaviour of magnetizing inductance measured on the stator side as a function of rotor position. The direct axis magnetizing inductance L_{md} is the measure of inductance along the direct axis. Respectively, quadrature axis magnetizing inductance L_{mq} is the measure of inductance along the quadrature axis.

The stator voltage is written as $u_s = R_s \cdot i_s + d\psi_s/dt$. Since, in general, $\psi = Li$, the terms dL/dt should be taken into account in the voltage equation of the stator during transients where L is changing; however, utilizing the differential of the inductance in the equations is difficult. Its inclusion easily leads to numerical instabilities. Furthermore, inductance behaviour should be accurately known to correctly calculate voltage. Therefore, synchronous coordinate systems are used instead.

Modelling the inductance variation as a function of rotor position can be avoided by working with a synchronously rotating reference frame, for example, the rotor reference frame. In general when modelling a synchronous machine, the reference frame that results in the simplest equations should be used. As explained earlier in Chapter 4, when transforming a stator reference frame vector into a dq reference frame that rotates fixed to the rotor or into a ψT flux-linkage reference frame fixed to the air-gap flux-linkage vector, the vector must be oriented according to the position angle of the rotating reference frame.

Figure 8.6 repeats the varying vector components involved in a coordinate transformation adapted to the synchronous machine. The reference frames employed here are the rectangular

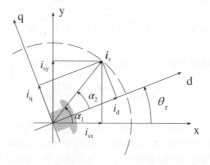

Figure 8.6 The components of the current vector for different reference frames. The xy reference frame is stationary and fixed to the stator, that is, the stator reference frame. The dq reference frame or rotor reference frame is fixed to the rotor. The d-axis of the reference frame aligns with the magnetic pole of the rotor. i_s is the stator current vector. i_{sx} and i_{sy} are its components in the stator reference frame, and i_{sd} and i_{sq} are its components in the rotor reference frame. θ_r is the rotor position angle, and α_1 and α_2 are the angles of the stator current vector in the stator and rotor reference frames.

xy reference frame fixed to the stator, that is, the stator reference frame, and the rectangular dq reference frame fixed to the rotor, that is, the rotor reference frame.

The voltage equation for the machine stator in its own (stator) frame of reference is as follows.

$$u_s^s = R_s i_s^s + \frac{d\psi_s^s}{dt} \tag{8.1}$$

The stator current vector is

$$i_s^s = i\, e^{j\alpha_1} \tag{8.2}$$

where i is the magnitude or length of the vector, and α_1 is its angle in the stator reference frame. Since the rotor of a synchronous machine is both magnetically and electrically anisotropic (only one winding on the d-axis), the rotor reference frame is used. First, the stator current is transferred to the rotor reference frame as follows.

$$i_s^r = i\, e^{j\alpha_2} = i\, e^{j(\alpha_1 - \theta_r)} \tag{8.3}$$

Here, α_2 is the vector angle in the rotor reference frame, and θ_r is the rotor position angle in the stator reference frame. The superscript r refers to the rotor reference frame.

$$i_s^r = i_s^s e^{-j\theta_r} \tag{8.4}$$

$$i_s^s = i_s^r e^{j\theta_r} \tag{8.5}$$

Applying the substitutions $i_s^s = i_s^r e^{j\theta_r}, u_s^s = u_s^r e^{j\theta_r}$ and $\psi_s^s = \psi_s^r e^{j\theta_r}$ makes it possible to transform (8.1) into the rotor reference frame.

$$u_s^r e^{j\theta_r} = R_s i_s^r e^{j\theta_r} + \frac{d(\psi_s^r e^{j\theta_r})}{dt} \tag{8.6}$$

$$u_s^r e^{j\theta_r} = R_s i_s^r e^{j\theta_r} + \frac{d\psi_s^r}{dt} e^{j\theta_r} + j\frac{d\theta_r}{dt}\psi_s^r e^{j\theta_r} \tag{8.7}$$

Finally, both sides are divided by the term $e^{j\theta_r}$, which yields

$$u_s^r = R_s i_s^r + \frac{d\psi_s^r}{dt} + j\frac{d\theta_r}{dt}\psi_s^r \tag{8.8}$$

The first derivative term is the voltage generated by the change in the magnitude of the flux linkage, in other words, the induction voltage. The latter derivative term is the rotating voltage caused by rotation. According to the latter term, rotating the stator windings induces voltage, which makes sense. In fact, explicitly writing the stator voltage equation yields a virtual rotating stator winding.

The vector model according to Equation (8.8) is a complex single-axis model. It cannot account for the magnetic anisotropy of the salient-pole machine. Therefore, it is advisable to divide the quantities into two components along the magnetic axes to produce a two-axis model. The two-axis model is represented in the rotor reference frame, because in that frame, the inductance parameters of the flux-linkage equations are not dependent on rotor position

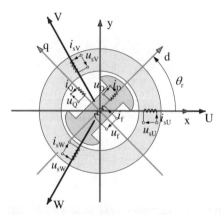

Figure 8.7 Representation of a synchronous machine with its three-phase stator windings illustrated as three concentrated phase windings. There is one field winding and two equivalent damper windings in the rotor. i_{sU}, i_{sV}, and i_{sW} are the stator currents. u_{sU}, u_{sV}, and u_{sW} are the stator voltages. i_D and i_Q are the damper winding currents. i_f and u_f are the field-winding current and voltage. θ_r is the rotor position angle. The UVW axes and the xy reference frame are fixed to the stator, and the dq reference frame is fixed to the rotor.

angle. Investigating the structure of the machine leads to the same result. Figures 8.7 through 8.9 illustrate various cases where the actual distributed windings are replaced by virtual concentrated windings. In other words, a virtual winding is depicted on the magnetic axis that has the same effect on overall model behaviour as the actual distributed winding.

Initially, there is an ordinary three-phase winding in the stator. See Figure 8.8. The direction of the magnetic pole of the rotor, the direct direction, is called the d-axis. The direction perpendicular to the d-axis is known as the quadrature direction or the

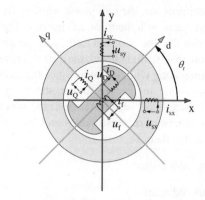

Figure 8.8 Representation of a synchronous machine with the three-phase stator winding replaced by a two-phase winding – There is one field winding and two equivalent damper windings in the rotor. i_{sx} and i_{sy} are the stator currents, and u_{sx} and u_{sy} are the stator voltages. i_D and i_Q are the damper-winding currents. i_f and u_f are the field-winding current and the voltage. θ_r is the rotor position angle.

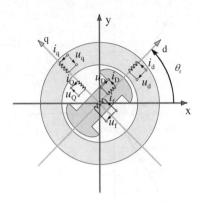

Figure 8.9 Representation of a synchronous machine with the three-phase winding replaced by a two-phase rotating winding. There is one field winding and two equivalent damper windings in the rotor. i_d and i_q are the stator currents, and u_{sd} and u_{sq} are the stator voltages. i_D and i_Q are the damper-winding currents. i_f and u_f are the field-winding current and the voltage. θ_r is the rotor position angle.

q-axis. The field winding magnetizes the magnetic circuit in the direction of the d-axis. The damper winding is illustrated by the two short-circuited equivalent windings, one of which magnetizes the machine together with the field winding in the d-direction, and the other which magnetizes in the q-direction. The equivalent damper windings are denoted D and Q. The angle θ_r between the d-axis of the rotor and the direction of the U-phase winding is the rotor angle with respect to the stator.

The rotating magnetic field generated by a three-phase stator winding can also be produced by a two-phase winding where the magnetic axes are perpendicular to each other as shown in Figure 8.8. In the two-phase case, the symbols x and y refer to the magnetic axes of the windings.

When the rotor turns, the magnetic connection between the stator and rotor windings changes, which means the inductance terms in the flux-linkage equations depend on rotor angle θ_r. To eliminate this dependence, the two-phase winding fixed to the stator is replaced by an imaginary winding that turns with the rotor, as shown in Figure 8.9. The d-direction of this winding is congruent with the d-axis of the rotor, and the direction of the q-winding is perpendicular to this direction, that is, aligned with the q-axis. There is another advantage with the dq reference frame rotating along with the rotor. Since the rotor rotates at the same speed as the magnetic field in steady state, the vectors remain stationary in the dq reference frame, whereas in the xy reference frame, the vectors rotate at the synchronous speed.

The following symbols represent the resistances and inductances present in the rotor reference frame that refer to the stator.

- L_d d-axis synchronous inductance

- L_q q-axis synchronous inductance

- L_{md} d-axis magnetizing inductance

- L_{mq} q-axis magnetizing inductance

- $L_{s\sigma}$ stator leakage inductance

- L_f total inductance of the field winding

- $L_{f\sigma}$ leakage inductance of the field winding

- L_{df} mutual inductance between the stator equivalent winding on the d-axis and the field winding (in practice L_{md})

- L_{dD} mutual inductance between the stator equivalent winding on the d-axis and the direct equivalent damper winding (in practice L_{md})

- L_{qQ} mutual inductance between the stator equivalent winding on the q-axis and the quadrature equivalent damper winding (in practice L_{mq})

- L_D total inductance of the d-axis damper winding

- $L_{D\sigma}$ leakage inductance of the d-axis damper winding

- L_Q total inductance of the q-axis damper winding

- $L_{Q\sigma}$ leakage inductance of the q-axis damper winding

- R_s stator resistance

- R_f resistance of the field winding

- R_D resistance of the d-axis damper winding

- R_Q resistance of the q-axis damper winding

Sometimes, the artificial Canay inductance (Canay, 1969) is used. Canay inductance may even have negative values. Therefore, it must be interpreted as a corrective factor in the model, which can in many cases be omitted altogether.

- $L_{k\sigma}$ mutual leakage inductance between the field winding and the direct damper winding, that is, the Canay inductance

Next, the same situation is approached using the vector model. Previously, the voltage equation of the synchronous machine in the rotor reference frame was expressed as follows.

$$u_s^r = R_s i_s^r + \frac{d\psi_s^r}{dt} + j\frac{d\theta_r}{dt}\psi_s^r \tag{8.9}$$

The current, voltage, and flux-linkage vectors can be decomposed into their real and imaginary parts on the d- and q-axes of the rotor reference frame.

$$u_s^r = u_d + ju_q; \; i_s^r = i_d + ji_q; \; \psi_s^r = \psi_d + j\psi_q \tag{8.10}$$

The equations for the real and imaginary parts of the voltage equation become

$$u_d = R_s i_d + \frac{d\psi_d}{dt} - \omega_r \psi_q \tag{8.11}$$

$$u_q = R_s i_q + \frac{d\psi_q}{dt} + \omega_r \psi_d \tag{8.12}$$

The voltage equations of the rotor circuits (the field winding and the d- and q-axis damper windings) referred to the stator are then expressed thusly.

$$u_f = R_f i_f + \frac{d\psi_f}{dt} \tag{8.13}$$

$$0 = R_D i_D + \frac{d\psi_D}{dt} \tag{8.14}$$

$$0 = R_Q i_Q + \frac{d\psi_Q}{dt} \tag{8.15}$$

As these equations are written in the natural reference frame of the windings no rotational voltage is needed. The inductances of the synchronous machine model are determined in the rotor reference frame.

$$L_d = L_{md} + L_{s\sigma} \tag{8.16}$$

$$L_q = L_{mq} + L_{s\sigma} \tag{8.17}$$

$$L_f = L_{md} + L_{f\sigma} + L_{k\sigma} \tag{8.18}$$

$$L_D = L_{md} + L_{k\sigma} + L_{D\sigma} \tag{8.19}$$

$$L_Q = L_{mq} + L_{Q\sigma} \tag{8.20}$$

The following equations may be written for stator flux linkages and other flux linkages; the inductances and currents refer to the stator.

$$\psi_d = L_d i_d + L_{df} i_f + L_{dD} i_D \tag{8.21}$$

$$\psi_q = L_q i_q + L_{qQ} i_Q \tag{8.22}$$

$$\psi_f = L_{df} i_d + L_f i_f + L_{fD} i_D \tag{8.23}$$

$$\psi_D = L_{dD} i_d + L_{fD} i_f + L_D i_D \tag{8.24}$$

$$\psi_Q = L_{qQ} i_q + L_Q i_Q \tag{8.25}$$

In the traditional two-axis model, the stator circuit (that is, the armature circuit), the damper windings, and the field winding are assumed to be magnetically interconnected only through the magnetizing inductances L_{md} and L_{mq}. However, measurements have shown that in transients, the alternating component of the field-winding current may be larger than the calculated value. Therefore, the Canay inductance parameter $L_{k\sigma}$ is added to the model, since the traditional model only correctly describes the armature. The Canay inductance takes into account the deviation of the magnetic connection of the damper winding and the field winding from the d-axis magnetizing inductance. Therefore, the different mutual inductances are

$$L_{dD} = L_{df} = L_{md} \tag{8.26}$$

$$L_{fD} = L_{md} + L_{k\sigma} \tag{8.27}$$

$$L_{qQ} = L_{mq} \tag{8.28}$$

Based on the above assumptions, the equations of the flux linkages can be expressed in the rotor (dq) reference frame. The sum of the d-axis currents produced flux linkage in the magnetizing inductance and the stator winding leakage flux linkage together define the stator flux linkage on the d-axis.

$$\psi_{\mathrm{d}} = L_{\mathrm{md}}(i_{\mathrm{d}} + i_{\mathrm{f}} + i_{\mathrm{D}}) + L_{\mathrm{s}\sigma}i_{\mathrm{d}} \tag{8.29}$$

Correspondingly, the sum of the q-axis currents define the q-axis flux linkage.

$$\psi_{\mathrm{q}} = L_{\mathrm{mq}}(i_{\mathrm{q}} + i_{\mathrm{Q}}) + L_{\mathrm{s}\sigma}i_{\mathrm{q}} \tag{8.30}$$

As was the case for d-axis stator flux linkage, there is a common part coming from the sum of all the d-axis currents plus the leakage components when defining the field-winding flux linkage.

$$\psi_{\mathrm{f}} = L_{\mathrm{md}}i_{\mathrm{d}} + L_{\mathrm{f}}i_{\mathrm{f}} + (L_{\mathrm{md}} + L_{\mathrm{k}\sigma})i_{\mathrm{D}} = L_{\mathrm{md}}(i_{\mathrm{d}} + i_{\mathrm{f}} + i_{\mathrm{D}}) + L_{\mathrm{f}\sigma}i_{\mathrm{f}} + L_{\mathrm{k}\sigma}i_{\mathrm{D}} \tag{8.31}$$

The damper-winding flux linkages are as follows.

$$\psi_{\mathrm{D}} = L_{\mathrm{md}}i_{\mathrm{d}} + (L_{\mathrm{md}} + L_{\mathrm{k}\sigma})i_{\mathrm{f}} + L_{\mathrm{D}}i_{\mathrm{D}} = L_{\mathrm{md}}(i_{\mathrm{d}} + i_{\mathrm{f}} + i_{\mathrm{D}}) + L_{\mathrm{k}\sigma}i_{\mathrm{f}} + L_{\mathrm{D}\sigma}i_{\mathrm{D}} \tag{8.32}$$

$$\psi_{\mathrm{Q}} = L_{\mathrm{mq}}i_{\mathrm{q}} + L_{\mathrm{Q}}i_{\mathrm{Q}} = L_{\mathrm{mq}}(i_{\mathrm{q}} + i_{\mathrm{Q}}) + L_{\mathrm{Q}\sigma}i_{\mathrm{Q}} \tag{8.33}$$

8.4 Equivalent circuits and machine parameters for a synchronous machine

The equivalent circuits of Figures 8.10 and 8.11 can be represented for a synchronous machine in the rotor reference frame, since in that frame, the inductance coefficients of the

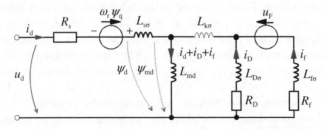

Figure 8.10 The equivalent circuit of the synchronous machine in the d-axis direction referred to the stator winding – i_{d} and u_{d} are the d-axis components of the stator current and voltage. ψ_{d} and ψ_{q} are the d- and q-axis components of the stator flux linkage. i_{D} is the current of the direct damper winding. i_{f} is the field-winding DC current. R_{s} is the stator resistance, R_{D} is the resistance of the direct damper winding, and R_{f} is the resistance of the field winding. $L_{\mathrm{s}\sigma}$ is the leakage inductance of the stator, L_{md} is the direct magnetizing inductance, $L_{\mathrm{k}\sigma}$ is the Canay inductance, $L_{\mathrm{D}\sigma}$ is the leakage inductance of the direct damper winding, and $L_{\mathrm{f}\sigma}$ is the leakage inductance of the field winding. u_{f} is the voltage of the field winding.

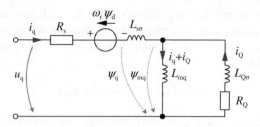

Figure 8.11   The equivalent circuit of the synchronous machine in the q-direction referred to the stator winding – i_q and u_q are the quadrature components of the stator current and voltage. ψ_d and ψ_q are the direct and quadrature components of the stator flux linkage. i_Q is the quadrature current of the damper winding. R_s is the stator resistance, R_Q is the resistance of the quadrature damper winding. $L_{s\sigma}$ is the leakage inductance of the stator, L_{mq} is the quadrature magnetizing inductance, and $L_{Q\sigma}$ is the leakage inductance of the quadrature damper winding.

flux-linkage equations no longer depend on rotor position. Therefore, the coefficients are constants. The equivalent circuits are given separately for the d- and q-directions, since the salient-pole machine is magnetically anisotropic. Although the nonsalient-pole machine is in principle magnetically isotropic, it also involves enough anisotropy to make employing the two-axis model advisable. Furthermore, the field winding is usually a single-phase construction, which also justifies the application of the two-axis model. Only a slip-ring asynchronous machine, which can also be used as a synchronous machine by supplying direct current to the rotor, is in principle magnetically completely isotropic. Therefore, it does not require a two-axis model fixed to the rotor.

The equivalent circuits of Figures 8.10 and 8.11 are DC-circuits. This is the result of representing the currents and voltages in the rotor reference frame. In steady state; voltages, currents, and flux linkages have constant DC-values. Only during transients do the variables have AC-components. For example, the DC d-axis stator current is steady only along the magnetizing inductance path of the d-axis equivalent circuit, because that path has no resistance. There is current in the damper winding only when air-gap flux-linkage changes and induces a variable air-gap voltage. The AC-portion of the air-gap voltage sees high impedance in the magnetizing inductance path, and therefore the damper winding route seems attractive. Similarly, any transient in the air-gap flux linkage results in current components via the damper winding, the magnetizing circuit, and the magnetizing inductance.

In principle, the appropriate parameters for a synchronous machine are provided by the manufacturer. These parameters are supplemented or adjusted based on measurements taken by the user of the machine. The provided parameters are not the most suitable values to apply to the construction of equivalent circuits, because they have been defined using traditional methods that represent different magnetic states of the machine. For instance, d-axis synchronous inductance is determined based on no-load DOL operation in the absence of field-winding current or in a sustained short circuit. The transient and subtransient inductances, on the other hand, are defined using data collected during short-circuit testing. Quadrature-axis synchronous inductance and q-axis transient inductances are determined using various methods in various loading situations.

As a result, all the parameters determined for a synchronous machine are not simultaneously valid, but instead represent different magnetic states. In their design software, the manufacturers apply their own experimentally defined coefficients or simulate transients with time-stepping FEM-based analyses. Frequency converter technology also provides some solutions for determining machine parameters. The parameters can be updated online, or they can be determined in the initial identification run of the drive.

An example will serve to identify the traditional parameters and time constants used for synchronous machine modelling and to illustrate how they are determined. A three-phase short-circuit test beginning from a no-load condition is a traditional parameter determination approach. Figure 8.12 depicts the single-phase short-circuit current measured during three-phase short circuit test. In the figure, the extracted subtransient, transient, permanent, and DC component parts of the short-circuit current are shown.

The test data plots clearly show the decay of each component. The subtransient component decays fastest. The decay of the transient component is slower. The test ends in a sustained short circuit determined by the internal emf and the d-axis synchronous inductance. From the equivalent circuits in Figures 8.10 and 8.11 and the short circuit behaviour, it is possible to define the machine parameters of the circuits shown in Figure 8.13. The figure clearly illustrates the parameters traditionally given by the synchronous machine manufacturer.

The mutual inductances between windings on the d-axis of the machine and the magnitude of the direct magnetizing inductance are usually assumed to be equal, that is, $L_{df} = L_{dD} = L_{fD} = L_{md}$.

- The d-axis synchronous inductance is the sum of the stator leakage inductance and the d-axis magnetizing inductance. This is, together with the q-axis synchronous inductance, the most important machine parameter in steady state.

$$L_d = L_{md} + L_{s\sigma} \tag{8.34}$$

- Correspondingly, the q-axis synchronous inductance can be written

$$L_q = L_{mq} + L_{s\sigma} \tag{8.35}$$

The stator leakage inductance values for both axes are assumed equal. This represents a simplification and a negligible error, as some of the leakage inductance components depend on air-gap size, and therefore the q- and d-axis leakages differ slightly.

- Direct transient inductance is the sum of the stator leakage inductance and the parallel connection of the direct magnetizing and field-winding leakage inductances as follows.

$$L_d' = L_{s\sigma} + \frac{L_{md} L_{s\sigma}}{L_{md} + L_{f\sigma}} \tag{8.36}$$

- The subtransient inductances L_d'' and L_q'' are important parameters that describe machine behaviour at the beginning of a transient. For example, in a sudden short

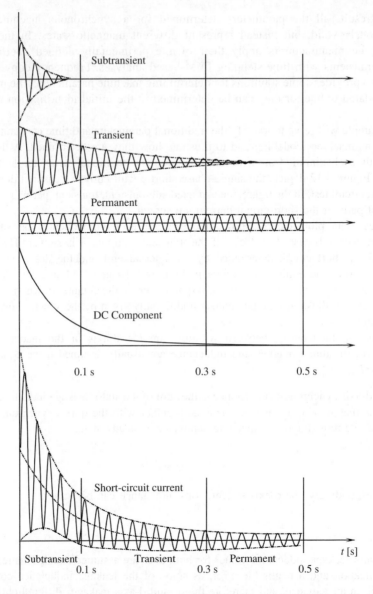

Figure 8.12 Result of a three-phase short-circuit test. The extracted subtransient, transient, permanent, and DC component regions of the single-phase short-circuit current are shown above the measured current.

circuit, subtransient inductances define the currents. If a synchronous machine is supplied from a PWM source, the pulses see L_d'', which filters the current ripple of the machine. The d-axis subtransient inductance L_d'' is the sum of the stator leakage inductance and the parallel connection of the d-axis magnetizing, damper-winding

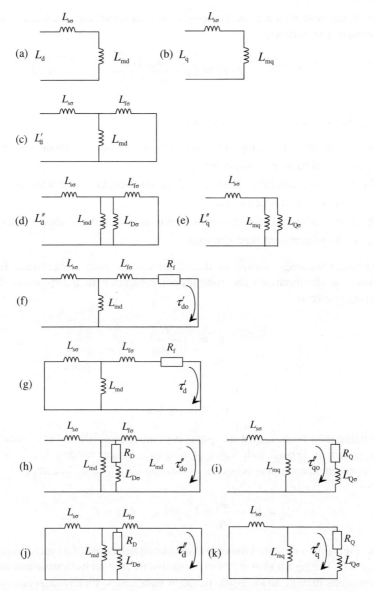

Figure 8.13 Traditional machine parameters for a synchronous machine and the respective equivalent circuits. *Source:* Pyrhönen et al. (2014). Reproduced with permission of John Wiley & Sons Ltd.

leakage, and field-winding leakage inductances.

$$L_d'' = L_{s\sigma} + \frac{L_{md}\dfrac{L_{D\sigma}L_{f\sigma}}{L_{D\sigma}+L_{f\sigma}}}{L_{md}+\dfrac{L_{D\sigma}L_{f\sigma}}{L_{D\sigma}+L_{f\sigma}}} \tag{8.37}$$

- There is no field winding on the q-axis, and therefore the quadrature subtransient inductance is as follows.

$$L''_q = L_{s\sigma} + \frac{L_{mq}L_{Q\sigma}}{L_{mq} + L_{Q\sigma}} \tag{8.38}$$

Figure 8.13 also shows a number of time constants

τ''_{do} is the d-axis subtransient time constant with an open-circuit stator winding,

τ''_d is the d-axis subtransient time constant,

τ'_{do} is the d-axis transient time constant with an open-circuit stator winding,

τ'_d is the d-axis transient time constant,

τ''_{qo} is the q-axis subtransient time constant with an open-circuit stator winding, and

τ''_q is the q-axis subtransient time constant.

All machine parameters shown in the figure can be easily determined first in the commissioning of the machine. The subtransient inductance of the q-axis is determined by the equivalent circuit

$$L''_q = L_q - \frac{L^2_{mq}}{L_Q} \tag{8.39}$$

where

$$L_Q = L_{mq} + L_{Q\sigma} \tag{8.40}$$

The magnetizing inductance L_{mq} of the q-axis is usually, especially in traditional machines, notably larger than the leakage inductance of the q-axis damper winding $L_{Q\sigma}$. Assuming that $L_Q \approx L_{mq}$, an acceptable simplification, the subtransient inductance becomes

$$L''_q = L_q - \frac{L^2_{mq}}{L_Q} \approx L_{s\sigma} + L_{mq} - L_{mq} = L_{s\sigma} \tag{8.41}$$

In principle, Figure 8.13 describes the equivalent circuit behaviour of a synchronous machine during a transient. During a transient, the machine first reacts with its subtransient inductances and time constants moving towards the transient parameters. Finally, it settles into steady state, where synchronous inductances dominate.

The time constants describe machine behaviour depending on stator connection status. A shorted stator means the stator is in short circuit from the point of view of the phenomenon observed, for example, at a certain frequency. The stator can be shorted during a real terminal short circuit or, for example, by applying a frequency converter zero vector.

If the supplying system has a significant inductance the leakage of the machine must be increased accordingly to find the drive subtransient parameters. The obtained subtransient estimate can be used to estimate the stator leakage inductance

$$L_{s\sigma} = k_{ls\sigma}L''_q \tag{8.42}$$

where the coefficient $k_{ls\sigma}$ varies typically between 0.4 and 0.6. The exact measurement of stator leakage inductance is carried out without the rotor in place, and therefore an operating synchronous machine cannot be used to make this inductance measurement. An estimate must suffice. The other parameters required for the current model are obtained by employing the equivalent circuits from Figure 8.13 as follows.

$$L_{md} = L_d - L_{s\sigma}$$
$$L_{mq} = L_q - L_{s\sigma}$$
(8.43)

$$L_{f\sigma} = \frac{L_d' L_{md} - L_{s\sigma} L_{md}}{L_d - L_d'}$$

$$L_{Q\sigma} = \frac{L_{mq}^2}{L_q - L_q''} - L_{mq}$$
(8.44)

$$L_{D\sigma} = \frac{L_d'' L_{md} L_{f\sigma} - L_{s\sigma} L_{md} L_{f\sigma}}{L_{s\sigma} L_{f\sigma} + L_{s\sigma} L_{md} + L_{md} L_{f\sigma} - L_d'' L_{f\sigma} - L_d'' L_{md}}$$

When all the parameters according to the two-axis model are known, the resistances R_D and R_Q of the damper windings can be calculated.

$$R_D = \frac{\left(L_{D\sigma} + \frac{L_{md} L_{f\sigma}}{L_{md} + L_{f\sigma}}\right) L_d''}{\tau_d''}$$
(8.45)

$$R_Q = \frac{\left(L_{Q\sigma} + \frac{L_{mq} L_{s\sigma}}{L_{mq} + L_{s\sigma}}\right)}{\tau_q''}$$
(8.46)

The current model can be determined by employing Equation (8.44) through Equation (8.46). The successful accomplishment of this task depends on the accuracy of the traditional motor parameters.

The main problem with the equivalent circuits in Figure 8.13 is that the parameters have different values in different situations, and therefore solving the system of equations based on the equivalent circuits and real parameters of the machine is not possible, because it results in irrelevant values for some parameters. The figure describes the principal behaviour of the machine and the components involved in certain states, but the values of the different parameters must be determined experimentally or from FEA.

If a rotor is at all conductive, even without a damper winding there will be subtransient parameters, at least in principle. For example, rotor surface permanent magnet machines without damper windings are common; however, they still possess some of the properties of machines that do have damper windings, because of eddy currents produced in the magnets.

Laboratory measurement technologies available for determining synchronous machine parameters are defined in the IEC 60034-4 (*Methods for Determining Synchronous Machine*

Table 8.1 The machine parameters of a test machine determined by standard measurement procedures – Test machine properties are: 14.5 kVA, 400 V/21A, field-winding current 10.5 A, $k_{ri} = 4.64$, 50 Hz/1500 rpm, and rated power factor as generator cos $\varphi = 0.8$ ind.

Parameter	Value	Notes
per-unit stator resistance R_s	0.05	
per-unit field-winding resistance R_f	0.0083*	*referred to the stator voltage level
referring factor k_{ri}	4	
per-unit d-axis synchronous inductance L_d	1.19	
per-unit q-axis synchronous inductance L_q	0.56	
per-unit d-axis transient inductance L'_d	0.33	
per-unit d-axis subtransient inductance L''_d	0.105	
per-unit q-axis subtransient inductance L''_q	–*	* could not be measured
d-axis transient time constant with open-circuited stator τ'_{do}	0.236 s	
transient time constant of the d-axis τ'_d	0.054 s	
subtransient time constant of the d-axis τ''_d	0.024 s	
subtransient time constant of the q-axis τ''_q	–*	* could not be measured

Quantities from Tests) and IEEE 115-1983 (*Test Procedures for Synchronous Machines*) standards. The measurements comprise a DC resistance measurement, a no-load test, a steady-state short-circuit measurement, a slip test, a short-circuit test of the field winding, a sudden three-phase short-circuit measurement, and the measurement of V-curves. Table 8.1 lists the test motor data obtained from these measurements.

The data provided by the machine manufacturer are summarized in Table 8.2.

8.5 Measuring motor parameters using a frequency converter

The DC resistance of the stator, the direct synchronous inductance at no load for different voltage steps, and the subtransient inductance, both in the d- and q-axis directions, can be measured using a frequency converter. A modern frequency converter has good measuring and computing capacity. Therefore, various measurements can be carried out automatically during the commissioning of the drive.

If the drive is being commissioned under load, and it can be run, that is, using both the current model and the model based on the voltage integral in the flux linkage and torque estimation, then the parameters of the current model can be updated both on the direct and quadrature axes. The torque values determined using the current and voltage models must be equal, and therefore the inductance parameters have to be selected to establish this condition.

The measurement of transient inductance is straightforward. The motor is supplied with short voltage pulses from the frequency converter: $dt \ll \tau_D$. By switching on the six active voltage vectors in their directions, a good picture of transient inductance can be determined for each direction of the machine. When supplying the machine in the direction

Table 8.2 Machine parameters given by the supplier: 14.5 kVA, 400 V/21 A, 50 Hz/1500 rpm

Parameter	Value	Notes
per-unit stator resistance R_s	0.048	
per-unit field-winding resistance R_f	0.00793*	*in the stator voltage level
referring factor k_{ri}	4.64	
per-unit d-axis synchronous inductance L_d	1.17	
per unit d-axis magnetizing inductance $L_{md,pu}$	1.05	
per unit q-axis synchronous inductance L_q	0.57	
per unit q-axis magnetizing inductance $L_{mq,pu}$	0.45	
stator leakage inductance $L_{s\sigma}$	0.12	
field-winding leakage inductance $L_{f\sigma}$	0.27	in the stator voltage level
d-axis damper leakage inductance $L_{D\sigma}$	0.07	
q-axis damper leakage inductance $L_{Q\sigma}$	0.14	
Canay inductance $L_{k\sigma}$	0	
d-axis damper resistance $R_{D\sigma}$	0.02	
q-axis damper resistance R_Q	0.03	
per-unit d-axis transient inductance L'_d	0.13	
per-unit d-axis subtransient inductance L''_d	0.09	
per-unit q-axis subtransient inductance L''_q	0.109	
d-axis transient time constant with open-circuited stator τ'_{do}	0.284 s	
transient time constant of the d-axis τ'_d	0.031 s	
subtransient time constant of the d-axis τ''_d	0.006 s	
subtransient time constant of the q-axis τ''_q	0.008 s	

of the d-axis of the machine, the d-axis subtransient inductance can be calculated as follows.

$$L''_d = \frac{u_d dt}{di_d} = \frac{d\psi_d}{di_d} \approx \frac{\Delta\psi_d}{\Delta i_d} \tag{8.47}$$

Correspondingly, supplying the machine with a voltage pulse in the direction of the q-axis of the machine, $dt \ll \tau_Q$, reveals the quadrature subtransient inductance.

$$L''_q = \frac{u_q dt}{di_q} = \frac{d\psi_q}{di_q} \approx \frac{\Delta\psi_q}{\Delta i_q} \tag{8.48}$$

In the laboratory, measurements can also be made in intermediate positions. Figure 8.14 illustrates the subtransient inductance of a synchronous machine as a function of rotor angle.

In this case, the subtransient inductance is larger in the q-axis direction than in the d-axis direction. Evidently, the damper-winding activity preventing flux penetration into the rotor is not as efficient on the q-axis as it is on the d-axis. Therefore, a short voltage pulse generates a larger flux linkage when the rotor is in the q-axis position than when it is in the d-axis position.

The d-axis synchronous inductance of a synchronous motor can also be measured when the motor is operating at no load. The measurement is based on flux linkage, obtained by

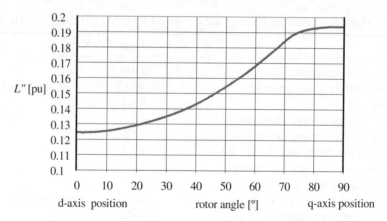

Figure 8.14 Measurement result of the per-unit subtransient inductance as a function of rotor angle. The angle is given in electrical degrees. *Source:* Adapted from Kaukonen (1999). Reproduced with permission of J. Kaukonen.

integrating from the voltages and comparing to the measured currents. In this inductance measurement, the motor must be run at a relatively high speed to reduce the uncertainty of the measurement. The measurement result cannot be recorded before all transients have settled to ensure the damper-winding currents are zero.

The synchronous inductance measurement can be performed by varying the current of the stator and rotor. Completely eliminating the direct current of the rotor provides the first value for the synchronous inductance. Since the absolute value of the terminal voltage is $|\omega_s \psi_s| \approx |\omega_s i_{s0} L_d| \approx |u_s|$, the no-load current can be measured and kept chiefly as inductive excitation current to calculate the d-axis synchronous inductance. As rotor current increases, stator current decreases until the stator becomes completely resistive.

Measuring q-axis magnetizing inductance is difficult in no-load operation. This measurement is best carried out in a laboratory using two similar machines – one that is driving while the other is being measured. The machine rotors must be connected with a 90 electrical degree phase shift.

To measure its parameters, a synchronous machine must be loaded. The magnetizing inductance and the stator leakage inductance can be calculated by applying the information on the stator flux linkage in the rotor reference frame.

$$L_{md} = \frac{\psi_d - i_d L_{s\sigma}}{i_d + i_f + i_D}\Big|_{(i_D \approx 0)}$$

$$L_{mq} = \frac{\psi_q - i_q L_\sigma}{i_q + i_Q}\Big|_{(i_Q \approx 0)} \tag{8.49}$$

$$L_\sigma = \frac{\psi_d}{i_d} - L_{md}\Big|_{(i_f = 0, i_D \approx 0)}$$

Figure 8.15 illustrates the measured no-load saturation curves in the d-direction.

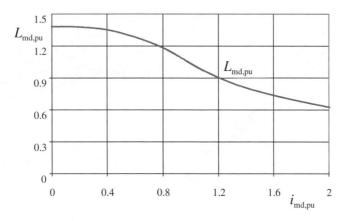

Figure 8.15 Measured no-load saturation curves for the d-axis magnetizing inductance.

These inductance measurements are made by setting the sum current of the d-axis, $i_{md} = i_d + i_f + i_D$, at the desired value (in practice with no damper-winding current, $i_D = 0$) and varying the torque, which in turn impacts the sum field-winding current of the q-axis; $i_{mq} = i_q + i_Q$ (with $i_Q = 0$). The measurements are repeated across the flux-linkage range of $\psi_{ref} = 0.3–1.3$ pu, with torque varying between $T_{ref} = 0–2.5$ pu. At each measured point, the measuring result is computed by the inverter control program. The result is the d-axis, $L_{md} = f(i_{md}(t), i_{mq}(t))$, and q-axis magnetizing, $L_{mq} = f(i_{md}(t), i_{mq}(t))$, inductances as a function of i_{md} and i_{mq}. Stator leakage inductance is assumed constant and similar for both axes even though air-gap affects the leakage inductance calculation. See Pyrhönen et al. (2014). Under heavy loads, the leakage may saturate, which should considered when designing, for example, a rolling mill drive subject to heavy overload conditions.

Figure 8.16 offers surface plots of the d-axis and q-axis magnetizing inductances as a function of the axis sum currents. The plots show that q-axis current affects the magnetizing

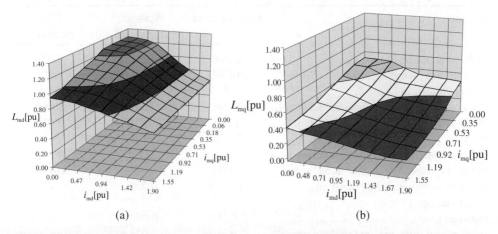

(a)   (b)

Figure 8.16 Surface plots of (a) the d-axis magnetizing inductance and (b) the q-axis magnetizing inductance as functions of the sum of the magnetizing currents. *Source:* Adapted and reproduced with permission of J. Kaukonen (1999).

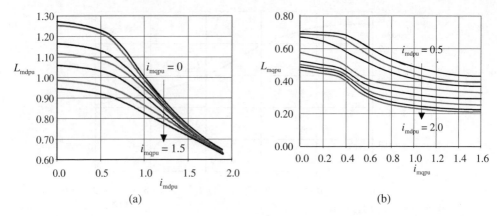

(a)   (b)

Figure 8.17 Curve plots of (a) the d-axis magnetizing inductance and (b) the q-axis magnetizing inductance as functions of the sum of the field-winding currents.

inductance of the d-axis and vice versa. The effect of so-called cross-saturation, therefore, is clearly visible. Figure 8.17 presents the same information as two-dimensional curve plots.

Table 8.3 presents a comparison of the results obtained using the different methods. The DTC measurement data have been gathered at the nominal operation point. The transient values in particular are notably different for the different methods.

Table 8.3 A comparison of results obtained using different methods – The DTC-converter measurement data was collected at the nominal operation point. The transient values, in particular, are notably different between the different methods.

Parameter	Manufacturer	Standard measured	DTC measured
Per-unit stator resistance R_s	0.048	0.048	0.051
Per-unit field-winding resistance R_f	0.00793	0.0083	$-^*$
Current referring factor k_{ri}	4.63	4	3.96
Per-unit d-axis synchronous inductance L_d	1.196	1.19	1.066
Per-unit q-axis synchronous inductance L_q	0.475	0.56	0.439
Per-unit d-axis transient inductance L'_d	0.129	0.33	$-^*$
Per-unit d-axis subtransient inductance L''_d	0.09	0.105	0.125
Per-unit q-axis subtransient inductance L''_q	0.109	$-^*$	0.194
direct-axis transient time constant	0.284 s	0.236 s	$-^*$
with an open-circuit stator winding τ'_{do}			
transient time constant of the d-axis τ'_d	0.031 s	0.054 s	$-^*$
subtransient time constant of the d-axis τ''_d	0.006 s	0.024 s	$-^*$
subtransient time constant of the q-axis τ''_q	0.008 s	$-^*$	$-^*$

*was not possible to measure.

8.6 Finite element analysis (FEA) for determining the synchronous machine inductances

With electromagnetic analysis, the preceding inductance surfaces of Figure 8.16 can be prepared early on in the machine design phase. FEM magnetic calculation is useful in determining the saturation behaviour of the direct and quadrature magnetizing inductances. The inductances can also be defined at a particular loading point. At present, however, FEM magnetic calculation is still rather laborious, and therefore not even the largest suppliers calculate all their machines using the method. Instead, simpler methods are applied to the machine calculation.

For example, consider a small synchronous test motor with a three-phase winding, a stator slot number $Q_s = 24$, and $q_s = 2$ slots per pole and per phase. The stator stack length is $l = 140$ mm, the inner diameter of the stator is $D_s = 196$ mm, and the number of turns in series in the stator winding per slot is $z_Q = 56$. With a four-pole field winding, the number of turns per pole is $N_{rp} = 220$. The d-axis magnetizing inductance $L_{md} = f(i_f)|_{i_d=i_q=0}$ can be calculated using different values of rotor current. In the calculation, complete quadrature magnetizing can be accomplished, so it is possible to determine $L_{mq} = f(i_q)|_{i_f=i_d=0}$. At each loading point, the inductances $L_{md} = f(i_d, i_f, i_q)$ and $L_{mq} = f(i_d, i_f, i_q)$ can be calculated. Figure 8.18 illustrates the machine running at no load and loaded at the nominal operating point.

The finite element method yields, for instance, the air-gap flux density, which is naturally distorted due to the slot openings and the armature reaction. A Fourier analysis of the curve is needed to calculate the magnetizing inductances, since space vector theory is based on the assumption of a sinusoidal curve form. Figure 8.19 depicts the air-gap flux density distribution and its fundamental harmonic at no load.

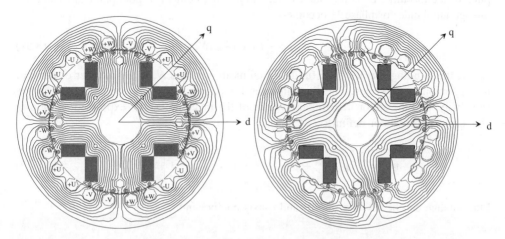

Figure 8.18 Flux diagrams of the machine at no load and at the rated point. *Source:* Adapted from Kaukonen (1999). Reproduced with permission of J. Kaukonen.

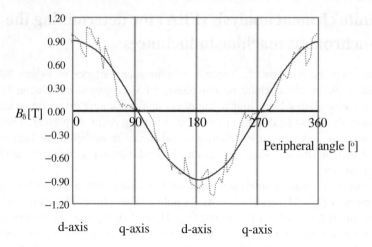

d-axis   q-axis   d-axis   q-axis

Figure 8.19 Air-gap flux density and its fundamental harmonic at nominal no-load operation calculated by FEM.

To calculate the inductances, the air-gap flux density fundamental $\hat{B}_{1\delta}$ is searched from the result of the Fourier analysis. Air-gap flux can then be determined from the following equation.

$$\hat{\Phi}_m = \int\limits_{0}^{l'} \int\limits_{-\frac{\tau_p}{2}}^{+\frac{\tau_p}{2}} \hat{B}_{1\delta} \cos\left(\frac{x}{\tau_p}\pi\right) dx dl' = \frac{2}{\pi}\hat{B}_{1\delta}\tau_p l' \tag{8.50}$$

where the effective machine equivalent length l' must be defined, for example, based on the information given in Pyrhönen et al. (2014). In the simplest case, there are no cooling ducts, and $l' \approx l + 2\delta_0$, where δ_0 is the air gap in the middle of the pole shoe, $\tau_p = \pi D_s/(2p)$ is the pole pitch of the machine, D_s is the stator bore, and p is the number of pole-pairs. Air-gap flux-linkage amplitude for phase U becomes

$$\hat{\psi}_{mU} = \frac{2}{\pi}\hat{B}_{1\delta}\tau_p l' k_{ws1} N_s \tag{8.51}$$

k_{ws1} is the winding factor of the fundamental harmonic, and N_s is the number of turns of the stator phase in series. The flux linkages for phases V and W are calculated in the same way, and the stator flux-linkage vector of the machine can be expressed as follows (see e.g. Chapter 4 for definition of γ).

$$\psi_m = \frac{2}{3}\left[\psi_{mU}(t) + \psi_{mV}(t)e^{j\frac{2\pi}{3}} + \psi_{mW}(t)e^{j\frac{4\pi}{3}}\right] = \hat{\psi}_{mU}e^{j\gamma} \tag{8.52}$$

Flux linkage can be divided into d- and q-axis components.

$$\begin{aligned}\psi_{md} &= \psi_m \cos\gamma \\ \psi_{mq} &= \psi_m \sin\gamma\end{aligned} \tag{8.53}$$

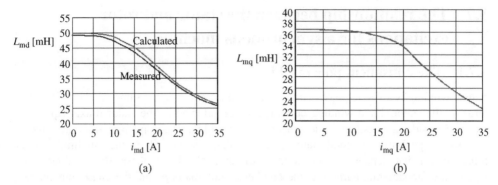

(a)   (b)

Figure 8.20 (a) The converter measured and calculated d-axis inductance, when there is magnetization only on the d-axis and (b) the calculated quadrature inductance (reference machine of Table 8.1).

Finally, the inductances can be calculated using these equations.

$$L_{md} = \frac{\Psi_{md}}{i_d + i_f + i_D}\bigg|_{(i_D=0)}$$

$$L_{mq} = \frac{\Psi_{mq}}{i_q + i_Q}\bigg|_{(i_Q=0)}$$

(8.54)

Figure 8.20a shows how the measured and calculated results for d-axis inductance compare. The agreement is acceptable. Figure 8.20b presents the calculated quadrature inductance.

The results of the FEM analysis for a machine running under load are interesting. Figure 8.21a depicts the behaviour of the d-axis inductance at nominal stator flux linkage for different loads. Figure 8.21b illustrates the behaviour of the q-axis inductance. Again, the results deviate somewhat from the measured ones.

This example clearly shows how the inductances vary as a function of load. Obtaining exact results for the synchronous machine, therefore, is difficult, and the machine parameters must be updated for the machine's working conditions.

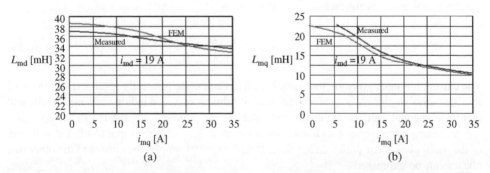

(a)   (b)

Figure 8.21 (a) The converter measured and calculated behaviour of the d-axis inductance and (b) the behaviour of the q-axis inductance when the torque varies between 0 % and 250 % (reference machine of Table 8.1).

8.7 The relationship between the stator and rotor excitations for a synchronous machine

8.7.1 The nonsalient pole machine

A field winding is a single-phase winding, and it does not have the same number of winding turns as the stator. Field-winding current must be referred to the stator d-axis equivalent circuit. Consider a nonsalient-pole synchronous machine. The polyphase armature winding produces a peak-to-peak total stator current linkage value ($\hat{\Theta}_s$ is the amplitude of the fundamental) that depends on the number of turns N_s of the phase, the winding factor k_{ws1} of the fundamental harmonic, the number of pole-pairs p, and the phase number m

$$2\hat{\Theta}_s = m\frac{4}{\pi}\frac{k_{ws1}N_s}{2p}\hat{i}_s \tag{8.55}$$

For a three-phase two-pole machine, the above is rewritten as

$$2\hat{\Theta}_s = 3\frac{4}{\pi}\frac{k_{ws1}N_s}{2}\hat{i}_s = 3\frac{4}{\pi}\frac{k_{ws1}N_s}{2}\sqrt{2}I_s = \frac{6\sqrt{2}}{\pi}k_{ws1}N_sI_s \tag{8.56}$$

In a synchronous machine, there is a single-phase field winding, which is supplied with direct current. The peak value of the total field-winding current linkage becomes for two poles magnetically in series

$$2\hat{\Theta}_r = \frac{4}{\pi}k_{wr1}N_rI_{rfDC} \tag{8.57}$$

To express armature current as a function of the direct current of the rotor, the current linkages are set to be equal.

$$I_s = \frac{k_{wr1}N_rI_{rfDC}}{k_{ws1}N_s}\frac{\pi}{6\sqrt{2}}\frac{4}{\pi} = \frac{2}{3\sqrt{2}}\frac{k_{wr1}N_r}{k_{ws1}N_s}I_{rfDC} \tag{8.58}$$

The relationship between the fundamental of the stator current and the magnetizing DC current is therefore as follows.

$$\frac{I_s}{I_{rfDC}} = k_{ri} = \frac{\sqrt{2}}{3}\frac{k_{wr1}N_r}{k_{ws1}N_s} \tag{8.59}$$

The current linkages given by the equations, if set equal, do not really result exactly in equal flux densities in the air gap since the leakage fluxes of the windings are not equal, and therefore unequal parts of the stator and rotor current linkages are applied in the air gap.

Assuming a time instant, as illustrated in Figure 8.22, yields the same result, but with one of the stator phases currentless. Therefore, for a sinusoidal supply, the current in the other two phases can be expressed

$$i = \frac{\sqrt{3}}{2}I_s\sqrt{2} \tag{8.60}$$

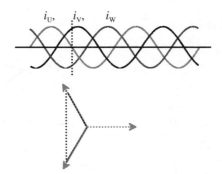

Figure 8.22 Currents of a symmetrical three-phase system, when the current of one phase is zero ($i_U = 0$) and the armature behaviour resembles single-phase behaviour with instantaneous currents at phase angle 60° or 300°, both 30° from peak.

The sum current of the instantaneous currents of the phases that corresponds to the common current linkage may therefore be written as follows.

$$i_{com} = 2\frac{\sqrt{3}}{2}I_s\sqrt{2}\cos 30° = \frac{3}{2}I_s\sqrt{2} \qquad (8.61)$$

The current linkage at time instant corresponding to single-phase stator (one phase current is zero) is therefore

$$2\hat{\Theta}_s = \frac{4}{\pi}\frac{3}{2}I_s\sqrt{2}k_{ws1}N_s \qquad (8.62)$$

and the current linkage of the field winding is the same as previously. Therefore, the relationship between the currents is realized as before.

8.7.2 The salient-pole machine

A salient-pole machine has three different magnetic air gaps: 1) the first is the one seen by the rotor current linkage, 2) the second is the one seen on the d-axis by the stator current linkage, and 3) the third is the one seen on the q-axis by the stator current linkage.

The air gap seen by rotor pole excitation is usually shaped by the pole shoes so the d-axis sees, as much as possible, a sinusoidal, no-load, rotor-excitation-produced flux density distribution. The pole surface must be shaped so that the length of the flux line is inversely proportional to the cosine of the electrical angle when the frame of reference is fixed in the middle of the pole shoe. This pole shoe shape generates a sinusoidally distributed magnetic flux density in the air gap.

The stator winding is constructed so its current linkage is also sinusoidally distributed in the air gap. This current linkage produces flux density in the air gap. Since the air gap is shaped to produce sinusoidal rotor excitation flux, the flux generated by the stator is clearly not sinusoidal. The three-phase winding of the stator produces its own current linkage, and correspondingly, its own flux density component to the air gap. In normal operation, the air-gap flux density comprises the flux components generated by all the machine's windings.

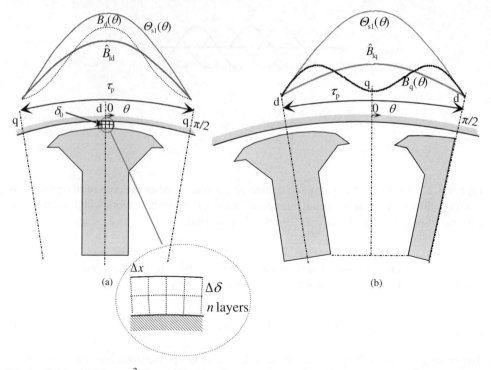

Figure 8.23   (a) The $\cos^2$-shaped flux density wave B_d produced in an air gap with stator current linkage shaped by the pole shoe. The flux density wave is generated by a cosinusoidal current linkage Θ_{s1} falling along the d-axis of the stator. The figure also illustrates the corresponding equivalent cosinusoidal fundamental harmonic $\hat{B}_{1d}$. (b) The cosinusoidal current linkage distribution Θ_{s1} of the stator occurring at the q-axis generates the curve B_q. The peak value of the corresponding equivalent cosinusoidal fundamental harmonic flux density curve is $\hat{B}_{1q}$.

To understand how the stator magnetizes a shaped air gap ($\delta = \delta_0/\cos\theta$), assume that $\hat{\Theta}'_d$ of the stator current linkage amplitude is exerted along centreline d of the pole. The magnetic voltage varies along the pole pitch, therefore, as follows.

$$\Theta'_d(\theta) = \hat{\Theta}'_d \cos\theta \tag{8.63}$$

where θ is the electrical angle along the air-gap periphery, see Figure 8.23. The permeance dΛ of the passage from this point to the rotor surface through surface d$S = Dxl$ and through n potential layers is

$$d\Lambda = \mu_0 \frac{dS}{n\Delta\delta} = \mu_0 \frac{Dld\theta}{2p} \frac{\cos\theta}{\delta_{0e}} \tag{8.64}$$

where δ_{0e} is the air gap in the middle of the pole, corrected with the Carter factor

The magnetic flux density at the point θ is therefore

$$B_d(\theta) = \frac{d\Phi}{dS} = \frac{\mu_0}{\delta_{0e}} \hat{\Theta}'_d \cos^2\theta \tag{8.65}$$

The proportion of the air-gap flux density produced by the stator current linkage is, therefore, proportional to the square of the cosine. However, this density function is often replaced by its fundamental harmonic, that is, by a cosine function with an equal flux. The condition for keeping the magnitude of the flux equal is expressed thusly.

$$\frac{\mu_0}{\delta_{0e}} \hat{\Theta}'_d \int\limits_{-\pi/2}^{+\pi/2} \cos^2 \theta d\theta = \hat{B}_{1d} \int\limits_{-\pi/2}^{+\pi/2} \cos\theta d\theta \tag{8.68}$$

The amplitude of the corresponding cosine function is therefore

$$\hat{B}_{1d} = \frac{\pi}{4} \frac{\mu_0}{\delta_{0e}} \hat{\Theta}'_d = \frac{\mu_0}{\delta_{de}} \hat{\Theta}'_d \tag{8.69}$$

In the latter form of Equation (8.69), the equivalent air gap δ_{de}, which the stator current linkage meets, is equivalent. Its theoretical value is (see more thorough analysis in Pyrhönen et al. (2014))

$$\delta_{de} = \frac{4\delta_{0e}}{\pi} \tag{8.70}$$

Figure 8.23a illustrates the effect of this air gap. Equation (8.70) describes the air gap experienced by the stator for a salient-pole machine when the air gap is shaped to produce a sinusoidal distribution in the presence of rotor magnetization. In reality, the distance at the pole edge from the stator to the rotor cannot be infinite, so the theoretical value of Equation (8.70) is not realized exactly. Precise values for this equivalent air gap are best determined using an FE method; however, it is also possible to calculate accurate results manually.

Figure 8.23b illustrates the quadrature air-gap definition. The magnetic voltage axis of the stator is imagined at the q-axis of the machine. The flux density curve on the q-axis is sketched, and the flux Φ_q is calculated. The flux density amplitude corresponding to this flux can be written as follows.

$$\hat{B}_{1q} = \frac{p\,\Phi_q}{Dl} = \frac{\mu_0}{\delta_{qe}} \hat{\Theta}'_q \tag{8.71}$$

where δ_{qe} is the equivalent quadrature air gap. All the air-gap current linkages are set equal. That is, $\hat{\Theta}_f = \hat{\Theta}'_d = \hat{\Theta}'_q$. For this case, the equivalent air gaps behave like inverses of the flux density amplitudes.

$$\hat{B}_\delta : \hat{B}_{1d} : \hat{B}_{1q} = \frac{1}{\delta_{0e}} : \frac{1}{\delta_{de}} : \frac{1}{\delta_{qe}} \tag{8.72}$$

Direct and q-axis equivalent air gaps are calculated using this expression of proportionality, and the direct and quadrature magnetizing inductances of the stator are inversely proportional to these air gaps. Consequently, the q-axis synchronous inductance of a traditional salient-pole machine is usually notably lower than the direct synchronous inductance.

In case of air gap producing sinusoidal excitation flux density, the q-axis equivalent air gap can be evaluated, also based on (Heikkilä, 2002), as follows.

$$\delta_{qe} \approx \frac{3\pi\delta_e}{4 \sin^2\left(\frac{\alpha\pi}{2}\right)} \tag{8.73}$$

In the equation, α is the per-unit value of the pole shoe width, and the $\delta = \delta_0/\cos\theta$ form can be assumed valid for the air gap. A generally suitable value for α is 0.9.

Therefore, the current linkage of the rotor produces a sinusoidal flux density distribution on the stator surface, the peak value of which converges with the smallest air gap δ_0. Stator slotting increases the effective air gap by an amount proportional to the Carter coefficient k_C (see Pyrhönen et al 2014).

$$\delta_0' = k_C \delta_0 \tag{8.74}$$

Observed from the stator, the fundamental harmonic of the direct air gap theoretically behaves according to Equation (8.70). The length of the air gap has therefore become $4/\pi$ times δ_0'. See Equation (8.70).

Since the field winding is positioned only along on the d-axis, determining the ratio of stator and rotor excitation current is only necessary in this direction. The ratio of currents is determined by setting the fundamental harmonics of the flux densities caused by the stator and rotor currents to be equal. It is assumed there are no leakage fluxes and that the reluctance of iron is zero. Therefore, all the current linkages are applied to the air gaps. Stator current linkage amplitude can then be written thusly.

$$\hat{\Theta}_s = \frac{3}{p\pi} k_{ws1} N_s \sqrt{2} I_s \tag{8.75}$$

The fundamental flux density can be calculated accordingly.

$$\hat{B}_{1s} = \frac{\frac{3}{p\pi} k_{ws1} N_s \sqrt{2} I_s}{\frac{4}{\pi} \delta_0'} \mu_0 = \frac{3 k_{ws1} N_s \sqrt{2} I_s}{4 p \delta_0'} \mu_0 \tag{8.76}$$

The square-wave current linkage of the rotor has a magnitude determined by the following.

$$\Theta_r = \frac{1}{2} \frac{N_r}{p} I_{rfDC} \tag{8.77}$$

Since the air gap is shaped to produce a sinusoidal density distribution, the peak value of the flux density becomes

$$\hat{B}_{1r} = \frac{\frac{1}{2} \frac{N_r}{p} I_{rfDC}}{\delta_0'} \mu_0 \tag{8.78}$$

The current ratio is revealed by comparing Equation (8.76) and Equation (8.78).

$$k_{ri} = \frac{I_s}{I_{rfDC}} = \frac{N_r}{\frac{3\sqrt{2}}{2} k_{ws1} N_s} \tag{8.79}$$

This result is equivalent to the one obtained previously. See Equation (8.59). The current ratio is the same for both a salient-pole and nonsalient-pole machine if the rotor pole shoe of the salient-pole machine is shaped to produce a sinusoidal distribution. Otherwise, the current ratio must be reconsidered by employing the partly empirical form factors given in the literature.

If it can be done, the current ratio can easily be determined via permanent short-circuit testing. In a short circuit, the current linkage of the stator current should cancel the current linkage of the rotor current. Therefore, a short-circuit test determines the current ratio directly. A certain magnetizing direct current of the rotor has a corresponding stator short-circuit current at steady state.

8.7.3 The referring factor

When referring the resistances from the single-phase rotor winding to the three-phase stator winding or vice versa, the single-phase rotor DC can be considered to form the current vector of an equivalent three-phase rotor. The rotor is represented by a nonsalient-pole sheet rotor with a three-phase winding. The three-phase current flowing in the winding generates an equal flux density in the air gap as the rotating pole wheel is magnetized by the direct current.

According to the energy principle, an equal power loss must occur in both the actual rotor and in the space vector equivalent circuit. When operating with equivalent three-phase rotor quantities suitable for the equivalent circuit, the rotor power is determined as P_{r3ph}. And, this power must be equal to the power calculated using actual rotor quantities.

$$P_{r3ph} = 3R_{r3ph}I_{r3ph}^2 = P_f = I_{fDC}^2 R_{fDC} \tag{8.80}$$

Therefore

$$R_{r3ph} = \frac{I_{fDC}^2}{3I_{r3ph}^2} R_{fDC} \tag{8.81}$$

Substituting the effective value I_{r3ph} of the rotor phase current for the respective stator current I_s results in the following expression.

$$R_{r3ph}' = \frac{I_{fDC}^2}{3I_s^2} R_{fDC} = \frac{1}{3k_{ri}^2} R_{fDC} \tag{8.82}$$

The same result can also be obtained by replacing the single-phase rotor winding with a three-phase winding that has an equal number of turns. The current linkages per pole-pair for the single-phase and the equivalent three-phase rotors are set equal to get the following expression for a two-pole machine.

$$\hat{\Theta}_{r3ph} = \hat{\Theta}_{fDC} \rightarrow \frac{3}{2}\frac{4}{\pi}\frac{k_{ws1}N_{rp}}{2}\sqrt{2}I_{r3ph} = \frac{4}{\pi}\frac{k_{ws1}N_{rp}}{2}I_{fDC} \tag{8.83}$$

N_{rp} is the number of turns in the rotor pole-pair. For the equivalent three-phase current of the rotor, this yields

$$I_{r3ph} = \frac{\sqrt{2}}{3}I_{fDC} \tag{8.84}$$

According to the energy principle, the power losses in the actual rotor and in the equivalent three-phase circuit must be equal, so when operating with three-phase quantities, the rotor power is determined as follows.

$$P_{r3ph} = 3R_{r3ph}I_{r3ph}^2 = 3R_{r3ph}\left(\frac{\sqrt{2}}{3}I_{fDC}\right)^2 = \frac{2}{3}R_{r3ph}I_{fDC}^2 \tag{8.85}$$

This power must be equal to the power calculated with the actual rotor quantities, and therefore the resistance of the equivalent three-phase rotor becomes

$$R_{r3ph} = \frac{3}{2}R_{fDC} \tag{8.86}$$

Since both the equivalent three-phase rotor and the stator have the same number of phases, the resistance is now referred to the stator by employing the current ratio determined by the turns and winding factors.

$$R'_{r3ph} = \frac{3}{2}R_{fDC}\left(\frac{k_{ws1}N_s}{k_{wr1}N_r}\right)^2 = \frac{1}{3}R_{fDC}k_{ri}^{-2} \tag{8.87}$$

8.7.4   Referring rotor quantities to the stator when applying space vectors

When operating with equivalent three-phase quantities in the rotor, the effective value of the three-phase current of the rotor was as follows. See Equation (8.84).

$$I_{r3ph} = \frac{\sqrt{2}}{3}I_{fDC} \tag{8.88}$$

Using sinusoidal quantities, the peak value of this current becomes

$$\hat{i}_{r3ph} = \frac{2}{3}I_{fDC} \tag{8.89}$$

If this equation is applied to construct the rotor current space vector, a rotor current vector length is determined as shown in Figure 8.24, for instance, when the phase U current is at its positive maximum, and the phase V and W currents are at half of their negative maximum values.

The magnitude of the space vector is, therefore, $(2/3)I_{fDC}$. Setting the space vector expressions of the stator and rotor current linkages to be equal results in the following (N_{sp} is the number of turns in stator phase per pole).

$$\frac{2}{3}I_{fDC}N_{rp} = |i'_r|k_{ws1}N_{sp} \tag{8.90}$$

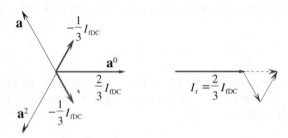

Figure 8.24   Formation of the electrical current space vector for an equivalent three-phase rotor with the phase U current at its positive maximum.

This yields the ratio of the magnitudes of the current vectors.

$$\frac{|i'_r|}{I_{fDC}} = \frac{2}{3}\frac{N_{rp}}{k_{ws1}N_{sp}} = k_{riav} \tag{8.91}$$

This result differs from the DC referring by $\sqrt{2}$. In this system, the phase resistance of the rotor is $R_{r3ph} = 3/2(R_{fDC})$. Power can then be calculated by applying space vectors. By setting this power equal to the effective magnitude of the direct current, the following expressions can be written

$$P = \frac{3}{2}ui^* = \frac{3}{2}\left(\frac{2}{3}\frac{N_{rp}}{k_{ws1}N_{sp}}I_{fDC}\right)^2 R'_r = I^2_{fDC}R_{rfDC}$$

$$= \frac{3}{2}k^2_{riav}I^2_{fDC}R'_r = I^2_{fDC}R_{fDC} \tag{8.92}$$

When operating with space vectors, Equation (8.92) can be simplified as follows.

$$R'_r = \frac{2}{3}\frac{1}{k^2_{riav}}R_{fDC} \tag{8.93}$$

8.8   The vector diagram for a synchronous machine

Often, a space-vector diagram (shortly vector diagram) is used to analyse the operation of a machine. A vector diagram is quite similar to an RMS value phasor diagram. However, unlike the phasor diagram, the vector diagrams rotate. Figure 8.25 is a vector diagram for a synchronous machine in the rotor oriented dq reference frame. The machine represented is operating as a motor that has a rotor spinning in the counterclockwise direction.

The stator flux linkage ψ_s of the machine can be calculated by integrating the stator voltage vector u'_s with the resistive voltage drop removed ($u'_s = u_s - i_sR_s$). The flux linkage, therefore, follows about 90° behind the stator voltage vector. Because damper-winding currents are present the Figure represents a dynamic state and the stator flux-linkage vector is not perpendicular to the voltage vector. Perpendicularity occurs only in steady state when operating with sinusoidal quantities. In PWM-supplied machines the stator voltage PWM signal fundamental may be about perpendicular to the stator flux-linkage vector. However, individual voltage vectors generated by the converter and the winding are rarely found in positions perpendicular to the stator flux linkage.

The air-gap flux linkage ψ_m differs from the stator flux linkage by the amount of the leakage flux linkage $\psi_{s\sigma} = L_{s\sigma}i_s$. The field-winding current produces an equivalent flux-linkage component $L_{md}i_f$, which is transformed towards the air-gap flux linkage by the armature reactions in the d- and q-axis directions $L_{md}i_d$ and $L_{mq}i_q$, respectively. Each flux linkage, including the leakage flux linkage, induces its own electromotive force (back emf).

For a motor control, the induced voltages $e_s = -j\omega_s\psi_s$ and $e_m = -j\omega_s\psi_m$ are important. The back emf e_s, which is induced by the stator flux linkage, is the induced voltage that almost totally opposes the supply voltage. When it is added to the supply voltage, the small remaining back emf determines machine current $i_s = (u_s + e_s)/R_s$. The induced voltage produced by the field-winding current $-j\omega_sL_{md}i_f$ is measurable only at no-load. Under load it is just imaginary, because the electric machine cannot usually generate the high flux-linkage magnitude

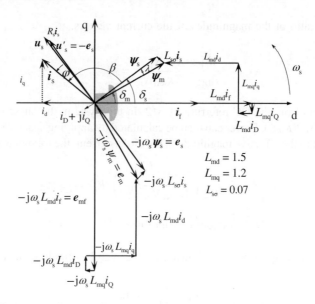

Figure 8.25 The vector diagram for a general case of a synchronous machine in the rotor reference frame. ω_s is the electrical angular speed. δ_s is the load angle of the stator flux linkage, δ_m is the load angle of the air-gap flux linkage, β is the angle between the stator flux linkage and the stator current, and φ is the angle between the stator current and stator voltage. i_s is the stator current vector, and u_s is the stator voltage vector. ψ_s is the vector of the stator flux linkage, and ψ_m is the vector of air-gap flux linkage. Finally, e_s is the back electromotive force induced by the stator flux linkage. The figure also shows a small damper-winding current $i_D + j\, i_Q$, which means machine operation is transient. The entire vector diagram rotates counterclockwise at angular speed ω_s.

represented by the nonsaturating value of $L_{md}i_f$. The induced voltages caused by the armature and damping reactions $-j\omega_s L_{md}i_d$, $-j\omega_s L_{mq}i_q$, $-j\omega_s L_{md}i_D$, and $-j\omega_s L_{mq}i_Q$ transform this into the actual air-gap flux linkage.

The back emf induced by the rotation of the air-gap flux linkage becomes observable if the stator of the machine suddenly becomes currentless. Particularly due to field weakening, the magnitude of the air-gap flux linkage can be notably larger than the magnitude of the stator flux linkage. If control of this type of machine is lost for some reason, the situation can become dangerous, because the terminal voltage of the machine can rise excessively – first to a value corresponding to the air-gap flux linkage and finally to the maximum value allowed by the machine iron, which corresponds to the field-winding current.

Figure 8.26 shows the vector diagram for the synchronous machine without damper currents, that is, at steady state.

For practical control purposes, the vector diagram can be simplified by omitting several induced voltages. See Figure 8.27.

Figure 8.28 depicts the synchronous machine currents in the different reference frames revealing that field-winding current i_f contributes to torque production. In the stator flux linkage to torque reference frame ψ_sT, field-winding current includes a component i_{fT} that clearly decelerates the machine.

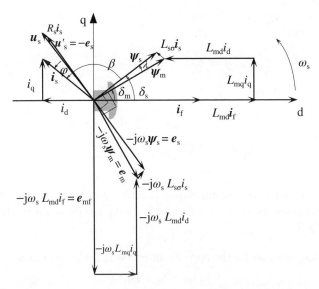

Figure 8.26 Steady-state vector diagram of a synchronous machine with no damper currents.

The different coordinate transformation equations must be presented in component form when applying the two-axis model. These component forms can be expressed by defining stator current as the subject quantity. However, similar equations hold also for stator voltages and stator flux linkages. When transferring from three-phase to two-phase quantities, the following equations apply.

$$i_{sx} = \frac{2}{3}\left[i_{sU} - \frac{1}{2}(i_{sV} + i_{sW})\right] \tag{8.94}$$

$$i_{sy} = \frac{1}{\sqrt{3}}[i_{sV} - i_{sW}] \tag{8.95}$$

For the normally nonexistent zero sequence current,

$$i_{s0} = \frac{1}{3}[i_{sU} + i_{sV} + i_{sW}] \tag{8.96}$$

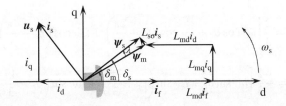

Figure 8.27 Simplified vector diagram for a synchronous machine. The machine runs at the power factor $\cos\varphi = 1$.

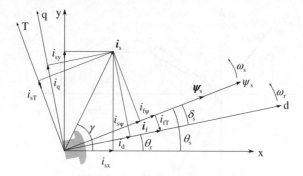

Figure 8.28 Stator and field-winding current in different reference frames. Represented are the stator reference frame xy, the rotor reference frame dq, and the stator flux linkage to torque reference frame ψ_sT.

The transformation formulas for the opposite direction of the transformation are as follows.

$$i_{sU} = i_{sx} + i_{s0} \tag{8.97}$$

$$i_{sV} = -\frac{1}{2}i_{sx} + \frac{\sqrt{3}}{2}i_{sy} + i_{s0} \tag{8.98}$$

$$i_{sW} = -\frac{1}{2}i_{sx} - \frac{\sqrt{3}}{2}i_{sy} + i_{s0} \tag{8.99}$$

As stated previously, when changing over from the stator to the rotor reference frame, the angle between the frames, which is rotor position angle θ_r, must be known. Since the zero component is not included in the current vector, it must be taken into account separately.

$$i_d = i_{sx}\cos\theta_r + i_{sy}\sin\theta_r \tag{8.100}$$

$$i_q = -i_{sx}\sin\theta_r + i_{sy}\cos\theta_r \tag{8.101}$$

$$i_{s0} = i_{s0} \tag{8.102}$$

For the opposite direction of the transformation, the equations are written

$$i_{sx} = i_d\cos\theta_r - i_q\sin\theta_r \tag{8.103}$$

$$i_{sy} = i_d\sin\theta_r + i_q\cos\theta_r \tag{8.104}$$

$$i_{s0} = i_{s0} \tag{8.105}$$

When transferring directly from the phase quantities to the rotor reference frame, the following equations apply.

$$i_d = \frac{2}{3}\left[i_{sU}\cos\theta_r + i_{sV}\cos\left(\theta_r - \frac{2\pi}{3}\right) + i_{sW}\cos\left(\theta_r - \frac{4\pi}{3}\right)\right] \tag{8.106}$$

$$i_q = -\frac{2}{3}\left[i_{sU}\sin\theta_r + i_{sV}\sin\left(\theta_r - \frac{2\pi}{3}\right) + i_{sW}\sin\left(\theta_r - \frac{4\pi}{3}\right)\right] \tag{8.107}$$

$$i_{s0} = \frac{1}{3}\left[i_{sU} + i_{sV} + i_{sW}\right] \tag{8.108}$$

For the opposite direction of the transformation

$$i_{sU} = i_d \cos \theta_r - i_q \sin \theta_r + i_{s0} \tag{8.109}$$

$$i_{sV} = i_d \cos\left(\theta_r - \frac{2\pi}{3}\right) - i_q \sin\left(\theta_r - \frac{2\pi}{3}\right) + i_{s0} \tag{8.110}$$

$$i_{sW} = i_d \cos\left(\theta_r - \frac{4\pi}{3}\right) - i_q \sin\left(\theta_r - \frac{4\pi}{3}\right) + i_{s0} \tag{8.111}$$

8.9   Torque production for a synchronous machine

A synchronous electric machine operates according to the cross-field principle. The air-gap flux linkage in particular is significant for torque production. It is the current component perpendicular to the air-gap flux linkage that produces torque. As previously presented, expressions for electrical torque vector T_e can be given in various forms. These are briefly recapitulated here. The general equation for the torque of any rotating-field machine can be given as the following vector representation.

$$T_e = \frac{3}{2} p \boldsymbol{\psi}_s \times \boldsymbol{i}_s \tag{8.112}$$

where p is the number of pole-pairs, $\boldsymbol{\psi}_s$ is the stator flux-linkage space vector, and $\boldsymbol{i}_s$ is the stator-current space vector.

The vector diagrams for a synchronous machine do not show any contradiction between this representation and the role of the aforementioned air-gap flux linkage. When the angle β between the vectors is known the torque equation can also be expressed in scalar form.

$$T_e = \frac{3}{2} p |\boldsymbol{\psi}_s| |\boldsymbol{i}_s| \sin(\beta) \tag{8.113}$$

According to these equations, the magnitude of torque depends on the magnitudes of the vectors and the angles between them. Moreover, the sign of the torque expressed in scalar form also depends on the angle between the vectors. Therefore, in the case of synchronous machines, changing over from motor operation to generator operation or vice versa is accomplished by simply changing the sign of the angle between the vectors.

When Equation (8.112) is represented by its components in the rotor reference frame, it too can be written in scalar form.

$$T_e = \frac{3}{2} p \left(\psi_d i_q - \psi_q i_d \right) \tag{8.114}$$

Substituting the stator flux-linkage components and applying the inductances and currents results in the following expression.

$$T_e = \frac{3}{2} p \left[\left(L_d - L_q \right) i_d i_q + L_{md} (i_f + i_D) i_q - L_{mq} i_Q i_d \right] \tag{8.115}$$

Finally, by decomposing the middle term, the equation can be rewritten as follows.

$$T_e = \frac{3}{2}p\left[(L_d - L_q)i_d i_q + L_{md}i_f i_q + L_{md}i_D i_q - L_{mq}i_Q i_d\right] \tag{8.116}$$

Equation (8.116) reveals that electrical torque comprises the following.

- the reluctance torque $(L_d - L_q)i_d i_q$ resulting from the magnetic anisotropy, which only occurs in salient-pole machines,

- a component related to the field-winding current and the quadrature stator current, which is the actual torque of the machine, and

- two terms describing the torque components of the damper-winding currents. These components occur only in transients.

8.10   Simulating an electrically excited salient-pole machine via constant parameters

All rotor quantities are referred to the stator voltage level before the per-unit values are constructed. The stator voltage equation is expressed in the rotor reference frame as

$$u_s^r = R_s i_s^r + \frac{d\psi_s^r}{dt} + j\omega_s\psi_s^r \tag{8.117}$$

When using per-unit values, the above can be divided by the base value of the voltage. And, if grouped appropriately, the expression becomes

$$\frac{u_s^r}{\hat{u}_N} = \frac{R_s \hat{i}_N}{\hat{u}_N} \frac{i_s^r}{\hat{i}_N} + \frac{d\psi_s^r}{d(\omega_N t)} \frac{\omega_N}{\hat{u}_N} + \frac{j\omega_s}{\omega_N} \frac{\omega_N \psi_s^r}{\hat{u}_N} \tag{8.118}$$

These equation values were introduced in Chapter 2, but are repeated here for convenience. Therefore, the voltage equation given in per-unit values becomes

$$u_{spu}^r = R_{spu}i_{spu}^r + \frac{d\psi_{spu}^r}{d\tau} + j\omega_{spu}\psi_{spu}^r \tag{8.119}$$

The form of this equation is the same as the original equation. However, the per-unit time may cause problems. If normal time is used in the per-unit value equations, the equation can be rewritten as

$$u_{spu}^r = R_{spu}i_{spu}^r + \frac{1}{\omega_N}\frac{d\psi_{spu}^r}{dt} + j\omega_{spu}\psi_{spu}^r \tag{8.120}$$

8.11   The current equations for a synchronous machine

The relationship between the flux linkages and currents of a synchronous machine must be investigated in the rotor reference frame to avoid the division of the direct and quadrature quantities into their components. The dependence between the flux linkages and currents is determined by an inductance matrix. Omitting the pu (per-unit value) notations, the equation takes the following form.

$$\psi = \mathbf{L} \cdot \mathbf{i} \tag{8.121}$$

or

$$
\begin{bmatrix}
\psi_d \\
\psi_q \\
\psi_D \\
\psi_Q \\
\psi_f
\end{bmatrix}
=
\begin{bmatrix}
L_d & 0 & L_{md} & 0 & L_{md} \\
0 & L_q & 0 & L_{mq} & 0 \\
L_{md} & 0 & L_D & 0 & L_{fD} \\
0 & L_{mq} & 0 & L_Q & 0 \\
L_{md} & 0 & L_{fD} & 0 & L_f
\end{bmatrix}
\cdot
\begin{bmatrix}
i_d \\
i_q \\
i_D \\
i_Q \\
i_f
\end{bmatrix}
\tag{8.122}
$$

where

$$L_d = L_{md} + L_{s\sigma} \tag{8.123}$$

$$L_q = L_{mq} + L_{s\sigma} \tag{8.124}$$

$$L_D = L_{md} + L_{D\sigma} + L_{k\sigma} \tag{8.125}$$

$$L_Q = L_{mq} + L_{Q\sigma} \tag{8.126}$$

$$L_f = L_{md} + L_{f\sigma} + L_{k\sigma} \tag{8.127}$$

$$L_{fD} = L_{md} + L_{k\sigma} \tag{8.128}$$

The currents can be obtained from the flux linkages by applying the inverse inductance matrix $\mathbf{K} = \mathbf{L}^{-1}$.

$$
\begin{bmatrix}
i_d \\
i_q \\
i_D \\
i_Q \\
i_f
\end{bmatrix}
=
\begin{bmatrix}
k_{sd} & 0 & k_{dD} & 0 & k_{df} \\
0 & k_{sq} & 0 & k_{qQ} & 0 \\
k_{dD} & 0 & k_D & 0 & k_{fD} \\
0 & k_{qQ} & 0 & k_Q & 0 \\
k_{df} & 0 & k_{fD} & 0 & k_f
\end{bmatrix}
\cdot
\begin{bmatrix}
\psi_d \\
\psi_q \\
\psi_D \\
\psi_Q \\
\psi_f
\end{bmatrix}
\tag{8.129}
$$

where

$$k_{sd} = \frac{L_f L_D - L_{fD}^2}{\Delta_1} \tag{8.130}$$

$$k_{sq} = \frac{L_Q}{\Delta_2} \tag{8.131}$$

$$k_D = \frac{L_f L_d - L_{md}^2}{\Delta_1} \tag{8.132}$$

$$k_Q = \frac{L_q}{\Delta_2} \tag{8.133}$$

$$k_f = \frac{L_d L_D - L_{md}^2}{\Delta_1} \tag{8.134}$$

$$k_{dD} = (L_{fD} - L_f)\frac{L_{md}}{\Delta_1} \tag{8.135}$$

$$k_{df} = (L_{fD} - L_D)\frac{L_{md}}{\Delta_1} \tag{8.136}$$

$$k_{qQ} = \frac{-L_{mq}}{\Delta_2} \tag{8.137}$$

$$k_{fD} = \frac{L_{md}^2 - L_{fD}L_d}{\Delta_1} \tag{8.138}$$

The denominator components Δ_1 and Δ_2 can be written

$$\Delta_1 = 2 \cdot L_{md}^2 \cdot L_{fD} - L_{md}^2 \cdot L_D - L_{md}^2 \cdot L_f + L_d \cdot L_D \cdot L_f - L_{fD}^2 \cdot L_d \tag{8.139}$$

$$\Delta_2 = L_Q \cdot L_q - L_{mq}^2 \tag{8.140}$$

8.12   Simulating a synchronous machine in a discrete-time system

To represent a synchronous machine mathematically, a simulation model can be prepared with the following inputs: stator voltage vector $\boldsymbol{u}_{spu}$, excitation voltage u_{fpu}, and mechanical angular speed n (to integrate the rotor angle). The outputs are stator current $\boldsymbol{i}_{spu}$, field-winding current i_{fpu}, damper-windings current i_{Dpu} and i_{Qpu}, and electric machine torque T_{epu}. Naturally, the simulation presumes a knowledge of the synchronous machine inductances and resistances.

The steps of the simulation are as follows.

0. Appropriate initial values are initialized for the flux linkage and current parameters, and the inductance matrix is calculated

1. The derivatives of the stator flux linkage $\psi_{spu}(k)$ are calculated in the stator reference frame in the x and y directions by applying the stator voltage $\boldsymbol{u}_{spu}(k)$

$$\frac{d\psi_{sxpu}(k)}{dt} = \omega_N\left[u_{sxpu}(k) - i_{sxpu}(k-1)R_{spu}\right] \tag{8.141}$$

$$\frac{d\psi_{sypu}(k)}{dt} = \omega_N\left[u_{sypu}(k) - i_{sypu}(k-1)R_{spu}\right] \tag{8.142}$$

2. The derivatives of the flux linkages of the damper winding and the field winding are calculated in the rotor reference frame

$$\frac{d\psi_{fpu}}{dt}(k) = \omega_N\left[u_{fpu}(k) - i_{fpu}(k-1)R_{fpu}\right] \tag{8.143}$$

$$\frac{d\psi_{Dpu}}{dt}(k) = -\omega_N i_{Dpu}(k-1)R_{Dpu} \tag{8.144}$$

$$\frac{d\psi_{Qpu}(k)}{dt} = -\omega_N i_{Qpu}(k-1)R_{Qpu} \tag{8.145}$$

3. New values are integrated for the flux linkages by applying the obtained derivatives with Δt as calculation time step

$$\psi_{sx}(k) = \psi_{sx}(k-1) + \frac{d\psi_{sx}}{dt}(k)\Delta t$$

$$\psi_{sy}(k) = \psi_{sy}(k-1) + \frac{d\psi_{sy}}{dt}(k)\Delta t$$

$$\psi_D(k) = \psi_D(k-1) + \frac{d\psi_D}{dt}(k)\Delta t \qquad (8.146)$$

$$\psi_Q(k) = \psi_Q(k-1) + \frac{d\psi_Q}{dt}(k)\Delta t$$

$$\psi_f(k) = \psi_f(k-1) + \frac{d\psi_f}{dt}(k)\Delta t$$

4. New values are determined for the currents by applying the inverse inductance matrix

$$i(k) = \mathbf{K\Psi}(k) \qquad (8.147)$$

5. The electrical torque is calculated for the machine (where superscripts "re" and "im" refer to real and imaginary components)

$$T_{epu}(k) = \psi_s^{re}(k)i_s^{im}(k) - \psi_s^{im}(k)i_s^{re}(k) \qquad (8.148)$$

6. The cycle repeats from item 1

8.13 The implementation of vector control for a synchronous machine

Figure 8.29 presents a general signal processing diagram for a vector-controlled synchronous machine drive. The control loop feedbacks in the diagram are typically proportional-integral (PI) controllers. First, the currents and the rotor position of the motor are measured. Next, a two-phase rotor-oriented representation is adopted. Finally, the motor model is applied to calculate the current components producing the flux linkage and torque. The speed controller establishes a torque reference, which in turn yields the respective current reference. The flux limiter is an essential element when considering field-weakening control. Cross-coupling-effects are eliminated by a circuit designed for the purpose. Decoupling is examined in more detail in Example 8.1. The angular speeds shown in the figure are electrical angular speeds.

First, the reference values for the field-winding current components and for the current components producing electrical torque must be determined. These current references are calculated in the stator flux-linkage–torque (ψ_sT) reference frame. The axes of this reference frame are known as the flux-linkage and torque axes. An appropriately expedient flux-linkage axis can and should be chosen.

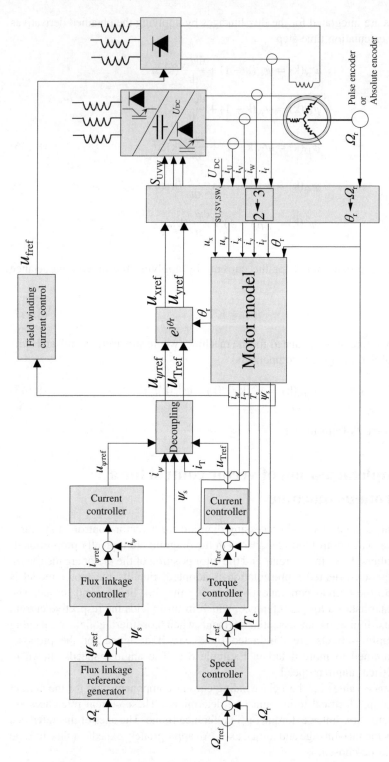

Figure 8.29 Current-vector control for a synchronous machine in a flux-linkage–oriented system.

For this discussion, the stator flux-linkage direction is selected as the flux-linkage axis, and torque will be produced by the current component normal to this axis. The equation for torque in an AC drive was previously expressed as $T_e = 3/2p(\psi_s \times i_s)$. This equation indicates that the current component perpendicular to the stator flux linkage, in principle, produces the torque. However, torque can also be calculated from $T_e = 3/2p(\psi_m \times i_s)$, and the highest torque per ampere comes from current perpendicular to the air-gap flux linkage. However, stator flux linkage is selected in this case as the reference for the vector control orientation. Therefore, the references for current will be designated as the stator flux-linkage–aligned current reference and the torque axis–aligned current reference.

The torque reference comes from the output of the speed controller; the input to the PI controller is the difference between the speed reference and the actual speed value. The speed reference is a setting that is input to the controller by the system controlling the overall process. Because this motor control is based on the stator flux-linkage reference frame, the electrical torque will be proportional to the magnitudes of the stator current and the stator flux linkage. The current reference aligned with the torque axis is simply obtained by dividing the torque reference by the stator flux-linkage magnitude reference value.

The stator flux-linkage axis–oriented current reference is obtained from the output of the flux-linkage controller. The flux-linkage controller is also a PI controller. The input consists of the difference between the field-winding current reference and the actual current value. The stator flux-linkage reference depends on the actual value for rotation speed. Stator flux linkage is kept at the rated value when operating well below the rated speed.

As discussed previously in Chapter 2, stator flux linkage must be decreased at an inversely proportional rate at speeds above rated. The latter speed range is known as the field weakening range. A suitable voltage reserve must be maintained. Therefore, field weakening begins taking effect in highly dynamic drives, for example, at 90% of motor rated speed to maintain a 10% voltage reserve if rated voltage corresponds to the full voltage output of the converter.

The absolute value of the rotation induced back emf vector in the stator reference frame is

$$e_s = \omega_s \psi_s \tag{8.149}$$

Bearing in mind that stator voltage magnitude u_s differs only slightly from the magnitude of the back emf e_s, a simplified expression can be written that neglects the resistive voltage drop.

$$u_s \approx \psi_s \omega_s \tag{8.150}$$

Taking only the proportional relationships into account and from the point of view of the stator flux-linkage absolute value this becomes

$$\psi_s \approx \frac{u_s}{2\pi f} \tag{8.151}$$

Flux linkage in drives remains constant if the ratio of the terminal voltage and the frequency is kept constant. Voltage also remains constant if the flux linkage is reduced inversely proportional to the frequency. Figure 8.30 illustrates, in more detail, the flux-linkage reference generator seen in Figure 8.29. The presented functional block produces a stator flux-linkage reference value for the drive controller.

Determining the field weakening point is an essential part of electrical drive design, because the speed at which field weakening is applied influences motor sizing. Not

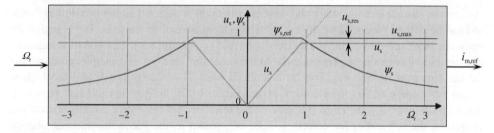

Figure 8.30 A stator flux-linkage absolute value reference generator. The functional block produces a magnetizing current reference in the stator flux-linkage oriented coordinate system. At angular velocities under 0.9 per unit, ψ_s will be held constant at its rated value. At speeds greater than 0.9 per unit, the magnitude of the stator flux linkage must be reduced to maintain the voltage reserve $u_{s,res}$.

considering the influence of field weakening results in a motor that is larger than it needs to be to reach the desired torques at lower rotational speeds. To appropriately dimension the machine, selecting a correct field weakening point is essential. However, there is no simple mathematical expression available to select rated speed. A techno-economic analysis must be carried out to determine the optimal field-weakening point.

The actual values for stator and air-gap flux linkages are calculated from the flux-linkage equations in real time during operation of the drive. To make these calculations, real-time values for phase current, field-winding current, rotor position angle, and damper-winding current must be measured and made available. The measured three-phase currents are transformed into current components in the rotor reference frame. For the transformation, rotor position angle must be known. As previously stated, the stator d- and q-axis flux-linkage components are as follows.

$$\psi_d = \psi_{md} + L_{s\sigma}i_d = L_{md}(i_d + i_D + i_f) + L_{s\sigma}i_d \tag{8.152}$$

$$\psi_q = \psi_{mq} + L_{s\sigma}i_q = L_{mq}(i_q + i_Q) + L_{s\sigma}i_q \tag{8.153}$$

The d-axis and q-axis components are used to calculate the magnitudes of the flux linkages and the corresponding load angles for the stator and air-gap flux linkages.

$$|\psi_s| = \sqrt{\psi_d^2 + \psi_q^2} \tag{8.154}$$

$$|\psi_m| = \sqrt{\psi_{md}^2 + \psi_{mq}^2} \tag{8.155}$$

$$\delta_s = \arctan\frac{\psi_q}{\psi_d} \tag{8.156}$$

$$\delta_m = \arctan\frac{\psi_{mq}}{\psi_{md}} \tag{8.157}$$

Since it is normally impossible to measure the damper-winding currents, their expressions must be eliminated from the equations and replaced by appropriate time constant representations. The first step for the rotor reference frame is to determine the direct and quadrature flux-linkage components.

$$\psi_{md} = i_{md}L_{md} = L_{md}(i_f + i_d + i_D) \tag{8.158}$$

$$\psi_{mq} = i_{mq}L_{mq} = L_{mq}(i_q + i_Q) \tag{8.159}$$

The voltage equation for the d-axis damper winding of the rotor can be expressed

$$u_D = 0 = R_D i_D + \frac{d\psi_D}{dt} = R_D i_D + \frac{d}{dt}(\psi_{md} + i_D L_{D\sigma}) \tag{8.160}$$

Next, the time constants describing the rotor damping are adopted.

$$\tau_{rD} = \frac{L_{md} + L_{D\sigma}}{R_D} \tag{8.161}$$

which equals the time constant τ''_{d0} in Figure 8.13. Another leakage flux time constant is defined as

$$\tau_{rD\sigma} = \frac{L_{D\sigma}}{R_D} \tag{8.162}$$

Then, the equation for the damper d-axis flux linkage is plugged into the voltage equation.

$$0 = R_D i_D + \frac{d}{dt}(L_{md}(i_f + i_d + i_D) + i_D L_{D\sigma}) \tag{8.163}$$

This expression can be developed somewhat further thusly.

$$0 = i_D + \frac{d}{dt}\left(\frac{L_{md}}{R_D}(i_f + i_d + i_D) + i_D \frac{L_{D\sigma}}{R_D}\right) \tag{8.164}$$

$$0 = i_D + \frac{d}{dt}\left(\frac{L_{md}}{R_D}(i_f + i_d) + i_D \frac{L_{D\sigma} + L_{md}}{R_D}\right) \tag{8.165}$$

Now, the following equations can be formulated.

$$\left[1 + \frac{d}{dt}\frac{L_{D\sigma} + L_{md}}{R_D}\right]i_D = -\frac{d}{dt}\frac{L_{md}}{R_D}(i_f + i_d) \tag{8.166}$$

$$\left[1 + \tau_{rD}\frac{d}{dt}\right]i_D = +\tau_{rD\sigma}\frac{d}{dt}(i_f + i_d) - \tau_{rD}\frac{d}{dt}(i_f + i_d) \tag{8.167}$$

$$i_D + \tau_{rD}\frac{d}{dt}(i_f + i_d + i_D) = \tau_{rD\sigma}\frac{d}{dt}(i_f + i_d) \tag{8.168}$$

Multiplying both sides of the equation by the magnetizing inductance and adding the equation for the air-gap flux linkage to both sides results in the following expression.

$$\tau_{\mathrm{rD}}\frac{d}{dt}L_{md}(i_f + i_d + i_D) + L_{md}(i_f + i_d + i_D) = L_{md}(i_f + i_d) + \tau_{\mathrm{rD}\sigma}\frac{d}{dt}L_{md}(i_f + i_d) \qquad (8.169)$$

The flux-linkage equation can then be rewritten as

$$\tau_{\mathrm{rD}}\frac{d}{dt}\psi_{md} + \psi_{md} = L_{md}(i_f + i_d) + \tau_{\mathrm{rD}\sigma}\frac{d}{dt}L_{md}(i_f + i_d), \qquad (8.169)$$

and the equation can be solved for the air-gap flux linkage in the real and Laplace domains.

$$\psi_{md} = L_{md}(i_f + i_d)\frac{1 + \tau_{\mathrm{rD}\sigma}\dfrac{d}{dt}}{1 + \tau_{\mathrm{rD}}\dfrac{d}{dt}} \Rightarrow \psi_{md}(s) = L_{md}(i_f(s) + i_d(s))\frac{1 + \tau_{\mathrm{rD}\sigma}s}{1 + \tau_{\mathrm{rD}}s} \qquad (8.170)$$

Correspondingly, it is possible to represent, dependent on the q-axis time constants of the damper winding, the q-axis air-gap flux-linkage component.

$$\psi_{mq} = L_{mq}i_q\frac{1 + \tau_{\mathrm{rQ}\sigma}\dfrac{d}{dt}}{1 + \tau_{\mathrm{rQ}}\dfrac{d}{dt}} \Rightarrow \psi_{mq}(s) = L_{mq}i_q(s)\frac{1 + \tau_{\mathrm{rQ}\sigma}s}{1 + \tau_{\mathrm{rQ}}s} \qquad (8.171)$$

Dividing the flux-linkage terms by the respective magnetizing inductances makes it possible to solve the equation for the damper-winding currents.

$$i_D + i_f + i_d = \frac{\psi_{md}}{L_{md}} = (i_f + i_d)\frac{1 + \tau_{\mathrm{rD}\sigma}\dfrac{d}{dt}}{1 + \tau_{\mathrm{rD}}\dfrac{d}{dt}} \qquad (8.172)$$

$$i_D = (i_f + i_d)\left(\frac{1 + \tau_{\mathrm{rD}\sigma}\dfrac{d}{dt}}{1 + \tau_{\mathrm{rD}}\dfrac{d}{dt}} - 1\right) \Rightarrow i_D(s) = (i_f(s) + i_d(s))\left(\frac{1 + \tau_{\mathrm{rD}\sigma}s}{1 + \tau_{\mathrm{rD}}s} - 1\right) \qquad (8.173)$$

and

$$i_Q + i_q = \frac{\psi_{mq}}{L_{mq}} = i_q\frac{1 + \tau_{\mathrm{rQ}\sigma}\dfrac{d}{dt}}{1 + \tau_{\mathrm{rQ}}\dfrac{d}{dt}} \qquad (8.174)$$

$$i_Q = i_q\left(\frac{1 + \tau_{\mathrm{rQ}\sigma}\dfrac{d}{dt}}{1 + \tau_{\mathrm{rQ}}\dfrac{d}{dt}} - 1\right) \Rightarrow i_Q(s) = i_q\left(\frac{1 + \tau_{\mathrm{rQ}\sigma}s}{1 + \tau_{\mathrm{rQ}}s} - 1\right) \qquad (8.175)$$

The time constants for the leakage components are so much shorter than the damping time constants in the denominators that these components can be neglected in certain cases. Therefore, since the damper-winding currents cannot be measured, they are calculated from

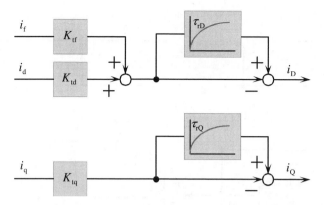

Figure 8.31   Determining the damper-winding currents from other currents in the dq plane. i_f is the field-winding current, i_d and i_q are the stator current components in the rotor reference frame, and i_D and i_Q are the calculated damper-winding currents. The time constants for the filter blocks are τ_{rD} and τ_{rQ}. K_{tf}, K_{td}, and K_{tq} are constant factors.

the other currents. The equations are transformed to the Laplace domain and the effect of Canay inductance is added yielding the following expressions.

$$i_D = (K_{td}i_d + K_{tf}i_f) \cdot \left(\frac{1}{\tau_{rD}s + 1} - 1 \right) \tag{8.176}$$

$$i_Q = K_{tq}i_q \cdot \left(\frac{1}{\tau_{rQ}s + 1} - 1 \right) \tag{8.177}$$

Figure 8.31 presents the block diagrams corresponding to these equations.

The parameters needed to determine the damper-winding currents, that is, the coefficients and time constants of the filters, can be calculated as a function of the inductance values.

$$K_{td} = \frac{L_{md}}{L_{md} + L_{k\sigma} + L_{D\sigma}} \tag{8.178}$$

$$K_{tf} = \frac{L_{md} + L_{k\sigma}}{L_{md} + L_{k\sigma} + L_{D\sigma}} \tag{8.179}$$

$$K_{tq} = \frac{L_{mq}}{L_{mq} + L_{Q\sigma}} \tag{8.180}$$

$$\tau_{rD} = \frac{L_{md} + L_{k\sigma} + L_{D\sigma}}{R_D} \tag{8.181}$$

$$\tau_{rQ} = \frac{L_{mq} + L_{Q\sigma}}{R_Q} \tag{8.182}$$

The following example investigates a control that is stator flux-linkage oriented. The current references, aligned with the stator flux linkage and perpendicular torque axes, and the sine and cosine of the stator flux-linkage load angle can be determined using the method presented in Figure 8.32.

The outputs from the block diagram are the current references in the ψT-coordinate system and the sine and cosine of the load angle, which are both needed in transforming the current references to the dq reference frame.

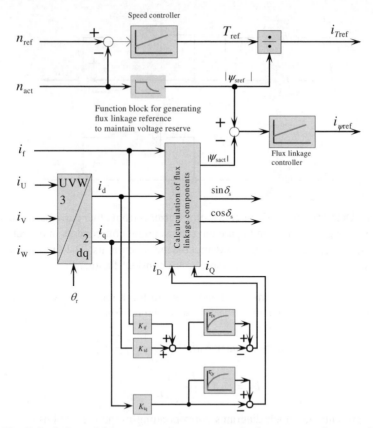

Figure 8.32 Calculation of the current references. i_{Tref} is the current reference aligned with the effective power axis, and $i_{\psi ref}$ is the current reference aligned with the stator flux-linkage axis. δ_s is the load angle of the stator flux linkage, $|\psi_s|$ is the magnitude, and $|\psi_{sref}|$ is the reference value. n_{ref} is the speed reference, and n is the actual value. I_f is the actual value of the field-winding current, and i_U, i_V, and i_W are the actual values of the stator phase currents. The block $3 \rightarrow 2$ transforms the phase currents into rotor reference frame currents. i_d and i_q are the actual stator current components in the rotor reference frame. θ_r is the rotor position angle.

Stator flux linkage can also be determined by applying the voltage model in which the stator flux linkage is integrated from the supply voltage, $\psi_s = \int (u_s - i_s R_s)dt$. The voltage model is not applicable to zero speed or low-speed operation where the voltage is too noisy to be correctly integrated. Furthermore, rotor angle must be known to transfer the flux linkage to the rotor reference frame, where its magnitude and load angle would be known.

The next phase of vector control is the construction of the current references in the rotor reference frame. In this step, the sine and cosine of the stator flux-linkage load angle must be known to change over from the stator flux-linkage reference frame to the rotor reference frame. See Figure 8.28. The transformation can be done using the following expressions.

$$i_{dref} = -i_{\psi ref} \cos \delta_s - i_{Tref} \sin \delta_s \tag{8.183}$$

$$i_{qref} = i_{\psi ref} \sin \delta_s + i_{Tref} \cos \delta_s \tag{8.184}$$

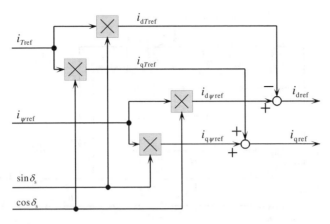

Figure 8.33 Transforming the $\psi_s T$ current references to the dq reference frame. i_{Tref} is the torque producing current reference, and $i_{\psi ref}$ is the stator flux-linkage producing current reference. i_{qTref} and i_{dTref} are the torque producing current components, $i_{q\psi ref}$ and $i_{d\psi ref}$ are the flux-linkage–producing current reference components. i_{dref} is the d-axis current reference, and i_{qref} is the q-axis current reference.

This information is expressed in the block diagram of Figure 8.33.

Establishing stator current control is the next step in implementing vector control. See Figure 8.34. The d- and q-axis current errors are fed to an integrating controller. Then, the feedforward from the rotating voltages is added to these current signal outputs. This, actually in practice, helps in decoupling the different axes current controls from each other. Since the rotating voltages of the machine are close to the terminal voltages, it is easier to adjust the output of the controller to within the correct range, and the current model must only correct the errors resulting from the nonidealities of the model and the converter.

Next, the reference values are transformed into phase values. Since integration is carried out in the rotor reference frame, the signals to be integrated are DC signals and phase errors, for example, are avoided. It is natural to perform the integration in the dq reference frame, because integrating AC-quantities is problematic from the control point of view. In principle, the result of integration is just a phase shifting of the AC-quantities. When operating with DC-quantities, it is easier to eliminate error using an integrating controller. The proportional DC controller could also be realized in the dq reference frame, but there are some benefits to modelling the DC controller in the stator reference frame.

The dq-current references are transformed into three-phase stator current references to perform the P-control for the motor phase currents. This is done because it is, in practice, more reliable to operate directly with measured quantities than the dq-quantities. This procedure e.g. prevents the occurrence of a zero sequence component. Ideally, it should be equal whether the control is realized with dq-components of xy-components. As was discussed previously, a zero-sequence component in the rotor reference frame currents is not detectable. The P-controllers yield the other components for the reference values. These components are summed up for different phases to provide the final phase current reference values, which are mainly voltage references for the different phases. These references are modified further to produce appropriate signals for the power electronics. In this case, the converter bridge is a

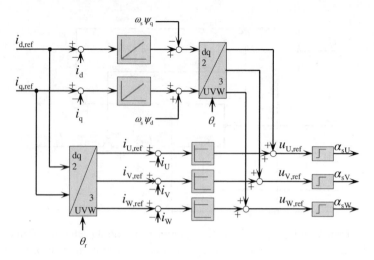

Figure 8.34 Stator current integrating control and the feedforwards in the dq reference frame with proportional control in the stator coordinate system for phase quantities. $i_{d\text{ref}}$ is the d-axis current reference, and $i_{q\text{ref}}$ is the q-axis current reference. θ_r is the measured rotor angle, and ψ_d and ψ_q are the direct and q-axis components of the stator flux linkage. $-\omega_s\psi_q$ is the d-axis rotation-caused voltage. $\omega_s\psi_d$ is the q-axis rotation-caused voltage. ω_s is the electrical angular velocity. $i_{U,\text{ref}}$, $i_{V,\text{ref}}$, and $i_{W,\text{ref}}$ are the references of the stator phase currents. i_d, i_q, i_U, i_V, and i_W are the respective actual values of the currents. The 2/3 blocks perform the transformation from the rotor reference frame to stator three-phase quantities. $u_{U,\text{ref}}$, $u_{V,\text{ref}}$, and $u_{W,\text{ref}}$ are the voltage references for the different phases, and α_{sU}, α_{sV}, and α_{sW} are the control angles for the bridges of the different phases.

cycloconverter, and the current reference signals must be converted, therefore, to appropriate ignition angle references for the thyristor bridges.

Field-winding current is an essential quantity that must be controlled when considering the operation of the synchronous machine. It has to compensate for the effect of the d-axis stator current and maintain the desired power factor of the stator. If the target is to keep the power factor of the stator at unity ($\cos\varphi = 1$), the field-winding current reference $i_{f,\text{ref}}$ can be calculated as follows.

$$i_{f,\text{ref}} = \frac{|\psi_{s,\text{ref}}|\cos\delta_s - L_d i_{d,\text{ref}}}{L_{md}} \tag{8.185}$$

where $|\psi_{s,\text{ref}}|$ is the stator flux-linkage reference value, δ_s is the load angle of the stator flux linkage, L_d is the d-axis synchronous inductance, and L_{md} is the d-axis magnetizing inductance.

A PI controller is used also in the control of field-winding current. The controller provides the voltage reference from which the control angle reference is modified for the excitation bridge as shown in Figure 8.35. Field-winding current control must be tuned correctly to avoid oscillation during transients. In the above-calculated reference value, magnetizing inductance does not remain constant, which is a problem. Therefore, magnetizing inductance should be estimated before the calculation process begins.

This vector control comprises the current model and the respective coordinate transformations of the currents, the flux-linkage controller, and the generation of the direct and quadrature current references. All other aspects are taken care of by the ordinary system controllers. The vector control is stator flux-linkage oriented. The control does not have to be stator flux-linkage

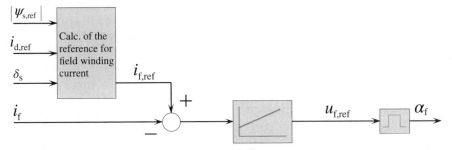

Figure winding current controller

Figure 8.35 The field-winding current controller. $|\psi_{s,\text{ref}}|$ is the reference value for the magnitude of the stator flux linkage, $i_{d,\text{ref}}$ is the reference value for the current i_d, δ_s is the load angle, and i_f is the field-winding current. $u_{f,\text{ref}}$ is the voltage reference, and α_f is the control angle for the excitation bridge.

oriented, and the leakage flux linkage of the stator flux linkage does not finally produce torque. However, voltage source drives—and the cycloconverter is of this type—must ensure sufficient voltage. Therefore, a control based on stator flux linkage is a justified choice.

Figure 8.32 depicts the current reference i_{Tref} in a stator flux-linkage oriented system. A control based on air-gap flux linkage would best correspond to the control of a fully compensated DC machine. However, because of its voltage dependence, this control is based on stator flux linkage. Figure 8.36 illustrates the difference between controls based on the stator flux linkage and air-gap flux linkage.

The figure shows that when the machine operates at rated per-unit voltage $u_s = 1$, $\psi_s = 1$ ($\omega_s = 1$) and the power factor and magnitude of the stator current is held at 1 ($i_s = 1$, $\cos\varphi = 1$), torque is given by $T_e = 1$. This is possible, because the air-gap flux-linkage value is slightly higher than 1 ($\psi_m = 1.03$), which compensates for the effect of the angle λ between ψ_s and ψ_m.

If the stator current vector in Figure 8.36b is kept perpendicular to the air-gap flux-linkage reference frame, and if $\psi_m = 1$, the stator flux linkage should grow ($\psi_s = 1.04$). With this higher flux linkage, the drive should shift into field weakening before reaching rated speed.

Figure 8.37 illustrates the comprehensive schematic control diagram of the cycloconverter-synchronous motor drive based on the stator flux-linkage orientation.

A cycloconverter drive can also operate as a generator, and therefore the vector diagram of a synchronous machine in generator operation is also introduced. The representation follows the same logic as in the case of motor operation. See Figure 8.38.

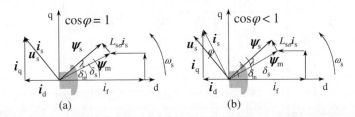

Figure 8.36 Vector diagrams for rated operation based on (a) stator flux linkage and (b) the air-gap flux linkage. The latter will lead to slightly higher torque values, lower power factor, and increased stator flux linkage. It should hit the voltage ceiling earlier and shift into field weakening earlier than the stator flux-linkage–oriented control.

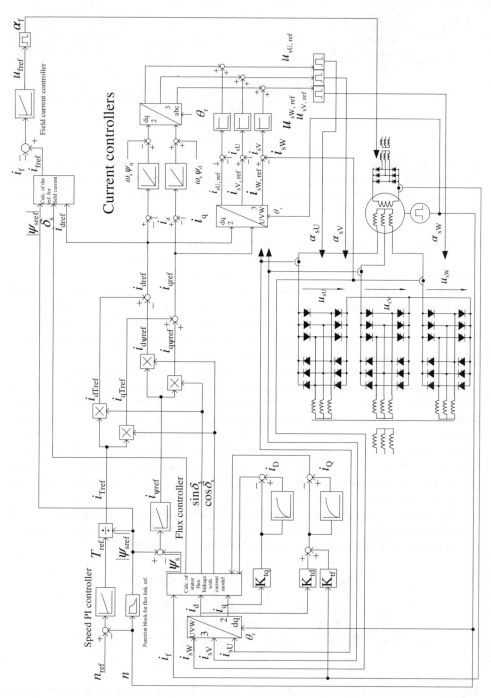

Figure 8.37 Total schematic diagram of a stator flux-linkage–based cycloconverter drive containing the blocks presented previously.

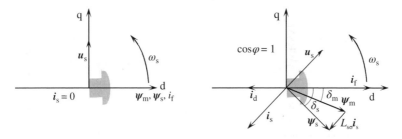

Figure 8.38 Operation at no-load and as a generator at a power factor of $\cos\varphi = 1$.

EXAMPLE 8.1: Tuning the control of an electrically excited synchronous machine

SOLUTION: (derived by Dr. Pasi Peltoniemi, LUT, 2015) This example reviews the control principles of the electrically excited synchronous machine (EESM) and considers EESM control system tuning.

Control system

Equations (8.11) through (8.28) are the equations in the d-q reference frame for an electrically excited synchronous machine rotating at rotor angular speed, that is, synchronous angular speed. This model can be used to derive model-based tuning rules for the controls used. From Figure E8.1, five different controllers are used to control the electrically excited synchronous machine (EESM) including the speed controller, the stator flux-linkage controller, the field current controller, and two current controllers.

For these controllers, the stator phase currents along with the field current are used for feedback signals. The angular speed of the machine could be measured to provide feedback, or sensorless schemes could be applied. It is assumed here that rotor electrical angular speed ω_r and the currents mentioned i_{UVW} and i_f are measured as depicted in Figure E8.1. The angular velocity ω_c is used in the control.

EESM drive controller tuning is presented for each controller separately.

Stator current control

EESM stator current control is based on a rotor-oriented reference frame that is aligned with the field winding produced flux-linkage component $L_{md}i_f$ as was shown previously in Figure 8.26. The frame is also referred to as the field-oriented control frame. Stator current control is carried out in this specific frame since the inductance values are not affected by the change in rotor angle. However, the inductance values of machine are still subject to saturation, especially when operating in the field-weakening region (Pyrhönen, 1998).

In this example, control laws are derived for stator current control in the rotor-oriented frame, and tuning rules are given for the controller parameters based on the system. Inductance values remain within limits tolerated by the feedback control.

Substituting the flux-linkage expressions of Equation (8.24) and Equation (8.25) into Equation (8.14) and Equation (8.15), respectively, gives the rotor voltage equations in the

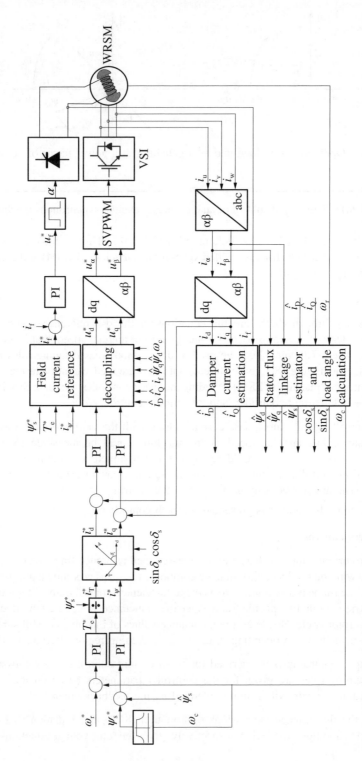

Figure E8.1 Stator flux-linkage oriented control scheme for a wound rotor synchronous machine (WRSM) *Source:* Reproduced with permission from Dr Pasi Peltoniemi.

rotor-oriented reference frame.

$$0 = R_D i_D + \frac{d}{dt}(L_{md} i_d + L_D i_D + L_{fD} i_f) \tag{E8.1}$$

$$0 = R_Q i_Q + \frac{d}{dt}\left(L_{mq} i_q + L_Q i_Q\right) \tag{E8.2}$$

Similarly, substituting Equation (8.21) and Equation (8.22) into Equation (8.11) and Equation (8.12) yields

$$u_d = R_s i_d + \frac{d}{dt}(L_d i_d + L_{md}(i_D + i_f)) - \omega_c \psi_q \tag{E8.3}$$

$$u_q = R_s i_q + \frac{d}{dt}\left(L_q i_q + L_{mq} i_Q\right) + \omega_c \psi_d \tag{E8.4}$$

The rotor damper currents i_D and i_Q are solved from Equation (8.24) and Equation (8.25).

$$i_D = \frac{\psi_D - L_{md} i_d - L_{fD} i_f}{L_D} \tag{E8.5}$$

$$i_Q = \frac{\psi_Q - L_{mq} i_q}{L_Q} \tag{E8.6}$$

The rotor currents can be eliminated from (E8.3) and (E8.4) using (E8.5) and (E8.6). After simplification, the result is as follows.

$$u_d = R_s i_d + L_{cc,d}\frac{d}{dt} i_d + \frac{L_{md}}{L_D}\frac{d}{dt}\psi_D + L_{md}\left(1 - \frac{L_{fD}}{L_D}\right)\frac{d}{dt} i_f - \omega_c \psi_q \tag{E8.7}$$

$$u_q = R_s i_q + L_{cc,q}\frac{d}{dt} i_q + \frac{L_{mq}}{L_Q}\frac{d}{dt}\psi_Q + \omega_c \psi_d \tag{E8.8}$$

$L_{cc,d} = L_d - \frac{L_{md}^2}{L_D}$ and $L_{cc,q} = L_q - \frac{L_{mq}^2}{L_Q}$. The subscript cc refers to current control.

The main idea of field-oriented current control is to enable the independent control of the d-axis stator current i_d component and the q-axis stator current i_q component. This makes it possible to independently control the flux linkage ψ and the electromagnetic torque T of the machine. However, the equations for the stator voltage components (E8.8) and (E8.9) are coupled. That is to say, the d-axis component u_d also depends on variables other than i_d. Similarly, the q-axis component u_q depends on other variables. Therefore, stator voltage components u_d and u_q cannot be considered decoupled. Stator currents i_d and i_q can only be independently controlled if the stator voltage equations are decoupled.

It is possible to derive a decoupling scheme that is valid for any operation point. First, (E8.5) and (E8.6) are substituted into (E8.1) and (E8.2), and then the following derivative terms are solved from the resulting equations.

$$\frac{d}{dt}\psi_D = -\frac{R_D}{L_D}\psi_D + \frac{L_{md}R_D}{L_D} i_d + \frac{L_{fD}R_D}{L_D} i_f \tag{E8.9}$$

$$\frac{d}{dt}\psi_Q = -\frac{R_Q}{L_Q}\psi_Q + \frac{L_{mq}R_Q}{L_Q} i_q \tag{E8.10}$$

Substituting Equation (E8.9) into Equation (E8.7) and Equation (E8.10) into Equation (E8.8) results in the following expressions.

$$u_d = R_{cc,d}i_d + L_{cc,d}\frac{d}{dt}i_d - \frac{L_{md}R_D}{L_D^2}\psi_D + \frac{L_{md}L_{fD}R_D}{L_D^2}i_f + L_{md}\left(1 - \frac{L_{fD}}{L_D}\right)\frac{d}{dt}i_f - \omega_c\psi_q$$

(E8.11)

$$u_q = R_{cc,q}i_q + L_{cc,q}\frac{d}{dt}i_q - \frac{L_{mq}R_Q}{L_Q^2}\psi_Q + \omega_c\psi_d$$

(E8.12)

$R_{cc,d} = R_s + \left(\frac{L_{md}}{L_D}\right)^2 R_D$ and $R_{cc,q} = R_s + \left(\frac{L_{mq}}{L_Q}\right)^2 R_Q$. Inserting Equation (8.32) and Equation (8.33) into Equation (E8.11) and Equation (E8.12), respectively, results in these voltage equations.

$$u_d = R_s i_d + L_{cc,d}\frac{d}{dt}i_d - \frac{L_{md}R_D}{L_D}i_D + L_{md}\left(1 - \frac{L_{fD}}{L_D}\right)\frac{d}{dt}i_f - \omega_c\psi_q$$

(E8.13)

$$u_q = R_s i_q + L_{cc,q}\frac{d}{dt}i_q - \frac{L_{mq}R_Q}{L_Q}i_Q + \omega_c\psi_d$$

(E8.14)

Expressed as follows, the voltage equations can be used as control laws for the EESM.

$$u_{d,ref} = R_s i_d + L_{cc,d}\frac{d}{dt}i_d + e_d,$$

(E8.15)

$$u_{q,ref} = R_s i_q + L_{cc,d}\frac{d}{dt}i_q + e_q,$$

(E8.16)

$u_{d,ref}$ and $u_{q,ref}$ represent the voltage references for the power electronics. The decoupling voltage terms e_d and e_q can be determined from

$$e_d = -\frac{L_{md}R_D}{L_D}i_D + L_{md}\left(1 - \frac{L_{fD}}{L_D}\right)\frac{d}{dt}i_f - \omega_c\psi_q,$$

(E8.17)

$$e_q = -\frac{L_{mq}R_Q}{L_Q}i_Q + \omega_c\psi_d.$$

(E8.18)

If the field current derivative term can be considered negligible, other decoupling terms may be either measured or estimated and then used for decoupling.

Cross-coupling between stator voltage equations can be eliminated by feeding the decoupling terms (E8.17) and (E8.18) as positive feedback (measured quantities are used to calculate e_d and e_q) or as feed-forward (reference values are used to calculate e_d and e_q) to the output of the current controllers. In both cases, the system behaves like two decoupled linear first order systems from the perspective of the current control. Therefore, current controllers can be designed assuming the system has a transfer function.

$$G(s) = \frac{I_{dq}(s)}{U_{dq}(s)} = \frac{\frac{1}{R_s}}{\tau_{cc,dq}s + 1}$$

(E8.19)

where $\tau_{cc,dq} = \frac{L_{cc,dq}}{R_s}$.

Simple PI controllers can be used as current controllers, because controlled variables appear as DC quantities in the dq reference frame. According to the Internal Model Control (IMC) principle, PI controller gains can be chosen as follows.

$$C_{PI}(s) = K_p + \frac{K_i}{s} \tag{E8.20}$$

$$K_p = \alpha_{cc}L_{cc,dq}, K_i = \alpha_{cc}R_s \tag{E8.21}$$

α_{cc} is the desired closed loop bandwidth of the current control. Choosing controller parameters using (E8.21), the closed loop system becomes

$$G_{cl}(s) = \frac{C_{PI}(s)G(s)}{1 + C_{PI}(s)G(s)} = \frac{\alpha_{cc}}{s + \alpha_{cc}} \tag{E8.22}$$

Therefore, using the (E8.17) and (E8.18) decoupling scheme and choosing controller parameters based on the (E8.13), (E8.14), and (E8.21) machine model results in good dynamic performance. Instead of using system bandwidth, it may be more convenient to specify desired control performance using rise time $t_{r,cc}$. For a first order system, rise time and bandwidth have the following relationship.

$$\alpha_{cc} = \frac{\ln(9)}{t_{r,cc}} \tag{E8.23}$$

Current rise time is limited by the voltage reserve of the inverter, so it cannot be set fast arbitrarily. Severe saturation of the controller should be avoided. Obviously, current rise time also determines the torque rise time. Therefore, $t_{r,cc}$ is an important design parameter.

Field current control

Another low level control in the EESM control concept is field current control. Field current control not only plays a crucial role in producing machine flux, but also in machine stability. Especially in the field-weakening range, the higher level field current reference calculation significantly impacts machine performance. This will be discussed in more detail later in Section 8.14. Here, the focus is on field current controller tuning.

As was done in the previous discussion of stator current control, the upcoming paragraphs will derive the model-based control law and consider the tuning of the appropriate controller for field current control. It is assumed that the variation in inductance values remains small enough to be manageable by the feedback control. Field current supply methods quite commonly suffer from delays within the control loop that limits EESM drive performance. There are compensation methods available, but as a minimum requirement the field current controller must be tuned so that field current rise time is higher than the maximum delay in the excitation system. Otherwise, instability is likely to occur. In this example delay is ignored.

In Figure E8.1, the field current controller produces the field voltage reference u_f^* on the control loop level, which is then produced by the excitation unit. A simplified field current control loop is shown in Figure E8.2 where the field current reference calculation is also depicted.

The dynamics of the field current are modelled as in Equation (8.13) and for field flux linkage as in Equation (8.23). Substituting Equation (8.23) into Equation (8.13) gives

$$u_f = R_f i_f + L_f \frac{d}{dt}i_f + L_{md}\frac{d}{dt}i_d + L_{fD}\frac{d}{dt}i_D \tag{E8.24}$$

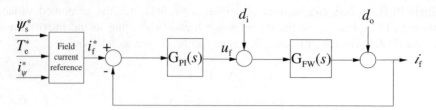

Figure E8.2 Field current control loop. *Source:* Reproduced with permission from Dr Pasi Peltoniemi.

The following expression describes $\frac{d}{dt}i_D$.

$$\frac{d}{dt}i_D = \frac{1}{L_D}\left(-R_D i_D - L_{md}\frac{d}{dt}i_d - L_{fD}\frac{d}{dt}i_f\right)$$
(E8.25)

Substituting (E8.25) into (E8.24) gives

$$u_f = R_f i_f + \left(L_f - \frac{L_{fD}^2}{L_D}\right)\frac{d}{dt}i_f + \left(L_{md} - \frac{L_{fD}L_{md}}{L_D}\right)\frac{d}{dt}i_d - \frac{L_{fD}R_D}{L_D}i_D$$
(E8.26)

If stator current is assumed constant during the field current transients, then

$$u_f = R_f i_f + \left(L_f - \frac{L_{fD}^2}{L_D}\right)\frac{d}{dt}i_f - \frac{L_{fD}R_D}{L_D}i_D$$
(E8.27)

Neglecting the damper current term, (E8.27) can be divided into a first-order system and feed-forward term. Neglecting the damper current term only affects the transient performance of the drive, because $i_D = 0$ in steady-state. A field-current controller can now be designed assuming that the system has a transfer function.

$$G(s) = \frac{I_f(s)}{U_f(s)} = \frac{\frac{1}{R_f}}{\tau_f s + 1}$$
(E8.28)

where $\tau_f = \dfrac{L_f - \frac{L_{fD}^2}{L_D}}{R_f}$.

Again, simple PI controllers can be used as current controllers, because the controlled variables are DC quantities. The PI controller gains can be determined as follows.

$$C_{PI}(s) = K_{pf} + \frac{K_{if}}{s}$$
(E8.29)

$$K_{pf} = \alpha_f\left(L_f - \frac{L_{fD}^2}{L_D}\right), \; K_{if} = \alpha_f R_f$$
(E8.30)

α_f is the desired closed loop bandwidth. Choosing controller parameters using (E8.30), the closed loop system becomes

$$G_{CL}(s) = \frac{C_{PI}(s)G(s)}{1 + C_{PI}(s)G(s)} = \frac{\alpha_f}{s + \alpha_f}$$
(E8.31)

Here, as well as with the stator current controllers, the desired control performance is specified using the $t_{r,f}$ rise time. For a first-order system, the rise time and bandwidth have

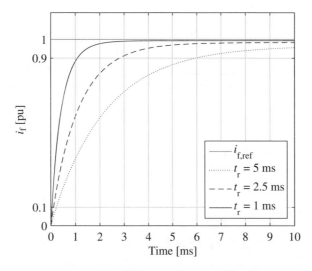

Figure E8.3 Step responses of the field current controller. *Source:* Reproduced with permission from Dr Pasi Peltoniemi.

the following relationship.

$$\alpha_f = \frac{\ln(9)}{t_{r,f}} \tag{E8.32}$$

A current rise time cannot be chosen fast arbitrarily, because of the limited voltage reserve of the inverter and the possible delay as previously mentioned. Generally, the excitation system saturation should be avoided.

Test simulations have been carried out for various design bandwidths. The results are shown in Figure E8.3. As expected, the responses are relatively accurate. However, significant parameter variations occur in EESM drives operating in demanding conditions. These variations must be taken into account when designing a control system.

Stator flux linkage control

Stator flux linkage is one of the primary control variables of the EESM drive. Flux linkage control is performed in a stator flux-linkage-oriented frame. In the stator flux-linkage-oriented frame, as was illustrated in Figure 8.28, stator flux linkage is controlled via the stator current component i_ψ, which is aligned with the stator flux linkage. Similarly, electromagnetic torque is controlled with the stator current component i_T, which is aligned with the stator voltage. To obtain the current references for current control operating in the field-oriented dq frame, a coordinate transformation is applied as was expressed in Equation (8.183) and Equation (8.184).

The stator flux-linkage control system can be presented as shown in Figure E8.4 (a). The figure illustrates that stator flux is controlled from both the stator and field winding. A significant consequence is that the power factor at the stator poles can be controlled. Flux control from the stator side is considered in the upcoming paragraphs. This flux control

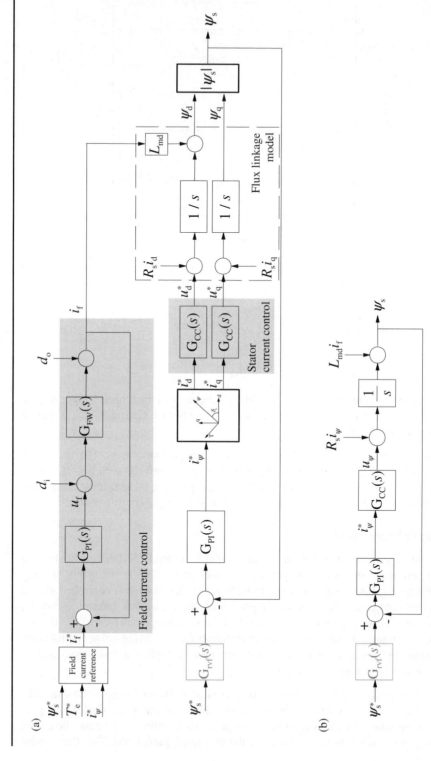

Figure E8.4 Stator flux-linkage control loop showing (a) the detailed stator flux-linkage control system structure and (b) a simplified diagram to illustrate controller tuning. *Source*: Reproduced with permission from Dr Pasi Peltoniemi.

Table E8.1 Matlab script for root locus based flux-linkage controller tuning

```
%% Stator flux linkage controller tuning

Kp_psi=1;Ti_psi=4*0.0129;tr_cc=5e-3;alpha_cc=log(9)/tr_cc; % Parameters
sys_psi_ol=tf([alpha_cc*Kp_psi*Ti_psi alpha_cc*Kp_psi],
   [Ti_psi alpha_cc*Ti_psi 0 0]);% (E8.33)

rlocus(sys_psi_ol) % plotting of root locus
```

operates in cascade structure with the stator current controller as shown in the figure. Assuming equivalent stator current controller dynamics, and if the coordinate transformation from the stator flux-linkage frame into the rotor flux-linkage frame can be considered a static multiplication presenting negligible delay, the stator flux-linkage control scheme can be simplified into the diagram shown in Figure E8.4 (b).

The voltage drop across the stator resistance becomes significant only at low speeds. Therefore, it is neglected here. The open loop transfer function of the stator flux-linkage control loop can be written as follows.

$$G_{\psi,ol}(s) = \frac{K_{p\psi}(T_{i\psi}s + 1)}{T_{i\psi}s} \frac{\alpha_{cc}}{s + \alpha_{cc}} \frac{1}{s} \tag{E8.33}$$

where $K_{p\psi}$ and $T_{i\psi}$ are the proportional gain and integrator time constant of the stator flux-linkage controller, respectively, and α_{cc} is the bandwidth of the current control loop from (E8.23).

In the symmetric optimum method, the integrator time constant of the cascade controller is equal to $T_{i,\psi} = 4T_{i,cc}$. When the parameters shown in Table 8.2 are applied, $T_{i,cc} = 0.0129$ is obtained for the current controllers. Since the bandwidth of the current control loop (α_{cc}) is known, setting $K_{p\psi} = 1$ and plotting the root locus curve (Table E8.1) reveals the effect of gain variation on the location of the system poles, and the desired system properties can be chosen.

In Figure E8.5, the root loci have been plotted for current control rise times of 5 ms and 1 ms. As the figure reveals, using faster current controllers allows bigger separation between the pole near the origin and the pole-pair that eventually becomes complex valued. Moreover, the root locus trajectories show that the current control system is robust, since the gain could be increased to infinity, and the system poles would remain in the left half plane (LHP). In practice, converter imposed limits on the performance of flux-linkage control are inevitable. Available voltage and current handling capability of the power electronics are limited. Furthermore, noise in the measurements could become a control problem if it increases at specific frequencies in the selected bandwidth.

A closed loop transfer function for the stator flux-linkage control loop can be derived from (E8.31).

$$\frac{\Psi_s(s)}{\Psi_{s,ref}(s)} = \frac{\alpha_{cc}(K_{p\psi}s + K_{i\psi})}{s^3 + \alpha_{cc}s^2 + \alpha_{cc}K_{p\psi}s + \alpha_{cc}K_{i\psi}} \tag{E8.34}$$

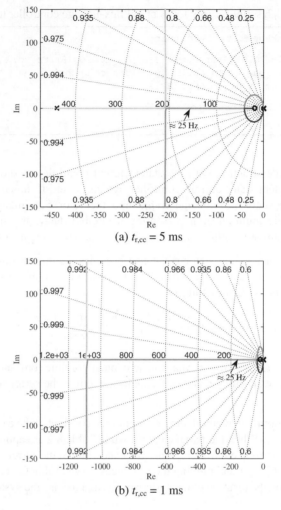

Figure E8.5 Root locus of stator flux-linkage control. In (a), the rise time of the current control loop t_r was selected as 5 ms and in (b) as 1 ms. *Source:* Reproduced with permission from Dr Pasi Peltoniemi.

where $K_{i\psi} = K_{p\psi}/T_{i\psi}$ is the integrator gain of the stator flux-linkage controller. As shown in Figure E8.5, selecting a pole location at 25 Hz, where the damping is equal to 1, results in gain values of $K_{p\psi,1ms} = 167$ and $K_{p\psi,5ms} = 116$.

In Figure E8.6, (a) a step response is plotted (Table E8.2) for both designs using a transfer-function-based step response. In both cases, there is response overshoot. This overshoot becomes negligible if a reference value filter (RVF) is applied as shown in Figure E8.4. In Figure E8.6, (b) the same responses are obtained from the Simulink model that contains the detailed model of the control system and machine. The responses are similar, but the settling times of the Simulink model responses are longer. The difference is due to simplifications made in the transfer function approach.

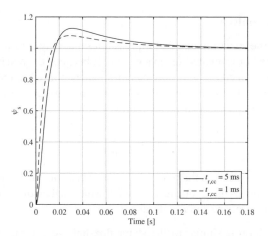

(a) Transfer function (E8.34) based response

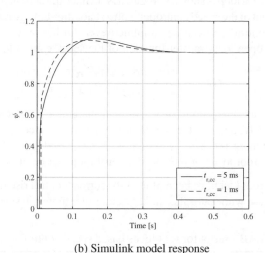

(b) Simulink model response

Figure E8.6   The step response of the stator flux-linkage control loop. (a) shows a transfer function response based on (E8.34), and (b) shows the response obtained using the Simulink model of the control system introduced in this chapter and illustrated in Figure E8.1.

Table E8.2   MATLAB® script for closed loop step response analysis of the stator flux-linkage control loop

```
%% Step response – Stator flux-linkage control loop
Kp_psi=116;Ti_psi=4*0.0129;tr_cc=5e-3;alpha_cc=log(9)/tr_cc; % Parameters
sys_psi_cl= tf([alpha_cc*Kp_psi alpha_cc*(Kp_psi/Ti_psi)],
  [1 alpha_cc alpha_cc*Kp_psi alpha_cc*(Kp_psi/Ti_psi)]); % (E8.32)
step(sys_psi_cl) % plotting of step response
```

Speed and torque control

In the control scheme, electromagnetic torque is not separately controlled, and as a result, there is no measured or estimated feedback. This is because the machine operates in the stator flux-linkage oriented frame, where electromagnetic torque can be controlled via the current component i_T. If the torque and the stator flux linkage are known, this torque-producing current component can be solved as follows.

$$i_{sT}^* = \frac{T_e^*}{\frac{3}{2} p \psi_s^*} \tag{E8.35}$$

where the superscript * denotes the reference value. Therefore, as seen in Figure E8.1, the speed controller produces the torque reference T_e^*, which is then divided by the stator flux-linkage reference to produce the i_T reference.

The speed control loop is similar to the stator flux-linkage control loop. Therefore, the control loop can be simplified as was done with the stator flux-linkage control loop. The simplified speed control loop is shown in Figure E8.5. Since the current reference is obtained by dividing the output of the speed controller with a stator flux-linkage reference value, it is a static value. The control loop can be simplified even further by not accounting for the division. Now, the open loop transfer function of the speed control loop can be written

$$G_{T,ol}(s) = \frac{K_{p\omega}(T_{i\omega}s + 1)}{T_{i\omega}s} \frac{\alpha_{cc}}{s + \alpha_{cc}} \frac{1}{Js} \tag{E8.36}$$

where $K_{p\omega}$ and $T_{i\omega}$ are the proportional gain and integrator time constant of the speed controller, respectively, and J is the inertia of the machine. Since α_{cc} is known from the current controller tuning process, by setting $T_{i\omega} = 4T_{i,cc}$ and using the machine inertia $J = 0.001 \text{ kgm}^2$, the root locus of the speed control loop can be plotted (Table E8.3).

In Figure E8.7, the root loci have been plotted for current control rise times of 5 and 1 ms. Again, as was the case with stator flux-linkage controller tuning, faster current controllers

Table E8.3 MATLAB® script for root locus based speed controller tuning

```
%% Speed controller tuning

Kpw=1;Tiw=4*0.0129;tr_cc=5e-3;alpha_cc=log(9)/tr_cc;J=0.001; % Parameters

sys_w_ol=tf([alpha_cc*Kpw*Tiw alpha_cc*Kpw],[J*Tiw alpha_cc*J*Tiw 0 0]); %
   (E8.34)

rlocus(sys_w_ol) % plotting of root locus
```

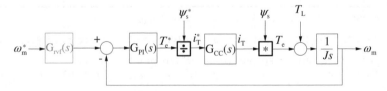

Figure E8.7 Simplified block diagram of the speed control loop. *Source:* Reproduced with permission from Dr Pasi Peltoniemi.

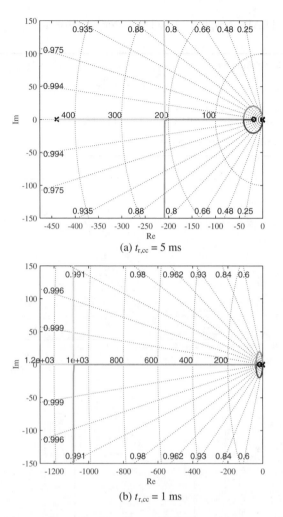

(a) $t_{r,cc}$ = 5 ms

(b) $t_{r,cc}$ = 1 ms

Figure E8.8 Root locus of speed control loop. Different current controller tunings were applied in (a) and (b). *Source:* Reproduced with permission from Dr Pasi Peltoniemi.

allow bigger separations between the pole near the origin and the pole-pair farther away in the left half plane. Similarly, the speed control loop is also robustly stable with respect to speed controller gain variation, since the poles of the control system remain on the left half plane despite the gain value. However, as mentioned earlier for the stator flux-linkage control, increasing the bandwidth could lead to measurement noise. In fact, as seen in Figure E8.7, the damping of the complex valued pole-pair drops continuously as speed controller gain increases.

For the speed control, the closed loop transfer function can be written in the following form.

$$\frac{\omega_m(s)}{\omega_{m,ref}(s)} = \frac{\frac{\alpha_{cc}}{J}\left(K_{p\omega}s + K_{i\omega}\right)}{s^3 + \alpha_{cc}s^2 + \frac{\alpha_{cc}}{J}K_{p\omega}s + \frac{\alpha_{cc}}{J}K_{i\omega}} \tag{E8.37}$$

Table E8.4   MATLAB® script for closed loop step response analysis of the speed control loop

```
%% Speed controller tuning
Kpw=1;Tiw=4*0.0129;tr_cc=5e-3;alpha_cc=log(9)/tr_cc;J=0.001; % Parameters
sys_w_cl=tf([(alpha_cc/J)*Kpw (alpha_cc/J)*(Kpw/Tiw)],[1 alpha_cc
  (alpha_cc/J)*Kpw (alpha_cc/J)*(Kpw/Tiw)]);% (E8.35)
step(sys_w_cl) % plotting of step response
```

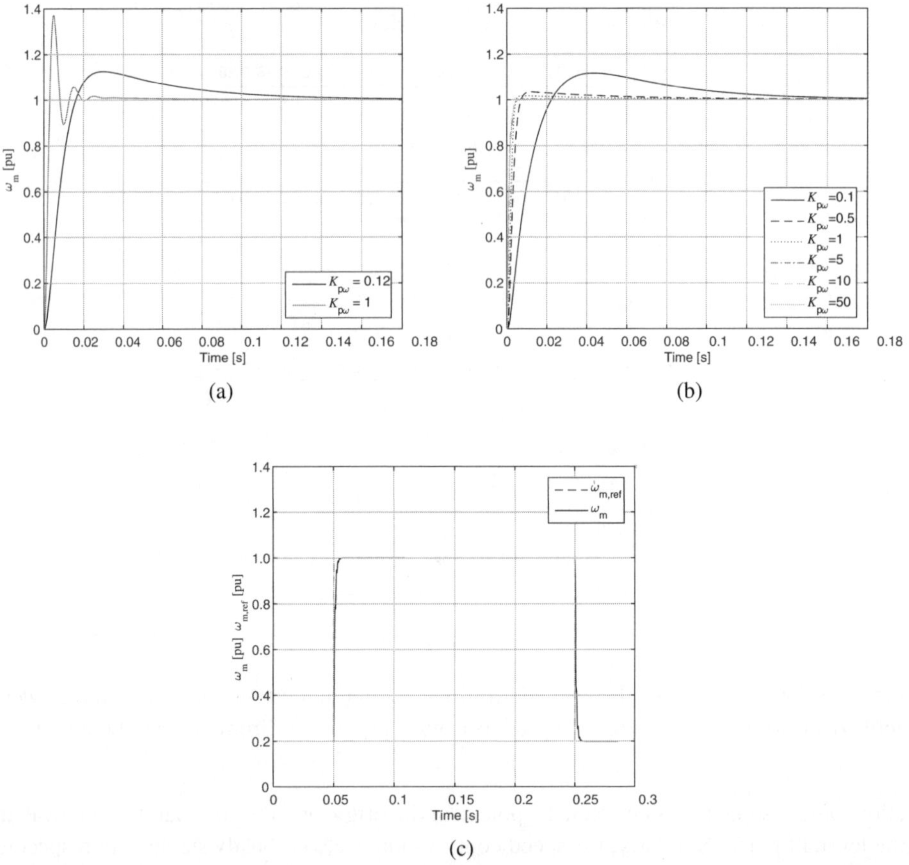

(a)

(b)

(c)

Figure E8.9   Step responses of the speed control loop. A transfer-function–based (E8.37) response is shown in (a). In (b), the step responses that result when an RVF is applied are shown. And, the speed control performance obtained from a detailed Simulink model is illustrated in (c). *Source:* Reproduced with permission from Dr Pasi Peltoniemi.

The script given in Table E8.4 produces a step response of the speed control loop. The plotted responses are shown in Figure E8.9 (a).

Despite the gain value, some overshoot exists in the response. This overshoot can be made negligible by applying an RVF as shown in Figure E8.7. The effect of the RVF on control performance can be illustrated using the following transfer function.

$$\frac{\omega_{\mathrm{m}}(s)}{\omega_{\mathrm{m,ref}}(s)} = \frac{1}{T_{\mathrm{rvf}}s + 1} \frac{\frac{\alpha_{\mathrm{cc}}}{J}\left(K_{\mathrm{p}\omega}s + K_{\mathrm{i}\omega}\right)}{s^3 + \alpha_{\mathrm{cc}}s^2 + \frac{\alpha_{\mathrm{cc}}}{J}K_{\mathrm{p}\omega}s + \frac{\alpha_{\mathrm{cc}}}{J}K_{\mathrm{i}\omega}} \qquad \text{(E8.38)}$$

where T_{rvf} is the time constant of the filter. Now, different speed controller gain values are applied when $T_{\mathrm{rvf}} = 1$ ms and the responses are shown in Figure E8.9 (b). The responses reveal the expected result. That is, response approaches the response of the RVF as the gain value is increased, and simultaneously, the overshoot becomes negligible. However, in practice, the limitations mentioned previously in the stator flux-linkage controller discussion are present, and performance is therefore constrained by these limitations.

Finally, the speed control responses are tested with the detailed Simulink model when RVF is used ($T_{\mathrm{rvf}} = 1$ ms). A high gain value is applied ($K_{\mathrm{pw}} = 5$) for the speed controller. The result of the simulation is shown in Figure E8.7 (c). The response obtained with Simulink is in line with expectations.

EESM drive simulation using unity power factor control

One simulation has been carried out with controller tunings determined as above and applied to the drive. In the simulation, the drive operates under the unity power factor principle, which is executed by calculating the field current reference as follows per (Bühler, 1997-II).

$$i_{\mathrm{f}}^* = \frac{\psi_{\mathrm{s}}^2 + L_{\mathrm{d}}L_{\mathrm{q}}|i_{\mathrm{s}}|^2}{L_{\mathrm{md}}\sqrt{\psi_{\mathrm{s}}^2 + L_{\mathrm{q}}^2|i_{\mathrm{s}}|^2}} \qquad \text{(E8.39)}$$

The machine parameters used correspond to those previous given in Table 8.2. The simulation results are shown in Figure E8.10.

In the simulation, the EESM drive rotates at nominal speed in no-load operation ($T_{\mathrm{L}} = 0$). At $t = 0.1$ s, a nominal torque step takes place and a change in rotational speed is evident from Figure E8.10 (a). The drive reaches new steady-state values within approximately 0.1 s. The new steady-state values can be calculated as shown in the following Table E8.5. The calculated steady-state values when $L_{\mathrm{d}} = 1.17$, $L_{\mathrm{q}} = 0.57$, $L_{\mathrm{md}} = 1.05$, $\psi_{\mathrm{s}}^* = 1$, and

Table E8.5 Steady-state solutions of an EESM drive using unity power factor control in pu values

$$\delta_{\mathrm{s}} = \operatorname{atan}\left(\frac{L_{\mathrm{q}}T_{\mathrm{e}}^*}{\psi_{\mathrm{s}}^{*2}}\right);$$

$$I_{\mathrm{f}}^* = \frac{\psi_{\mathrm{s}}^2 + L_{\mathrm{d}}L_{\mathrm{q}}\left(\frac{T_{\mathrm{e}}^*}{\psi_{\mathrm{s}}^*}\right)^2}{L_{\mathrm{md}}\sqrt{\psi_{\mathrm{s}}^2 + L_{\mathrm{q}}^2\left(\frac{T_{\mathrm{e}}^*}{\Psi\psi_{\mathrm{s}}^*}\right)^2}}; \quad I_{\mathrm{d}} = -\frac{L_{\mathrm{q}}\left(\frac{T_{\mathrm{e}}^*}{\psi_{\mathrm{s}}^*}\right)^2}{\sqrt{\psi_{\mathrm{s}}^2 + L_{\mathrm{q}}^2\left(\frac{T_{\mathrm{e}}^*}{\psi_{\mathrm{s}}^*}\right)^2}}; \quad I_{\mathrm{q}} = \frac{T_{\mathrm{e}}}{\sqrt{\psi_{\mathrm{s}}^2 + L_{\mathrm{q}}^2\left(\frac{T_{\mathrm{e}}^*}{\psi_{\mathrm{s}}^*}\right)^2}};$$

$$\psi_{\mathrm{d}} = L_{\mathrm{d}}I_{\mathrm{d}} + L_{\mathrm{md}}I_{\mathrm{f}}; \quad \psi_{\mathrm{q}} = L_{\mathrm{q}}I_{\mathrm{q}}$$

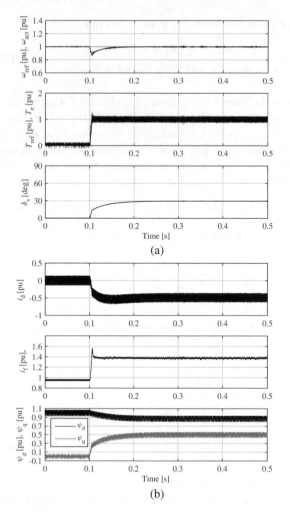

(a)

(b)

Figure E8.10 EESM drive simulation via unity power factor control. *Source:* Reproduced with permission from Dr Pasi Peltoniemi.

$T_e^* = 1$ are $\delta_s = 29.7°$, $I_f = 1.38$, $I_d = -0.495$, $I_q = 0.869$, $\psi_d = 0.869$, and $\psi_q = 0.495$. Figure E8.10 reveals these values to be where the drive settles after the torque transient. The electric current values in the stator flux-oriented frame can be calculated from the inverse transformation of Equation (8.183) and Equation (8.184) as follows.

$$I_\psi = I_d \cos\delta_s + I_q \sin\delta_s = 0 \text{ and } I_T = -I_d \sin\delta_s + I_q \cos\delta_s = 1$$

The machine is only magnetized from the rotor when operated with unity power factor. The result also shows that minimum stator current operation is achieved.

8.14   Field-winding current, reactive power, and the dynamics of a synchronous machine drive

The vector control of a synchronous machine based on the stator flux-linkage and rotor reference frames was presented in the preceding analysis. The control of field-winding current was not studied in detail even though it can have a big influence on drive behaviour and can offer significant improvements in dynamic performance, especially in the field weakening area.

8.14.1   The dynamic response of a synchronous machine in a vector control drive to different field-winding current control methods

The vector control case above was based on maintaining a unity power factor, and therefore the dynamics of field-winding current control were not studied in detail. In the following text, synchronous machine behaviours using different principles of the field-winding control will be examined. In steady state, the control of field-winding current is often based on keeping the power factor at unity in the base speed region. In the field weakening region, the control strategy can vary a great deal based on boundary and load conditions. For transients, it is more critical to understand behaviours and carefully manage the control of the field-winding currents.

According to the load angle equation, torque is directly proportional to e_f and the stator flux linkage ψ_s. And, the synchronous inductances L_d and L_q define the maximum static torque. Figure 8.39 depicts the trajectories of the tips of the stator current vector i_s and the stator flux-linkage vector ψ_s with differing load angles, $\delta_s \in [0,\pi]$ and with constant field-winding current. The field-winding current has a certain fixed value enabling near rated torque despite the low stator flux-linkage modulus ($\psi_s = 0.3$ in the field weakening at high speed, $\Omega_r = 3$). Therefore, the machine is first heavily over-excited, and a large demagnetizing current i_{s0} is applied to keep the stator flux linkage small despite the large i_f.

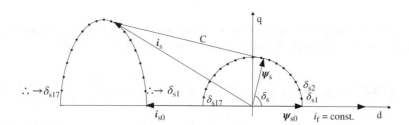

Figure 8.39   Stator current and flux-linkage vectors i_s and ψ_s at different static torques with field-winding current i_f and $|\psi_s|$ held constant. Vectors i_{s0} and ψ_{s0} correspond to no load, and i_s and ψ_s correspond to the maximum torque point $\psi_s \times i_s$. The machine properties listed in Table 8.1 have been used. *Source:* Adapted from Pyrhönen (1998). Reproduced with permission of O. Pyrhönen.

The maximum torque T_{max} producing static load angle $\delta_{s,max}$ can be determined given i_f, ψ_s, and the appropriate machine parameters. First, the stator flux linkage and stator current components must be found.

$$\psi_d = \psi_s \cos \delta_s,$$

$$\psi_q = \psi_s \sin \delta_s,$$

$$i_d = \frac{1}{L_d}(\psi_s \cos \delta_s - L_{md}i_f), \tag{8.186}$$

$$i_q = \frac{1}{L_q}\psi_s \sin \delta_s.$$

Per-unit torque is then

$$|T_e| = |\psi_s \times i_s| = \psi_s \cos \delta_s \frac{1}{L_q}\psi_s \sin \delta_s - \psi_s \sin \delta_s \frac{1}{L_d}(\psi_s \cos \delta_s - L_{md}i_f)$$

$$= \frac{1}{L_d}\psi_s L_{md}i_f \sin \delta_s + \psi_s^2\left(\frac{1}{L_q} - \frac{1}{L_d}\right)\sin \delta_s \cos \delta_s \tag{8.187}$$

$$= \left(\frac{\psi_s L_{md}i_f}{L_d} + \psi_s^2\left(\frac{1}{L_q} - \frac{1}{L_d}\right)\cos \delta_s\right)\sin \delta_s,$$

which includes the excitation and reluctance torque components and corresponds to the salient pole synchronous machine power equation. The per-unit power P expression is the more traditional load angle equation.

$$P = \omega_s T_e = \left(\frac{u_s e_f}{L_d \omega_s}\sin \delta_s + u_s^2\left(\frac{L_d - L_q}{2L_d L_q \omega_s}\right)\sin 2\delta_s\right) \tag{8.188}$$

where

$$u_s = \omega_s \psi_s, \, e_f = \omega_s L_{md}i_f \tag{8.189}$$

The maximum torque producing load angle can be found by differentiating Equation (8.187) with respect to the load angle δ_s and finding the zero for the differential.

$$\frac{\partial T_e}{\partial \delta_s} = \left(\frac{\psi_s L_{md}i_f}{L_d} + \psi_s^2\left(\frac{1}{L_q} - \frac{1}{L_d}\right)\cos \delta_s\right)\cos \delta_s + \sin \delta_s\left(-\psi_s^2\left(\frac{1}{L_q} - \frac{1}{L_d}\right)\sin \delta_s\right) \tag{8.190}$$

The zero of Equation (8.190) can be found by rewriting it in polynomial form with $\cos\delta_s$ as a variable.

$$2\psi_s^2\left(\frac{1}{L_q} - \frac{1}{L_d}\right)\cos^2 \delta_s + \frac{\psi_s L_{md}i_f}{L_d}\cos \delta_s - \psi_s^2\left(\frac{1}{L_q} - \frac{1}{L_d}\right) = 0 \tag{8.191}$$

Equation (8.191) represents a quadratic equation, which is expressed

$$\cos \delta_s = \frac{-\dfrac{L_{md}i_f}{L_d} \pm \sqrt{\dfrac{(L_{md}i_f)^2}{L_{sd}^2} + 8\psi_s^2\left(\dfrac{1}{L_q} - \dfrac{1}{L_d}\right)^2}}{4\psi_s\left(\dfrac{1}{L_q} - \dfrac{1}{L_d}\right)} \tag{8.192}$$

For a motor, the maximum static torque will be reached using a load angle $\delta_s \in (-\pi/2, \pi/2)$. A root must be selected from Equation (8.192) that gives a positive value to the load angle cosine.

$$\cos \delta_s = \frac{-\dfrac{L_{md}i_f}{L_d} + \sqrt{\dfrac{(L_{md}i_f)^2}{L_d^2} + 8\psi_s^2\left(\dfrac{1}{L_q} - \dfrac{1}{L_d}\right)^2}}{4\psi_s\left(\dfrac{1}{L_q} - \dfrac{1}{L_d}\right)} \tag{8.193}$$

Substituting (193) into (191) yields maximum torque as a function of the field-winding current i_f and the stator flux-linkage modulus ψ_s.

$$T_{e\,max} = \sqrt{1 - \left[\frac{\dfrac{L_{md}i_f}{L_d} - \sqrt{\dfrac{(L_{md}i_f)^2}{L_d^2} + 8\psi_s^2\left(\dfrac{1}{L_q} - \dfrac{1}{L_d}\right)^2}}{4\psi_s\left(\dfrac{1}{L_q} - \dfrac{1}{L_d}\right)}\right]^2}$$

$$\cdot \left[\psi_s^2\left(\dfrac{1}{L_q} - \dfrac{1}{L_d}\right)\frac{-\dfrac{L_{md}i_f}{L_d} + \sqrt{\dfrac{(L_{md}i_f)^2}{L_{sd}^2} + 8\psi_s^2\left(\dfrac{1}{L_q} - \dfrac{1}{L_d}\right)^2}}{4\psi_s\left(\dfrac{1}{L_q} - \dfrac{1}{L_d}\right)} + \frac{\psi_s L_{md}i_f}{L_d}\right] \tag{8.194}$$

Figure 8.40 depicts torque and load angle as a function of field-winding current for the motor deep in the field weakening region. The figure illustrates two cases: maximum torque with a particular field-winding current and the torque resulting from a unity power factor with the same current.

8.14.2 Maximum dynamic torque

According to the load angle equation, in steady state, synchronous machines with normal saliency ($L_d \geq L_q$) produce maximum torque with a load angle range of $\delta_s \in [-\pi/2, \pi/2]$.

However, because the damper winding works to stabilize the air-gap flux linkage ψ_m, in machines undergoing rapid transients, that is, when the transient time constant is so short the damper windings cannot respond quickly enough, maximum dynamic torque can be achieved with a load angle δ_s significantly larger than $\pi/2$. Niiranen (1993) introduced the following equation for the torque of a synchronous machine during a transient torque step.

$$T_e = \frac{\psi_s^2}{L_d}\left[(k_m - k_d)\sin \delta_s + \frac{k_d - k_q}{2}\sin 2\delta_s + (k_q - k_{q2})\sin \delta_{s0} \cos \delta_s\right] \tag{8.195}$$

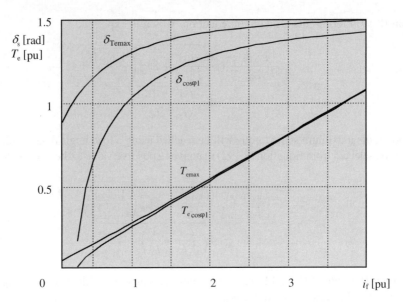

Figure 8.40 The maximum torque T_{emax} and the torque producing load angle δ_{Temax} as a function of field-winding current i_f in the field-weakening region compared to the same quantities for a unity power factor $\delta_{cos\varphi 1}$ and $T_{ecos\varphi 1}$, $\psi_s = 0.3$. *Source:* Adapted from Pyrhönen (1998). Reproduced with permission of O. Pyrhönen.

where

$$k_d = \frac{\Delta\psi_{md}}{\Delta\psi_d}, \quad k_q = \frac{\Delta\psi_{mq}}{\Delta\psi_q}, \quad k_m = \frac{\psi_{md0}}{\psi_{d0}}, \quad \text{and} \quad k_{q2} = \frac{L_{mq}}{L_{s\sigma} + L_{mq}}$$

$\Delta\psi_{md}$, $\Delta\psi_{mq}$, $\Delta\psi_d$, and $\Delta\psi_q$ represent transient changes in the flux-linkage components and ψ_{md0}, ψ_{sd0} and δ_{s0} are the initial values of the variables.

Synchronous machine transients can be investigated in the rotor reference frame. The angular velocity of the stator flux linkage in the rotor reference frame depends on the voltage reserve available.

$$u_{res} \approx |u_s|_{max} - |e_s| = |u_s|_{max} - \omega_s|\psi_s|, \quad R_s \approx 0 \tag{8.196}$$

The angular velocity $\omega_{r\psi}$ of the stator flux-linkage vector can be determined as follows.

$$\omega_{r\psi} = \frac{1}{\omega_b}\frac{\partial\delta_s}{\partial t} = \frac{\partial\delta_s}{\partial\tau} \approx \frac{u_{res}}{|\psi_s|} \tag{8.197}$$

If the voltage reserve u_{res} is assumed constant, the control can rotate the flux-linkage vector at a constant speed in the rotor reference frame during a torque step. The behaviour of the stator current, at the same time, depends on the control of the field-winding current.

Figure 8.41 depicts the behaviour of the stator current and the stator flux linkage during a torque step in a vector-controlled synchronous motor when the DC voltage supplying the field winding is kept constant. This is a nonpractical case, because field-winding current should also be controlled. With constant DC voltage, however, the final transient-time load angle is significantly larger than $\pi/2$ meaning the working point cannot be stable and is reached only because the damper-winding current compensates for the large change in stator current.

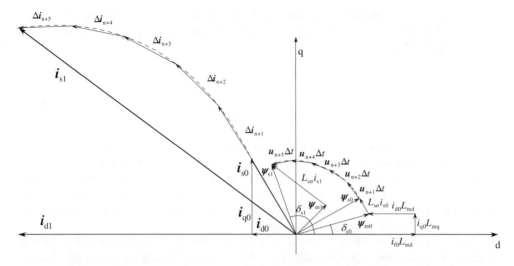

Figure 8.41   The step-by-step behaviour of stator flux linkage and stator current during a fast torque transient. If the stator voltage is held constant, the initial and final stator flux linkages ψ_{s0}, ψ_{s1} have the same amplitude values – The machine per-unit parameters are according to Table 8.1. $L_{md} = 1.05$, $L_{mq} = 0.45$, and $L_{s\sigma} = 0.12$. The behaviour of the stator current depends on the control of the field-winding current. Constant field-winding voltage is assumed that enables a small dynamic increase in field-winding current. Because of the increasing demagnetizing armature reaction, the air-gap flux linkage ψ_m becomes continually smaller as the transient progresses until a final level ψ_{m1} is reached. The d-axis damper and extra field-winding currents must compensate for the large negative d-axis current that occurs when the transient ends. *Source:* Adapted from Pyrhönen (1998). Reproduced with permission of O. Pyrhönen.

If field-winding current is kept sufficiently high following the transient, the stator flux linkage can be driven with the synchronous angular speed, and the new torque value can be maintained. Load angle must also remain stable. Stator flux linkage in the rotor reference frame can be expressed

$$\psi_s^r = |\psi_s| e^{j(\delta_{s0} + \omega_{r\psi}\tau)} \tag{8.198}$$

where δ_{s0} is the load angle before the torque step. The accelerating voltage reserve is

$$u_{res}^r = u_{res} e^{j(\delta_{s0} + \omega_{r\psi} \cdot \tau + \frac{\pi}{2})} \tag{8.199}$$

With constant voltage reserve the stator flux linkage rotates in the rotor reference frame with a constant angular speed during a torque step. The corresponding stator current vector trajectory depends on the motor parameters, Figure 8.42. Approximate transient analysis can be done by

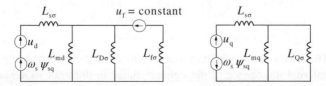

Figure 8.42   Simplified two-axis inductance transient model of a salient pole synchronous machine for transient analysis without rotor resistances.

neglecting the rotor resistances, if the transient is fast compared to the time constants of the damper windings. The damper-winding per-unit time constants for the d-axis τ_{Dpu} and for the q-axis τ_{Qpu} are defined as follows.

$$
\begin{aligned}
\tau_{\text{Dpu}} &= \frac{L_{\text{D}}}{\omega_{\text{b}} R_{\text{D}}}, \\
\tau_{\text{Qpu}} &= \frac{L_{\text{Q}}}{\omega_{\text{b}} R_{\text{Q}}}.
\end{aligned}
\tag{8.200}
$$

Both τ_{Dpu} and τ_{Qpu} are scaled to real time by the factor $1/\omega_{\text{b}}$, where ω_{b} is the base value of the angular speed of the per-unit system. In large synchronous machines, the damper-winding time constants τ_{D} and τ_{Q} are typically higher than 100 ms (except in our small test machine with time constants in the range of 10 ms). The stator current change is proportional to the voltage time integral $\int u_{\text{n}} dt \approx u_{\text{n}} \Delta t$, and it is also dependent on the motor inductance parameters. A constant field-winding voltage u_{f} is assumed.

The machine reacts differently in the d- and q-axis directions. On the q-axis, the transient inductance is the sub-transient inductance L_{q}'', which is formed by the stator leakage inductance $L_{\text{s}\sigma}$ and the q-axis damper-winding leakage and magnetising inductances $L_{\text{Q}\sigma}$ and L_{mq}, which are connected in parallel.

$$
L_{\text{q,tr}} = L_{\text{q}}'' = L_{\text{s}\sigma} + \frac{L_{\text{mq}} L_{\text{Q}\sigma}}{L_{\text{mq}} + L_{\text{Q}\sigma}}.
\tag{8.201}
$$

On the d-axis, to achieve the lowest value transient inductance, the sub-transient inductance and the field-winding leakage inductance $L_{\text{f}\sigma}$ is coupled in parallel with the magnetising inductance L_{md} and the damper-winding leakage inductance $L_{\text{D}\sigma}$.

$$
L_{\text{d,tr}} = L_{\text{d}}'' = L_{\text{s}\sigma} + \frac{L_{\text{D}\sigma} L_{\text{md}} L_{\text{f}\sigma}}{L_{\text{md}} L_{\text{D}\sigma} + L_{\text{md}} L_{\text{f}\sigma} + L_{\text{D}\sigma} L_{\text{f}\sigma}}
\tag{8.202}
$$

Field-winding current control has a big influence on the d-axis transient. The basic different methods include constant field-winding voltage control, constant field-winding current control, and d-axis damper-winding compensation control, which is also known as reaction control (Mård et al., 1990).

If the case of constant field-winding voltage, the d-axis stator subtransient current uses all three inductance paths. See Figure 8.43. The d-axis current time differential is inversely proportional to the subtransient inductance in Equation (8.202).

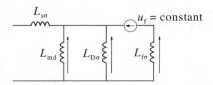

Figure 8.43 The d-axis stator current transient paths when the field-winding voltage is kept constant. The internal impedance of the voltage source u_{f} does not significantly resist fast changes, and therefore the field winding may take part in the transient. In a transient moving faster than τ_{D}, the stator transient meets the d-axis subtransient inductance. The arrows indicate the transient current directions.

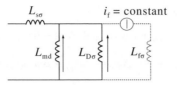

Figure 8.44 d-axis components during a transient if the field-winding current is held constant. The arrows describe the transient currents.

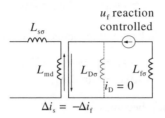

Figure 8.45 The paths of the d-axis component of the stator current if the d-axis component of the air-gap flux is kept constant. There will be no damper-winding current. Changing the field-winding current is also easy, because there is no change in the d-axis air-gap flux energy when the air-gap flux linkage is held constant. Therefore, this is the fastest possible field-winding control method for a synchronous machine.

If field-winding current instead of voltage is kept constant (Figure 8.44), there is no leakage flux path available in the field winding, because the current source has infinite impedance for changes, and the transient will be dominated by a new type of transient inductance.

$$L_{\text{d,tr,ifconst}} = L_{\text{s}\sigma} + \frac{L_{\text{D}\sigma}L_{\text{md}}}{L_{\text{md}} + L_{\text{D}\sigma}} \tag{8.203}$$

And, if the d-axis component of the air-gap flux is kept constant, an approach referred to as reaction control (Figure 8.45), the transient inductance is expressed as follows. The fastest transients in a synchronous machine are made possible by the reaction control.

$$L_{\text{d,tr,react contr}} = L_{\text{s}\sigma} \tag{8.204}$$

8.14.3 Maximum dynamic torque in the case of a fast transient

When analysing a transient taking place faster than τ_{D}, rotor resistances can be neglected, and rotor phenomena can be observed by monitoring rotor inductances alone. The dependencies between stator flux linkage and the stator current changes can be expressed separately on the different axes.

$$\Delta i_{\text{d}} = \frac{1}{L_{\text{d,tr}}} \Delta \psi_{\text{d}},$$

$$\Delta i_{\text{q}} = \frac{1}{L_{\text{q,tr}}} \Delta \psi_{\text{q}}. \tag{8.205}$$

Torque during a transient can be expressed as

$$
\begin{aligned}
T_e &= \psi_d i_q - \psi_q i_d = (\psi_{d0} + \Delta\psi_d)(i_{q0} + \Delta i_q) - (\psi_{q0} + \Delta\psi_q)(i_{d0} + \Delta i_d) \\
&= (\psi_{d0} + \Delta\psi_d)\left(i_{q0} + \frac{1}{L_{q,tr}}\Delta\psi_q\right) - (\psi_{q0} + \Delta\psi_q)\left(i_{d0} + \frac{1}{L_{d,tr}}\Delta\psi_d\right)
\end{aligned}
\tag{8.206}
$$

In the following, no load is assumed. The initial q-axis stator current and stator flux-linkage components can be set equal to zero, and Equation (8.206) can be simplified.

$$
T_e = (\psi_{d0} + \Delta\psi_d)\frac{1}{L_{q,tr}}\Delta\psi_q - \Delta\psi_q \frac{1}{L_{d,tr}}\Delta\psi_d
\tag{8.207}
$$

The changes in the stator flux-linkage components are as follows.

$$
\begin{aligned}
\Delta\psi_d &= (\cos\delta_s - 1)|\psi_s|, \\
\Delta\psi_q &= \sin\delta_s|\psi_s|.
\end{aligned}
\tag{8.208}
$$

Therefore, the torque expression becomes

$$
T_e = \psi_s^2 \sin\delta_s \left(\frac{1}{L_{d,tr}} + \left(\frac{1}{L_{q,tr}} - \frac{1}{L_{d,tr}}\right)\cos\delta_s\right)
\tag{8.209}
$$

The maximum value for the torque is found by differentiating Equation (8.209) with respect to the load angle δ_s.

$$
\frac{\partial T_e}{\partial\delta_s} = \psi_s^2\left(2\cdot\left(\frac{1}{L_{q,tr}} - \frac{1}{L_{d,tr}}\right)\cos^2\delta_s + \frac{1}{L_{d,tr}}\cos\delta_s - \frac{1}{L_{q,tr}} + \frac{1}{L_{d,tr}}\right)
\tag{8.210}
$$

The load angle cosine, which corresponds to the maximum dynamic torque, is determined by solving for the root of the second order Equation (8.210).

$$
\cos\delta_s = \frac{-\dfrac{1}{L_{d,tr}} + \sqrt{\dfrac{1}{L_{d,tr}^2} + 8\left(\dfrac{1}{L_{q,tr}} - \dfrac{1}{L_{d,tr}}\right)^2}}{4\left(\dfrac{1}{L_{q,tr}} - \dfrac{1}{L_{d,tr}}\right)}
\tag{8.211}
$$

Table 8.4 shows the effective transient inductances, the optimal load angles and the maximum torques for different field-winding current control methods in the case of a reduced stator flux linkage $\psi_s = 0.3$ pu.

Table 8.4 Effective transient inductances, maximum dynamic torque $T_{ed,max}$ and corresponding load angle $\delta_{d,max}$ for different field-winding current-control methods. The stator flux-linkage modulus $\psi_s = 0.3$, and the machine parameters are according to Table 8.1. Motor load is initially zero. In all cases, the initial rotor field-winding current corresponds to the no-load voltage before the transient.

Field-winding current control method	$L_{d,tr}$ [pu]	$L_{q,tr}$ [pu]	$T_{ed,max}$ [pu]	$\delta_{d,max}$ [deg]
A. Constant voltage	0.17	0.23	0.54	103
B. Constant current	0.19	0.23	0.49	100
C. Reaction control	0.12	0.23	0.82	111

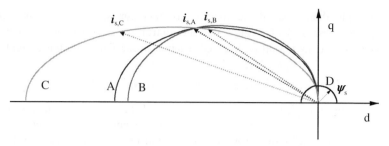

Figure 8.46 Stator flux linkage and stator current trajectory curves for a fast transient with the different excitation methods and the load angle $\delta_s \in [0,\pi]$ in field weakening where $\psi_s = 0.3$ – Curve A is the constant field-winding voltage, curve B is the constant field-winding current trajectory, curve C is the reaction control trajectory, and curve D is the stator flux-linkage vector trajectory. The current vectors shown correspond to the respective maximum dynamic torque points. Motor parameters are according to Table 8.1. *Source:* Adapted from Pyrhönen (1998). Reproduced with permission of O. Pyrhönen.

Figures 8.46 and 8.47 illustrate the three different transient trajectory curves resulting from the different field-winding current control methods. The comparison helps to explain the meaning of each field-winding current response. During a fast transient, the good dynamics of the field-winding current results in a much higher peak value for dynamic torque driven by a

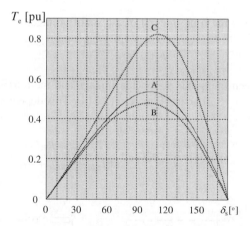

Figure 8.47 Fast per-unit torque response T_e in field weakening where $\psi_s = 0.3$ as a function of the load angle δ_s for the different field-winding current control methods. Curve A is for constant field-winding voltage with both the d-axis damper and field windings acting as damper windings. Curve B is for constant field-winding current with the field winding not taking part in the transient. Curve C is for a fully compensated d-axis damper winding with the d-axis air-gap flux linkage ψ_{md} held constant. The highest dynamic torque is achieved in case C. *Source:* Adapted from Pyrhönen (1998). Reproduced with permission of O. Pyrhönen.

higher load angle than is available for the constant current- or voltage-based control approaches. The constant field-winding current control gives the smallest dynamic torque response, because it does not allow induced reaction in the field-winding current.

8.14.4   Transient analysis with operating inductances

The previous analysis neglected the rotor resistances. If there is low voltage reserve when operating in the field weakening area, transient time constants may be longer than damper-winding time constants. Moreover, rotor resistances must be taken into account. In this case, the transient inductances are not constant, but are, instead, frequency dependent transfer functions. This scenario was examined, for example, by Bühler (1977).

The frequency-dependent coefficients are referred to as *operator inductances*. For a vector controlled drive, the resistance and inductance parameters of the two-axis equivalent circuit are used. The equivalent circuits can be updated with resistances as follows.

The voltage reserve components become

$$u_{d,res} = u_d - i_d R_s + \omega_{s0}\psi_q, \quad \text{and}$$
$$u_{q,res} = u_q - i_q R_s - \omega_{s0}\psi_d. \tag{8.212}$$

The angular speed ω_{s0} is the stator flux-linkage vector angular velocity before the transient. If the voltage drop in the stator resistance is assumed to be small and the voltage reserve is assumed constant, the transfer function $G_q(s)$ between the q-axis current i_q and the q-axis voltage reserve $u_{q,res}$ is defined as

$$G_q(s) = \frac{U_{q,res}(s)}{I_q(s)} = \frac{s^2 \cdot L_{Q\sigma}L_{mq} + sR_Q L_{mq}}{s(L_{Q\sigma} + L_{mq}) + R_Q} + sL_{s\sigma} \tag{8.213}$$

The stator flux linkage is the time integral of the stator voltage.

$$\psi_q(s) = \frac{U_{q,res}(s)}{s} \tag{8.214}$$

The dependence between the q-axis stator flux linkage and the q-axis stator current can be solved by using Equation (8.213) and Equation (8.214). This dependence defines the q-axis operator inductance as

$$L_q(s) = \frac{G_q(s)}{s} = L_{s\sigma} + \frac{sL_{Q\sigma}L_{mq} + R_Q L_{mq}}{s(L_{Q\sigma} + L_{mq}) + R_Q} \tag{8.215}$$

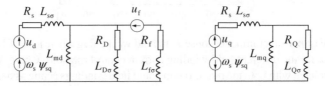

Figure 8.48 Two-axis parameter model of a salient pole synchronous machine for the transient analysis. Resistance parameters are included in the transient analysis. The stator resistance is assumed to be small and has been neglected.

In a fast transient, $sL_{Q\sigma}L_{mq} \gg R_Q L_{mq}$, and $s(L_{Q\sigma} + L_{mq}) \gg R_Q$, so the result of Equation (8.215) approaches the value determined for the transient inductance $L_{q,tr}$ defined in Equation (8.201). In steady state, where all the differentials are zero, Equation (8.215) results in the q-axis synchronous inductance L_q.

The analysis of the d-axis for the transient inductances above also can be carried out using the three field-winding current control methods previously discussed. The added inclusion of the rotor resistive components in the observations is the only difference.

If the field-winding voltage is kept constant during the transient (see previous Figure 8.43), the stator current transient will use all three paths. The operator inductance becomes

$$L_d(s) = L_{s\sigma} + \frac{L_{md}(sL_{D\sigma} + R_D)(sL_{f\sigma} + R_f)}{sL_{md}(sL_{D\sigma} + R_D) + (sL_{D\sigma} + R_D)(sL_{f\sigma} + R_f) + sL_{md}(sL_{f\sigma} + R_f)}$$

$$= L_{s\sigma} + \frac{s^2 L_{md}L_{D\sigma}L_{f\sigma} + sL_{md}(L_{D\sigma}R_f + L_{f\sigma}R_D) + L_{md}R_D R_f}{s^2(L_{md}L_{D\sigma} + L_{D\sigma}L_{f\sigma} + L_{md}L_{f\sigma}) + s(L_{md}(R_D + R_f) + L_{D\sigma}R_f + L_{f\sigma}R_D) + R_D R_f}$$

$$(8.216)$$

If instead the field-winding current is kept constant (see previous Figure 8.44), the operator inductance looks like the operator inductance on the q-axis, because no transient takes place in the field winding.

$$L_{difconst}(s) = L_{s\sigma} + \frac{sL_{D\sigma}L_{md} + R_D L_{md}}{s(L_{D\sigma} + L_{md}) + R_D} = \frac{s(L_{D\sigma}L_{md} + L_{s\sigma}L_{md} + L_{s\sigma}L_{D\sigma}) + R_D(L_{md} + L_{s\sigma})}{s(L_{D\sigma} + L_{md}) + R_D}$$

$$(8.217)$$

Finally, if reaction control is used (see previous Figure 8.45), no rotor resistive path is used by the transient current, and the effective operator inductance equals the stator leakage, that is to say, $L_{dreactcontr}(s) = L_{s\sigma}$.

In the ideal case, there is linear dependence between the changes to stator flux linkage and stator current. See Equation (8.205). When operator inductances are used, frequency dependency becomes obvious. The difference between the analysis with the transient inductances and the more realistic analysis using the operator inductances depends on the speed of the transient. In field weakening with a small voltage reserve, a transient cannot be as fast as it is in the nominal speed range and the analysis using operator inductances is preferred.

Equation (8.197) defined the angular velocity of the stator flux-linkage vector in a transient in the rotor reference frame. The per-unit time constant describes the time for a full stator flux-linkage vector rotation. However, a full rotation does not correspond to a real transient, since the maximum torque is achieved approximately with the load angle of $\pi/2$. Therefore, it is more convenient to define the dimensionless transient time constant to be the time corresponding to the stator flux-linkage vector rotation from zero to the load angle of $\pi/2$.

$$\tau_{trpu} = \frac{1}{4}\frac{2\pi}{\omega_{r\psi}} = \frac{\pi}{2}\frac{\psi_s}{u_{res}}$$

$$(8.218)$$

Since time constants are normally expressed in seconds, converting the transient time constant into real time is convenient. The time constant depends on the relation between $|\psi_s|$ and the voltage reserve u_{res}.

$$\tau_{tr} = \frac{1}{\omega_b}\tau_{tr,pu} = \frac{1}{2\pi f_b}\frac{\pi}{2}\frac{\psi_s}{u_{res}} = \frac{1}{4 f_b}\frac{\psi_s}{u_{res}}$$

$$(8.219)$$

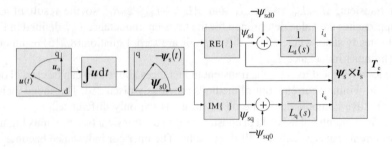

Figure 8.49 Transient simulation algorithm block presentation. *Source:* Adapted from Pyrhönen (1998). Reproduced with permission of O. Pyrhönen.

Numerical simulation is needed in transient analysis. A change in the stator flux-linkage component results in a change to the stator current component.

$$
\begin{aligned}
i_d(s) - i_{d0} &= \frac{1}{L_d(s)} \left(\psi_d(s) - \psi_{sd0} \right), \\
i_q(s) - i_{q0} &= \frac{1}{L_q(s)} \left(\psi_q(s) - \psi_{sq0} \right).
\end{aligned}
\tag{8.220}
$$

A change in the relative position of the stator flux linkage in the rotor reference frame can only be caused by a voltage reserve, and transient angular velocity is dependent on the voltage reserve. Figure 8.49 shows the symbolic block diagram of the simulation scenario starting from no load.

The damper-winding time constant has a big impact on the transient behaviour of a synchronous machine. Therefore, the direct relationship between transient time and the damper-winding time constant is particularly interesting. Figure 8.50 offers simulation results with two different ratios between the transient and damper-winding time constants.

Curves A2, B2, and C2 were determined by reducing the damper-winding resistances to one-fifth of their original values to get $\tau_D = \tau_Q = 5\tau_{tr} = 100$ ms. Neglecting resistive losses, both constant field-winding voltage and constant field-winding current result in a 20% smaller peak value (A2, B2) than determined for the transient inductances of Figure 8.48.

With reaction control, the dynamic torque curves (C1, C2) are close to each other, and the effect of the transient time constant on the dynamic peak torque is small. The torque rise time is, however, directly dependent on the transient time constant. Figure 8.51 shows two different torque transients with different transient time constants: $\tau_{tr} = 20$ ms (A) and $\tau_{tr} = 40$ ms (B). The damper-winding time constants in both cases are set to be equal, that is, $\tau_D = \tau_Q = 40$ ms. The torque rise time occurs to be approximately the same as the transient time constant.

In the first case, the transient and damper-winding time constants are equal; $\tau_D = \tau_Q = \tau_{tr} = 20$ ms. These time constants are somewhat larger than actual. The curves A1, B1, and C1 are determined with these time constants. The comparison between the torque peak values produced by the operator inductance analysis (Figure 8.50) and those of the transient

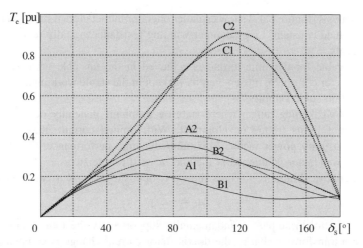

Figure 8.50 Torque development during a transient using various field-winding current control methods and different damper-winding time constants with a reduced stator flux-linkage modulus of $\psi_s = 0.3$ pu. The time corresponding to the transient time constant is 40 ms (time to rotate load angle to $\pi/2$). Curves A represent constant field-winding voltage, curves B represent constant field-winding current, and curves C represent reaction control. The number 1 corresponds to $\tau_D = \tau_Q = \tau_{tr} = 20$ ms, and the number 2 corresponds to $\tau_D = \tau_Q = 5\tau_{tr} = 100$ ms. Reference motor parameters are used, and only the damper-winding resistances are modified. *Source:* Adapted from Pyrhönen (1998). Reproduced with permission of O. Pyrhönen.

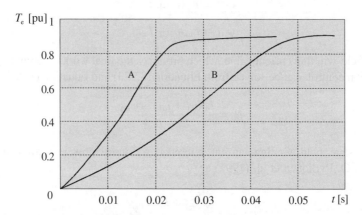

Figure 8.51 Torque rise with different transient time constants and with reaction control. (a) $\tau_{tr} = 20$ ms. (b) $\tau_{tr} = 40$ ms. In both cases, $\tau_D = \tau_Q = 40$ ms. The stator flux-linkage modulus is reduced, that is, $\psi_s = 0.3$ pu. Reference motor parameters are used, and only the damper-winding resistances are modified. *Source:* Adapted from Pyrhönen (1998). Reproduced with permission of O. Pyrhönen.

inductance analysis (Figure 8.47) shows smaller operator inductances if field-winding current or voltage are held constant. Large damper-winding resistances rapidly reduce the damper-winding currents.

The peak value for dynamic torque is achieved in the unstable working area, where $|\delta_s| > \pi/2$. In vector control, δ_s must generally be kept in the stable working area or where $|\delta_s| < \pi/2$ to maintain static stability.

In the field-weakening range, where stability problems generally occur, the available converter current limits torque development. With limited current, maximum torque is achieved with a unity power factor. Maximum dynamic performance, therefore, requires that a unity power factor is attained at the end of each torque step.

Reaction control (Mård, 1990) uses the exact opposite d-axis current components.

$$\Delta i_f = -\Delta i_d \tag{8.221}$$

The reaction control principle is fundamentally supported by the Lenz's law. The d-axis operates like a transformer, that is, the d-axis stator current change is compensated by the field-winding current.

If the post-transient d-axis currents are i_{f1} and i_{d1}, and the initial values are i_{f0} and i_{d0}, the d-axis current components change as follows.

$$\begin{aligned} \Delta i_f &= i_{f1} - i_{f0}, \\ \Delta i_d &= i_{d1} - i_{d0}. \end{aligned} \tag{8.222}$$

The changes in the d-axis current components should have the same modulus and the opposite sign.

$$i_{f1} - i_{f0} = -(i_{d1} - i_{d0}) \tag{8.223}$$

The initial and final d-axis stator flux-linkage components are

$$\psi_{d0} = L_{md} i_{f0} + L_d i_{d0} \tag{8.224}$$

$$\psi_{d1} = L_{md} i_{f1} + L_d i_{d1} \tag{8.225}$$

The initial field-winding current value, which produces the final working point (i_{f1}, i_{d1}) when using reaction control, can be solved from Equation (8.223) and Equation (8.224) as follows.

$$i_{f0} = \frac{L_d(i_{f1} + i_{d1}) - \psi_{d0}}{L_d - L_{md}} \tag{8.226}$$

The steady state field-winding current reference for unity power factor can be solved according to O. Pyrhönen et al. (1997) as

$$i_{fref} = \frac{\psi_{s\,ref}^2 + L_d L_{sq} \frac{T_{e\,ref}^2}{\psi_{s\,ref}^2}}{L_{md} \sqrt{\left[\psi_{s\,ref}^2 + L_q^2 \frac{T_{e\,ref}^2}{\psi_{s\,ref}^2} \right]}} \tag{8.227}$$

The reaction control law of Equation (8.221) and the unity power factor control of Equation (8.227) result in different field-winding current references. When maximum drive performance is desired, the minimum stator current-to-torque ratio must be reached when applying the maximum available stator current. This is especially important when using reduced stator flux linkage and the maximum torque of a drive is defined by the maximum inverter current. The initial field-winding current reference should be selected so Equation (8.219) satisfies Equation (8.227) at the end of each torque step. At partial loads, the field-winding current reference, in practice, should be above the value given by Equation (8.227). The machine should operate over-excited if the highest possible torque output is wanted in a transient.

The following example clarifies the principle of partial load over excitation. In the previous example illustrated by Figure 8.50, the dynamic peak torque of 0.9 pu was achieved using reaction control. Let the same value be the torque reference in this example. Bühler (1977) has shown the load angle value of a salient pole synchronous machine with unity power factor to be

$$\delta_s = \text{atan}\left(\frac{T_e L_q}{\psi_s^2}\right) \tag{8.228}$$

EXAMPLE 8.2: Calculate the final values of field-winding current, load angle, stator d-axis flux linkage, and initial field-winding current for a torque step from zero to 0.9 per unit at a reduced stator flux-linkage modulus of 0.3 pu using the reference machine parameters from Table 8.1.

SOLUTION: According to Equation (8.227), the field-winding current should be

$$i_{f1} = \frac{\psi_{s\,\text{ref}}^2 + L_d L_q \frac{T_{e\,\text{ref}}^2}{\psi_{s\,\text{ref}}^2}}{L_{md}\sqrt{\left[\psi_{s\,\text{ref}}^2 + L_q^2 \frac{T_{e\,\text{ref}}^2}{\psi_{s\,\text{ref}}^2}\right]}} = \frac{0.3^2 + 1.17 \cdot 0.57 \cdot \frac{0.9^2}{0.3^2}}{1.05\sqrt{\left[0.3^2 + 0.57^2 \cdot \frac{0.9^2}{0.3^2}\right]}} = 3.35$$

The load angle will be as follows.

$$\delta_{s1} = \text{atan}\left(\frac{T_{e\,\text{ref}} L_q}{\psi_{s\,\text{ref}}^2}\right) = \text{atan}\left(\frac{0.9 \cdot 0.57}{0.3^2}\right) = 80$$

The d-axis stator flux-linkage final component can be solved according to the load angle and the stator flux-linkage modulus.

$$\psi_{d1} = \cos\delta_{s1} \cdot \psi_{s\,\text{ref}} = \cos 80° \cdot 0.3 = 0.052$$

Finally, the d-axis current components can be solved from Equation (8.226) and Equation (8.227).

$$i_{d1} = \frac{\psi_{d1} - L_{md}i_{f1}}{L_d} = \frac{0.052 - 1.05 \cdot 3.35}{1.17} = -2.96$$

$$i_{f0} = \frac{L_d(i_{f1} + i_{d1}) - \psi_{d0}}{L_d - L_{md}} = \frac{1.17 \cdot (3.35 - 2.96) - 0.3}{1.17 - 1.05} = 1.3$$

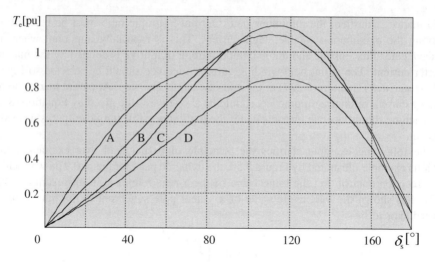

Figure 8.52 Static or dynamic torque as a function of the load angle δ_s. Curve A represents the static torque with fixed $\psi_s = 0.3$ pu and $i_s = 3$. Curve B illustrates fast dynamic torque when the initial value of the field-winding current has the value of the above example. Curve C is the same as B, but the q-axis is assumed to follow the steady-state equation. And, curve D is the dynamic torque with the initial value $i_d = 0$. $\tau_{tr} = \tau_D = \tau_Q = 40$ ms, and the reference motor parameters are used. *Source:* Adapted from Pyrhönen (1998). Reproduced with permission of O. Pyrhönen.

Figure 8.52 shows four different dynamic results. Curve B illustrates the dynamic torque with reaction control and the above presented initial field-winding current. Curve A represents the static torque with $|\psi_s| = 0.3$ pu and $|i_s| = 3$. In steady state, these values produce the desired torque when the unity power factor is wanted. The required transient torque will be reached at approximately the same load angle, which corresponds to the unity power factor working point for steady state. Curve C shows the dynamic torque that results when the q-axis components are assumed to conform to the steady state equations during the transient also (no damping effect on the q-axis). In this case, the desired torque is reached exactly with unity power factor load angle, that is, at $\delta_{s, \cos\varphi1} = 80.05°$.

When the maximum converter current and the stator flux-linkage modulus are known, a solution for initial field-winding current can be found. The maximum torque is then

$$T_{e\ max} = \psi_s i_{s\ max} \tag{8.229}$$

The required field-winding current $i_{f,ref}$ and the d-axis current i_d realizing a power factor of 1 are found from Equation (8.227) and Equation (8.228). Reaction control requires d-axis air-gap flux linkage to remain constant. According to Equation (8.221), changes to d-axis current must be compensated for with field-winding current. Starting from the final working point with maximum available torque and stator current, the values for d-axis stator and field-winding current can be calculated as a function of the decreasing torque by conforming to

Equation (8.221). Using the following expressions

$$\psi_d^2 + \psi_q^2 = \psi_s^2$$

$$\psi_d = \psi_{d1} + \Delta i_d L_d$$

$$\psi_q = L_q i_q \tag{8.230}$$

$$T_e = \psi_d i_q - \psi_q(i_{d1} + \Delta i_d)$$

a polynomial equation form of fourth order can be found for Δi_d.

$$k_4 \Delta i_d^4 + k_3 \Delta i_d^3 + k_2 \Delta i_d^2 + k_1 \Delta i_d + k_0 - T_e^2 = 0 \tag{8.231}$$

where

$$k_4 = -L_{s\sigma}^2 \left(1 - \frac{L_{s\sigma}}{L_q}\right)^2,$$

$$k_3 = -2\psi_{d1} L_{s\sigma} \left(1 - \frac{L_{s\sigma}}{L_q}\right)^2 - 2L_{s\sigma}^2 \left(1 - \frac{L_{s\sigma}}{L_q}\right)\left(i_d^* - \frac{\psi_{d1}}{L_q}\right),$$

$$k_2 = -L_{s\sigma}^2 \left(\frac{\psi_{d1}}{L_q} - i_{d1}\right)^2 - 4\psi_{sd1} L_{s\sigma} \left(1 - \frac{L_{s\sigma}}{L_q}\right)\left(i_{d1} - \frac{\psi_{d1}}{L_q}\right) + \left(\psi_s^2 - (\psi_{d1})^2\right)\left(1 - \frac{L_{s\sigma}}{L_q}\right)^2,$$

$$k_1 = -2\psi_{sd1} L_{s\sigma} \left(\frac{\psi_{d1}}{L_q} - i_{d1}\right)^2 + 2\left(\psi_s^2 - (\psi_{d1})^2\right)\left(1 - \frac{L_{s\sigma}}{L_q}\right)\left(i_{d1} - \frac{\psi_{d1}}{L_q}\right),$$

$$k_0 = \left(\psi_s^2 - (\psi_{d1})^2\right)\left(\frac{\psi_{d1}}{L_q} - i_{d1}\right)^2.$$

Using another numerical example to demonstrate, Table 8.5 lists a number of parameter values for a maximum torque working point in the field weakening range.

Table 8.5 A maximum torque working point in the field weakening range with a given stator flux-linkage modulus and maximum stator current

Variable		Description
δ_{s1}	81°	load angle with the maximum torque and unity power factor
ψ_s	0.3 pu	stator flux-linkage reference
i_{smax}	3.33 pu	maximum inverter current modulus
T_{emax}	1.0 pu	maximum torque with unity power factor
ψ_{d1}	0.047 pu	d-axis stator flux-linkage component in maximum torque point
i_{d1}	−3.29 pu	d-axis stator current component in maximum torque point
I_{f1}	3.71 pu	field-winding current component in maximum torque point

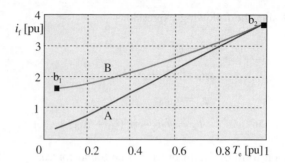

Figure 8.53 Reaction control. Curve B starts with overexcitation and ends at $\cos \varphi = 1$. The d-axis component of the air-gap flux linkage is kept constant irrespective of the torque. Also, the curve (A) for constant $\cos \varphi = 1$ is illustrated. *Source:* Adapted from Pyrhönen (1998). Reproduced with permission of O. Pyrhönen.

At every working point then, field-winding current can be solved for when the final value i_{f1} and the required change $\Delta i_f = -\Delta i_d$ are known. Figure 8.53 represents the field-winding current value for the constant d-axis air-gap flux-linkage component. For comparison, the unity power factor field-winding current as a function of the torque is also shown.

The transition from working point b_1 to working point b_2 shown in Figure 8.53 is represented by space vectors in the rotor reference frame in Figure 8.54. Over excitation demands large demagnetizing d-axis stator currents at lower load conditions. However, this is necessary to achieve the maximum torque defined by maximum converter current immediately following a load transient.

Field-winding current control plays an important role in the operation of synchronous motor drives, especially, when the drive is working with a high load angle in the field weakening range. By using partial load over excitation and reaction control the available

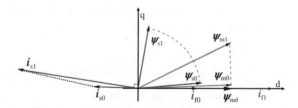

Figure 8.54 The initial and final stator flux-linkage vector (ψ_{s0}, ψ_{s1}), the air-gap flux-linkage vector (ψ_{m0}, ψ_{m1}), the stator current vector (i_{s0}, i_{s1}), and the field-winding current modulus (i_{f0}, i_{f1}) before and after a torque step from $T_e = 0.05$ to $T_e = 1.0$ corresponding to points b_1 and b_2 in Figure 8.53. Reaction control keeps the d-axis air-gap flux-linkage component ψ_{md} constant. The maximum torque point with unity power factor is achieved at the end of the transient. The current vectors are not to scale but should be significantly larger. *Source:* Adapted from Pyrhönen (1998). Reproduced with permission of O. Pyrhönen.

stator current can be fully utilised in each torque step and maximum torque is achieved without stability problems.

8.14.5   The field-winding current control of synchronous machine DTC drive

To explore the operation of a synchronous machine DTC drive, it must be observed separately in the constant flux range and in field weakening.

In the constant flux range

In direct torque control (DTC), stator current is not a control parameter. It is merely one of the outputs of the system. The key control parameters for DTC are stator flux linkage and torque. As such, traditional field-winding current control methods that link with the control of the stator current components do not apply.

A DTC control can hold torque at any set point value within the stable operation range of the synchronous machine. Dynamic field-winding current control errors do not affect its torque control accuracy. However, the field-winding current must keep the drive within the stable operating range by increasing the field-winding current as a function of torque. A field-winding reference that produces unity power factor can be calculated as a function of torque and flux linkage when the inductances of the synchronous machine are known. This was shown previously in Equation (8.227).

Equation (8.227) links field-winding current control to direct torque control so magnetic energy oscillations between the stator and rotor can be avoided. Due to saturation, inductance values also become inaccurate. A correction term can be defined by investigating the angle between stator flux linkage and stator current, which must be perpendicular at unity power factor. The vector dot product will be zero when the power factor is 1.

$$\boldsymbol{\psi}_s \cdot \boldsymbol{i}_s = 0 \rightarrow \cos \varphi = 1 \tag{8.232}$$

If the scalar product differs from zero, its output can be used as a correction term for the field-winding control. In that case, the field-winding current control as a whole can be implemented as shown in Figure 8.55. This system maintains unity power factor in the constant flux-linkage area.

In field weakening

In the field weakening range, drive performance is constrained by two factors: low voltage reserve and the current limit of the converter. Previously, field-winding current reaction control was shown to provide optimal torque response. Reaction control contradicts the field-winding current reference of Equation (8.227), because it does not keep the d-axis component of the air-gap flux-linkage constant. Figure 8.56 shows two examples of the amplitudes of the stator flux linkage as a function of torque.

However, field-winding current can be controlled so the d-axis component of the air-gap flux linkage remains constant and a power factor $\cos \varphi = 1$ and the current limit is

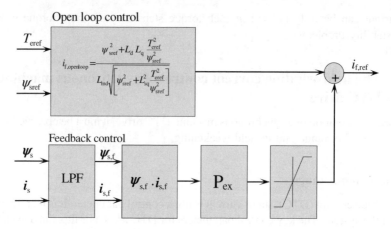

Figure 8.55 Field-winding current control of a DTC synchronous motor drive in the constant flux range. Suitable amplification P_{ex} and limitation to the correction term has been adopted. *Source:* Adapted from Pyrhönen (1998). Reproduced with permission of O. Pyrhönen.

reached simultaneously. This result is good dynamics, and the maximum available stator current produces a theoretical maximum torque. The resulting excitation curve and the vector representations are similar to those previously presented in Figures 8.53 and 8.54.

High stator and rotor currents at small loads, which increase drive losses considerably, is the disadvantage of the method. Another problem related to the high over excitation is high

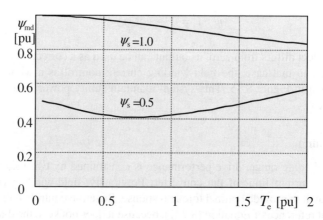

Figure 8.56 Changes to the d-axis components of the air-gap flux linkage when $\cos \varphi = 1$ at different flux-linkage levels. The field-winding current reference is determined according to Equation (8.224). *Source:* Adapted from Pyrhönen (1998). Reproduced with permission of O. Pyrhönen.

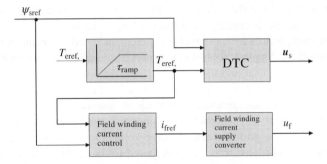

Figure 8.57 Reaction excitation control block diagram for a DTC synchronous machine drive. The torque reference rise time must be limited. The time constant is adjusted to achieve the best torque response. *Source:* Adapted from Pyrhönen (1998). Reproduced with permission of O. Pyrhönen.

overvoltage in a fault situation. Figure 8.57 illustrates how reaction control can be implemented in a DTC synchronous machine drive.

The fastest possible field-winding current rise is constrained by the available excitation voltage reserve. At low speeds, large changes can be made rapidly on the stator side, and therefore, the allowed torque rise speed must be limited depending on the field-winding voltage reserve available. The voltage reserve is naturally

$$u_{f,res} = u_{f,max} - i_f R_f \qquad (8.233)$$

For good dynamic performance, significant voltage reserve must be used. If the field winding is supplied by a thyristor bridge from a 400 V AC network, the typical rated field-winding current uses about 100 V DC resulting in about a 400 V DC reserve for the current control. The field-winding leakage inductance is, however, large and therefore a high voltage reserve is needed.

8.15   The DOL synchronous machine and field-winding current supply

8.15.1   The synchronous machine and the network

Synchronous machines are largely used in DOL applications both as generators and motors. In addition to their main function, large motors have been used for reactive power compensation tasks. In an industrial plant, running a large synchronous motor overexcited easily compensates for the reactive power of tens to hundreds of small induction motors maintaining overall an acceptable reactive power status. The most extreme use of a synchronous machine has been as a synchronous compensator without any mechanical task functionality. Within its electric current limits, the synchronous machine can be used as an effective stepless reactive power compensator. From the network point of view, it is possible to run the machine either as a reactor (under excited) or as a capacitor (overexcited) by just controlling the field-winding current accordingly. Today, this is rare, because static compensators with capacitors are more energy efficient and need less maintenance than rotary machine compensators. Stepless control can, however, be achieved only with synchronous machines.

The following text examines the interaction with the network of both motors and generators. Both in motor and generator applications, generator-island operation excluded, and the excitation state of a synchronous machine is defined chiefly by the stator terminal voltage determined by the grid. If stator resistive voltage loss is neglected, and if motor or generator power are considered to be low compared to the short-circuit power of the grid,

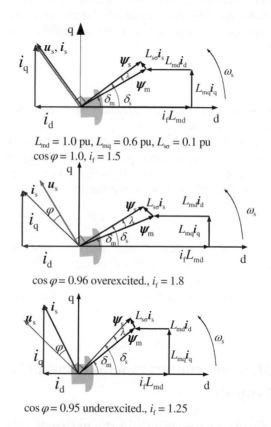

$L_{md} = 1.0$ pu, $L_{mq} = 0.6$ pu, $L_{s\sigma} = 0.1$ pu
$\cos\varphi = 1.0$, $i_f = 1.5$

$\cos\varphi = 0.96$ overexcited., $i_f = 1.8$

$\cos\varphi = 0.95$ underexcited., $i_f = 1.25$

Figure 8.58 Effect of field-winding current on the excitation state of a motor when the machine operates on a rigid network (or the terminal voltage is determined by an inverter). The angle between stator flux linkage and voltage is always 90° electrical. In the upper graph, the motor is running at its rated stator current. No-load behaviours are shown in the lower graph. See Figure 8.59. The upper graph clearly shows that machine load angle depends on field-winding current. An overexcited machine has the lowest load angle δ_s. In each representation, voltage and stator flux linkage have pu values equal to 1. The value for field-winding current must be greater than 1 to compensate for the armature reaction. In each representation, $i_f = 1.5$pu, $i_f = 1.8$pu, $i_f = 1.25$pu. $L_{md} = 1$pu, $L_{mq} = 0.6$pu, and $L_{s\sigma} = 0.1$pu.

stator flux linkage can be determined directly from the terminal voltage. Large generators in power plants can impact local line voltage, and therefore their terminal voltages can be regulated by a PI controller. However, the PI controller is not suitable for controlling the voltage of small machines. The integrator in the controller would probably saturate at either control extreme, that is, when there is no field-winding current or when there is maximum field-winding current. The machine cannot, at its terminals, significantly affect network voltage. Therefore, reactive power and not voltage is controlled in small DOL machines.

One might question if this can even be considered synchronous machine excitation control. While it is true that field-winding current impacts the magnetic state of the machine, it cannot determine stator flux linkage. Instead, the ratio of stator flux linkage and air-gap flux linkage, and thus the reactive power of the machine, is determined on the basis of field-winding current. Therefore, it would be advisable to consider this case as the control of field-winding current.

Figure 8.58 depicts a salient-pole synchronous motor in different excitation states with constant terminal voltage $u_s = 1$ and current $i_s = 1$. The per-unit magnetizing inductance of the illustrated machine in the d-direction is $L_{md} = 1$ pu. In the q-direction, it is $L_{mq} = 0.6$ pu. The leakage inductance is $L_{s\sigma} = 0.1$ pu. The machine operates at its rated current. The power and the power factor, therefore, depend on field-winding current level. High field-winding currents overexcite the machine, and correspondingly, low field-winding currents under excite. An overexcited motor supplies reactive power to the network and compensates, for example, for the reactive power of induction motors connected to a common coupling point. Overexcited synchronous motors have traditionally been used to compensate for the reactive powers needed by induction motors that share the same system.

Figure 8.59 illustrates the synchronous machine at no load with different field-winding currents. The diagrams of the machine will be shown in three conditions, under excited, at ideal no-load and overexcited.

Figure 8.60 shows synchronous generators illustrated in accordance with the motor logic that best suits the idea of terminal voltage determining stator flux linkage. According to the

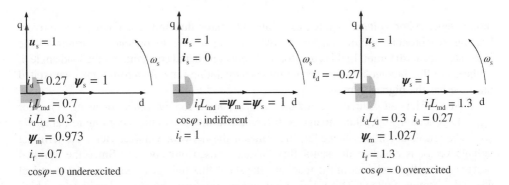

Figure 8.59   At no load, $i_f = 1.0$ suffices to excite the machine so it absorbs no reactive power from the network, and the current is zero. If the field-winding current is decreased to $i_f = 0.7$, the no-load becomes under excited, and the machine looks like an inductor. If the field-winding current is increased to $i_f = 1.3$, the no-load becomes overexcited and the machine looks like a capacitor. $i_f = 0.7$, $i_f = 1.0$, $i_f = 1.3$. $L_{md} = 1$ pu, $L_{mq} = 0.6$ pu, and $L_{s\sigma} = 0.1$ pu.

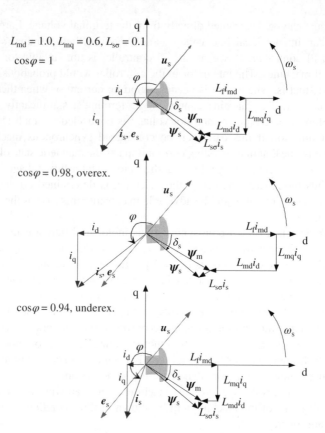

Figure 8.60 Vector diagrams of a salient-pole synchronous generator in different excitation states illustrated in accordance with the motor mode.

motor logic, stator voltage is integrated into the stator flux linkage. Each regeneratively braking synchronous motor can be smoothly changed over to operate according to the illustrated vector diagram. In this case, torque is not considered constant, but the load angle δ_s of the stator flux linkage is kept constant for each operating state. As a result, the overexcited machine has the highest torque.

Figure 8.61 is an ordinary vector diagram of a synchronous generator using generator logic. In generator logic, flux linkage is differentiated into back emf according to Faraday's law. The terminal voltage, however, must match the network voltage. Here, the terminal voltage vector is equal to the stator flux-linkage–caused induction e_s. Since the terminal voltage equals the back emf of the machine, the stator flux linkage ψ_s must be 90° electrical ahead the terminal voltage. This rotating flux linkage produces the induced voltage e_s, which equals the terminal voltage of the network. Stator current is produced as an effect of the difference of the terminal voltage and the induction caused by the stator flux linkage in the stator resistance of the machine. In Figure 8.61, the difference is zero, since $R_s = 0$.

Figure 8.62 illustrates the behaviour of a nonsalient pole generator at constant power with various power factors. At constant voltage, the loci plotted by the points of the current vector

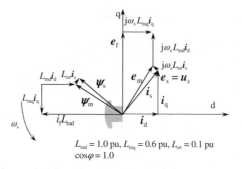

$L_{md} = 1.0$ pu, $L_{mq} = 0.6$ pu, $L_{s\sigma} = 0.1$ pu
$\cos\varphi = 1.0$

Figure 8.61   Vector diagram of a salient-pole synchronous generator in generator mode. The illustration includes the reactive voltage drops $j\omega_s L_{md}i_d$ and $j\omega_s L_{mq}i_q$ corresponding to the armature reaction and the leakage flux linkage $j\omega_s L_{s\sigma}i_s$. The induced voltage e_f is the imaginary voltage induced by the rotor current i_f, e_m is the voltage induced by the air-gap flux linkage, and e_s is the voltage induced by the stator flux linkage.

and the imaginary emf vector e_f produced by field-winding current, are straight lines when operating with different power factors and at constant active power. The power of the machine, expressed by applying the torque equation, is written as

$$P = \left|\frac{\omega_s}{p}T_e\right| = \left|\frac{\omega_s}{p}\frac{3}{2}p\boldsymbol{\psi}_s \times \boldsymbol{i}_s\right| = \left|\omega_s\frac{3}{2}\boldsymbol{\psi}_s \times \boldsymbol{i}_s\right| = \left|\frac{3}{2}j\boldsymbol{u}_s \times \boldsymbol{i}_s\right| = \frac{3}{2}\boldsymbol{u}_s \cdot \boldsymbol{i}_s = \frac{3}{2}u_s i_s \cos\varphi$$

$$(8.234)$$

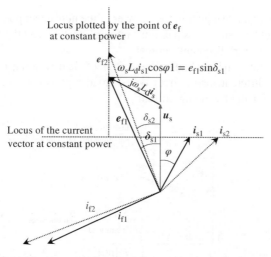

Figure 8.62   Vector diagram of a nonsalient pole generator with two power factor values at a constant terminal voltage u_s. Increasing the excitation raises the emf e_f (full and dotted vectors), reduces the load angle δ_s, and increases the power factor angle φ. At increased internal voltage, the machine can supply more reactive power to the network. If the active power of the generator remains the same, the power factor gets increasingly lower with increasing field-winding current.

However, apparent power is expressed as follows.

$$\underline{S} = \frac{3}{2}\boldsymbol{u}_s\boldsymbol{i}_s{}^* = \frac{3}{2}u_si_s\cos\varphi + j\frac{3}{2}u_si_s\sin\varphi \qquad (8.235)$$

When the power is kept constant at a constant terminal voltage, $i_s\cos\varphi$ remains constant, and the point of the current vector plots a straight line as depicted as the field-winding current of the machine varies.

The total voltage drop $j\omega_sL_d\boldsymbol{i}_s$ caused by the armature reaction, and leakage flux can be expressed as

$$\begin{aligned} j\omega_sL_d\boldsymbol{i}_s = jX_d\boldsymbol{i}_s &= jX_d(i_s\cos\varphi) - jX_d(ji_s\sin\varphi) \\ &= X_d(i_s\sin\varphi) + jX_d(i_s\cos\varphi). \end{aligned} \qquad (8.236)$$

If X_d is constant, the component $X_d(i_s\cos\varphi)$ remains constant, and as the excitation state changes, the emf vector $\boldsymbol{e}_f$ produced by field current plots the straight line depicted above. This component can also be expressed with the load angle and the emf vector.

$$X_d(i_s\cos\varphi) = e_f\sin\delta_s \qquad (8.237)$$

Solving for current from the above yields the following expression.

$$i_s = \frac{e_f\sin\delta_s}{X_d(\cos\varphi)} \qquad (8.238)$$

Substituting the expression into the power equation gives

$$P = \frac{3}{2}|\boldsymbol{u}_s||\boldsymbol{i}_s|\cos\varphi = \frac{3}{2}|\boldsymbol{u}_s|\frac{|e_f|\sin\delta_s}{X_d} \qquad (8.239)$$

This is the familiar load angle equation for a power of the nonsalient pole machine. When the terminal voltage is held constant, $|e_f|\sin\delta_s$ must remain constant, and therefore the emf vector $\boldsymbol{e}_f$ produced by field current becomes vertical.

Figure 8.63 illustrates the qualitative behaviour of field-winding current at a constant terminal voltage for different load current types. Load type is decisive in determining the generator's demand for field-winding current.

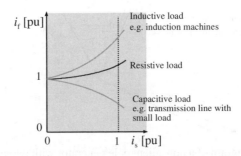

Figure 8.63 Effect of load on the demand for field-winding current in a synchronous generator. Inductive load produces a demagnetizing armature reaction that must be compensated for by increased field-winding current. Resistive load mostly affects the q-axis, and therefore does not demand much extra field-winding current. Capacitive load produces exciting armature reaction, and therefore field-winding current can be reduced.

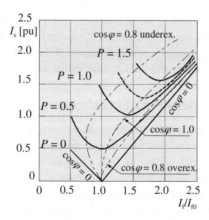

Figure 8.64 V-curves of a synchronous machine. The current axes as well as the parameter (power) are given in per-unit values.

This phenomenon can be more accurately observed using a V-curve. When the machine is operating on a constant-voltage, constant frequency network, the armature current is a function of the field-winding current with constant active power as parameter. If stator resistance is neglected and considering a nonsalient pole machine, V-curves can be graphed when the stator current is illustrated as a function of the field-winding current. The continuous lines of Figure 8.64 show the V-curves of an unsaturated synchronous machine when the pu value of the synchronous reactance is $X_d = 1.0$. The dashed lines indicate the constant cos φ values. Synchronous motors usually operate overexcited, which produces magnetizing current for induction motors running in parallel with them, connected in the network.

A synchronous machine can be loaded momentarily with a power much higher than its rated value. In controllable motor drives in particular, the machine is often briefly over-loaded—depending on the dimensioning of the power electronics—to a torque value up to three times its rating. To represent continuous operation, a PQ diagram is typically presented for a DOL machine. Figure 8.65 is such a diagram expressing the load capacity of a synchronous machine based on differing limitations.

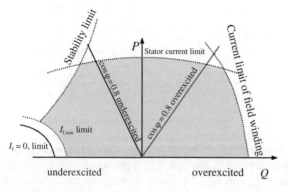

Figure 8.65 Limits set for the power plant generator in continuous operation. The "stability limit" is supplied by the manufacturer.

The machine in Figure 8.65 can reach a stability limit if the field-winding current of the machine is too low. In that case, the load angle equation indicates that e_f decreases, and therefore the load angle δ_s increases correspondingly. If the emf e_f is too low, the machine is pulled out of synchronism when the load torque increases.

8.15.2   The parallel operation of synchronous machines

In power plants, synchronous generators must often run in parallel, and load sharing needs must be considered. Figure 8.66 illustrates the reactive power division between two parallel operating generators. Terminal voltage is held constant by a rigid network. The power of generator 1 accounts for 40% of the power of generator 2. During changes in reactive power, both machines operate at constant power. In the initial state, the machines have currents i_{s1} and i_{s2}. The field-winding current of generator 2 is raised, and the field-winding current of generator 1 is reduced at the same time. Consequently, new currents i'_{s1} and i'_{s2} result, which are indicated by the dashed lines. The changes in the currents Δi_{s1} and Δi_{s2} are equal but opposite, and thus the situation remains unaltered with respect to the network.

If the two generators are operating in parallel and the power of the first generator is increased without changing its field-winding current, the rigid network maintains the constant voltage and the other machine must adjust its reactive power to maintain the original situation. When the supply torque of the first machine is increased, the generator accelerates momentarily and then settles back down to its original speed. Its load angle δ_s increases until a new power balance is reached. Since the field-winding current i_f is held constant, the amplitude of the emf e_f for the first generator remains constant, and the locus plotted by the emf vector is a circle. The situation is illustrated in Figure 8.67. As the tip of the vector of the armature reaction and leakage follows the same path as e_f, the tip of the current vector must also follow a circle as the figure illustrates.

To maintain the original sum apparent power, corresponding changes ($-\Delta i_P$ and $+\Delta i_Q$) must be made in the other generator. In other words, its power must be reduced, and the field-winding current must be increased. If the common grid current vector is i_{net}, the currents for generator 2 in the initial and final state become

$$i_{s2} = i_{net} - i_s,$$
$$i'_{s2} = i_{net} - i'_s.$$

(8.240)

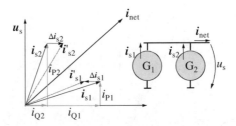

Figure 8.66   Reactive power division between two parallel operating generators. Generator 1 produces power with current i_{P1}, and generator 2 produces power with i_{P2}. The corresponding reactive currents are i_{Q1} and i_{Q2}. Opposite changes in the field-winding currents would generate the opposite respective changes in the reactive powers of the machines. Δi_{s1} and Δi_{s2} are perpendicular to the terminal voltage and thus represent reactive power.

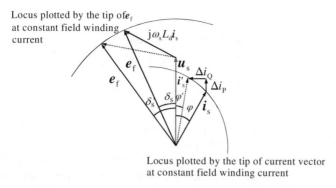

Locus plotted by the tip of e_f at constant field winding current

Locus plotted by the tip of current vector at constant field winding current

Figure 8.67 Behaviour of a nonsalient pole generator at a constant field-winding current when the power increases. Since the tip of e_f moves in a circle (constant i_f), the stator current vector tip must also move in a circle. The power of the generator increases according to Δi_P, and the reactive power decreases according to Δi_Q.

The emf of the generator 2 must be adjusted to match.

$$e'_{f2} = u_s + jX_{d2}i'_{s2} \tag{8.241}$$

To maintain the original apparent power of the two generator system, field-winding current must be increased based on the increase in the internal induced voltage of generator 2 when e_{f2} increases to e'_{f2}.

Voltage droop control

As previously mentioned, a simple PI controller is not suitable for controlling low-power generator terminal voltage, because the controller integrator is likely to saturate and the generator would not be capable of affecting the network voltage. Instead, reactive power of the machines must be controlled. This can be accomplished using a voltage controller that has droop capability. A droop controller ensures adequate reactive power control and an appropriate division of the reactive power between parallel operating machines.

Reactive power droop, as shown in Figure 8.68, is a property of the voltage controller. The voltage reference $u_{s,ref}$ is reduced (drooped) as generator reactive power increases. The value of the reactive power droop is typically of the scale of $\Delta u_s = 3\%$.

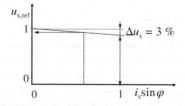

Figure 8.68 Principle of reactive power droop. When the reactive current $i_s \sin\varphi$ of the generator increases the reference voltage $u_{s,ref}$ of the generator voltage controller is reduced according to the droop. Appropriate droop control prevents saturation of the voltage PI controller. A typical droop value is 3%.

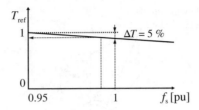

Figure 8.69 Torque controller droop. When the turbine operates at the generator rated frequency, the torque reference is slightly lower than the capacity of the turbine. If the network frequency drops, the torque reference is raised according to the droop. If the network frequency increases, the torque reference is lowered. This system ensures the appropriate torque control of the prime mover for the parallel operation of a synchronous generator drive system. If the frequency drops by 5%, the torque of the prime mover is raised by 5%.

Generator prime mover torque control

A generator is an electromechanical energy converter, which when connected to an electrical network, settles at a load angle that corresponds to the incoming mechanical torque of the prime mover, for example, a turbine or diesel engine, if the preconditions for continuous operation are met. The torque control of the prime mover is based on the measurement of the network (supply) frequency. Torque changes induce slight variations in the network frequency. In this type of multi-machine system, integral rotation speed control cannot be used. Instead, and as was done with voltage control, power management for the prime mover is based on droop control. As frequency decreases, higher torque is requested from the driving machine.

However, unlike integral control, droop control can allow a continuous control error. If there were an integral controller in the turbine of a small generator operating in parallel with the network, the turbine would always give its maximum power below the 50 Hz supply frequency. Correspondingly, at a supply frequency above 50 Hz, it would deliver zero power. This is impractical, so droop control is used instead of integral control. Figure 8.69 illustrates the frequency control droop principle.

Droop speed control is an easy first step in solving a load-sharing problem between generators. Droop is expressed as % speed droop when the torque is at 100%. The droop amount can be set by experiment. A 5% speed droop at 100% torque is a good starting point.

Large power plant voltage control

If the synchronous machine is large and its excitation state can be used to influence network voltage, a PI-type voltage controller is used that can adjust field-winding current to appropriate levels. The voltage control of an actual power plant generator functions differently for different operating conditions. Table 8.6 offers a summary.

The primary function of the voltage control is to keep the terminal voltage of the generator at the desired level in normal operation where slight variations in power and reactive power are continual. A secondary function is to appropriately divide reactive power between parallel operating generators, thereby improving network stability. Table 8.7 lists requirements that can be set for generator voltage control.

Table 8.6 Operating conditions and functions of the voltage control of a large power plant generator

operating mode	network event	transient state	functions of the voltage controller
static	several nearly static large machines in an extensive network	slight changes in power or reactive power	keeping the voltage constant, damping of the transients
dynamic	a switching situation	changes in power, reactive power, terminal voltage, current, field-winding current	as steady transition to new operating state as possible
	a momentary fault situation, fault switch-off	significant temporary changes in power, reactive power, terminal voltage, current, and field-winding current. Damper windings assist the control	maintenance of stability and damping of the pole angle fluctuation
	a failed fault switch-off, high-speed automatic reclosing, permanent fault	strong consequent temporary changes in power, reactive power, terminal voltage, current, and field-winding current. Damper windings assist the control	smooth transition to a new operating state, damping, control of reactive power and voltage, rapid shutdown, protection of generator and turbine

Table 8.7 Requirements for generator voltage control

Parameter	Requirement
Static accuracy: permitted deviation of the voltage from the rated value as the voltage has settled after a transient	ca. 1%
Dynamic accuracy: permitted deviation of the voltage from the rated value as the rated load is connected to the generator or disconnected from the generator	ca. 15%
Control speed: time in which the voltage has to return to the value required by the static accuracy after the rated load has been disconnected (switched off)	ca. 1 . . . 3 s
Permitted deviation of the reactive power compared with the rated power between the parallel operating generators	ca. 10 %

8.15.3   Field-winding current supply methods

The previous discussion shows clearly that field-winding current plays a central role in the operation of a synchronous machine. When connected to a rigid network, field-winding current determines the power factor and the load angle of the machine. Proper field-winding current control also improves the stability of the machine through its dynamic states.

Power electronic controlled drives, in particular, must have the ability to rapidly adjust field-winding current. Therefore, drives subjected to demanding dynamic conditions use a separately magnetized slip-ring machine. When supplying the rotor of the machine from a 400 V network, the resistance of the field winding is dimensioned so the voltage at the nominal operating point is on the order of 100 V. The thyristor bridge can instantaneously supply a 540 V potential across the field winding, thus forcing rapid change. In a large machine, changing air-gap flux linkage takes a long time. The time constant τ'_{d0} of a synchronous machine can be several seconds. The larger the voltage reserve of the converter feeding the field winding, the easier it is to change the air-gap flux linkage. And with a thyristor bridge, it is possible to apply high negative voltage to the field winding to reduce electric current, for instance, if the machine's load state drops abruptly.

Figure 8.70 presents a typical supply arrangement for the field-winding current of a synchronous machine. In this arrangement, the field winding is fed by a thyristor bridge between the winding and the grid. This is the best approach, because it also makes it possible to rapidly adjust field-winding current in the event of a high overvoltage condition in a dynamic event. If necessary, the thyristor bridge also guarantees negative excitation voltage in the field winding to rapidly bring the current down.

Brushless excitation is often required for a machine intended for continuous operation. Figure 8.71 illustrates the excitation system of a fully independent generator. A permanent magnet generator produces the excitation power for the external pole synchronous machine, the armature winding of which is mounted on the same axis with the rotor of the main machine. Also, the rectifier rotates along with the rotor. This yields a brushless solution;

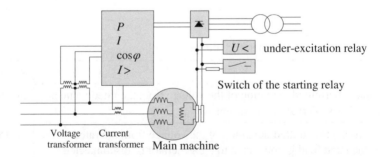

Figure 8.70   Thyristor bridge excitation of a separately magnetized slip-ring machine. If the machine is started as an asynchronous motor DOL with damper winding, a short circuit of the field winding or the depicted resistance is required during the startup to avoid excessively high voltage stresses. The control block has current, voltage, power, and power factor measurement capabilities. The under-excitation relay releases the drive if field-winding current disappears. The same control method applies for both synchronous motors and generators; however, the startup of a generator is normally executed by the prime mover machine.

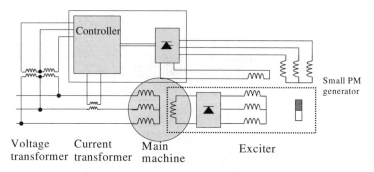

Voltage   Current   Main   Exciter
transformer transformer machine

Figure 8.71 Brushless excitation system for a synchronous generator in island operation. Island operation means the generator alone produces the system voltage and no other generators are present. The field-winding current is changed via the time constants of the three generators. If required, the time constants of a permanent magnet machine can be shortened, and therefore using a secondary permanent magnet machine does not significantly retard the control of the field-winding current of the main machine.

however, the dynamics of the field-winding current suffers considerably. The method is not applicable to dynamically demanding drives.

Figure 8.72 offers another brushless excitation system that is primarily designed for DOL operation. It is a modification of the previous example. The energy required by the field winding of the excitation machine is taken from the grid by using a thyristor bridge.

Figure 8.73 illustrates the somewhat out-of-date use of a DC generator for the excitation of a synchronous generator. A separately excited DC generator is mounted on the same axis as the main generator. However, both slip rings and a commutator are required in this case. Applications of this type are probably no longer being produced.

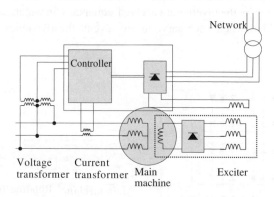

Voltage   Current   Main   Exciter
transformer transformer machine

Figure 8.72 Brushless excitation system for a synchronous generator. The field-winding current is altered via the time constants of the two generators. In principle, system response can be faster. The control method is also commonly applied in continuous motors; however, an auxiliary system must always be used to start up the machine. To start the machine with damper windings, some form of protection (for example, a resistance) against the high induced voltages of startup must be provided for the field winding. This protection may also be required for generator fault conditions.

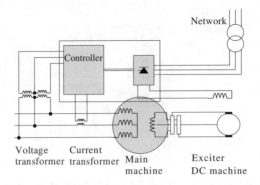

Figure 8.73 Formerly traditional, but now out-of-date, synchronous machine DC generator excitation.

The field-winding current arrangements of brushless generators are in principle also applicable to DOL motors. However, if a power-electronics–controlled brushless motor is required, field-winding current must also be supplied when the machine rotor is not turning, that is, at zero speed. A converter-drive-related solution is required. In this solution, the field-winding current is produced by a rotating-field machine that is equipped with polyphase windings on both the stator and rotor. When the main machine rotor is held stationary, field-winding current can be supplied to the machine via the rotating excitation of the exciter.

The direction of excitation should be opposite to the rotation direction of the main machine, so when the main machine starts up, exciter frequency tends to increase and can be decreased as required. Once the main machine has accelerated to its rated speed, direct current can be fed to the polyphase winding of the exciter. In dynamic transients, the "field winding" of the exciter can be fed by the inverter at varying frequencies, in which case induction can, in principle, be intensified when necessary. In any event, the dynamics resulting from this

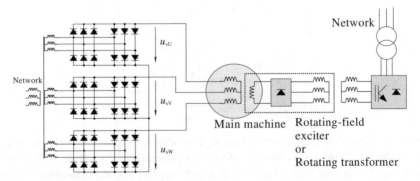

Figure 8.74 Method for supplying field-winding current to a brushless synchronous motor so the converter-supplied drive can be operated at full torque when the rotor is held stationary. A cycloconverter drive is illustrated here, but any brushless drive needs a similar field-winding supply system.

method are slower than the dynamics made available using the slip ring approach. Since brushless motors are nonwearing, they are often used in marine drives.

References

Bühler, H. (1977). *I, Einführung in die Theorie Geregelter* Drehstromantriebe, Band 1 (p. 268) Grundlagen: Birkhäuser Verlag Basel und Stuttgart, p. 268.

Bühler, H. (1997). *II, Einführung in die Theorie Geregelter* Drehstromantriebe, Band 2 (p. 348) Anwendungen: Birkhäuser Verlag Basel und Stuttgart.

Canay, I. M. (1969). Causes of discrepancies on calculation of rotor quantities and exact equivalent circuit diagrams of the synchronous machine. *IEEE Transactions on Power App. Syst.*, 88 (7),1114–1120.

Heikkilä, T. (2002). Permanent magnet synchronous motor for industrial inverter applications – analysis and design. Dissertation Lappeenranta University of Technology, ISBN 951-764-699-2.

Kaukonen, J. (1999). Salient-pole synchronous machine modelling in an industrial direct torque controlled drive. Dissertation Lappeenranta University of Technology, ISBN 951-764-305-5.

Mård, M., Niiranen, J., & Vauhkonen, V. (1990). The control properties of synchronous motors within load angles of 90 degrees. *ICEM* 434–439.

Pyrhönen, O. (1998). Analysis and control of excitation, field weakening and stability in direct torque controlled electrically excited synchronous motor drives. Research Papers 74. Dissertation, Lappeenranta University of Technology. ISBN 951-764-274-1.

Pyrhönen, J., Niemelä, M., Kaukonen, J., Luukko, J., Pyrhönen, O., Tiitinen, P., & Väänänen, J. (1997). Synchronous motor drives based on direct flux linkage control. *EPE Conference Proceeding in Trondheim*, 1, 1434–1439.

Pyrhönen, O., Kaukonen, J., Luukko, J., Niemelä, M., Pyrhönen, J., Tiitinen, P., & Väänänen, J. (1997). Salient pole synchronous motor excitation control in direct flux linkage control based drives. *EPE Conference Proceeding, Trondheim*, 3, 3678–3682.

Pyrhönen, O., Kaukonen, J., Niemelä, M., Pyrhönen, J., & Luukko, J. (1998). Salient pole synchronous motor excitation and stability control in direct torque control drives. *International Conference on Electric Machines in Istanbul Proceedings*, 1, 83–88.

9

Permanent magnet synchronous machine drives

The increasing commercial availability of high-quality permanent magnet (PM) materials has encouraged manufacturers to introduce various PM synchronous machines (PMSMs). For some time now, PMSMs have been used in servo drives and, currently, large PMSMs are seeing use in industrial applications and distributed generation, especially as wind turbine generators. Megawatt-range PMSMs are successfully operating even as low-speed, direct-drive wind turbine generators. Moreover, electric and hybrid car manufacturers often use PM technology in their products. The number of electromagnetic structures, such as stator teeth or rotor poles, in a PMSM can be increased without incurring significant degradation in machine performance. This makes PM machines popular in all kinds of special applications. In addition, using PM motors or generators in electric drives makes it possible to attain the highest possible efficiencies. In a well-designed PMSM, rotor losses can be insignificant, making the architecture superior in efficiency. Low-power machines using PM rather than induction rotors, in particular, see significant improvements in efficiency.

Comparing PMSM losses to the losses of an induction motor (IM) reveals that efficiency can be significantly improved by eliminating rotor Joule losses, which in a low-power IM amounts to 25–30 % of total losses. Because the power factor of the PMSM is often slightly higher than the corresponding IM value, stator Joule losses are also smaller. Assuming a power factor improvement from 0.8 to 0.9, stator Joule losses drop by 20 %. Stator Joule losses make up between 30 % and 40 % of all losses in low-power industrial IMs, and therefore PMSM losses can be up to 40 % smaller than IM losses for a similarly sized motor.

Because of the nonisotropic magnetic structure of the PM rotor, a vector control approach must usually be used with PMSMs. However, direct online (DOL) PMSMs do also exist. Historically, poor magnet material quality has restricted the effective

Electrical Machine Drives Control: An Introduction, First Edition. Juha Pyrhönen, Valéria Hrabovcová and R. Scott Semken.
© 2016 John Wiley & Sons, Ltd. Published 2016 by John Wiley & Sons, Ltd.

implementation of PMSM control. The poor demagnetization characteristics of AlNiCo magnets encouraged the development of so-called $i_d = 0$ control. The lack of demagnetizing d-axis current in these drives makes it possible to ensure the stability of the PM polarization. However, this is only a partial explanation. In rotor surface magnet machines, $i_d = 0$ control is the best at low speeds.

When operating a rotor surface magnet machine, demagnetizing current can be used if the magnets are NdFeB or SmCo, because of the properties of the magnetic materials. This is particularly helpful when implementing field weakening. However, at higher power factors, there will be a negative current aligned with the d-axis in the constant flux range. The selection of best current vector direction is also influenced by PMSM rotor construction. In many cases, to achieve optimal performance, negative d-axis current is needed, which results in magnet stresses that are higher than for $i_d = 0$ control.

PMSM control principles differ from those for other AC machines, because of the properties of the PM material, and because the material is part of the magnetic circuit of the machine, which significantly influences reluctance. For example, the relative permeability of the PM material μ_r is approximately equal to one, and therefore the effective d-axis air gap of the PMSM often becomes very large. Machine inductances usually remain low, particularly in machines with rotor surface magnets. In addition, when the rotor magnets are embedded, the value of the d-axis synchronous inductance can be less than the q-axis value. For a separately excited salient-pole synchronous machine, the d-axis value exceeds the value of the q-axis synchronous inductance.

Demagnetizing stator current must be used for field weakening in PM machines. This is not appropriate if the machines are low in inductance. For rotor surface magnet servomotors, the synchronous inductance per unit (pu) value L_d is typically between 0.2 and 0.4. An adequate speed range is usually achieved by raising the machine's rated frequency. However, in machines with embedded rotor magnets, the rotation speed range can be increased more simply by increasing machine inductances via machine design.

Often when staying within rated current limits, the maximum upper speed limit is about twice the rated speed. However, the back emf produced by the PMs is directly proportional to the rotation speed of the machine. If demagnetizing current is lost for some reason, the inverter must be able to withstand this back emf. Therefore, when applying field weakening, the relationship between rotation speed and back emf must be taken into account to minimize the risk of damage to the converter, which uses DC link capacitors that are only capable of withstanding overvoltage levels of about 125 % rated voltage.

Since a PMSM can be designed with the PMs of the rotor arranged in many different configurations, adopting a generalized control methodology is not practical. Instead, the optimal control approach depends on the rotor's magnetic configuration. Most control methods are based on modelling the PMSM in the rotor reference frame; however, these control methods require knowledge of rotor position angle. Therefore, pulse-encoder–based speed data and initial rotor angle must be known. In servo drives, it is often necessary to get rotor position feedback information from an absolute encoder.

Much research activity has been devoted to the development of PM drive control methods that do not require position feedback. Several of the introduced methods are based on the use of computationally intensive estimators. Good results have been achieved in several studies for operation at moderate supply frequencies. However, operating at close to zero speed remains somewhat problematic.

9.1 PMSM configurations and machine parameters

The performance characteristics of a PM machine are highly dependent on rotor structure. PMSM rotor poles can be implemented in various ways. The simplest is to mount magnets directly onto the rotor yoke surface. See Figure 9.1. The yoke can be a simple steel tube if the stator harmonics are at sufficiently low frequencies. However, in machines with higher-frequency stator harmonics, it is better to use a laminated structure to minimize the eddy current losses that can lead to excessive rotor temperatures.

Rotor surface magnet machines are very low in magnetizing inductance. If they also have a small number of poles, leakage inductance also remains low. Therefore, a voltage source inverter with a high switching frequency is required for best current behaviour. The configuration is applied to servomotors, for example, which require minimal inertia. The d- and q-axis inductances of rotor surface magnet machines are approximately equal, so the machine is nonsalient.

Adjusting machine inductances can be accomplished by appropriately shaping the electrical steel components. Pole shoes can be mounted on the magnets to produce sinusoidal air-gap flux densities. At the same time, they protect the magnets against both electric and magnetic stresses. Moreover, if appropriately shaped, the pole shoes can protect the brittle sintered magnet material from mechanical damage during manufacturing assembly. Although PMs can tolerate significant pressure, they have very low tensile strength.

A possible solid-steel rotor configuration with pole shoes is illustrated in Figure 9.2. The configuration can be used to implement a multipole, low-speed machine. The solid pole shoes, which should be shaped to produce a sinusoidal flux density distribution in the air gap, function also as weak damper windings. Therefore, smooth and quiet operation is achieved.

The relative permeability of current hard PM materials is approximately 1, the same as with air, so the effective air gap of a PMSM is relatively large. Because of the large air gap, d-axis armature reaction effects remain low, and the harmonics resulting from the small

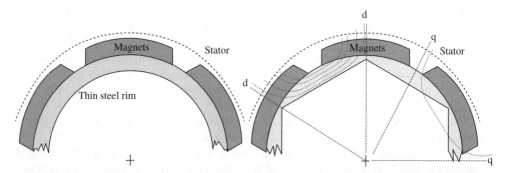

Figure 9.1 The low-inductance rotor configuration of a PMSM. The image on the left illustrates a nonsalient pole configuration. In the right image, the steel rim is made as thin as possible on the d-axes to reduce machine inertia. The flux path at the d-axis is sufficient for the excitation flux (see the magnetic flux lines at the d-axis), and magnetically the machine is slightly asymmetric, since there is higher reluctance on the quadrature axis than on the direct axis (see the flux lines at the q-axis). The q-axis path on the right shows higher reluctance than the d-axis path. However, because of the large magnetic air gap, $L_d \approx L_q$, which is typical of rotor surface magnet PMSMs.

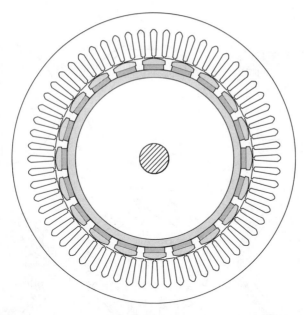

Figure 9.2 Cross section of a 20-pole, solid-pole shoe, three-phase machine with one slot per pole and phase $q = 1$.

number slots per pole and phase in the stator do not produce significant torque ripple, a fact that gives the machine certain special characteristics.

In solid pole-shoe machines, the expensive PM material is used fairly efficiently, because the magnetic flux flows almost completely through the air gap, and the PM magnetic leakage flux is small. However, some eddy-current losses are produced in the solid-pole shoes. Pole shoes offer the PMs good protection against demagnetization, because most of the demagnetizing forces do not pass through the magnets but are instead conducted by the pole shoes.

As with the rotors of asynchronous motors, the rotor yokes of PM machines can be constructed of electrical steel sheet metal. There are a number of alternative laminated rotor configurations available to achieve the desired PMSM performance characteristics. The PMs can be glued to the rotor outer diameter surface as is done with a solid rotor yoke. However, the magnets can also be embedded, in part or completely. Using the embedded magnet approach, the magnets can be mounted in different positions and orientations. Some basic configurations for laminated PM rotor structures producing different machine properties illustrated in Figure 9.3.

For the configurations illustrated by Figure 9.3a–9.3c, the physical air gap is approximately constant, and the PMs produce an approximately trapezoidal air-gap flux density. Depending on winding arrangement, the voltage induced in the stator of such a machine may include harmonics. These harmonics may affect the torque output of the machine resulting in vibration and noise. Since smooth torque production is normally required, the stator and rotor current linkages and the fundamental must not include same order harmonics. This would result in harmonics-generated torque components or torque ripple. The d-axis inductance of the machine for the a–c configurations will be significantly lower than the q-axis inductance. Therefore, when suitably controlled, the machines produce reluctance torque.

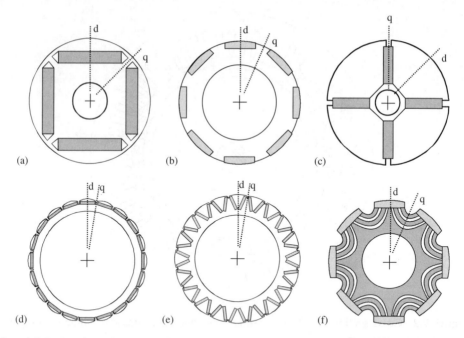

Figure 9.3 Laminated rotor configurations of PM machines with embedded or rotor-surface magnets include (a) embedded magnets, (b) inset magnets, (c) I-type, (d) embedded magnets with laminated pole shoes and magnet pockets, (e) flux concentrating version of version (d), and (f) synchronous reluctance machine with rotor surface magnets on d-axis. Constructions (a) through (e) provide saliency, and $L_q > L_d$. With construction (f), $L_d > L_q$.

To be an attractive drives alternative, especially for low-speed operation, a synchronous machine must offer smooth torque production; the aim of the configurations is illustrated by Figures 9.3d and 9.3e. For these constructions, the rotor laminations are shaped to produce poles that resemble the poles of a salient-pole machine achieving a sinusoidal no-load flux density in the air-gap. These configurations also include reluctance differences that may produce a torque harmonic at a frequency that is six times the supply frequency. This phenomenon takes place with the fifth and seventh stator harmonics travelling in different directions in the air gap. However, by skewing either the stator or the rotor, this torque ripple can be mostly avoided. A more-detailed analysis of harmonics can be found, such as in Pyrhönen et al. (2014).

PMSMs have different performance characteristics depending on rotor configuration. The rotor structures illustrated by Figures 9.3a and 9.3c are for hybrid machines, which could operate as synchronous reluctance machines, in principle, with the magnets removed. In these machines, some torque is normally produced by the reluctance differences, that is, the difference in the direct and quadrature inductance directions. In these machines, the q-axis inductance L_q is larger than the d-axis inductance L_d. The addition of magnets in these hybrid machines makes their behavioural characteristics considerably better than the behavioural characteristics of a synchronous reluctance machine. In particular, efficiency and power factor can be notably improved.

The rectangular magnet configuration shown in Figure 9.3c requires flux barriers near the rotor shaft to inhibit through-shaft PM leakage flux. This is a challenging mechanical design problem. Furthermore, the structure offers a good q-axis armature reaction path, which is not always desirable. The configuration does offer, however, large flux density concentration in multiple pole-pair machines.

In the Figure 9.3b configuration, the magnets are surface mounted. This construction provides some reluctance difference between the direct and quadrature axes. Because of the reluctance difference, maximum torque is produced at a load angle well above 90°. In PM machines, maximum torque is often produced at load angles greater than 90°, since inductance in the q-direction is often slightly higher than in the d-direction.

The configurations illustrated in Figures 9.3d and 9.3e were developed to produce smooth and quiet operation at low rotation speeds with a stator with the number of slots per pole and phase $q \geq 1$. Since a PM machine can operate at a low speeds with good efficiency and power factor, mechanical gearing is not necessarily needed to produce a low-speed PM electrical drive. Although the fundamental frequency of the machine can be set at any desired level using an inverter supply, a relatively high pole-pair number should still be selected so that the relative stator yoke thickness is minimized, which in turn maximizes acting rotor diameter within a predefined outer diameter.

The number of slots per pole and phase q can be equal to or even less than one, resulting in excess harmonic content in the stator current linkage. One way to smooth the resulting torque production of the machine is to make the rotor current linkage sinusoidal. In general, unless appropriate design steps are taken to minimize its production, machines with a low number of stator slots per pole and phase and with rotors configured as shown in Figures 9.3d and 9.3e will also be prone to torque rippling. A surface magnet rotor with sinusoidal flux density should be used in machines with low values of q to produce sinusoidal rotor current linkage with no inductance differences.

In all of the first five configurations, Figures 9.3a through 9.3e, the q-axis inductance L_q is larger than the d-axis inductance L_d (that is, $L_q > L_d$). In each case, the machine must be driven by a current vector that has negative d-axis current. A PMSM with $L_d > L_q$ is also possible. Figure 9.3f illustrates a rotor configuration that combines the characteristics of a synchronous reluctance machine and a rotor surface magnet PMSM. A machine configured in this way should be driven by a current vector with positive d-axis current. The PM-produced machine flux will be strengthened by the armature reaction. The machine designer, in this case, must be careful to avoid PM material hysteresis losses, which can occur if external magnetic field strength varies between positive and negative values.

Laminated rotor machines are subject to magnetic flux leakage, which can be reduced by integrating leakage flux guides as shown in Figure 9.3a. The flux guide can be air or some other poor conductor. The poles shown in Figures 9.3d and 9.3e produce a sinusoidally shaped flux pattern that simultaneously reduces magnetic leakage. The utilization of the magnets is still lower in an embedded magnet machine than in a salient pole machine, for example, in which the magnetic flux is almost completely in the air gap. A laminated rotor structure can be used to increase air-gap flux density by using two magnets per pole (Figure 9.3e), in which case the PM cross-sectional area increases in proportion to the machine pole area. More magnetic material is needed to implement this approach, however, so it results in a higher-cost machine. For higher pole-pair numbers p, the I-type configuration (Figure 9.3c) can also be a flux-concentrating construction.

Laminated rotor structures with clearly shaped pole shoes (Figures 9.3d and 9.3e) can easily be equipped with damper windings, which fit well into the pole shoes. The configuration type enables the production of direct online (DOL) machine versions. However, DOL-starting may be difficult in PM machines, and mostly they are designed for variable frequency drives.

The pu parameter values for PM machines differ from the pu values of traditional induction machines and synchronous machines in industrial use. Although the pu value of magnetizing inductance is typically greater than three for power induction machines rated higher than 100 kW and between one and two for synchronous machines, it may be a tenth of these values for servomotors with rotor surface–mounted magnets. The minimum value for stator flux leakage is typically close to 0.1 for all AC machine types with small pole numbers. The synchronous inductance of a servomotor with a surface-magnet rotor can be $L_s \approx 0.3$, typically.

Correspondingly, in multipole machines with embedded magnets, synchronous inductances are typically about $L_d = 0.4 - 0.6$ and $L_q = 0.6 - 0.9$. In PM machines with a larger number of poles, the stator flux leakage proportion can increase excessively, becoming up to a half of the total synchronous inductance. In tooth-coil machines, stator leakage inductance can be significantly higher than magnetizing inductance.

The RMS load angle (δ_s) equation, which is important in the analysis of a synchronous machine, is also important in the analysis of a PM machine. Neglecting stator resistance R_s and the corresponding Joule losses, the RMS load angle equation can be written as follows.

$$P = 3 \left(\frac{U_{sph} \, E_{PMph}}{\omega_s L_d} \sin \delta_s + U_{sph}^2 \frac{L_d - L_q}{2 \, \omega_s L_d \, L_q} \sin 2\delta_s \right) \tag{9.1}$$

Correspondingly, electromagnetic torque can be expressed in terms of the RMS phase values.

$$T_e = 3p \left(\frac{U_{sph} \, E_{PMph}}{\omega_s^2 L_d} \sin \delta_s + U_{sph}^2 \frac{L_d - L_q}{\omega_s^2 2 \, L_d \, L_q} \sin 2\delta_s \right) \tag{9.2}$$

For a nonsalient pole machine, the pull-out torque depends on the inverse of the synchronous inductance L_d. Therefore, the inductances should be low if high torque production is the goal. This requirement is even more important for PM machines, since the interior emf E_{PM} cannot be altered by adjusting the field-winding current as can the corresponding interior E_f in synchronous machines. Often and for practical reasons, the pu value of e_{PM} must be kept close to one, that is, $e_{PM} \in (0.7, 1)$. Since the supply voltage u_s is also one, the pu value of the synchronous inductance L_d must be considerably less than one. For a nonsalient pole machine and if $e_{PM} = u_s = 1$, synchronous inductance must be no more than $L_d = 0.625$ pu to achieve the commonly required 160 % pull-out torque (maximum) at rated speed and voltage.

Therefore, the synchronous inductances must be kept relatively low to reach the desired pull-out torque for PM machines. In a surface magnet machine, this precondition is easily achieved; however, getting there is more difficult for embedded magnet machines where pu values approach one and the pu value of e_{PM} cannot be increased considerably above one.

Therefore, the synchronous inductances must be kept relatively low to reach the desired pull-out torque for PM machines. In a surface magnet machine, this precondition is easily achieved; however, getting there is more difficult for embedded magnet machines where pu values approach one and the pu value of e_{PM} cannot be increased considerably above one.

The fractional-slot, nonoverlapping concentrated winding, that is, the tooth-coil winding machine configuration is a recent arrival to the electric drives marketplace. In a tooth-coil

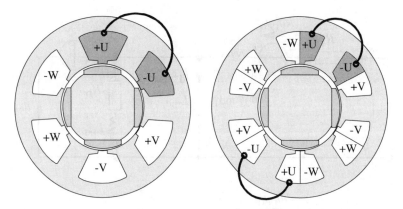

Figure 9.4 PM three-phase synchronous machines with single-layer and double-layer tooth-coil windings: $Q_s = 6$, $p = 2$, and $q = 0.5$.

machine, the number of slots per pole and phase varies between $q \in [0.25, 0.5]$. In particular, the machines are intended for low-speed direct drive machine applications where maximum torque is needed in a smaller volume. Tooth coils have smaller end windings and can work with a smaller stator yoke, so the architecture accommodates a larger maximum rotor diameter and length relative to the outer dimensions of the machine. Figure 9.4 gives cross-sectional views illustrating tooth-coil PM synchronous machines.

The best tooth-coil machines produce a completely sinusoidal voltage, and therefore smooth torque can be achieved with sinusoidal stator currents. The control of this kind of a machine does not differ in principle from the control of any other ordinary rotating-field PM machine. Only the machine parameters are of interest. In principle, tooth-coil machines can also produce some reluctance torque. However, the proportion normally stays low, and the machines behave mostly like rotor surface magnet nonsalient-pole PM machines. In tooth-coil machines, stator leakage inductance can be much higher than stator magnetizing inductance, which facilitates field weakening in some tooth-coil machines.

9.2 The equivalent circuit and space-vector diagram for a PMSM

As are separately excited synchronous machines, PM synchronous machines are normally treated in the dq reference frame fixed to the rotor. See Figure 9.5. The equivalent circuit of the machine is almost the same as for a separately excited synchronous machine.

If damper windings are included in the model, the equations of a PM machine differ from a separately excited synchronous machine only by the fact that the field winding current is replaced by the PM equivalent current. Therefore, the voltage equations of the PM machine are given in the rotor reference frame in the following familiar form for the stator.

$$u_d = R_s i_d + \frac{d\psi_d}{dt} - \omega_r \psi_q \tag{9.3}$$

$$u_q = R_s i_q + \frac{d\psi_q}{dt} + \omega_r \psi_d \tag{9.4}$$

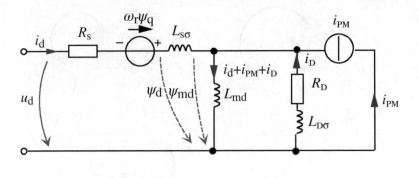

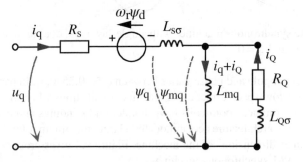

Figure 9.5 Equivalent circuits of a PMSM in the d- and q-axis directions showing R_s stator resistance, $L_{s\sigma}$ stator leakage inductance, L_{md} d-axis magnetizing inductance, L_{mq} q-axis magnetizing inductance, $L_{D\sigma}$ d-axis damper leakage inductance, $L_{Q\sigma}$ q-axis damper leakage inductance, R_D d-axis damper resistance, and R_Q q-axis damper resistance. The PM can be represented by a current source i_{PM} in the rotor circuit. In the magnetizing inductance, this current source produces the PM's share of the air-gap flux linkage $\psi_{PM} = i_{PM}L_{md}$.

For the possible rotor damper windings, the equations are

$$0 = R_D i_D + \frac{d\psi_D}{dt} \qquad (9.5)$$

$$0 = R_Q i_Q + \frac{d\psi_Q}{dt} \qquad (9.6)$$

The flux linkage components in the equations are determined by these equations. The direct axis stator flux linkage is

$$\psi_d = L_d i_d + L_{md} i_D + \psi_{PM} \qquad (9.7)$$

The quadrature axis stator flux linkage is

$$\psi_q = L_q i_q + L_{mq} i_Q \qquad (9.8)$$

The stator flux linkage is

$$\psi_s = \sqrt{\psi_d^2 + \psi_q^2} \qquad (9.9)$$

The d-axis damper winding flux linkage is

$$\psi_D = L_{md}i_d + L_{D}i_D + \psi_{PM} \tag{9.10}$$

And, the q-axis damper winding flux linkage is

$$\psi_Q = L_{mq}i_q + L_{Q}i_Q \tag{9.11}$$

where the damper total inductances $L_D = L_{md} + L_{D\sigma}$ and $L_Q = L_{mQ} + L_{Q\sigma}$. If the PM flux linkage ψ_{PM} is considered to be a function of a virtual current i_{PM}, then ψ_{PM} is equivalent to the product of i_{PM} and the magnetizing inductance L_{md}, and the following expression defines the virtual current.

$$i_{PM} = \frac{\psi_{PM}}{L_{md}} \tag{9.12}$$

The resulting flux linkage definitions now look similar to the equations for a separately excited synchronous machine. However, taking saturation of the magnetizing inductance L_{md} into account, i_{PM} does not remain constant, and it must be altered accordingly.

The power factor of a PMSM can be written

$$\cos \varphi = \frac{u_d i_d + u_q i_q}{u_s i_s} \tag{9.13}$$

The space-vector diagram of the PMSM is a modification of the space-vector diagram of a synchronous machine. PMs produce the flux linkage ψ_{PM} in the stator winding.

EXAMPLE 9.1: Draw the space-vector diagram for a nonsalient pole PMSM in pu values operating according to the following parameters: $\omega_s = 1$ pu, $u_s = 1$ pu, $i_s = 1$ pu, $\psi_s = 1$ pu. The machine is operating in light field weakening, that is, $i_d = 0.24$ pu in the negative direction. Resistance is neglected, the pu inductances are $L_d = L_q = 0.5$, and rotor position angle is $\theta_r = 40°$ measured from the x- to d-axis, in which ψ_{PM} is located.

Determine the load angle δ_s, the PM linkage flux ψ_{PM}, the d- and q-components of current and voltage, the power factor $\cos \varphi$, the electric current angle γ measured from the d-axis to the stator current space vector, and the electromagnetic torque T_e.

SOLUTION: Figure 9.6 shows the space-vector diagram. There is a stator reference frame xy and rotor reference frame dq shifted by rotor angle θ_r from the x-axis. The PM linkage flux ψ_{PM}, the value of which will be determined later, is along the d-axis.

The stator flux linkage ψ_s is shifted from the d-axis by a load angle that will now be calculated. Applying the Pythagorean theorem, ψ_d can be written as follows.

$$\psi_d = \psi_s \cos \delta_s = \sqrt{\psi_s^2 - (L_q i_q)^2} = \sqrt{1^2 - (0.5 \cdot 0.97)^2} = 0.874 \text{pu}$$

where

$$i_q = \sqrt{i_s^2 - i_d^2} = \sqrt{1^2 - 0.24^2} = 0.97 \text{pu}$$

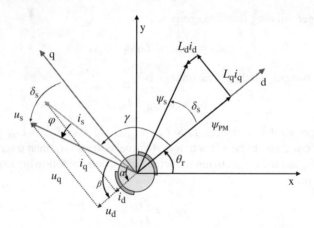

Figure 9.6 Space-vector diagram of a nonsalient-pole PMSM in pu values. $L_d = L_q = 0.5$. There is a stator reference frame (xy) and a rotor reference frame (dq). The machine operates as a motor at its nominal operating point in light field weakening (small negative $i_d = 0.24$). $\omega_s = 1$, $u_s = 1$, $i_s = 1$, and $\varphi \cong 15°$. The load angle $\delta_s \cong 29°$. Power factor and torque have equal values, that is, $\cos\varphi_s \cong 0.965$ and $T_e = 0.965$. $\gamma = 104°$ is the electric current angle measured from the d-axis.

Since $\omega_s = 1$ pu, the cosine function can be calculated to determine the load angle.

$$\cos \delta_s = \frac{\psi_s \cos \delta_s}{\psi_s} = \frac{0.874}{1} = 0.874 \Rightarrow \delta_s = 29°$$

For nonsalient pole machines, the load angle can also be measured between the q-axis and u_s space vector. This knowledge will be used later.

The PM flux linkage is

$$\psi_{PM} = \psi_s \cos \delta_s + L_d i_d = 0.874 + 0.5 \cdot 0.24 = 0.994.$$

The current components are given: $i_d = 0.24$ pu and $i_q = 0.97$ pu.

The direction of the current i_s will be located under the angle α from the d-component i_d.

$$\cos \alpha = \frac{i_d}{i_s} = \frac{0.24}{1} = 0.24 \Rightarrow \alpha = 76°$$

The voltage components in the d- and q-axis directions are as follows.

The stator voltage vector u_s has a pu value of one and is perpendicular to the stator linkage flux ψ_s, which also has a pu value of one. The angle between u_d and u_s is β, and

$$\beta + \delta_s = 90° \Rightarrow \beta = 90° - \delta_s = 90° - 29° = 61°.$$

Then

$$u_d = u_s \cos \beta = -1 \cdot \cos 61° = -0.484 \text{ pu}$$

$$u_q = u_s \sin \beta = 1 \cdot \sin 61° = 0.874 \text{ pu}$$

The phase angle between voltage u_s and current i_s can be used to find the power factor.

$$\varphi = \alpha - \beta = 76 - 61 = 15°$$

$$\cos \varphi = \cos 15° = 0.965$$

Note: From Equation (9.13): $\cos \varphi = \dfrac{u_d i_d + u_q i_q}{u_s i_s} = \dfrac{-0.484 \cdot (-0.24) + 0.874 \cdot 0.97}{1 \cdot 1} = 0.965 \text{ pu}$

The current angle can be determined from the angles φ and δ_s.

$$\gamma = 90° + \delta_s - \varphi = 90° + 29° - 15° = 104°.$$

Finally, torque in pu value is

$$T_e = \frac{P}{\omega_s} = \frac{u_s i_s \cos \varphi}{\omega_s} = \frac{1 \cdot 1 \cdot 0.965}{1} = 0.965 \text{ pu}$$

Commonly, PMSMs do not include separate damper windings, in which case the flux linkages are simply written as follows.

$$\psi_d = L_d i_d + \psi_{PM} \tag{9.14}$$

$$\psi_q = L_q i_q \tag{9.15}$$

In addition to the damping added by actual damper windings, damping in an electrical machine can result from several other factors. For example, conductive machine parts such as solid pole shoes will increase damping, and a solid steel rotor frame adds a significant amount of damping. Since magnet materials are resistive, their damping effects can normally be neglected, and a laminated rotor frame, which provides few paths for the eddy currents, does not significantly add to machine damping.

Many other damping factors are difficult to identify and quantify. Determining their overall effect on damping is best done by taking measurements. For example, a short-circuit test at rated current reveals, in principle, the torque produced by a damper winding. However, the presence of the PMs in a PMSM makes taking asynchronous measurements considerably more difficult.

The expression for torque, according to the cross-field principle, is the same for the PMSM as for a separately excited synchronous machine. It can be written as follows.

$$T_e = \frac{3}{2} p \left[\psi_{PM} i_q - (L_{mq} - L_{md}) i_d i_q + L_{md} i_D i_q + L_{mq} i_Q i_d \right] \tag{9.16}$$

One can see the main torque component of the PMSM is $\psi_{PM} i_q$, which corresponds to the SM torque $i_f L_{md} i_q$. As for synchronous machines, the PMSM torque is produced by the four terms determined according to the cross-field principle. The first term most important, and in many PM machines, it is the only term. It depends on the flux linkage of the PMs and on the stator current perpendicular to the flux linkage. The second term, which results from the inductance difference, is significant in machines where the difference between the d-axis and q-axis inductances large. The damper current terms are only significant during transients and in machines with significant asynchronous-mode damper currents. The torque equation is used as a development starting point for the various PM machine control principles.

9.3   Equations based on the electric current angle

In stator current control, using electric current angle γ is easier than using the load angle δ_s, because electric current angle is a vector control parameter. Section 9.5 will show load angle δ_s to be a direct torque control (DTC) parameter. Per-unit power can be expressed pu voltage and current as follows.

$$P = u_s i_s \cos \varphi \tag{9.17}$$

The power-factor angle φ can be written in terms of γ and δ_s on the basis of Figure 9.6 to become $\left(\left(\frac{\pi}{2} + \delta_s \right) - \gamma \right)$. Therefore

$$P = u_s i_s \cos \left(\left(\frac{\pi}{2} + \delta_s \right) - \gamma \right) \tag{9.18}$$

The pu value of power is as follows.

$$P = u_s i_s \cos \left(\left(\frac{\pi}{2} + \delta_s \right) - \gamma \right) = u_s i_s \sin(\gamma - \delta_s) \tag{9.19}$$

Equation (9.19) can also be written

$$P = u_s i_s \sin(\gamma - \delta_s) = u_s i_s (\sin \gamma \cos \delta_s - \cos \gamma \sin \delta_s) \tag{9.20}$$

Figure 9.6 reveals that

$$i_d = i_s \cos \gamma \tag{9.21}$$

$$i_q = i_s \sin \gamma \tag{9.22}$$

$$u_d = -u_s \sin \delta_s = -\omega_s L_q i_q \tag{9.23}$$

$$u_q = u_s \cos \delta_s = \omega_s(\psi_{PM} + L_d i_d) \tag{9.24}$$

The pu power can be expressed

$$P = \omega_s \left[\psi_{PM} i_s \sin \gamma - i_s^2 \sin 2\gamma \left(\frac{L_q - L_d}{2} \right) \right] \tag{9.25}$$

Correspondingly, the torque becomes

$$T_e = \frac{P}{\omega_s} = \left(i_s \psi_{PM} \sin\gamma - i_s^2 \sin 2\gamma \left(\frac{L_q - L_d}{2} \right) \right) \tag{9.26}$$

A so-called characteristic current i_x, which differs slightly from i_{PM}, has been introduced and is preferred by some researchers, for example, Soong et al. 2007a and 2007b.

$$i_x = \frac{\psi_{PM}}{L_d} \tag{9.27}$$

This value of this characteristic current determines one aspect of the motor drive's nature. If the value of i_x is close to 1, then it is easy to implement motor field weakening. If the value is equal to one, the field weakening range of the machine, in principle, is infinite.

PMSM drive behaviour is naturally limited by the maximum current and voltage of the converter, and voltage is proportional to rotor speed and stator flux linkage.

$$i_s = \sqrt{i_d^2 + i_q^2} \leq i_{s,max} \tag{9.28}$$

$$u_s = \sqrt{u_d^2 + u_q^2} \leq u_{s,max} \approx \frac{u_{DC}}{\sqrt{3}} \tag{9.29}$$

EXAMPLE 9.2: To illustrate the constraints given by Equations (9.28) and (9.29), draw a figure in the electric current plane for the rated machine speed using the parameters: $L_d = 0.4$ pu, $L_q = 0.67$ pu, $\psi_{PM} = 0.62$ pu.

SOLUTION: The calculation is performed based on Equations (9.23) and (9.24). Figure 9.7 shows how the voltage and current limits can be drawn as an ellipse and circle. The current circle is natural as if the current vector amplitude maximum is set to unity. A circle with radius $i_s = 1$ is drawn with its centre at the origin.

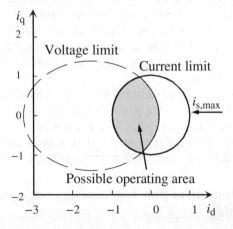

Figure 9.7 Possible operating area of a PMSM in $i_d i_q$-plane with current limit $i_{s,max} = 1$ at the rated speed $\omega_s = 1$. The machine parameters are $L_d = 0.4$ pu, $L_q = 0.67$ pu, and $\psi_{PM} = 0.62$ pu.

The possible zone of operation in Figure 9.7 is dictated by the converter current limit ($i_{s,max} = 1$) and the back electromotive force $\psi_s \omega_s$, which is heavily influenced by the armature reaction according to Equations (9.23) and (9.24). Naturally, the current limiting circle can be made bigger if a larger converter is used. However, motor rated current can only be exceeded temporarily. The voltage maximum caused limitation is speed dependent. At low speeds, the ellipse is larger than in the figure, and in field weakening, the ellipse becomes smaller as the internal induced voltage, $e_{PM} = \psi_{PM}\omega_s$, grows.

Considering stator resistance, the pu value of the stator voltage can be written in this form.

$$u_s^2 = \left(R_s i_d - \omega_s L_q i_q\right)^2 + \left(R_s i_q + \omega_s L_d i_d + \omega_s \psi_{PM}\right)^2 \tag{9.30}$$

By introducing the electric current components as absolute values, the equation can be reexpressed.

$$u_s = \sqrt{\left(R_s i_s \cos\gamma - \omega_s L_q i_s \sin\gamma\right)^2 + \left(R_s i_s \sin\gamma + \omega_s L_d i_s \cos\gamma + \omega_s \psi_{PM}\right)^2} \tag{9.31}$$

The equations reveal the effect of negative d-axis current as it reduces machine terminal voltage.

9.4   PMSM current vector control

Current vector control is a version implemented in the rotor reference frame. In current vector control, torque control generates the current references i_{dref} and i_{qref} for the d-axis and q-axis current components and then implements these references by suitably adjusting the voltage. As mentioned earlier, the current references depend on machine construction and on the constraints set by the drive system. For a nonsalient-pole machine without dq-axis inductance differences, the machine is normally driven with $i_d = 0$ before entering field weakening. With inductance differences, both i_d and i_q get nonzero values. In field weakening, $i_d < 0$, typically, for all PM machine types.

In low-speed operation, the current component references are formed directly from the torque reference. Therefore, by transforming two-phase current to three-phases, references are formed for the phase currents from the dq references. A current controller, for example a hysteresis controller, implements the current references. Figure 9.8 offers the principle block diagram of such a control system.

Current vector control is a widely adopted approach to the control of PM synchronous machines. Electric current vector control works well for PM machines, because unlike other machine types, the control parameter values for PM machines are not subject to significant variation during operation. Particularly in rotor surface magnet motors, inductances remain quite constant, and therefore the effect of armature reaction is small.

9.4.1   Control of nonsalient pole machines, $i_d = 0$ control, and field weakening

If the PMs are mounted on the outer diameter surface of the rotor and its internal steel is isotropic, the d-axis and q-axis machine inductances are approximately equal, that is, $L_d \approx L_q$.

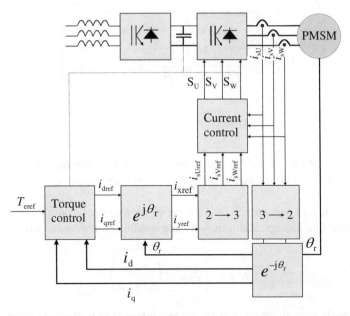

Figure 9.8 Simplified block diagram of the current vector control of a PMSM. The drive is equipped an active network bridge to enable effective regenerative braking common for PMSM drives. Alternatively a DC-link brake chopper could be used.

In steady state, the torque equation then simplifies to the following.

$$T_e \approx \frac{3}{2} p \left[\psi_{PM} i_q \right]$$ (9.32)

Therefore, d-axis current does not affect torque, and stator current at constant torque reaches a minimum when $i_d = 0$. This forms the basis for $i_d = 0$ control. In principle, nonsalient pole PMSMs are most energy efficient when $i_d = 0$, so if there is enough voltage to drive the motor without d-axis current, the $i_d = 0$ state should be preferred.

For $i_d = 0$ control, reference torque T_{eref} can be written in terms of the q-axis component i_{qref} of the reference current that should be realized by the converter.

$$i_{qref} = \frac{T_{eref}}{\frac{3}{2} p \psi_{PM}}$$ (9.33)

$$i_{dref} = 0$$ (9.34)

The $i_d = 0$ control method is easy to implement as long as rotor angle is known in real time. The torque control methodology is similar to that used to control a fully compensated DC machine – the torque is directly proportional to stator current. The $i_d = 0$ control method is best adapted to a machine with low inductances and almost insignificant armature reaction. For a rotor surface magnet machine with a small number of pole pairs, the inductance pu values are typically in the range of 0.2–0.6. Most low-power servo machines fall into this category.

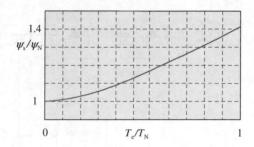

Figure 9.9 Movement of ψ_s/ψ_N (the ratio of stator flux linkage to rated flux-linkage amplitudes) as a function of torque T_e/T_N when $L_q = 1$ and $\psi_{PM} = 1$.

If the inductances are significant, the problem with this control method is rapidly increasing stator flux linkage. Stator flux-linkage amplitude increases as a function of torque.

$$|\psi_s| = \sqrt{\psi_{PM}^2 + \left(\frac{T_e L_q}{\frac{3}{2} p \psi_{PM}}\right)^2} \tag{9.35}$$

Figure 9.9 depicts the movement of ψ_s/ψ_n if the values for pu q-axis inductance and PM flux linkage are held at one, that is, $L_q = 1$ and $\psi_{PM} = 1$.

EXAMPLE 9.3: Produce a space-vector diagram for the PMSM condition at nominal voltage if $L_q = 1$ pu, $\psi_{PM} = 1$ pu, and $i_d = 0$. Calculate pu stator current i_s, voltage u_s, stator linkage flux ψ_s, load angle δ_s, angular speed for a two-pole machine ω_s, power factor $\cos\varphi$, electric current angle γ, and electromagnetic torque T_e.

SOLUTION: Figure 9.10 shows the space-vector diagram. The ψ_s/ψ_N behaviour shown previously in Figure 9.9 can easily be seen in the figures vector diagram.

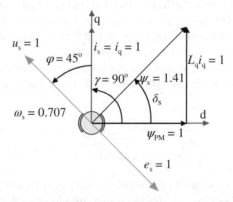

Figure 9.10 A machine implementing $i_d = 0$ control. The pu inductance is $L_q = 1$ for the machine at the pu speed $\omega_s = 0.707$. Armature reaction results in a stator flux-linkage value of $\psi_s = 1.41$ at the rated current $i_q = 1$. Already at a speed of $\omega_s = 0.707$, this value induces a back emf of $e_s = 1$. The maximum machine drive speed is 0.707. For $i_d = 0$ control, the power factor remains low if synchronous inductance is high. It stays at $\cos\varphi = 0.707$ at rated torque operation.

At nominal voltage, $u_s = 1$. If $i_d = 0$, $i_q = i_s = 1$, and i_q will be drawn in the q-axis. Stator linkage flux and load angle are calculated as follows.

$$\psi_s = \sqrt{\psi_{PM}^2 + (L_q i_q)^2} = \sqrt{1^2 + 1^2} = \sqrt{2} = 1.4141$$

$$\cos \delta_s = \frac{\psi_{PM}}{\psi_s} = \frac{1}{1.4141} = 0.707 \rightarrow \delta_s = 45°$$

$$u_s \cong e_s = \omega_s \psi_s = 1 \rightarrow \omega_s = \frac{1}{1.4141} = 0.707$$

The stator voltage space-vector is 90° ahead of ψ_s. Stator current $i_s = i_q$ is located on the q-axis. Therefore, the angle between the voltage and current is $\varphi = 45°$, and power factor is $\cos \varphi_s = 0.707$.

The electric current angle is between ψ_{PM} and the stator current i_s. ψ_{PM} is on the d-axis, and i_s is on the q-axis. Therefore, $\gamma = 90°$. Per unit, the electromagnetic torque is

$$T_e = \frac{u_s i_s \cos \varphi}{\omega_s} = \frac{1 \cdot 1 \cdot 0.707}{0.707} = 1$$

This is obvious based on the cross-field principle.

$$T_e = \psi_s \times i_s = \psi_d i_q + \psi_q i_d = \psi_{PM} i_q = 1$$

The voltage needed to supply the power to the machine has the same amplitude as the stator back emf.

$$u_s \approx -e_s = j\omega_s \psi_s \Rightarrow u_s \approx e_s$$

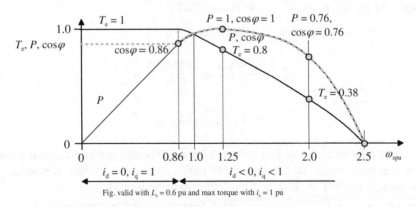

Fig. valid with $L_s = 0.6$ pu and max torque with $i_s = 1$ pu

Figure 9.11 Rotor surface magnet PMSM behaviour at different speeds starting with $i_d = 0$ control and entering into field-weakening from a small negative i_d until $i_d = -0.925$ and $i_q = 0.38$ at $\omega_s = 2.0$. At $\omega_s = 2.5$, $i_d = -1.0$, and $i_q = 0$, the torque and power are zero. $\psi_{PM} = 1$, $i_s = 1$, and $u_s = 1$. Synchronous inductance $L_s = 0.6$, and converter maximum current is 1 pu.

For machine speeds that do not call for field weakening, $i_d=0$ control is most energy efficient, and the low power factor is not significant. As $|\psi_s|$ increases, stator voltage increases and maximum inverter-voltage-limited speed drops. In principle, there is no field weakening with $i_d=0$ control so the speed of the drive is limited. To achieve higher speeds, field weakening must be engaged, and the d-axis current must be given a negative reference. In so doing, power factor begins to increase to its maximum at $\cos\varphi=1$ before going capacitive at higher speeds. This behaviour is illustrated in Figure 9.11 for a rotor surface magnet PMSM with pu values of $L_s=0.6$, $\psi_{PM}=1$, $i_s=1$, and $u_s=1$.

EXAMPLE 9.4: Produce a diagram T_e, P, $\cos\varphi=\mathrm{f}\,(\omega_s)$ that shows the field weakening dependence for a rotor surface magnet PMSM with $L_d=L_q=L_s=0.6$, $\psi_{PM}=1$, $i_s=1$, and $u_s=1$.

SOLUTION: The space-vector diagram is similar to that of Figures 9.6 and 9.10, however, in Figure 9.6, the different inductance results in changes in value for stator linkage flux and other parameters. For an $i_d=0$ control, $i_q=i_s=1$. The voltage increases to the nominal value when the pu value u_s is 1.

$$\psi_s = \sqrt{\psi_{PM}^2 + (L_s i_q)^2} = \sqrt{1^2 + (0.6 \cdot 1)^2} = 1.16$$

$$\cos\delta_s = \frac{\psi_{PM}}{\psi_s} = \frac{1}{1.16} = 0.86 \rightarrow \delta_s = 30.68°$$

$$u_s \cong e_s = \omega_s\psi_s = 1 \rightarrow \omega_s = \frac{1}{1.16} = 0.86$$

With $i_d=0$, $i_s=i_q=1$. In a nonsalient-pole machine using $i_d=0$ control, the angle shift between the voltage and current is identical with $\delta_s=30.7°$, because the q-axis is perpendicular to the d-axis, and the voltage vector u_s is perpendicular to the stator linkage flux ψ_s. Then, $\varphi=30.7°$ and $\cos\varphi=0.86$.

The pu electromagnetic power and torque are

$$T_e = \frac{u_s i_s \cos\varphi}{\omega_s} = \frac{1 \cdot 1 \cdot 0.86}{0.86} = 1$$

$$P_e = u_s i_s \cos\varphi = 0.86$$

This can be seen on the left in Figure 9.11. With $i_d=0$ control and at maximum available torque, the power factor stays at $\cos\varphi=0.86$ until a speed of $\omega_s=0.86$ is reached. At this point, voltage reaches its limit, and the torque must be limited.

To increase the speed, increasingly negative values of the d-axis current component must be applied, for example, at $i_d=-0.6$. Other parameter values are calculated as follows.

$$i_q = \sqrt{i_s^2 - i_d^2} = \sqrt{1 - 0.6^2} = 0.8$$

$$\psi_s = \sqrt{(\psi_{PM} - L_s i_d)^2 + (L_s i_q)^2} = \sqrt{(1 - 0.6 \cdot 0.6)^2 + (0.6 \cdot 0.8)^2} = 0.8$$

$$\cos \delta_s = \frac{(\psi_{PM} - L_d i_d)}{\psi_s} = \frac{(1 - 0.6 \cdot 0.6)}{0.8} = 0.8$$

$$\delta_s = 36.86°$$

If the angle between the q-axis and the stator current vector is α, then

$$\cos \alpha = \frac{i_q}{i_s} = \frac{0.8}{1} = 0.8 \rightarrow \alpha = 36.86°$$

Moreover, the phase angle between the voltage and current is

$$\varphi_s = \delta_s - \alpha = 36.86 - 36.86 = 0 \rightarrow \cos \varphi_s = 1$$

$$\omega_s = \frac{u_s}{\psi_s} = \frac{1}{0.8} = 1.25$$

The electromagnetic pu power and torque is as follows.

$$P_e = u_s i_s \cos \varphi = 1 \cdot 1 \cdot 1 = 1$$

$$T_e = \frac{P_e}{\omega_s} = \frac{1}{1.25} = 0.8$$

Power reaches a value of $P = 1$ at speed $\omega_s = 1.25$, and the power factor is $\cos\varphi = 1$. The torque at this point is $T_e = 1/1.25 = 0.8$. As the speed increases further, increasingly more current must be applied to the negative d-axis to weaken the stator flux linkage.

At $i_d = -0.925$, the other quantities are as follows.

$$i_q = \sqrt{i_s^2 - i_d^2} = \sqrt{1 - 0.925^2} = 0.38$$

$$\psi_s = \sqrt{(\psi_{PM} - L_s i_d)^2 + (L_s i_q)^2} = \sqrt{(1 - 0.6 \cdot 0.925)^2 + (0.6 \cdot 0.379)^2} = 0.5$$

$$\cos \delta_s = \frac{\psi_{PM} - L_s i_d}{\psi_s} = \frac{1 - 0.6 \cdot 0.925}{0.5} = 0.89 \rightarrow \delta_s = 26.9°$$

$$\alpha = \cos^{-1} \frac{i_q}{i_s} = \cos^{-1} \frac{0.38}{1} = 67.7°$$

$$\varphi = \delta_s - \alpha = 26.9° - 67.7° = -40.8° \rightarrow \cos \varphi = 0.76$$

$$\omega_s = \frac{u_s}{\psi_s} = \frac{1}{0.5} = 2.0$$

Electromagnetic pu power and torque become

$$P_e = u_s i_s \cos \varphi = 1 \cdot 1 \cdot 0.76 = 0.76$$

$$T_e = i_q \psi_{PM} = 0.38 \cdot 1 = 0.38$$

Figure 9.11 reveals the pu values of power $P_e = 0.75$ and torque $T_e = 0.378$ at twice the rated speed. At $i_d = -1.0$, the other quantities are as follows.

$$i_q = \sqrt{i_s^2 - i_d^2} = 0$$

$$\psi_s = \sqrt{(\psi_{PM} - L_s i_d)^2 + (L_s i_q)^2} = \sqrt{(1 - 0.6 \cdot 1.0)^2} = 0.4$$

$$\cos \delta_s = \frac{\psi_{PM} - L_s i_d}{\psi_s} = \frac{1 - 0.6 \cdot 1.0}{0.4} = 1.0 \rightarrow \delta_s = 0°$$

$$\alpha = \cos^{-1} \frac{i_q}{i_s} = \cos^{-1} \frac{0.0}{1} = 90°$$

$$\varphi_s = \delta_s - \alpha = 0° - 90° = -90° \rightarrow \cos \varphi_s = 0$$

$$\omega_s = \frac{u_s}{\psi_s} = \frac{1}{0.4} = 2.5$$

Electromagnetic pu power and torque are

$$P_e = u_s i_s \cos \varphi = 1 \cdot 1 \cdot 0 = 0$$

$$T_e = i_q \psi_{PM} = 0 \cdot 1 = 0$$

When machine inductances are so low that field weakening via stator current is not practical, $i_d = 0$ control is the approach often applied. Relatively high switching frequencies must be used to accommodate the high rates of change for electric current.

EXAMPLE 9.5: Produce a space-vector diagram for the $i_d = 0$ control of a rotor surface magnet PMSM with small pu inductances. $L_s = L_d = L_q = 0.2$. $u_s = 1, i_q = i_s = 1,$ and $\psi_{PM} = 1$.

SOLUTION:

$$\psi_s = \sqrt{\psi_{PM}^2 + (L_s i_q)^2} = \sqrt{1^2 + (0.2 \cdot 1)^2} = 1.02$$

$$\cos \delta_s = \frac{\psi_{PM}}{\psi_s} = \frac{1}{1.02} = 0.98 \rightarrow \delta_s = 11.36°$$

$$u_s \cong e_s = \omega_s \psi_s = 1 \rightarrow \omega_s = \frac{1}{1.02} = 0.98$$

With $i_d = 0$, $i_s = i_q = 1$ and the angle shift between voltage and current is $\delta = 11.36°$, because the q-axis is perpendicular to the d-axis, and the voltage vector $\boldsymbol{u}_s$ is perpendicular to the stator linkage flux $\boldsymbol{\psi}_s$. Therefore $\varphi = 11.36°$, and $\cos\varphi = 0.98$.

The electromagnetic pu power and torque are as follows.

$$T_e = \psi_{PM} i_q = 1 \cdot 1 = 1$$

$$P_e = T_e \omega_s = 1 \cdot 0.98 = 0.98$$

Figure 9.12 illustrates the implementation of the $i_d = 0$ control for a rotor surface magnet machine in which the synchronous inductance is $L_q = 0.2$. The load angle remains small and the power factor is high. Speed is restricted only by a few percent.

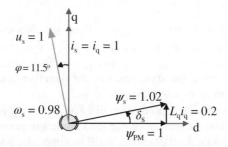

Figure 9.12   Rotor surface magnet machine implementing $i_d = 0$ control – $L_q = 0.2$ for the machine at the pu speed $\omega_s = 0.98$. Because of the armature reaction, the stator flux linkage reaches $\psi_s = 1.02$ at the rated current. At $\omega_s = 0.98$, this value induces a back emf of $e_s = 1$. The maximum speed of the machine drive in this control mode is 0.98. The $i_d = 0$ control no longer significantly affects power factor, and the power factor becomes $\cos\varphi = 0.98$.

Therefore, $i_d = 0$ control is best adapted to machines that have minimal armature reaction, because field weakening, in practice, is not available. A rated machine speed must be selected that suits the needs of the drive.

9.4.2   Torque production in machines with reluctance torque

The space vector torque Equation (9.16) can also be written for a steady state in pu values.

$$T_e = \left[\psi_{PM} i_q - \left(L_{mq} - L_{md} \right) i_d i_q \right] \tag{9.36}$$

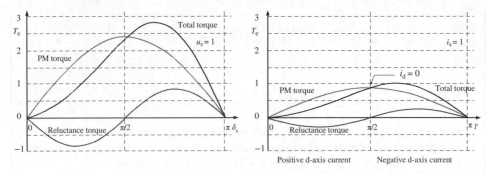

Figure 9.13 Per-unit load angle δ_s and electric current angle γ for a machine with $u_s = 0.9$, $\psi_{PM} = 0.8$, $e_{PM} = 0.8$, $L_d = 0.3$, and $L_q = 0.8$ – The effect of reluctance torque is significant and must be taken into account in the control. Unlike in rotor surface magnet machines, the peak torque is not found at a load angle of $\delta_s = \pi/2$, but instead at a significantly higher load angle. Also, the electric current angle must be higher than $\gamma = \pi/2$ to get maximum torque for a given current value. In this case, $i_s = 1$. The maximum reluctance torque is found with an electric current angle $\gamma = 3\pi/4$. With this angle and stator current value $i_s = 1$ and inductance difference $(L_d - L_q) = -0.5$ the reluctance torque gets value $T_{erel} = i_d i_q (L_d - L_q) = -1/\sqrt{2} \times 1/\sqrt{2} \times -0.5 = 0.25$. At electric current angle $\gamma = \pi/2$, the PM torque is $T_{ePM} = i_q$ and $\psi_{PM} = 0.8$.

The reluctance torque term $3p/2(L_{mq} - L_{md})i_d i_q$ is of significance only if the difference between magnetizing inductances is large. In a PM machine, this term has the same sign as the quadrature axis current i_q when $L_{md} < L_{mq}$, and $i_d < 0$. The current references for the dq reference frame that minimize the stator current are therefore achieved when $i_{dref} < 0$. If $L_{md} > L_{mq}$, then $i_{dref} > 0$ must be used instead.

Figure 9.13 illustrates the torque production of a PMSM having pu parameters $\psi_{PM} = 0.8$, $L_d = 0.3$, and $L_{qu} = 0.8$. The pu version of the load angle Equation (9.2) is illustrated as a function of the load angle with $u_s = 0.9$ and $\omega_s = 1$. The figure also gives the pu torque version of Equation (9.36) as a function of the electric current angle γ with $i_s = 1$.

In pu presentation, the term $3p/2$, which is known in space vector theory, is not needed. The pu presentations of the equations are repeated here.

Load angle equation

$$T_e = \frac{u_s\, e_{PM}}{\omega_s^2 L_d} \sin \delta_s + u_s^2 \frac{L_d - L_q}{\omega_s^2 2 L_d L_q} \sin 2\delta_s \tag{9.37}$$

Cross-product equation

$$T_e = \psi_{PM} i_q - (L_{mq} - L_{md})i_d i_q \tag{9.38}$$

Maximum torque per ampere – control

Machines producing both PM and reluctance torques should be controlled using the so-called maximum torque per ampere (MTPA) control method, which minimizes stator current at a certain torque, that is, the drive optimally combines PM and reluctance torques.

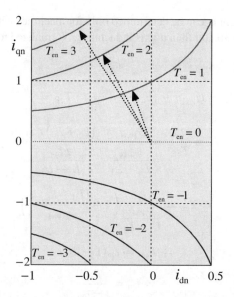

Figure 9.14 Examples of constant torque loci for the normalized torque T_{en} in the $i_d i_q$ plane for a machine with $L_d < L_q$. If $L_d > L_q$, the hyperbola should be mirrored with respect to the q-axis. The corresponding hyperbola moves farther from the origin with increasing torque.

Applying the cross-field principle, Equation (9.38) forms constant torque hyperbolas in the $i_d i_q$ plane.

$$i_q = \frac{T_e}{\psi_{PM} - (L_{mq} - L_{md})i_d} \tag{9.39}$$

The hyperbolas, which will be illustrated in Figure 9.14, have asymptotes.

$$i_q = 0 \tag{9.40}$$

$$i_d = \frac{\psi_{PM}}{(L_{mq} - L_{md})} = \frac{\psi_{PM}}{(L_q - L_d)} \tag{9.41}$$

Two different methods have been presented in the literature for implementing MTPA control based on the minimization of stator current (Jahns et al., 1986; Kim & Sul, 1997). The following normalizations are based on the method introduced by Jahns et al., which differ by definition from the pu system.

$$T_{en} = \frac{T_e}{T_{eB}}, i_{qn} = \frac{i_q}{i_B}, i_{dn} = \frac{i_d}{i_B} \tag{9.42}$$

where

$$i_B = \frac{\psi_{PM}}{L_q - L_d} \tag{9.43}$$

$$T_{eB} = \psi_{PM} i_b \tag{9.44}$$

The basic value for current i_B can be determined only for machines with saliency. Quadrature-axis inductance must be greater than d-axis inductance. Normalized torque is derived from pu torque as follows.

$$T_e = \psi_{PM} i_q - (L_q - L_d) i_q i_d \quad : T_{eB} = \frac{\psi_{PM}^2}{L_q - L_d}$$

$$\Leftrightarrow \frac{T_e}{T_{eB}} = \frac{\psi_{PM} i_q}{\frac{\psi_{PM}^2}{L_q - L_d}} - \frac{(L_q - L_d) i_q i_d}{\frac{\psi_{PM}^2}{L_q - L_d}}$$

$$\Leftrightarrow T_{en} = \frac{i_q}{\frac{\psi_{PM}}{L_q - L_d}} \left(1 + \frac{(L_d - L_q) i_d}{\psi_{PM}} \right)$$

$$\Rightarrow T_{en} = i_{qn}(1 - i_{dn}) \tag{9.45}$$

$$\Rightarrow i_{qn} = \frac{T_{en}}{1 - i_{dn}} \tag{9.46}$$

The square of the normalized current modulus can be expressed

$$|i_n|^2 = i_{dn}^2 + i_{qn}^2 = i_{dn}^2 + \left(\frac{T_{en}}{1 - i_{dn}} \right)^2 \tag{9.47}$$

The electric current minimum for a given torque T_{en} is found by differentiating Equation (9.47) and setting the differential to zero.

$$\frac{d|i_n|^2}{di_n} = 2i_{dn} + 2\frac{T_{en}^2}{(1 - i_{dn})^3} = 0 \tag{9.48}$$

$$\Leftrightarrow T_{en}^2 = i_{dn}(1 - i_{dn})^3 \tag{9.49}$$

Equation (9.49) can be used to determine the d-axis current reference. The quadrature axis current reference is found similarly with Equation (9.45) by eliminating i_{dn} and differentiating.

$$T_{en}^2 - T_{en} i_{qn} - i_{qn}^4 = 0 \tag{9.50}$$

Equation (9.50) can be solved for T_{en} as follows.

$$T_{en} = \frac{i_{qn}}{2} \left(1 \pm \sqrt{1 + 4i_{qn}^2} \right) = 0 \tag{9.51}$$

While the root in Equation (9.51) is always larger than one, only a positive root is allowed.

$$T_{en} = \frac{i_{qn}}{2} \left(1 + \sqrt{1 + 4i_{qn}^2} \right) = 0 \tag{9.52}$$

Solving both i_{dn} and i_{qn} requires iteration. However, a simplification can be made. From Equation (9.45)

$$i_{dn} = 1 - \frac{T_{en}}{i_{qn}}$$

(9.53)

From Equation (9.52)

$$\frac{T_{en}}{i_{qn}} = \frac{1}{2}\left(1 + \sqrt{1 + 4i_{qn}^2}\right)$$

(9.54)

Combining Equations (9.53) and (9.54) results in the following.

$$i_{dn} = \frac{1}{2}\left(1 - \sqrt{1 + 4i_{qn}^2}\right)$$

(9.55)

The resulting constant torque hyperbolas facilitate solving for the minimum stator current at a given constant torque. Minimum current corresponds to the hyperbola's minimum distance from the origin. The point from which this minimum value is obtained gives the required references for the normalized currents i_{dn} and i_{qn}. The current references i_{dnref} and i_{qnref} that produce minimum stator current are the obtained solutions. These references are also given by the equations based on the earlier study.

$$T_{enref} = \sqrt{i_{dnref}(i_{dnref} - 1)^3}$$

(9.56)

$$T_{enref} = \frac{i_{qnref}}{2}\left(1 + \sqrt{1 + 4i_{qnref}^2}\right)$$

(9.57)

The current references i_{dnref} and i_{qnref} must be computed iteratively, or the functions $i_{dnref} = f(T_{enref})$ and $i_{qnref} = f(T_{enref})$ must be computed in advance. Figure 9.15 shows the curves for the functions.

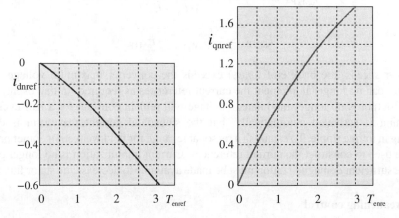

Figure 9.15 Current reference functions minimizing the stator current $i_{dnref} = f(T_{enref})$ and $i_{qnref} = f(T_{enref})$ for a machine with $L_d < L_q$.

Returning to the normal pu values gives the following.

$$T_e = \frac{\psi_{PM} i_q}{2} \left(1 + \sqrt{1 + 4\frac{i_q^2 (L_q - L_d)^2}{\psi_{PM}^2}} \right) \tag{9.58}$$

$$i_d = i_B i_{dn} = \frac{\psi_{PM} i_q}{2(L_q - L_d)} \left(1 - \sqrt{1 + 4i_q^2} \right) = \frac{\psi_{PM} i_q}{2(L_q - L_d)} - \sqrt{\frac{\psi_{PM}^2}{4(L_q - L_d)} + i_q^2} \tag{9.59}$$

The current references for the d- and q-axis currents can now be solved using Equations (9.58) and (9.59), respectively. If the machine is nonsalient Equation (9.59) becomes indifferent and an $i_d = 0$ control results.

Another way to formulate MTPA control by minimizing stator current was introduced by (Kim & Sul, 1997a). The pu torque equation is modified as follows.

$$T_e = \psi_{PM} i_s \sin \gamma - \frac{(L_q - L_d)}{2} i_s^2 \sin 2\gamma \tag{9.60}$$

To find the minimum, T_e is differentiated as a function of the electric current angle γ and set to zero.

$$\frac{dT_e}{d\gamma} = \psi_{PM} i_s \cos \gamma - (L_q - L_d) i_s^2 \cos 2\gamma = 0 \tag{9.61}$$

Because $\cos 2\gamma = 2\cos^2 \gamma - 1$, the cosine of the electric current angle γ can be solved from the following quadratic equation.

$$-2(L_q - L_d) i_s^2 \cos^2 \gamma + \psi_{PM} i_s \cos \gamma + (L_q - L_d) i_s^2 = 0 \tag{9.62}$$

Solving Equation (9.62) makes it possible to find the MTPA current pu components.

$$i_{dMTPA} = \frac{\psi_{PM} - \sqrt{\psi_{PM}^2 + 8(L_q - L_d)^2 |i_{sMTPA}^2|}}{4(L_q - L_d)} \tag{9.63}$$

$$i_{qMTPA} = \sqrt{|i_{sMTPA}^2| - i_{dMTPA}^2} \tag{9.64}$$

At higher speeds, the back emf ($\omega_s \psi_s$) exceeds the converter maximum voltage, and the converter can no longer implement the current references as required. In that event, inverter voltage output is no longer modulated. Voltage will remain constant for a half cycle. Field weakening is activated automatically, but the sinusoidal current waveform is distorted resulting in torque ripple. To establish a reasonable range for field weakening, other principles must be used to construct the current references. Current minimization is no longer possible, because sufficient demagnetization must be made available to decrease the stator flux linkage.

Field weakening control

At higher speeds, the MTPA control strategy is no longer appropriate, and field-weakening (FW) control should be considered. The d-axis current must be controlled so it weakens the stator flux linkage. Per-unit current components can be calculated using a maximum voltage u_{smax} constraint (Haque et al., 2003). The field-weakening current components are as follows.

$$i_{dFW} = \frac{L_d\psi_{PM} - \sqrt{(L_d\psi_{PM})^2 + (L_q^2 - L_d^2)(\psi_{PM}^2 + L_q^2 i_{sFW}^2 - u_{smax}^2/\omega_s^2)}}{L_q^2 - L_d^2} \tag{9.65}$$

$$i_{qFW} = \sqrt{|i_{sFW}^2| - i_{dFW}^2} \tag{9.66}$$

Current component values can be calculated from these equations at different speeds and current levels i_{sFW}. Figure 9.16 demonstrates the application of both the MTPA and the FW control strategies.

Figure 9.16 illustrates how torque must be reduced as speed increases. An increasing share of the electric current resources must be used to demagnetize the stator flux linkage. The centre of the voltage ellipses is located by the characteristic current i_x on the negative d-axis. In this machine, $i_x = 0.62/0.4 = 1.55$.

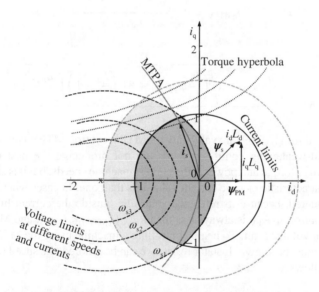

Figure 9.16 MTPA pu current trajectory, maximum current circles ($i_s = 1$ and $i_s = 1.5$), torque hyperbola, and voltage ellipses (with centre x) at different speeds ($\omega_{s1} = 1$, $\omega_{s2} = 2$, $\omega_{s3} = 3$). The voltage ellipses are drawn taking approximately 15 % voltage reserve into account. The machine pu parameters are $L_d = 0.4$, $L_q = 0.67$, and $\psi_{PM} = 0.62$. The current vector i_s is 0.9 pu at 110° producing 0.59 pu PM torque and 0.06 pu reluctance torque with the stator flux-linkage vector $\psi_s = 0.77$ at 45°. The dark gray area is the normal operating area of the drive, and the light gray area is the temporary operating area with 50 % increased current. With the rated current, stator flux linkage is reduced by 0.4 pu for a remaining $0.62 - 0.4 = 0.22$, which results in a maximum speed that is 4.5 times the rated speed at no load in field weakening. Greater torque locates the corresponding hyperbola further from the origin, while a higher speed produces a smaller voltage ellipse.

When the voltage limit is reached, the control method must be changed from MTPA to FW control. In field weakening, torque drops automatically with increasing electric current angle γ despite maintaining stator current at its rated value. Speed is further increased via FW control, or in some special cases, more torque can be produced using maximum torque per volt (MTPV) control.

Maximum torque per volt control

When the voltage limit is reached, armature current must be controlled accordingly. MTPV can be used when the characteristic current of the machine $i_x \le 1$. Differentiating the pu torque equation with respect to i_d and using Equation (9.65), the armature current vector producing the maximum amount of the output torque under voltage constraint can be derived (Illioudis & Margaris, 2010). The pu current references are

$$i_{\mathrm{dMTPV}} = -\frac{\psi_{\mathrm{PM}}}{L_d} - \Delta i_d \tag{9.67}$$

$$i_{\mathrm{qMTPV}} = \frac{\sqrt{\left(\dfrac{u_s}{\omega_s}\right)^2 - (L_d \Delta i_d)^2}}{L_q} \tag{9.68}$$

with

$$\Delta i_d = \frac{-\frac{L_q}{L_d}\psi_{\mathrm{PM}} + \sqrt{\left(\dfrac{L_q}{L_d}\psi_{\mathrm{PM}}\right)^2 + 8\left(\left(\dfrac{L_q}{L_d}-1\right)\left(\dfrac{u_{\mathrm{smax}}}{\omega_s}\right)\right)^2}}{L_q} \tag{9.69}$$

Equations (9.67) through (9.69) are valid only for MTPV control. MTPV control is used when MTPA and field-weakening control methods are not producing the best results. MTPV control can improve torque and power at the highest machine speeds, but it is suitable only for drives with a characteristic electric current value less than one. In cases where $i_x > 1$, MTPV cannot be considered, because its reference values lie outside the current limit circle.

In MTPV control, i_{sMTPV} is always less than i_{smax} ($i_{\mathrm{sMTPV}} < i_{\mathrm{smax}}$), and MTPV control is carried out at the voltage limit. To begin controlling a machine using the MTPV approach, a speed reference must be known. If stator resistance can be ignored, the speed reference can be calculated as follows.

$$\omega_{\mathrm{sMTPV}} = \frac{u_s}{\sqrt{(\psi_{\mathrm{PM}} + L_d i_{\mathrm{dMTPV}})^2 + (-L_q i_{\mathrm{qMTPV}})^2}} \tag{9.70}$$

In Equation (9.70), the d- and q-axis current components combine to produce a total stator current $i_s = 1$. At this point, the machine control method switches from FW to MTPV. Considering both the voltage and the current limits make it possible to produce maximum output power at all speeds.

Vector control mode selection

MTPA and current limit trajectories are independent of machine speed. These trajectories are determined by motor or converter properties alone. The voltage limit ellipse, however, varies

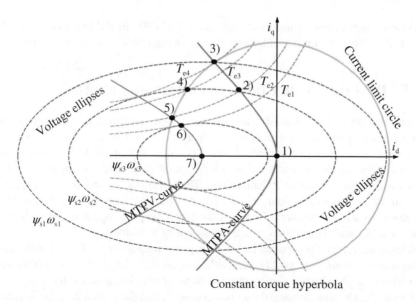

Figure 9.17 Controlling a PMSM using different control methods. Greater torque locates the corresponding hyperbola further from the origin, while a higher speed produces a smaller voltage ellipse.

with speed. The ellipse grows smaller as speed increases (Haque et al., 2002). Theoretically, a motor can reach a very high speed as long as its characteristic current ($i_x = \psi_{PM}/L_d$) remains less than the rated current. In MTPV, the L_d pu value is greater than the ψ_{PM} pu value ($L_d \geq \psi_{PM}$), which is not necessarily common.

Figure 9.17 shows the torque hyperbola and voltage ellipse plots for control mode selection between MTPA, field weakening, and MTPV.

The drive begins operation at point 1 where the machine is at rest. Next, the drive increases torque along the MTPA curve between points 1, 2, and 3. At point 3, maximum torque is produced as the curve intersects the current limit circle and intersects the maximum voltage ellipse at speed ω_{s1} and flux linkage ψ_{s1}. From points 3 to 4 and on to 5, the drive must follow the maximum current and voltage limits as the machine accelerates to higher speeds in field weakening. At point 5, where it is crossing the T_{e2} curve, the drive produces T_{e2} torque. For this machine configuration, the MTPV curve falls inside the maximum current circle, so further acceleration is accomplished by following the MTPV control curve to point 6. Continuing on to point 7, the voltage ellipse has collapsed to a zero point. At point 7, the machine can achieve, in principle, infinite no-load speed. Between points 6 and 7, the drive can produce torque at very high speeds.

With this understanding, it is now possible to implement a PMSM vector control methodology. The reference machine currents for the controller represented in the previous Figure 9.8 can be selected according to $i_d = 0$ control for a nonsalient-pole machine or MTPA control for a salient pole machine. At higher speeds, field-weakening control is enabled. Finally, current references can be produced via MTPV control.

As in all vector control applications, a machine saturation model should help provide control that is more accurate. In rotor surface magnet machines, saturation does not play as significant a role as in machines with embedded magnets, which facilitates control.

The upcoming section examines the application of DTC in PMSM drives. DTC brings benefits and at the same time enables machine modelling based on the drive system alone.

9.5 PMSM direct flux linkage and torque control

The method proposed by Takahashi and Nokuchi (1986) for direct flux linkage and torque control was initially investigated for IMs in particular. The control principle, known as DTC, is also commonly applied to commercial products, especially by ABB. The control method has been applied to the sensorless closed-loop speed control of synchronous reluctance machines and to the control of PM machines. The integral $\boldsymbol{\psi}_{s,est} = \int (\boldsymbol{u}_{s,est} - \boldsymbol{i}_{s,meas} R_s) dt$ is used to estimate the stator flux linkage in the stator reference frame. Because of this approach, only stator resistance is required for the stator flux-linkage integration step, which is the strength of DTC, and the estimation is independent of machine type.

However, the integration approach also leads to the weaknesses of DTC. Firstly, errors in the integration of constants quickly accumulate to produce large errors in the stator flux-linkage estimates. Usually, measuring stator voltage directly is not necessary. It is estimated, instead, from the DC-link voltage, the switch positions, and the switch models. Secondly, an accurate stator resistance estimate is important to accurate flux-linkage estimation. The resistive voltage drop that shows up in the integral at low speed becomes more dominant than stator voltage, and therefore errors are easily introduced in the flux-linkage estimate, which leads to making incorrect voltage vector choices.

Estimates of existing stator flux linkage and continual rotor angular position feedback are required to implement the electric current model. Machine stator flux-linkage estimates can be improved by computing stator flux linkage using all machine parameters, which can be accomplished using the equations $\psi_d = L_d i_d + L_{md} i_D + \psi_{PM}$ and $\psi_q = L_q i_q + L_{mq} i_Q$. Combining these flux-linkage estimates with available rotor angle feedback and the voltage integral makes it possible to implement a control method based on an electric current model that applies to all synchronous machine rotation speeds. Figure 9.18 offers a block diagram for DTC control with rotor position feedback for a PMSM.

To implement a DTC control system for a very low-inductance servo PM machine, a sufficiently high sampling frequency for the stator flux-linkage computation must be provided to maintain an adequate average switching frequency. On the other hand, a high sampling frequency is not needed to compute flux linkage with a position feedback current model.

If combined with drift correction methods proposed by Niemelä (1999) and Luukko (2000), DTC without rotor position feedback is capable of adequately controlling industrial PM machines. However, to carry out the machine identification run during startup, the angular position of the rotor is needed. Rotor angle can be determined, for example, by applying the method proposed by Luukko (2000), at least for all machines with an observable difference between the d- and q-axis inductances. The rotor position measurement is carried out by measuring the inductances produced by the electric currents that develop when the inverter is minimally pulsed in all possible voltage vector directions. Next, the measurement results are adjusted to plot a measurement curve that shows the placement of the d- and q-axes.

DTC may be well adapted for PM machine drives, because almost any reasonable reference value for stator flux linkage can be selected. The $i_d = 0$, MTPA, and MTPV control methods can be adapted for PMSM drives by selecting the appropriate flux-linkage reference. Naturally, this adds computational burden, because the current references must be converted

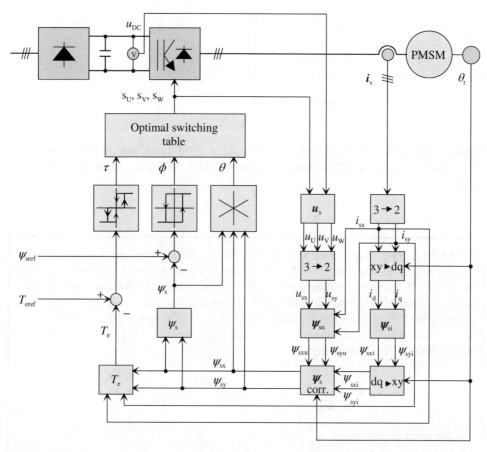

Figure 9.18 Block diagram for the DTC control of a PMSM with rotor position feedback; ψ_{sxu} and ψ_{syu} are the stator flux-linkage estimates calculated by integrating stator voltage. These estimates are periodically corrected using the stator flux-linkage estimates ψ_{sxi} and ψ_{syi} calculated from the electric current model. θ_r is the angular position of the rotor. (Luukko, 2000).

to flux-linkage references, which also require the machine parameters. For example, to implement the $i_d = 0$ control method in DTC, the stator flux-linkage reference must be set according to Figure 9.9, where the reference increases as a function of the torque.

Because a DTC drive does not need a machine model to rotate the rotor, flexible identification runs can be carried out. However, information about the flux linkage generated by the PMs must be available. ψ_{PM} can be easily found based on the voltage integral by rotating the motor at no load and adjusting the stator flux-linkage reference $\psi_{s,ref}$ of the inverter over a sufficiently wide range. The stator flux-linkage reference that produces the minimum no-load current is equal to the flux linkage ψ_{PM} generated by the PMs. See Figure 9.19.

A commercial converter applying DTC may input a 25-μs minimum pulse. This minimum pulse width is appropriate for industrial PM motors but may be too long for small servomotors with rotor surface magnets.

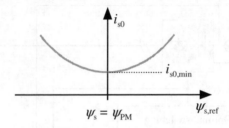

Figure 9.19 Behaviour of the no-load current i_{s0} of a PM motor as a function of the stator flux-linkage reference. The flux linkage ψ_{PM} of the PMs is where current reaches its minimum value.

EXAMPLE 9.6: Consider a 2 kW, 400 V, 150 Hz, Y-connected servo motor with a pu synchronous inductance of $L_d = 0.2$. Both the rated power factor and the efficiency of the motor are 0.9. Solve for the rated current, rated impedance, rated inductance, and synchronous inductance. Calculate the current rise for the machine if the commercial converter applying the DTC gives it a minimum pulse of 25 μs at a DC-link voltage of 540 V.

SOLUTION: The rated current is

$$I_N = \frac{P_N}{\sqrt{3}U_N\cos\varphi_N\eta_N} = \frac{2000}{\sqrt{3}400 \cdot 0.9 \cdot 0.9} = 3.56 \text{ A},$$

and the rated impedance is

$$Z_N = \frac{U_N/\sqrt{3}}{I_N} = \frac{U_N^2}{S_N} = \frac{U_N^2\cos\varphi_N\eta_N}{P_N} = \frac{400^2}{2000}0.9 \cdot 0.9 = 64.8 \ \Omega.$$

The rated inductance of the motor is therefore

$$L_N = \frac{Z_N}{\omega_N} = \frac{64.8}{150 \cdot 2\pi} = 69 \text{ mH}.$$

The synchronous inductance of the motor is as follows.

$$L_d = 0.2L_N = 0.2 \cdot 69 = 13.7 \text{ mH}$$

Because the inverter supplies a 25-μs pulse to the machine when the rotor is stationary, a DC-link potential of 540 V acts on the servomotor windings to put one phase winding in series with the parallel connection of the two other phases. The inductance is therefore $L' = 20.58$ mH.

The current rise rate becomes

$$\frac{di}{dt} = \frac{U_{DC}}{L'} = \frac{540}{20.58 \cdot 10^{-3}} = 26.23 \qquad \text{kA/s} = 0.65\text{A}/25 \text{ ms}.$$

The rated current of the motor is 3.56 A. The electric current changes about 18 % from rated current during a single minimum pulse, a seemingly large percentage. Therefore, the duration of the minimum pulse should be less than 25 µs in a low inductance machine.

9.5.1 Selecting the stator flux-linkage absolute value in a PMSM DTC drive in a practical case

For a PMSM in particular, correct analysis of the flux-linkage reference and the inductances is necessary. In the following examples, consider the responses of a real PM machine to the application of different control methods. Table 9.1 gives the nameplate values and relevant parameters for the subject machine.

Several different vector diagrams are illustrated in the upcoming examples for nominal operating points corresponding to rated load torque. Saturation of the inductances is not taken into account. In each example, the basic value for the flux linkage is always determined with respect to the stator voltage.

Table 9.1 Machine parameters of an industrial PMSM having small saliency

$E_{PMph} = 186$ V (phase voltage), no-load voltage	$U_{s,phN} = 202$ V (phase voltage), stator rated voltage used as the base voltage U_b corresponding to 1 pu
$I_{sN} = 115$ A, 1 pu	$f_N = 100$ Hz, 1 pu, $\omega_N = 2\pi f_N = 200\pi$ s$^{-1} = 628.318$ s^{-1}, 1 pu
Rated impedance $Z_N = U_{s,phN}/I_{sN} = 1.76\,\Omega$ Input apparent power $S_N = 3U_{sph}\,I_{sph} = 3\cdot202\cdot$ $115 = 69690$ VA, 1 pu	Rated inductance $L_N = Z_N/\omega_N = 1.76\,\Omega s/(200\pi) = 2.8$ mH
$P_N = 64$ kW, 0.92 pu, $T_{en} = 0.92$ pu	$n_N = 2000$ rpm, 1 pu (six-pole machine, converter treats all as two-pole machines)
$\cos\varphi_N = 0.96$ efficiency is 0.96: $P_N = S_N \cos\varphi_N\eta_N = $ $69.69 \cdot 0.96 \cdot 0.96$ kW $= 64$ kW	$R_s = 53$ mΩ, 0.03 pu
$L_d = 1.12$ mH, 0.401 pu	$L_q = 1.16$ mH, 0.415 pu
$\psi_{PM} = 186 \times \sqrt{2}/(200\pi)$ Vs = 0.418 Vs, 0.92 pu	$\psi_{sN} = 202 \times \sqrt{2}/(200\pi)$Vs = 0.455 Vs, 1 pu

EXAMPLE 9.7: Produce a vector diagram based on the Table 9.1 nameplate values with the d- and q-axis currents equal to 0.11 and 0.994, respectively. Note that stator resistance is 3% pu. The solution can be simplified by assuming the rated voltage is reduced, therefore, by 3%.

SOLUTION: Figure 9.20 is the space-vector diagram.

$$\psi_s = \sqrt{(\psi_{PM} - L_d i_d)^2 + (L_q i_q)^2} = \sqrt{(0.92 - 0.401 \cdot 0.11)^2 + (0.415 \cdot 0.994)^2} = 0.97$$

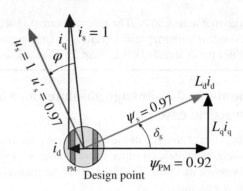

Figure 9.20 Space-vector diagram corresponding to $U_{\text{sph}} = 202$ V, 1 pu; $E_{\text{PMph}} = 186$ V, 0.92 pu; $I_s = 115$ A, 1 pu; $\cos\varphi = 0.95$; and $\delta_s = 24.5°$. The d- and q-axis currents are 0.11 and 0.994 respectively. The rotor symbol indicates embedded magnets. $T_{\text{em}} = 0.92$.

This corresponds well with the presumption of a 3 % reduced voltage.

Stator pu current for the given d-axis and q-axis current components is

$$i_s = \sqrt{i_d^2 + i_q^2} = \sqrt{0.11^2 + 0.994^2} = 1.0$$

The rated impedance as a base value is calculated as follows.

$$Z_N = \frac{U_{\text{sNph}}}{I_{\text{sNph}}} = \frac{202}{115} = 1.756 \ \Omega$$

The rated inductance is

$$L_N = \frac{\psi_{\text{sN}}}{\sqrt{2} I_N} = \frac{0.4546}{\sqrt{2} \cdot 115} = 2.795 \ \text{mH}$$

The pu inductance values in d- and q-axis are as follows.

$$L_{\text{d,pu}} = \frac{L_d}{L_N} = \frac{1.12 \ \text{mH}}{2.795 \ \text{mH}} = 0.415$$

$$L_{\text{q,pu}} = \frac{L_q}{L_N} = \frac{1.16 \ \text{mH}}{2.795 \ \text{mH}} = 0.415$$

The load angle is calculated trigonometrically.

$$\cos \delta_s = \frac{(\psi_{\text{PM}} - L_d i_d)}{\psi_s} = \frac{(0.92 - 0.11 \cdot 0.401)}{0.97} = 0.9 \rightarrow \delta_s = 25.4°$$

The voltage vector u_s is from ψ_{PM} under angle $\delta_s + 90° = 25.4 + 90 = 115.4°$ and the current vector under angle $90° + \cos^{-1} i_q/i_s = 90° + \cos^{-1} 0.994 = 90° + 6.3° = 96.3°$.

The phase angle between the voltage and current and the corresponding power factor are

$$\varphi_s = 115.4 - 96.3 = 19.1 \rightarrow \cos\varphi_s = 0.94$$

The machine is designed to operate about at $i_d = 0$ control. Therefore, the internal induced voltage must be slightly low (0.92 pu). Furthermore, the designer has included some voltage reserve for the machine drive. A converter connected to a 400 V network easily supplies a phase voltage fundamental of 220 V RMS, which corresponds to 1.09 V pu with the base voltage fixed to the motor rated voltage in this case. The base voltage could also be selected to be the converter maximum voltage when $U_b = 230$ V corresponding to 1 pu. In that case, the 202 V rated voltage should correspond to 0.88 pu.

EXAMPLE 9.8: Produce a vector diagram based on the Table 9.1 nameplate values if $i_d = 0$ and $i_q = i_s = 1$ at the rated frequency 100 Hz, 1 pu, $E_{PMph} = 186$ V, $I_s = 115$ A, 1 pu, and $L_q = 0.415$ pu. The stator resistive voltage drop can be ignored.

SOLUTION: Figure 9.21 is the space-vector diagram.

The pu inductances of the motor are rather low, which is a good result. Supply voltage is higher than in the previous example. The stator flux-linkage reference in the DTC has been raised to the operating point shown in the figure. The power factor is smaller, and the voltage reserve is very small, which is typical of $i_d = 0$ control.

$$\psi_{PM} = \frac{E_{PMph}}{2\pi f_N} = \frac{186 \cdot \sqrt{2}}{2\pi \cdot 100} = \frac{263.04}{628.31} = 0.418 \text{ Vs and } 0.92 \text{ pu}$$

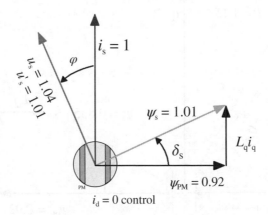

Figure 9.21 Space-vector diagram according to the $i_d = 0$ control at a rated torque and a rated frequency. $U_{sph} = 210$ V, 1.04 pu when the resistive voltage drop is taken into account. $E_{PMph} = 186$ V, 0.92 pu; $I_s = 115$ A, 1 pu; $\cos\varphi = 0.91$; $\delta_s = 24.4°$; and $T_e = 0.92$ pu.

$$\psi_s = \sqrt{\psi_{PM}^2 + (L_s i_s)^2} = \sqrt{0.92 + (0.415 \cdot 1)^2} = 1.01 \text{ pu}$$

$$\cos \delta_s = \frac{\psi_{PM}}{\psi_s} = \frac{0.92}{1.01} = 0.91 \rightarrow \cos^{-1} 0.91 = \delta_s = 24.4°$$

The power factor $\cos \varphi$ is calculated from the base of phase angle φ, which is now identical to the load angle, because the current space vector is perpendicular to the flux linkage ψ_{PM} and the voltage space vector is perpendicular to ψ_s. Referring to the space-vector diagram, the power factor is

$$\cos \varphi = 0.91$$

The electromagnetic pu power and torque are as follows.

$$P_e = u_s i_s \cos \varphi = 1 \cdot 1 \cdot 0.91 = 0.91 \text{ pu}$$

$$T_e = \frac{u_s i_s \cos \varphi}{\omega_s} = \frac{1.0 \cdot 1 \cdot 0.91}{1} = 0.91 \text{ pu}$$

The same motor can be driven using a unity power factor. Figure 9.22 shows the resulting vector diagram. The current vector points in the negative d-axis direction resulting in smaller stator flux linkage, because of the armature reaction. Moreover, the voltage level is lower. The figure clearly indicates that to obtain an appropriate stator current at the rated torque, the flux

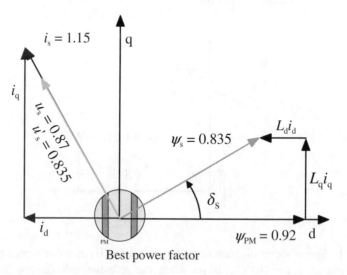

Best power factor

Figure 9.22 Vector diagram according to the $\cos \varphi = 1$ control at a rated torque and a rated frequency. $U_{sph} = 169$ V, $E_{PMph} = 186$ V, $I_s = 132$ A, $\cos \varphi = 1$, and $\delta_s = 30.3°$. The d- and q-axis currents are 0.57 and 0.99, respectively. $T_e = 0.92$ pu.

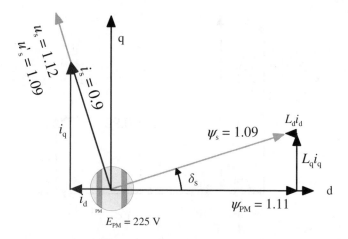

Figure 9.23 Vector diagram as a result of increased magnet material or greater coercive forces. The machine is redesigned to have $E_{PMph} = 225$ V. It operates at rated torque and frequency. $U_{sph} = 220$ V; 1.11 pu; $I_s = 103$ A, 0.9 pu; $\cos\varphi = 1$; and $\delta_s = 17.6°$.

linkage produced by the PMs should be higher. Here, the supply voltage is low, and as a result, the actual current demand exceeds the rated value by a considerable margin. There is plenty of voltage reserve.

Because machine saliency is low in this case, torque is driven mainly by the q-axis current ($T_{PM} = i_q \times \psi_{PM} = 0.99 \times 0.92 = 0.91$ pu), and the reluctance torque, expressed as $T_{rel} = (L_{md} - L_{mq})i_d i_q$ or $T_{rel} = (0.401 - 0.415) \times (-0.57) \times 0.99 \approx 0.01$, is insignificant.

With increased PM flux linkage, the motor could be made to operate with a low voltage reserve and high power factor at its nominal operating point. However, operating the motor with a small load or at no load at the rated speed would prove problematic. In this case, $E_{PM} = 225$ V. At small loads, demagnetization would be necessary to ensure sufficient voltage reserve.

Figure 9.23 illustrates a machine at its nominal operating point that has been designed using a different method. In practice, this machine requires more magnetic material or magnets with higher coercive force. A design of this kind is particularly adapted for machines that do not operate in field weakening and do not demand dynamic performance. Designing a high-power high-efficiency blower using this method would be appropriate, for example.

Figure 9.24 depicts the drive of the original motor, in which the stator flux linkage and the flux linkage of the PMs are regulated to be of equal magnitude in the DTC supply. In general, this is a good compromise, if good dynamic performance is required.

The machine in this example has significant voltage reserve, the power factor is good, and the motor current is slightly higher than rated.

Overall, these examples reveal how important stator flux linkage is to PMSM drive performance. The controller in a DTC application must pay special attention to the selection of the stator flux-linkage reference value. In steady state, a DTC converter may well find the MTPA operation by adjusting the stator flux-linkage reference to a value where the current is minimized at a given power level.

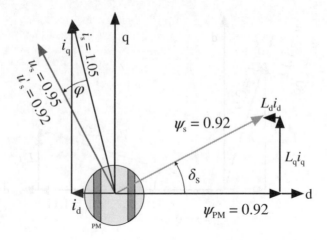

Figure 9.24 Control method, in which $\psi_{s,ref} = \psi_{PM}$ at a rated torque and at a rated frequency. $U_{sph} = 193$ V, $E_{PMph} = 186$ V, $I_s = 121$ A, $\cos\varphi = 0.975$, and $\delta_s = 27.4°$.

EXAMPLE 9.9: Consider a motor drive with respect to field weakening. The drive makes use of a 400 V frequency converter with $I_N = 147$ A. The maximum output phase voltage is 230 V. For DC-link protection, $U_{DCmax} = 730$ V. Observe the field weakening of the example motor.

SOLUTION: The upper safe operating limit for field weakening can be approximated based on the maximum DC-link voltage that occurs when the rotor runs freely at its highest speed.

The maximum line-to-line voltage tolerated by the DC link is

$$U_{max} \approx \frac{U_{DC}}{\sqrt{2}} = 516 \text{ V}$$

This corresponds to the following phase voltage.

$$U_{maxph} \approx \frac{U_{DC}}{\sqrt{2}\sqrt{3}} = 298 \text{ V}$$

Because the rated phase voltage of the machine is 186 V and the rated frequency is 100 Hz, the maximum theoretical safe field-weakening-point frequency f_{fwp} of the machine is

$$f_{fwp,limit} = \frac{730}{186 \cdot \sqrt{3} \cdot \sqrt{2}} \cdot 100 \text{ Hz} = 160 \text{ Hz}$$

At 160 Hz, the no-load voltage is approximately $E_{PMph} = 298$ V, which when rectified by the inverter freewheeling diodes suffices to produce 730 V in the DC link if the control-produced d-axis demagnetizing current disappears for some reason.

On the other hand, if the inverter is able to operate in field weakening, and if the motor can mechanically tolerate the high speeds, the drive can be run even faster than 160 Hz. If the rated current of the motor is used to demagnetize the motor, its flux can be reduced by i_d $L_d = 1 \times 0.401$ pu $= 0.401$ pu. Therefore, 60 % of the original PM produced flux remains, and the frequency can be raised to 1.67 times the rated frequency (a 1 : 0.6 ratio). Since E_{PM} at the rated frequency (100 Hz) is only 186 V, even a higher frequency can be reached by raising the voltage from the no-load voltage of 186 V to 230 V, in which case the frequency is $230/186 \times 167\ \text{Hz} = 206\ \text{Hz}$.

If the maximum converter current (1.28 pu) is used to demagnetize the motor at a full 230 V phase voltage, the maximum frequency of the motor in field weakening becomes $206\ \text{Hz} \times I_{Ninv}/I_{Nmotor} = 206\ \text{Hz} \times 147\ \text{A}/115\ \text{A} = 264\ \text{Hz}$. At this frequency, the stator flux linkage would be $\psi_{PM} - i_d L_d = 0.418\ \text{Vs} - 147\sqrt{2}\ \text{A} \times 1.12\ \text{mH} = 0.185\ \text{Vs}$ (41 % of the stator flux linkage at the nominal operating point). If for some reason, the demagnetizing current of the inverter should disappear at this operating point, the stator flux linkage would increase to 0.418 Vs, a value determined by the PMs. Correspondingly, the terminal phase RMS voltage of the motor would increase to $264\ \text{Hz} \times 2\pi \times 0.418/\sqrt{2} = 490\ \text{V}$. The line-to-line voltage should be 849 V, and the peak voltage of the intermediate DC link should be 1200 V, which would inevitably destroy the capacitors.

A 230 V stator phase voltage can be produced by the inverter, so running at no load with no voltage reserve, the maximum practical field weakening for the subject machine is $f_{fw} = 230/186 \cdot 100\ \text{Hz} = 123\ \text{Hz}$.

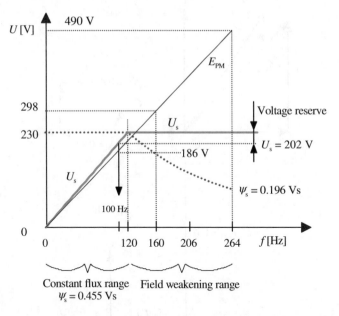

Figure 9.25 Motor voltage and back emf at different frequency points; 160 Hz corresponds to the induced voltage generated by the motor $E_{PM} = 298\ \text{V}$, which is the maximum voltage the DC-link capacitor can take without failing if the motor control malfunctions at 160 Hz. This is the maximum safe field weakening speed. If the rated current of the motor is used to demagnetize the machine, 206 Hz can be reached for no-load operation. If all the rated current available from the inverter is used to demagnetize the machine, the no-load operating point can be increased to 264 Hz. At that point, the internal back emf is $E_{PM} = 490\ \text{V}$, the stator voltage is 230 V, and the stator flux linkage is 0.196 Vs ($0.196\ \text{Vs} \times 2\pi 264\sqrt{3}/\sqrt{2}/\text{s} = 400$ V), which corresponds to the converter maximum output line-to-line voltage.

In field weakening, some and, in some cases finally, all of the stator current is used to reduce stator flux linkage. Consequently, in the constant flux range, more stator current is needed for field weakening than to produce torque. Figure 9.25 shows how different voltages behave for this motor at different supply frequency ranges.

Figure 9.25 shows that field weakening in a PM machine is a complex matter. The major problem is the fixed flux linkage ψ_{PM}, which can quickly lead to overvoltage risk to the inverter in field weakening. In this example case, frequency can be raised from the rated value of 100 Hz to 160 Hz without risking the inverter, because of the machine's rather low no-load voltage. If the no-load voltage is closer to the maximum voltage supplied by the inverter, failure risk moves closer to the nominal operating point.

Figure 9.26 offers a space-vector diagram that represents operation at a safe upper limit of field weakening (160 Hz) when the inverter operates at its own rated current and the stator current corresponds to a stator voltage of 230 V. The stator flux linkage is $\psi_s = 0.323$ Vs, 0.71 pu.

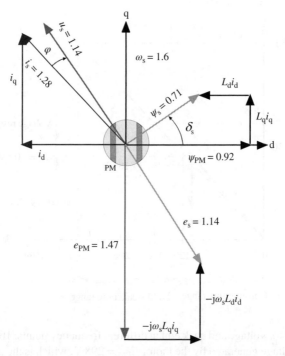

Figure 9.26 High-power operating point of the motor at the inverter rated current for 160% rated speed. $U_{sph} = 230$ V, 1.14 pu; $E_{PMph} = 298$ V, 1.47 pu; $I_s = 147$ A, 1.28 pu; $\cos\varphi = 0.987$; $\delta_s = 34°$; $P = 97$ kW, 1.39 pu; $f = 160$ Hz, 1.6 pu; $T = 0.915$; and $T_N = 0.84$ pu. Therefore, the motor runs close to its rated torque even at 60% overspeed, producing much more than its rated power. The rated current of the inverter, which is higher than the rated current of the motor, and the slightly low PM flux linkage of the motor makes this possible.

EXAMPLE 9.10: Prepare a space-vector diagram of a PMSM using the parameters and rated values given in Table 9.1, but with the frequency increased to 160 Hz, $U_{sph} = 230$ V, $E_{PMph} = 298$ V, converter maximum current $I_{smax} = 147$ A, and 1.28 pu as shown in Figure 9.25. Calculate the current components, load angle, power factor, power, and electromagnetic torque.

SOLUTION: Figure 9.26 showed the relationships between voltage, current, and flux-linkage vectors. The pu values can be calculated as follows.

$$u_s = \frac{230}{202} = 1.138$$

$$e_{PM} = \frac{298}{202} = 1.47$$

$$\psi_s = \frac{230 \cdot \sqrt{2}V}{2\pi \cdot 160\frac{1}{s}} = 0.323 Vs \rightarrow \psi_{s,pu} = \frac{0.323}{0.455} = 0.71$$

$$\psi_{PM} = \frac{298 \cdot \sqrt{2}V}{2\pi \cdot 160\frac{1}{s}} = 0.419 Vs \rightarrow \psi_{PM,pu} = \frac{0.419}{0.455} = 0.92$$

The current components are solved based on the space-vector diagram.

$$\psi_s^2 = (\psi_{PM} - L_d i_d)^2 + (L_q i_q)^2 = (0.92 - 0.401 \cdot i_d)^2 + (0.415 \cdot i_q)^2 = 0.71^2$$

where

$$i_q^2 = i_s^2 - i_d^2 = 1.278^2 - i_d^2$$

Solved this equation returns the i_d and i_q values.

$$i_d = 0.86$$
$$i_q = 0.945$$

The load angle is calculated as before.

$$\cos \delta_s = \frac{(\psi_{PMpu} - L_{dpu} i_d)}{\psi_{spu}} = \frac{0.92 - 0.401 \cdot 0.86}{0.71} = 0.821 \rightarrow \delta_s = 34°$$

or

$$\sin \delta_s = \frac{L_q i_q}{\psi_s} = \frac{0.415 \cdot 0.945}{0.71} = 0.56 \rightarrow \delta_s = 34°$$

The angle α between i_s and d-axis is given by

$$\sin \alpha = \frac{i_q}{i_s} = \frac{0.945}{1.278} = 0.738 \rightarrow \alpha = 47.7°$$

And, according the vector diagram

$$\alpha + \varphi + \delta_s = 90° \rightarrow \varphi = 90° - 47.7° - 34° = 8.316 \rightarrow \cos\varphi = 0.989$$

From this, it is possible to calculate the pu power and torque and the electromagnetic pu torque.

$$P_e = u_s i_s \cos\varphi = 1.14 \cdot 1.278 \cdot 0.989 = 1.44$$

$$T_e = \frac{u_s i_s \cos\varphi}{\omega_s} = \frac{1.44}{1.6} = 0.9$$

9.6 Torque estimation accuracy in a PMSM DTC drive

This book has considered that the stator flux-linkage space vector ψ_s can be estimated by integrating the stator voltage vectors u_s to stator flux-linkage estimate $\psi_{s,est} = \int(u_s - i_{s,meas}R_s)$. After that the torque T may be calculated by using the stator current vector $i_{s,meas}$ which is calculated based on measured phase currents. $T_{s,est} = \frac{3}{2}p(\psi_{s,est} \times i_{s,meas})$.

This assumption can be checked by reviewing the practical accuracy of the converter estimation. Figure 9.27 shows measured torque accuracy for a PM machine DTC drive based on ABB ACS 800 hardware. In the measurements, a high accuracy torque sensor was used to measure the motor output torque and this value was compared to the torque estimate given by the converter. According to this measurement result the torque estimate accuracy seems good. Only at the lowest speeds the estimate is somewhat more erroneous. This results from the

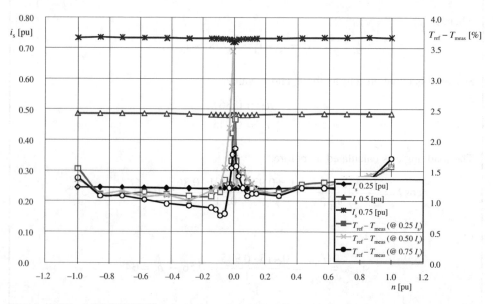

Figure 9.27 ACS 800 DTC converter torque estimation error in a PMSM drive measured at different torque and speed levels at LUT. *Source:* Minav et al. (2011). Reproduced by permission of the Institution of Engineering & Technology.

large proportion of the switch conduction losses when operating at the lowest voltages leading to somewhat erroneous stator flux-linkage estimate.

9.7 Speed and position sensorless control methods for PM machines

Induction motors can easily be run without rotor position encoders. Naturally, there is also a demand in less demanding applications to drive PM machines without position encoders. Position encoders are normally used with servomotors, but sensorless operation is favoured for high-power industrial drives. Several control methods have been introduced. They are based on calculating flux linkage, estimating machine state, using Kalman filters, or obtaining inductance variation information. Rotor angle estimates are needed for vector control methods, because the current references cannot be calculated without knowledge of rotor position.

DTC is inherently sensorless, because with DTCs there is, in principle, no need to calculate the electric current model. The electric current model is used only in demanding applications to stabilize and refine control. Furthermore, DTC is independent of rotating-field machine type. The methods developed by Niemelä (1999) and Luukko (2000) are well suited to the implementation of sensorless PMSM drives. Several of the proposed algorithms apply different methods depending of speed.

Initial rotor angle must always be determined for sensorless control of a PM machine. To accomplish this, Luukko (2000) uses the two-level converter active vectors as very short pulses to measure PMSM inductance in the directions of the voltage vectors. These inductance components are then fit to an inductance curve and compared to measured data to find initial rotor position. For a salient pole machine, there is a clear inductance difference in the d- and q-axis directions. Whether the machine is configured with magnets in the higher permeance direction must be known in advance. If the machine is a rotor surface magnet or salient-pole type with embedded magnets, because of saturation, the positive d-axis will be in the direction of the lowest inductance. Figure 9.28 illustrates the difference between positive and negative d-axis inductances based on the preexcitation caused by the PMs at $i_d = 0$.

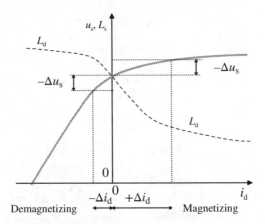

Figure 9.28 No-load curve and schematic behaviour of the stator d-axis inductance. The positive d-axis has the lowest inductance while the negative d-axis direction has the largest inductance. The q-axis inductance is somewhere in the middle. This phenomenon can be used to find initial rotor position in rotor surface magnet PMSMs.

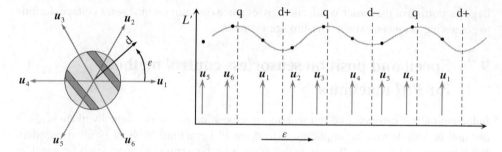

Figure 9.29 Finding the rotor initial position of a traditional PMSM. Voltage vectors are applied with minimum pulses to measure the transient inductances L' (dots), and curve fitting reveals the position of minimum inductance at the rotor positive d-axis.

Figure 9.29 illustrates testing a machine with six active voltage vector pulses.

Using the method of Figure 9.29, it is easy to identify the rotor position of a spinning rotor by applying two short successive short circuits, selecting zero voltage vectors, and then measuring the short-circuit current produced by the original back emf. The spinning rotor induces a voltage that is 90 electrical degrees behind ψ_{PM}, and the short-circuit current starts growing in the direction of the induced voltage vector. Using two successive pulses, rotor speed and ψ_{PM} position can be found, and the drive can be started. See Figure 9.30.

When a short circuit is triggered by a zero voltage vector at time t_1, the following current develops in the direction of e_{PM1} ($e_{PM1} = \psi_{PM}\omega_r e^{-j\pi/2}$).

$$i_1 \approx \int_t \frac{1}{L} e_{PM1} dt = \int_t \frac{1}{L} \psi_{PM1}\omega_r e^{-j\pi/2} dt \tag{9.71}$$

If test time t is sufficiently short, the vector direction of i_1 for an arbitrary angle θ_1 can be obtained.

$$i_1 = ie^{j\theta_1} \tag{9.72}$$

Following the second short circuit, i_2 can be determined.

$$i_2 = ie^{j\theta_2} \tag{9.73}$$

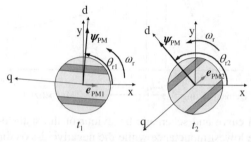

Figure 9.30 Sensorless start-up of a rotating PMSM. The rotation angle, ψ_{PM}, and e_{PM} are described at time instants t_1 and t_2.

The rotor electrical angular velocity is

$$\omega_r = \frac{\vartheta_2 - \vartheta_1}{t_2 - t_1} \tag{9.74}$$

Rotor position angle is as follows.

$$\theta_r = \vartheta_2 + \frac{\pi}{2} + \Delta t \omega_r \tag{9.75}$$

Here Δt is the time following the second short circuit that is needed for signal processing.

With information on rotor d-axis position, the motor can start producing torque when suitable current is fed on the q-axis. For a DTC inverter, the initial stator flux linkage will be $\psi_s = \psi_{PM}$ and modulation can begin with a voltage vector producing suitable torque and a suitable change to ψ_s according to DTC principles.

Several sensorless methods use scalar control to spin the rotor up to a speed sufficient for rotor angle detection using the state estimator. Currently, there is no general way of determining initial rotor angle that would encompass all PM machine types. Methods based on the detection and identification of the difference between the d-axis and q-axis inductance components were introduced, for example, by Schroedl (1990) and Östlund and Brokemper (1996).

As described, the inductance difference methods may also be used in nonsalient pole machines, where PM-induced saturation can be detected as a difference in inductance.

Ertugrul and Acarnley (1994) introduced a method for nonsalient pole PMSMs, which was presented in simplified form by Östlund and Brokemper (1996). In the method, stator flux linkage is estimated using the equation $\psi_{s,est} = \int (u_{s,est} - i_{s,meas} R_s) dt$. As a function of the integration time Δt, discrete flux linkage can be expressed in terms of xy components.

$$\psi_{sx\ est}(k) = \Delta t[u_{sx}(k) - R_s i_{sx}(k)] + \psi_{sx\ est}(k-1) \tag{9.76}$$

$$\psi_{sy\ est}(k) = \Delta t\left[u_{sy}(k) - R_s i_{sy}(k)\right] + \psi_{sy\ est}(k-1) \tag{9.77}$$

Integration is improved by comparing the stator current estimates $i_{sx\ est}$ and $i_{sy\ est}$ with the measured currents i_{sx} and i_{sy}.

$$\Delta i_{sx} = i_{sx} - i_{sx\ est} \tag{9.78}$$

$$\Delta i_{sy} = i_{sy} - i_{sy\ est} \tag{9.79}$$

Stator flux linkage is determined in terms of rotor angle θ_r as follows.

$$\psi_{sx} = L_s i_{sx} + \psi_{PM}\cos \theta_r \tag{9.80}$$

$$\psi_{sy} = L_s i_{sy} + \psi_{PM}\sin \theta_r \tag{9.81}$$

This division into components is illustrated in Figure 9.31.

Next, stator current estimates can be computed from the rotor angle estimate $\theta_{r\ est}(k)$.

$$i_{sx\ est} = \frac{1}{L_s}\left[\psi_{sx\ est} - \psi_{PM}\cos \theta_{r\ est}(k)\right] \tag{9.82}$$

$$i_{sy\ est} = \frac{1}{L_s}\left[\psi_{sy\ est} - \psi_{PM}\sin \theta_{r\ est}(k)\right] \tag{9.83}$$

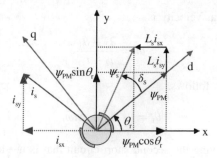

Figure 9.31 The components of Equations (9.80) and (9.81).

The direct and quadrature inductances are assumed to be equal $L_d = L_q = L_s$. If inductance is assumed constant, flux linkage becomes a function of current and rotor angle. Therefore, a total differential for the flux linkage can be written.

$$\Delta \psi_s = \left(\frac{\partial \psi}{\partial i}\right)\Delta i + \left(\frac{\partial \psi}{\partial \theta_r}\right)\Delta\theta_r \tag{9.84}$$

The rotor angle estimate $\theta_{r\,est}$ is corrected by an amount equal to $\Delta\theta_{r\,est}$ until the flux-linkage error $\Delta\psi_s$ becomes zero, and $\Delta\theta_{r\,est}$ can be calculated as follows.

$$\Delta\theta_{r\,est} = \Delta i\left(\frac{\partial \psi}{\partial i} \Big/ \frac{\partial \psi}{\partial \theta_r}\right) \tag{9.85}$$

The corrected rotor angle estimate is now

$$\theta_{cr\,est}(k) = \theta_{r\,est} + \Delta\theta_{r\,est} \tag{9.86}$$

The correction terms in xy components can be determined from the partial derivatives.

$$\Delta\theta_{xr\,est} = \Delta i_{sx}\frac{L_s}{\psi_{PM}\sin\theta_{r\,est}} \tag{9.87}$$

$$\Delta\theta_{yr\,est} = -\Delta i_{sy}\frac{L_s}{\psi_{PM}\cos\theta_{r\,est}} \tag{9.88}$$

The drawback of these equations is possibility of zeros in the denominators. Observation of the error and correction terms in an estimated rotor reference frame provides a solution to this problem. The correction term of the rotor angle estimate becomes

$$\Delta\theta_{r\,est} = \Delta i_q\frac{L_s}{\psi_{PM}} \tag{9.89}$$

Finally, new values are computed for the flux linkage with the updated rotor angle estimate

$$\psi_{sx} = L_s i_{sx} + \psi_{PM}\cos\theta_{cr\,est}(k) \tag{9.90}$$

$$\psi_{sy} = L_s i_{sy} + \psi_{PM}\sin\theta_{cr\,est}(k) \tag{9.91}$$

Table 9.2 Comparison of the control methods for a PM machine: 1 − low, 5 − high

Method	$i_d = 0$	MTPA	Position Sensorless DTC	DTC with a Position Sensor	Position Sensorless VC
Parameters required	$\psi_{PM}, (L_q)$	L_d, L_q, ψ_{PM}	R_s	L_d, L_q, ψ_{PM}, R_s	L_d, L_q, ψ_{PM}, R_s
Parameter sensitivity	1	2	3	2	5
Required computation capacity	1	2	3	4	5
Possibility for field weakening	No	No/yes	Yes	Yes	Depends on method
Applicability to different machine types	Dedicated for nonsalient pole machines, works with small L_d. Works nonideally with salient pole machines	$L_d < L_q$	No restrictions	No restrictions	Depends on method In signal injection large inductance difference $L_d - L_q$ helps
Other			Initial rotor position must be found before starting		Initial rotor position must be found before starting

According to Östlund and Brokemper (1996), the algorithm functions well both dynamically and in steady state from approximately 3 Hz to rated speed. However, like several other estimation algorithms, this method is parameter sensitive. It is particularly sensitive to stator resistance estimate effects. In addition, the saturation of inductances must be taken into account. Another drawback of the method is its assumption that direct and quadrature axis inductances are equal.

The Table 9.2 provides a brief comparison of the control methods for PMSMs. The application of field weakening depends not only on method, but also on the characteristics of the electrical machine. Only the $i_d = 0$ control, by definition, eliminates field weakening.

References

Aydin, M., Huang, S., & Lipo, T. A. (2010). Design, analysis and control of a hybrid field-controlled axial-flux permanent-magnet motor. *IEEE Trans. Ind.* Vol. 57(1), 78–87.

Consoli A., Musumeci S., Raciti A., & Testa A. A. (1994). Sensorless vector and speed control of brushless motor drives. *IEEE Transactions on Industrial Electronics*, IE-41(1), 91–96.

Chy, M. I., & Uddin, N. (2007). Analysis of flux control for wide speed range operation of IPMSM drive. IEEE, Ontario, Canada, pp. 256–260.

Dhaouadi R., & Shigyo M. (1991). Design and implementation of an extended Kalman filter for the state estimation of a PM synchronous motor. *IEEE Transactions on Power Electronics*, PE-6(1), 491–497.

Ertugrul N., & Acarnley P. (1994). A new algorithm for sensorless operation of PM motors. *IEEE Transactions on Industry Applications*, IA-30(1), 126–133.

Hall, E., & Balda, J. C. (2002). Permanent magnet synchronous motor drive for HEV propulsion: Optimum speed ratio and parameter determination. *IEEE*, pp. 1500–1504.

Haque, M. E., Zhong, L., & Rahman, M. F. (2002). Improved trajectory for an interior permanent magnet synchronous motor drive with extended operating limit. *Journal of Electrical and Electronic Engineering*, 22(1), 49–57.

Ilioudis, V. C., & Margaris, N. I. (2010). Flux weakening method for sensorless PMSM control using torque decoupling technique. Conf. Rec. IEEE, Thessaloniki, Greece, July 9–10, pp. 32–39.

Jahns T. M., Kliman G. B., & Neumann T. W. (1986). Interior permanent-magnet synchronous motors for adjustable-speed drives. *IEEE Transactions on Industry Application*, IA-22(4), 738–747.

Kaukonen, J. (1999). Salient pole synchronous machine modelling in an industrial direct torque controlled drive application. Dissertation. Lappeenranta University of Technology. ISBN 951-764-305-5.

Kim J.-M., & Sul S.-K. (1997a). Speed control of interior PM synchronous motor drive for the flux weakening operation. *IEEE Transactions on Industry Applications*, IA-33(1), 43–48.

Kim J.-S., & Sul S.-K. (1997b). New approach for high-performance PMSM drives without rotational position sensors. *IEEE Transactions on Power Electronics*, PE-12(5), 904–911.

Kulkarni A. B., & Ehsani M. (1992). A novel position sensor elimination technique for the interior permanent-magnet synchronous motor drive. *IEEE Transactions on Industry Applications*, IA-28(1), 144–150.

Lagerquist R., Boldea I., & Miller T. J. E. (1994). Sensorless control of the synchronous reluctance motor. *IEEE Transactions on Industry Applications*, IA-30, 673–682.

Lindh, P., Rilla, M., Jussila, H., Nerg, J., Tapia-Ladino, J. A., & Pyrhönen, J. (2011). Interior PM motors for traction application with nonoverlapping concentrated windings and with integer slot windings. *IREE*, 6(4).

Luise, F., Odorico, A., & Tessarolo, A. (2010). Extending the speed range of surface permanent-magnet axial-flux motors by flux-weakening characteristic modification. Conf. Rec. ICEM, Rome, Italy, September 6–8.

Luukko J. (2000). Direct torque control of PM synchronous machines – analysis and implementation. Acta Universitatis Lappeenrantaensis. Dissertation, Lappeenranta University of Technology.

Luukko J., Kaukonen J., Niemelä M., Pyrhönen O., Pyrhönen J., Tiitinen P., & Väänänen, J. (1997). Permanent magnet synchronous motor drive based on direct flux linkage control. *Proceedings of the 7th European Conference on Power Electronics and Applications*, 3, 683–688.

Meyer, M., Grote, T., & Böcker, J. (2007). *Direct torque control for interior permanent magnet synchronous motors with respect to optimal efficiency. Power electronic and applications*. Aalborg, Denmark, pp. 1–9. ISBN: 978-92-75815-10-8.

Minav, T., Immonen P., Laurila, L., Vtorov, V., Pyrhönen, J., & Niemelä, M. (2011). Electric energy recovery system for a hydraulic forklift – theoretical and experimental evaluation. *IET Electr. Power Appl.*, 5(4), 377–385. doi: 10.1049/iet-epa.2009.0302

Mohan, N., Undeland, T. M., & Robbins, W. P. (2003). *Power electronics: Converters, applications and design* (3rd. ed.). Chichester, UK: John Wiley & Sons, Ltd. ISBN: 0-471-22693-9.

Moynihan, J. F., Egan, M. G., & Murphy, J. M. D. (1994). The application of state observers in current regulated PM synchronous drives. *IEEE IECON*, 20–25.

Niemelä, M. (1998). Position sensorless electrically excited synchronous motor drive for industrial use, based on direct flux linkage and torque control. Acta Universitatis Lappeenrantaensis. Dissertation, Lappeenranta University of Technology.

Östlund, S., & Brokemper, M. (1996). Sensorless rotor-position detection from zero to rated speed for a PM synchronous motor drive. *IEEE Transactions on Industry Applications*, IA-32, 1158–1165.

Pyrhönen, J., Jokinen, T., & Hrabovcova, V. (2014). *Design of rotating electrical machines*. Chichester, UK: John Wiley & Sons, Ltd. ISBN: 978-1-118-58157-5.

Ruoho, S. (2011). *Modeling demagnetization of sintered NdFeB magnet material in time-discretized finite element analysis*. Aalto University publication series. ISBN 978-952-60-4000-4.

Schroedl, M. (1990). Operation of the PM synchronous machine without a mechanical sensor. Int. Conf. on Power Electronics and Variable Speed Drives, pp. 51–55.

Schiferl, R. F., & Lipo, T. A. (1990). Power capability of salient pole permanent magnet synchronous motors in variable speed drive applications. *IEEE Trans. Ind. Appl.* Vol. 26(1), 115–123.

Soong, W. L., Han, S., & Jahns, T. M. (2007a). Design of interior PM machines for field-weakening applications. Conf. Rec. ICEM, Seoul, Korea, October 8–11, pp. 654–664.

Soong, W. L., Reddy, P. B., El-Refaie, A. M., Jahns, T. M., & Ertugrul, N. (2007b). Surface PM machine parameter selection for wide field-weakening. Applications. *IEEE* 882–889.

Takahashi, I., & Noguchi, T. (1986). A new quick-response and high-efficiency control strategy of an induction motor. *IEEE Transactions on Industry Applications*, IA-22, 820–827.

Tiitinen, P., Pohjalainen, P., & Lalu, J. (1995). The next generation motor control method: Direct torque control DTC. *EPE Journal*, 5, 14–18.

Wu, R., & Slemon, G. R. (1990). A PM motor drive without a shaft sensor. Conference Record of IEEE IAS Annual Meeting, pp. 553–558.

Zhong, L., Rahman, M., Hu, W., & Lim, L. (1997). Analysis of direct torque control in PM synchronous motor drives. *IEEE Transactions on Power Electronics*, PE-12, 528–536.

Zhu, Z. Q., Li, Y., Howe, D., Bingham, C. M., & Stone, D. (2007). Influence of machine topology and cross-coupling magnetic saturation on rotor position estimation accuracy in extended back-emf based sensorless PM Brushless AC drives. Industry Applications Conference 2007, 42nd IAS Annual Meeting, Conference Record of the 2007 IEEE.

Zhu, Z. Q., Qi, G., Chen, J. T., Howe, D., Zhou, L. B., & Gu, C. L. (2009). Influence of skew and cross-coupling on flux-weakening performance of permanent-magnet Brushless AC machines. *IEEE Transactions on Magnetics*, 45(5), 2110–2117.

10

Synchronous reluctance machine drives

Traditionally, induction machines have been the least expensive industrial motor option. They continue to be the workhorse of the industry, and induction machine manufacturing has been fully automated for nearly a century. However, the slip inherent in an induction motor lowers its efficiency and makes it difficult to control speed without position feedback (see Chapter 11). As a result, other drive types with better characteristics, such as the synchronous reluctance motor drives, are gaining ground in the industrial drives arena.

The modern high-performance synchronous reluctance machine (SynRM) drive is the indcution machine's latest challenger. Especially at low power, the efficiency of a SynRM can be substantially higher than induction machine efficiency. In principle, the SynRM is the simplest rotating-field machine. It is typically equipped with a simple, salient-pole, laminated, unwound rotor. The simpler structure makes the machine type ideal for many purposes. Using a motor with four-pole construction, the SynRM drive can be more efficient than an induction machine drive. Moreover, a SynRM offers excellent speed control, easily establishing and maintaining synchronous speed without encoder feedback. Overall, the SynRM has the potential to overtake the popularity of the induction motor as a controlled drive.

Since the SynRM runs synchronously, rotor losses are minimized, which improves efficiency. However, the power factor of a SynRM drive is generally lower than that of a corresponding induction machine drive. Efficiency and power factor are directly proportional to saliency ratio, so the stator of a SynRM motor must use a distributed winding configuration. The concentrated (tooth-coil) windings that have become popular in PMSMs do not provide enough saliency. The high excitation current needed to magnetize the SynRM compromises overall performance, and as a result, SynRMs generally have lower power factors than IE3 induction motors.

A high-performance SynRM usually cannot operate without vector control, but a traditional low-performance SynRM with damper windings and a low inductance ratio

Electrical Machine Drives Control: An Introduction, First Edition. Juha Pyrhönen, Valéria Hrabovcová and R. Scott Semken.
© 2016 John Wiley & Sons, Ltd. Published 2016 by John Wiley & Sons, Ltd.

can operate direct online. Cage-equipped SynRM drives are often used in DOL applications where synchronous running is more important than overall performance. For example, several synchronously running DOL motors can be used to manage belt tensioning in long conveyers. Because of slip, which is the difference between synchronous and operating speeds, induction motors cannot be used for this purpose – the load sharing should become too uneven and speed differences should cause stretching of the conveyor.

The SynRM requires a cage winding for DOL startup. However, since the cage winding usually results in a lower saliency ratio, DOL SynRMs have lower power factors and machine efficiencies than do cageless rotor machines. In addition, the higher saliency ratio in a high-performance machine can adversely affect DOL startup, because the asymmetric damper winding, which has a half-speed synchronous operating point, may cause an incorrectly designed DOL machine to start at half-speed.

As with all electrical machines, a SynRM can also be operated as a generator. Reluctance generators are well adapted to low- and medium-power applications. Their low-loss rotor is a remarkable advantage when compared, for instance, with an asynchronous generator. Furthermore, their simpler brushless construction makes the SynRM generator a potential alternative in applications such as small-scale hydropower plants or wind power plants. As was the case for the SynRM motor drive, the efficiency and power factor of the SynRM generator are also directly proportional to its saliency ratio.

Similar methods are used to define the d- and q-axes when modelling a SynRM or a salient-pole synchronous machine. A different approach is used for the PMSM. In PMSMs, the d-axis always coincides with the PM flux axis regardless of the inductance ratio. For SynRMs (and salient-pole SMs), the highest inductance direction defines the d-axis position. Current angle is measured from the d-axis with minimum reluctance. Therefore, κ is used to symbolize the current angle instead of the γ used in PMSM modelling where the d-axis may have the maximum reluctance.

10.1 The operating principle and structure of a SynRM

A SynRM is a salient-pole rotating-field synchronous machine with a nonexcited rotor. Consequently, the RMS synchronous machine load angle equation can be simplified to the following form.

$$P = 3U_{\text{sph}}^2 \frac{L_\text{d} - L_\text{q}}{2\omega_s L_\text{d} L_\text{q}} \sin 2\delta_s = 3U_{\text{sph}}^2 \frac{1 - \dfrac{L_\text{q}}{L_\text{d}}}{2\omega_s L_\text{q}} \sin 2\delta_s \tag{10.1}$$

The equation suggests that $L_\text{d} - L_\text{q}$, the difference of the d-axis and q-axis synchronous inductances, should be as large as possible, and $L_\text{d}L_\text{q}$ should be as small as possible to produce maximum power. In practice, the target is to maximize the d-axis inductance and to minimize the q-axis synchronous inductance. In principle, the machine can yield its maximum torque at the load angle $\delta_s = 45°$. Saturation and other phenomena may cause apparent deviation from this value.

The corresponding current-based space-vector expression for absolute torque is of this familiar form.

$$T_\text{e} = \frac{3}{2} p (L_\text{d} - L_\text{q}) i_\text{d} i_\text{q} \tag{10.2}$$

Using κ to symbolize electric current angle, this equation takes the form shown previously in Chapter 5. However, Equation (5.48) was derived ignoring the possibility of having several pole pairs p. Taking this into account we get the following form.

$$T_e = \frac{3}{4}pi_s^2(L_d - L_q)\sin 2\kappa \tag{10.3}$$

As was the case with Equation (10.1), $L_d - L_q$ should be as large as possible to provide maximum torque.

Conventionally, a damper winding accompanies the rotor of a SynRM for direct on line start-up operation. The stator uses the same poly-phase (e.g., three-phase) winding as an ordinary rotating-field machine. A SynRM is, in principle, a salient-pole synchronous machine with no field winding current. Cutting away equal segments from opposite sides of an induction machine rotor is the simplest way to produce a SynRM drive. In practical applications however, the ferromagnetic rotor is shaped or laminated so the phase inductance variation with respect to the rotation of the rotor is as large as possible. The iron parts of the magnetic circuit of the SynRM are not allowed to saturate magnetically under normal operating conditions, since the target is to keep the operating range of the machine linear without sacrificing the inductance difference.

SynRM motor drives are often used as servo drives and in pumps, conveyors, and devices producing synthetic fibres. They are used in packaging and wrapping machines and are beginning to see use as drives in vehicles and robots. SynRM generator drives are being introduced in both wind power and mini-hydro power plants.

The rotor of a speed-controlled SynRM should be designed with as large a saliency ratio as possible. In other words, the d-axis inductance L_d should be maximized, and the q-axis inductance L_q should be minimized. This saliency ratio largely determines the characteristics of the SynRM drive, that is, what the peak torque of the machine is, how fast the machine responds to dynamic changes, and what power factor and efficiency can be reached. To be competitive with an induction motor of equal size, the saliency ratio of the SynRM must be at least in the range of 6 to 10.

The simplest rotor configuration for a SynRM is produced by removing teeth from the rotor of a conventional squirrel-cage induction motor as shown in Figure 10.1a. However, this geometry yields a relatively low saliency ratio ($L_d/L_q < 3$) making for a poor solution. Figure 10.1d shows the rotor of an ordinary salient-pole synchronous machine with the excitation windings removed. This rotor can also be manufactured of notched segments made of solid iron. Nevertheless, the saliency ratio for a SynRM motor drive with a conventional salient-pole rotor is too low for it to compete with an induction motor equipped with a similar stator. A typical saliency ratio for this type rotor has been determined to be 3 to 4 (Staton et al., 1993).

In the so-called single-layer flux-barrier rotors illustrated in Figures 10.1b and 10.1e, permanent magnets (PMs) can be mounted in the insulation spacer to keep the flux from progressing in the q-axis direction. Doing so improves machine characteristics and reduces the size of the inverter needed for the power supply. To do this, the magnet polarities must be chosen to oppose the q-axis armature reaction. This will be discussed further in Chapter 10.4. The rotor geometry illustrated in Figure 10.1e corresponds to that of a PM machine with internal rotor magnets. The Figure 10.1b geometry is a mixture of a salient pole and a PM excited rotor structure. The flux barrier in Figure 10.1b is usually made of a nonmagnetic

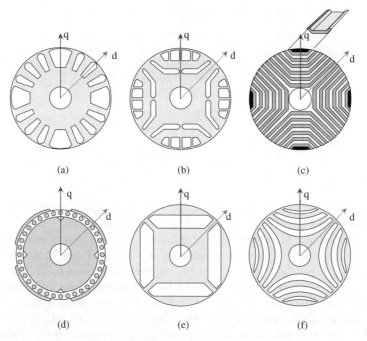

(a)   (b)   (c)

(d)   (e)   (f)

Figure 10.1 Four-pole rotors for a SynRM. The simplest construction (a) takes the rotor geometry of an induction motor and removes some teeth from the punched rotor laminations. Image (b) shows a laminated flux-barrier cage rotor, and (c) shows an axially laminated rotor with alternating U-shaped magnetic and nonmagnetic laminations fixed on the shaft by bolts and nonmagnetic wedges. Image (d) depicts a salient-pole rotor with its cage winding used also as a damper winding, (e) represents a cageless flux-barrier rotor, and (f) shows a punched flux-barrier rotor. For a synchronous solid rotor, it is important to configure the iron geometry to follow the magnetic flux lines as closely as possible. Rotors (a), (b), (d), (e), and (f) have round punched laminations. Rotor (c) uses laminations with axial gutter-like penetrations.

material such as aluminium, copper, slot insulation material, or plastic, or it can even be air. A SynRM equipped with the rotor of Figure 10.1b is suitable for direct on line start-p, since the bars on the rotor surface comprise a cage winding and the saliency ratio is low. Small bar cross sections are used for the cage windings to produce high rotor resistance during startup when machine startup torque is higher. The saliency ratio of this rotor type is in the range of 3 to 4.

The geometry shown in Figure 10.1c produces the largest saliency ratios. Its laminations are oriented axially, but are stacked in the radial direction. The L_d/L_q values for this rotor geometry are normally greater than 10 and may even exceed 15. Figure 10.1f shows a multilayer flux-barrier rotor. Its rotor saliency is lower, because of the rotor supporting steel parts in its round laminations. L_d/L_q values for the rotor geometry shown in Figure 10.1f are in the range of 6 to 8.

In a multilayer flux-barrier rotor, several curved sections are cut away from the rotor plate. When assembling the rotor stack, these sections can be filled with a desired nonmagnetic material via pressure casting. This way, the construction may become mechanically stronger than it would be if the sections were left unfilled. If a light material is used, rotor inertia can

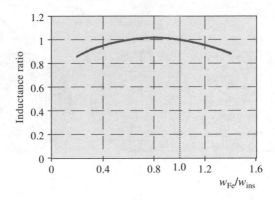

Figure 10.2 Normalized saliency ratio L_d/L_q for a low-power SynRM with the rotor geometry shown in Figure 10.1c. w_{ins} is the thickness of the insulation layer, and w_{Fe} is the thickness of the iron layer. The number of layers (electrical steel + insulator) is 24. The optimal inductance ratio is close to one resulting in 50 % steel and 50 % insulation. Flux density in the steel laminations is sensitive to w_{Fe} and, therefore, to iron losses. The w_{Fe} and w_{ins} thicknesses are key parameters for the number of the laminations in one stack of the rotor. One stack should occupy two-thirds of half of the pole pitch on the rotor periphery. This type of machine should be built using very high resistivity laminations to limit the iron losses caused by the heavy flux pulsation induced by the slot openings in the narrow electrical steel sheets.

also be kept low. The ratio of conductor and insulator layer thicknesses can significantly affect the saliency ratio of the machine.

Figure 10.2 shows how the saliency ratio in a small SynRM with the rotor geometry illustrated in Figure 10.1c depends on the ratio of the thicknesses of the iron and insulator layers w_{Fe}/w_{ins}. When the proportion of the iron layer w_{Fe} increases, the area of the cross section of the magnetic circuit of the d-axis flux increases, and therefore L_d increases. However, as the insulator layer becomes thinner, its preventive effect on the progression of the flux diminishes, and the quadrature inductance L_q increases. According the Figure 10.2, the optimum inductance ratio is approximately one. Beyond 10 layers, the total number of insulator and conductor layers does not significantly influence saliency ratio (Matsuo & Lipo, 1994; Staton et al., 1993).

In addition to rotor structural factors, the d- and q-axis inductances depend on the currents i_d and i_q, stator slotting, and the shape of the end windings. Accurately determining saliency ratio is a complex task. Often, it is necessary to resort to the estimation and determination of inductances using motor models.

10.2   Model, space-vector diagram, and basic characteristics of a SynRM

SynRM rotor design concentrates on maximizing the saliency ratio L_d/L_q and optimizing synchronous operation. In the absence of a damper winding, there will be, in principle, no rotor currents. The dq-model for a damper-less SynRM is simple with high saliency for transient inductances, which are then the same as synchronous inductances. The existence of a large inductance difference in a damper-less machine makes it possible, such as to efficiently define rotor position angle using signal injection. However, adding a damper winding makes the transient inductances of a SynRM more similar on both axes, which makes, for example, the dot-product–based stabilizing of the DTC drive more efficient.

A SynRM is similar to a nonexcited salient-pole machine. Because of their rotor geometries, both machines exhibit different machine characteristics in the d- and q-axis directions. The two-axis model equivalent circuit for a SynRM shown in Figure 10.3 is based on the corresponding model for a salient-pole synchronous machine. The SynRM circuit just leaves out the components representing the rotor excitation winding.

The voltage equations with supply angular velocity ω_s for the equivalent circuit of Figure 10.3 are written as follows.

$$u_d = R_s i_d + \frac{d\psi_d}{dt} - \omega_r \psi_q \tag{10.4}$$

$$u_q = R_s i_q + \frac{d\psi_q}{dt} + \omega_r \psi_d \tag{10.5}$$

The flux-linkage components are

$$\psi_d = L_d i_d = (L_{md} + L_{s\sigma})i_d + L_{md}i_D \tag{10.6}$$

$$\psi_q = L_q i_q = (L_{mq} + L_{s\sigma})i_q + L_{mq}i_Q \tag{10.7}$$

As for all rotating field machines, the stator flux linkage of a SynRM comprises the stator leakage flux linkage and the air-gap flux linkage.

$$\boldsymbol{\psi}_s = \boldsymbol{\psi}_{s\sigma} + \boldsymbol{\psi}_m \tag{10.8}$$

$\boldsymbol{\psi}_s$ is the stator flux linkage, $\boldsymbol{\psi}_{s\sigma}$ is the stator leakage flux linkage, and $\boldsymbol{\psi}_m$ is the air-gap flux linkage. The stator leakage flux-linkage vector can be expressed by the stator leakage inductance as follows.

$$\boldsymbol{\psi}_{s\sigma} = L_{s\sigma}\boldsymbol{i}_s \tag{10.9}$$

where $\boldsymbol{i}_s$ is the stator current space vector.

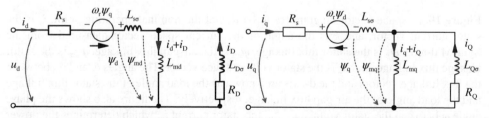

Figure 10.3 Equivalent d- and q-axis circuits of a SynRM with damper windings. The stator voltage and current components are indicated by the subscripts d and q. Damper windings are indicated by subscripts D and Q. The subscript σ refers to the leakage component. As usual, iron losses are neglected.

The magnitude of the space-vector cross product is proportional to the area of the parallelogram defined by these vectors. The torque produced by the machine is, therefore, proportional to the triangle delimited by the vectors ψ_s, ψ_m, and $\psi_{s\sigma}$ shown in Figure 10.4. In simplest terms, control is determining the optimum triangle for the required torque according to the specific control method. During a transient, control merely transitions from one triangle to another.

The load angle is also an indicator of torque. In a SynRM, load angle affects machine flux, and in steady state, the load angle producing the maximum torque is determined in terms of the stator flux linkage d- and q-axis components (not shown in Figure 10.4) as follows.

$$\psi_d = \psi_s \cos \delta_s = \psi_m \cos \delta_s + i_d L_{s\sigma} = i_d L_d \tag{10.10}$$

$$\psi_q = \psi_s \sin \delta_s = \psi_m \sin \delta_s + i_q L_{s\sigma} = i_q L_q \tag{10.11}$$

$$i_d = \frac{1}{L_d} \psi_s \cos \delta_s \tag{10.12}$$

$$i_q = \frac{1}{L_q} \psi_s \sin \delta_s \tag{10.13}$$

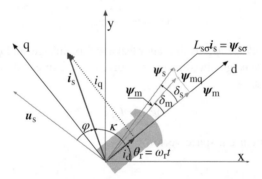

Figure 10.4 Space-vector diagram for a SynRM at the nominal operating point. i_s is the stator current vector, and κ is the angle of the current space vector measured from the SynRM d-axis. ψ_s is the stator flux linkage, ψ_m is the air-gap flux linkage, $\psi_{s\sigma}$ is the stator leakage flux linkage, and u_s is the stator voltage space vector. The angles δ_s and δ_m between the flux linkage vectors and the d-axis are known as the load angle of the stator flux linkage and the load angle of the air-gap flux linkage, respectively. The figure also shows the phase angle φ between the stator voltage u_s and the stator current i_s, which determines the power factor $\cos\varphi$, the rotor angle θ_r, and the xy reference frame fixed to the stator. The q-axis inductance must be very low to prevent the stator flux linkage and air-gap flux linkage from turning significantly away from the d-axis. $L_{md} = 3$, $L_{mq} = 0.2$, $L_{s\sigma} = 0.1$, $u_s = 1$, $i_s = 1$, $\omega_s = 1$, $\psi_s = 1$, $\psi_{md} = 0.95$, $\psi_{mq} = 0.18$, $\psi_{s\sigma} = 0.1$, $\cos\varphi = 0.82$, and $L_d/L_q = 3.1/0.3 = 10.3$.

Torque as a function of load angle can be calculated using space vectors.

$$
\begin{aligned}
T_e &= \frac{3}{2}p(\boldsymbol{\psi}_s \times \boldsymbol{i}_s) = \frac{3}{2}p\left(\psi_s \cos\delta_s \cdot \frac{1}{L_q}\psi_s \sin\delta_s - \psi_s \sin\delta_s \cdot \frac{1}{L_d}\psi_s \cos\delta_s\right) \\
&= \frac{3}{2}p\psi_s^2\left(\frac{1}{L_q} - \frac{1}{L_d}\right)\sin\delta_s \cos\delta_s \\
&= \frac{3}{2}p\psi_s^2\left(\frac{L_d - L_q}{L_q L_d}\right)\frac{1}{2}\sin 2\delta_s
\end{aligned}
\tag{10.14}
$$

Per-unit output power (with losses neglected) P_{out} can be expressed according to the following equation. Since space vectors, in principle, are defined only for two-pole machines, the converter does not usually "know" the number of poles unless it has speed feedback and can differentiate between mechanical speed Ω_r and electrical angular velocity ω_s.

$$
P_{out} = \omega_s T_e = u_s^2 \frac{L_d - L_q}{2 L_d L_q \omega_s}\sin 2\delta_s
\tag{10.15}
$$

when

$$
u_{s,pu} = \frac{\omega_s \psi_s}{U_b}
\tag{10.16}
$$

EXAMPLE 10.1: Draw a space-vector diagram of a SynRM, which has the following per-unit parameters. $L_{md} = 3$, $L_{mq} = 0.2$, $L_{s\sigma} = 0.1$, $u_s = 1$, $i_s = 1$, $\omega_s = 1$, $\psi_s = 1$, $\psi_{md} = 0.95$, $\psi_{mq} = 0.18$, and $\psi_{s\sigma} = 0.1$. Calculate $\cos\varphi$, L_d/L_q, the electric current components, load angle, current angle, and pu electromagnetic torque and power.

SOLUTION: The SynRM space-vector diagram was shown in Figure 10.4. The space vectors comprise d-axis and q-axis components.

The per unit values of the magnetizing flux linkage for the d- and q- components can be expressed as follows.

$$
\psi_m = \sqrt{\psi_{md}^2 + \psi_{mq}^2} = \sqrt{0.95^2 + 0.18^2} = 0.97
$$

$$
\cos\delta_m = \frac{\psi_{md}}{\psi_m} = \frac{0.95}{0.967} = 0.983 \rightarrow \delta_m = 10.7°
$$

The angle between ψ_s and ψ_m can be calculated using the cosine law.

$$
(L_{s\sigma}i_s)^2 = \psi_s^2 + \psi_m^2 - 2\psi_s\psi_m\cos(\delta_s - \delta_m)
$$

$$
\begin{aligned}
\cos(\delta_s - \delta_m) &= \frac{\psi_s^2 + \psi_m^2 - (L_{s\sigma}i_s)^2}{2\psi_s\psi_m} = \frac{1^2 + 0.967^2 - (0.1 \cdot 1)^2}{2 \cdot 1 \cdot 0.967} = 0.995 \rightarrow (\delta_s - \delta_m) \\
&= 5.5°
\end{aligned}
$$

$$
(\delta_s - \delta_m) = 5.5° \rightarrow \delta_s = 5.5° + \delta_m = 5.5° + 10.7°
$$

$$
\delta_s = 16.2°.
$$

The electric current per-unit components according to Equations (10.12) and (10.13) are

$$i_d = \frac{1}{L_d} \psi_{spu} \cos \delta_s = \frac{1}{3.1} 1 \cdot \cos 16.2° = 0.31$$

$$i_q = \frac{1}{L_q} \psi_s \sin \delta_s = \frac{1}{0.31} 1 \cdot \sin 16.2 = 0.9$$

Per-unit torque as a function of load angle, calculated with space vectors is

$$T_e = \psi_s^2 \left(\frac{L_d - L_q}{L_q L_d} \right) \frac{1}{2} \sin 2\delta_s = 1^2 \frac{3.1 - 0.3}{2 \cdot 0.3 \cdot 3.1} \sin 2 \cdot 16.2° = 0.81$$

Neglecting losses, the per-unit output power P_{out} becomes

$$P_{out} = \omega_s T_e = u_s^2 \frac{L_d - L_q}{2 L_d L_q \omega_s} \sin 2\delta_s = 1^2 \frac{3.1 - 0.3}{2 \cdot 0.3 \cdot 3.1 \cdot 1} \sin 2 \cdot 16.2° = 0.81$$

Because the q-axis is perpendicular to the d-axis, u_s is perpendicular to ψ_s. Therefore, the angle between the voltage and q-axis is the load angle δ_s. The angle between i_s and the q-axis must be added to get the phase angle φ between the voltage and current.

$$\sin^{-1} \frac{i_d}{i_s} = \sin^{-1} \frac{0.309}{1} = 18°$$

$$\varphi = 16.18° + 18° = 34.18° \rightarrow \cos\varphi = 0.827$$

The electric current angle is shown at the base of the Figure 10.4 diagram.

$$\kappa + \varphi = 90° + \delta_s \rightarrow \kappa = 90° + \delta_s - \varphi = 90° + 16.18° - 34.18° = 72°$$

Expressed more simply, the statement becomes

$$\kappa = \arccos \frac{i_d}{i_s} = \arccos \frac{0.309}{1} = 72°$$

Per-unit torque can be checked using the electric current angle.

$$T_e = \frac{1}{2} i_s^2 (L_d - L_q) \sin 2\kappa = \frac{1}{2} \cdot 1^2 (3.1 - 0.3) \sin 2 \cdot 72 = 0.82$$

This result is close to the 0.81 calculated above (within a small rounding error). The saliency ratio is $L_d/L_q = 3.1/0.3 = 10.33$.

EXAMPLE 10.2: Calculate the per-unit electromagnetic torque versus load angle for a saliency ratio between 5 and 50. $L_q = 0.27$ at rated voltage and frequency.

SOLUTION: Equation (10.15) can be rearranged in terms of per-unit electromagnetic torque. See also Equation (10.1). The result can be the basis for diagramming $T_e = f(\delta_s)$ with saliency ratio as a parameter.

$$T_e = \frac{P_{out}}{\omega_s} = \frac{u_s^2}{\omega_s} \frac{L_d - L_q}{2L_dL_q\omega_s} \sin 2\delta_s = \frac{u_s^2}{\omega_s^2} \frac{1 - \frac{L_q}{L_d}}{2L_q} \sin 2\delta_s = 1 \frac{1 - \frac{1}{L_d/L_q}}{2L_q} \sin 2\delta_s$$

For a saliency ratio $L_d/L_q = 10$ and a load angle of $20°$

$$T_e = \frac{1 - \frac{1}{L_d/L_q}}{2L_q} \sin 2\delta_s = \frac{1 - \frac{1}{10}}{2 \cdot 0.27} \sin 2 \cdot 20° = 1.666 \sin 40° = 1.07$$

$$T_{emax10} = \frac{1 - \frac{1}{L_d/L_q}}{2L_q} \sin 2\delta_s = \frac{1 - \frac{1}{10}}{2 \cdot 0.27} \sin 2 \cdot 45° = 1.666$$

Therefore, the maximum value for $\kappa = 45°$ is 1.67. Using the same approach, the calculated per-unit torques versus load angles for saliency ratios of $L_d/L_q = 5$, 10, and 50 are illustrated in Figure 10.5. The maximum values for a saliency ratio of 5 and 50 are as follows.

$$T_{emax5} = \frac{1 - \frac{1}{L_d/L_q}}{2L_q} \sin 2\delta_s = \frac{1 - \frac{1}{5}}{2 \cdot 0.27} \sin 2 \cdot 45° = 1.48$$

$$T_{emax50} = \frac{1 - \frac{1}{L_d/L_q}}{2L_q} \sin 2\delta_s = \frac{1 - \frac{1}{50}}{2 \cdot 0.27} \sin 2 \cdot 45° = 1.81$$

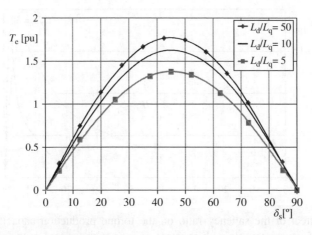

Figure 10.5 Torque production capacity of a SynRM at different saliency ratios. The q-axis per-unit inductance is constant at $L_q = 0.27$.

Maximum torque is produced when the load angle is $\delta_s = \pi/4$. Figure 10.5 illustrates how the torque production capacity of a SynRM changes with load angle for the saliency ratios of 5, 10, and 50. The q-axis inductance is held constant at $L_q = 0.27$ pu.

A saliency ratio of $L_d/L_q = 50$ is not achievable in practice. Because stator leakage alone comprises almost half of its value, it is difficult to reach or go below an $L_q = 0.27$ value for q-axis synchronous inductance. For $L_d/L_q = 50$ and $L_q = 0.27$, L_d must be a practically impossible $L_d = 13.5$. With a practically reachable d-axis inductance value of $L_d = 2.7$, a saliency ratio of $L_d/L_q = 10$ can be achieved. In large induction machines, magnetizing inductance values are approximately $L_d = 3.5$, and the magnetizing inductances for the SynRM architecture are about the same.

Exceeding a saliency ratio of 10 in a SynRM machine with $L_q = 0.27$ pu will result in the required 160% breakdown torque. However, this value is lower than that produced by an induction machine. In industrial asynchronous machines, breakdown torque is typically above 300%. To produce higher breakdown torque, q-axis inductance especially should be less than $L_q = 0.27$, which was the value use here.

Figure 10.6 shows how torque behaves as function of current angle κ for a constant voltage supply with $L_{qpu} = 0.2$. High L_d values result in low i_d values. When producing torque, i_q values are increasing. The higher the L_d and the lower the L_q, the closer i_s will be to the q-axis. As torque increases, stator leakage begins to reduce air-gap flux linkage. As a result, torque drops off with higher current angles.

Neglecting the stator resistance effect and based on Figure 10.4, the power factor angle φ can be determined as follows.

$$\varphi \approx \delta_s + \frac{\pi}{2} - \kappa \qquad (10.17)$$

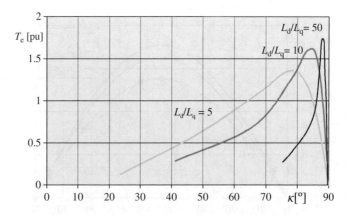

Figure 10.6 Effect of the saliency ratio on the torque production capacity of a SynRM and on the electric current angle with constant voltage supply $u_s = 1$ at the rated speed $\omega_s = 1$ – The calculation uses the rated current of an induction motor with a corresponding stator geometry. Values are calculated for a 30 kW, four-pole, 50 Hz machine with $L_q = 0.2$ per unit.

At the maximum torque-producing load angle δ_{Tmax}, the corresponding power factor angle φ_{Tmax} is approximately

$$\varphi_{\text{Tmax}} \approx \delta_{\text{Tmax}} + \frac{\pi}{2} - \kappa = \frac{\pi}{4} + \frac{\pi}{2} - \kappa = \frac{3\pi}{4} - \kappa \tag{10.18}$$

The current angle is always in the range

$$\kappa \in \left[0, \frac{\pi}{2}\right] \tag{10.19}$$

Therefore, the power factor of the SynRM at maximum power is always

$$\cos \varphi_{\text{Pmax}} < \cos\left(\frac{\pi}{4}\right) = 0.707 \tag{10.20}$$

Neglecting stator winding and iron losses, the SynRM power factor, in absolute and per-unit values, becomes

$$\cos \varphi = \frac{2T_e \omega_s}{3 p u_s i_s} = \frac{2T_e \Omega_r}{3 u_s i_s} = \frac{T_{e,\text{pu}} \Omega_{r,\text{pu}}}{u_{s,\text{pu}} i_{s,\text{pu}}} \tag{10.21}$$

In steady state, ω_s is the electric angular speed of the stator rotating field and rotor, u_s is the absolute value of the stator voltage space vector, and i_s is the absolute value of the stator current space vector. By applying the cross-field principle, the torque can be expressed as

$$T_e = \frac{3}{2} p \left(L_d - L_q\right) i_d i_q \tag{10.22}$$

Flux linkage magnitude can be rewritten as follows.

$$\psi_s = \sqrt{\psi_d^2 + \psi_q^2} = \sqrt{\left(L_d i_d\right)^2 + \left(L_q i_q\right)^2} \tag{10.23}$$

Plugging the identity from Equation (10.21) into Equations (10.22) and (10.23) and incorporating the following relations

$$u_s = \omega_s \psi_s \tag{10.24}$$

$$i_s = \sqrt{i_d^2 + i_q^2} \tag{10.25}$$

results in the following equation for power factor. Its value depends on the electric current components and therefore on the current angle.

$$\cos \varphi = \frac{\left(L_d - L_q\right) i_d i_q}{\sqrt{\left(L_d i_d\right)^2 + \left(L_q i_q\right)^2} \sqrt{i_d^2 + i_q^2}} \tag{10.26}$$

Differentiating Equation (10.26) with respect to i_d/i_q and setting the resulting expression to zero produces the following condition for the maximum power factor.

$$\frac{i_q}{i_d} = \sqrt{\frac{L_d}{L_q}} \tag{10.27}$$

Therefore, the power factor reaches its maximum at this electric current angle.

$$\kappa = \arctan \sqrt{\frac{L_d}{L_q}} \tag{10.28}$$

The maximum power factor is therefore

$$\cos \varphi_{max} = \frac{\dfrac{L_d}{L_q} - 1}{\dfrac{L_d}{L_q} + 1} \tag{10.29}$$

EXAMPLE 10.3: Calculate power factor versus current angle for saliency ratios between 5 and 50 and $L_q = 0.2$ at rated voltage, current, and frequency. Calculate the current angle at which the power factor reaches its maximum value and the maximum power factor at that current angle.

SOLUTION: Equation (10.26) is the basis for the power factor calculation. The calculation is carried out for saliency ratio values of 5, 10, and 50.

$$\cos \varphi = \frac{(L_d - L_q) i_d i_q}{\sqrt{(L_d i_d)^2 + (L_q i_q)^2}\sqrt{i_d^2 + i_q^2}}$$

Electric current angle changes with changes in the i_d and i_q components. The diagram in Figure 10.4 shows that current angle is linked to the current components by means of Equation (10.28).

$$\kappa = 90° - \arctan \frac{i_d}{i_q}$$

or more simply

$$\kappa = \arccos \frac{i_d}{i_s}$$

If i_d changes from 0.9 to zero, the current angle will move from 25.8° to 90°. For example, with a saliency ratio of 10 at $i_d = 0.2$, the i_q component will be as follows.

$$i_q = \sqrt{i_s^2 - i_d^2} = \sqrt{1^2 - 0.2^2} = 0.979$$

At this same saliency ratio of 10 and with $L_q = 0.2$ and $L_d = 2$, the calculation of power factor for the current angle given by this expression,

$$\kappa = 90° - \arctan \frac{0.2}{0.979} = 90° - 11.54° = 78.45°$$

results in

$$\cos \varphi_{10} = \frac{(L_d - L_q)i_d i_q}{\sqrt{(L_d i_d)^2 + (L_q i_q)^2}\sqrt{i_d^2 + i_q^2}} = \frac{(2 - 0.2)0.2 \cdot 0.979}{\sqrt{(2 \cdot 0.2)^2 + (0.2 \cdot 0.979)^2}\cdot 1} = \frac{0.352}{0.445} = 0.79$$

So, the maximum value results at this current angle

$$\kappa_{10} = \arctan\sqrt{\frac{L_d}{L_q}} = \arctan\sqrt{10} = \arctan 3.16 = 72.45°$$

and from Equation (10.29) will reach the value

$$\cos \varphi_{max10} = \frac{\dfrac{L_d}{L_q} - 1}{\dfrac{L_d}{L_q} + 1} = \frac{10 - 1}{10 + 1} = \frac{9}{11} = 0.818$$

At the electric current angle of 72.45°, which corresponds to current components $i_d = 0.3$ and $i_q = 0.954$, the power factor reaches its maximum of 0.818. This can be confirmed using Equation (10.27).

$$\frac{i_q}{i_d} = \sqrt{\frac{L_d}{L_q}} = \frac{0.954}{0.3} = 3.18 = \sqrt{10} = 3.16$$

Figure 10.7 illustrates for the range of the current angles κ between 25° and 90° and for saliency ratios 5, 10, and 50.

For a saliency ratio of $L_d/L_q = 5$, the maximum power factor value will be reached at the following current angle.

$$\kappa_5 = \arctan\sqrt{\frac{L_d}{L_q}} = \arctan\sqrt{5} = \arctan 2.23 = 65.9°$$

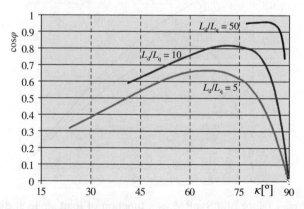

Figure 10.7 Power factor of a SynRM as a function of current angle at different saliency ratios with $L_q = 0.2$ pu

The maximum power factor will be

$$\cos \varphi_{max5} = \frac{\frac{L_d}{L_q} - 1}{\frac{L_d}{L_q} + 1} = \frac{5 - 1}{5 + 1} = \frac{4}{6} \approx 0.67$$

For a saliency ratio of $L_d/L_q = 50$, the maximum power factor value will be reached at

$$\kappa_{50} = \arctan \sqrt{\frac{L_d}{L_q}} = \arctan \sqrt{50} = \arctan 7.07 = 81.95°$$

and will be

$$\cos \varphi_{max50} = \frac{\frac{L_d}{L_q} - 1}{\frac{L_d}{L_q} + 1} = \frac{50 - 1}{50 + 1} = \frac{49}{51} = 0.96$$

For $L_d/L_q = 10$, the current angle becomes $\kappa = 72.7°$, and the maximum power factor will be $\cos \varphi_{max} = 0.82$. A large saliency ratio leads to a higher operating current angle, and therefore a good power factor.

EXAMPLE 10.4: Figure 10.8 shows power factor versus load angle for a SynRM motor drive at rated voltage and speed and saliency ratios of 5, 10, and 50. Check the operating point power factor versus load angle for the saliency ratio of 50 with $L_q = 0.2$ pu at the rated voltage, current, and frequency.

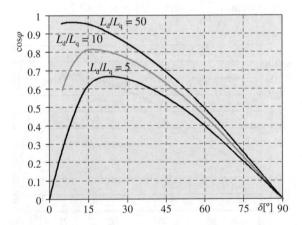

Figure 10.8 Power factor of a SynRM as a function of load angle at different saliency ratios with $L_q = 0.2$ pu.

SOLUTION: The RMS load angle Equation (10.1) can be written in per-unit form as follows.

$$P = u_s^2 \frac{L_d - L_q}{2\omega_s L_d L_q} \sin 2\delta_s = u_{spu}^2 \frac{1 - \dfrac{L_q}{L_d}}{2\omega_s L_q} \sin 2\delta_s$$

At the $L_d/L_q = 50$ saliency ratio with a load angle of $\delta_s = 10°$, per-unit power is

$$P = 1^2 \frac{1 - \dfrac{0.2}{10}}{2 \cdot 1 \cdot 0.2} \sin(2 \cdot 10°) = 0.83$$

This expression can be simplified by assuming there is no leakage. With $\psi_s = \psi_m = 1$, $\psi_{md} = \psi_m \cos \delta_s = 0.98$. The current i_d needed to excite the machine is as follows.

$$i_d = \frac{\psi_{md}}{L_{md}} = \frac{0.98}{10} = 0.098$$

Correspondingly, q-axis current is

$$i_q = \sqrt{i_{spu}^2 - i_{dpu}^2} = \sqrt{1^2 - 0.098^2} = 0.995$$

So the electric current angle becomes $\kappa = \arcos(0.995/.098) = 84.4°$, and the power factor is

$$\cos \varphi = \frac{(L_d - L_q) i_d i_q}{\sqrt{(L_d i_d)^2 + (L_q i_q)^2}\sqrt{i_d^2 + i_q^2}} = \frac{(10 - 0.2)0.098 \cdot 0.995}{\sqrt{(10 \cdot 0.098)^2 + (0.2 \cdot 0.995)^2}\sqrt{1}} = 0.955$$

Figure 10.8 illustrates the power factor of a SynRM as a function of load angle with rated voltage supply at the rated speed. You will see the agreement between the result of the previous example and the $L_d/L_q = 50$ curve.

Figure 10.9 illustrates the power factor of a SynRM as a function of shaft output power. The figure clearly indicates how essential a large saliency ratio is for the good machine characteristics.

SynRM machine efficiency can also be investigated at different saliency ratio values. Efficiency is calculated with RMS values as follows.

$$\eta = \frac{P_{out}}{P_{out} + P_{loss}} = \frac{T_e \Omega_r}{T_e \Omega_r + m I_s^2 R_s + P_{Fe+mech}} \tag{10.30}$$

$$\eta = \frac{m U_s^2 (L_d - L_q) I_d I_q \sin 2\delta_s}{m U_s^2 (L_d - L_q) I_d I_q \sin 2\delta_s + m I_s^2 R_s + P_{Fe+mech}} \tag{10.31}$$

The so-called no-load losses term $P_{Fe+mech}$ accounts for iron losses, mechanical losses, and additional losses. Figure 10.10 depicts the idealized efficiencies of a 30 kW, four-pole SynRM

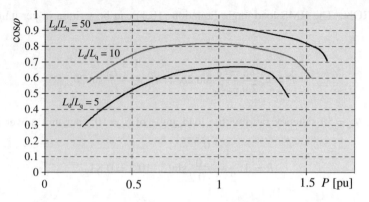

Figure 10.9 Power factor of a SynRM as a function of shaft output power at rated speed with $L_q = 0.2$ pu.

at different saliency ratios and load angles. Because, in reality, the iron circuit and air gap must be changed to affect saliency ratio; motor parameters do not remain constant, and the family of curves presented in Figure 10.10 does not represent actual behaviour. Nonetheless, the qualitative information offered is useful.

10.3 The control of a SynRM

Although the theory and operating principles of the SynRM machine were documented early in the 20th century, the first SynRMs were nothing more than converted induction-machine architectures offering poor torque-current ratios, poor power factors, poor efficiencies, and

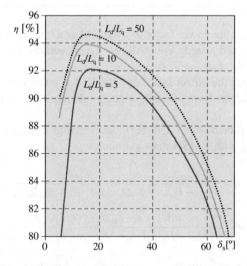

Figure 10.10 Efficiency of a 30 kW machine as a function of load angle with saliency ratio as a parameter and $L_q = 0.2$ pu, $P_{Fe+mech} = 0.02$, and $R_s = 0.028$.

excessive torque ripple. These performance weaknesses prevented the SynRM from becoming a serious alternative among electrical drives.

Converter drives make it possible to use a cageless rotor for efficient synchronous operation. The cageless rotor enables a higher saliency ratio, which improves machine efficiency and reduces inertia. Lower inertia improves transient response time. The SynRM stator is identical to that of an asynchronous machine, and its control system comprises similar components; however, SynRM implementation is simpler, and the SynRM drive costs less than a corresponding induction machine or PMSM drive.

The general control theories for the SynRM drive are similar to the control theories of other AC machines that were introduced previously in this book. For this reason, only the most essential and relevant control principles are discussed in the upcoming discussion. These control principles are based either on DTC or on vector control using the two-axis model of a SynRM in the rotor reference frame. Vector control requires rotor angle information such as position or speed feedback. Sensorless control methods make use of a number of different numerical estimators, each requiring substantial calculation capacity.

10.3.1 Vector control

As references, vector control uses the direct- and q-axis electric current components i_{dref} and i_{qref}. Usually these references are formed directly from the torque reference T_{eref}; however, in certain cases, such as in field weakening, the current references are formed based on the measured rotation speed or the torque reference given by the rotation speed controller. The torque control block produces the current references according to the selected control strategy. These strategies can be categorized as either constant current angle (κ) control or constant i_d control. The latter is better adapted to operation below the rated speed.

The phase current reference values are generated from the current references of the two-axis model using two-to-three-phase ($2 \rightarrow 3$) transformations. A hysteresis controller, for example, implements the current control using the measured stator phase currents and the phase current references. Figure 10.11 offers a block diagram of the operating principle of this kind of a control system.

10.3.2 Constant i_d control

This approach is most natural for a SynRM drive, because suitable armature current must be used to excite the machine. The SynRM remains safely excited if d-axis current is held at a constant value. For AC drives, the magnitude of the flux linkage using space-vector absolute values is

$$\psi_s = \sqrt{\psi_d^2 + \psi_q^2} = \sqrt{(L_d i_d)^2 + (L_q i_q)^2} \tag{10.33}$$

Solving for the q-axis current yields

$$i_q = \frac{\sqrt{\psi_s^2 - (L_d i_d)^2}}{L_q} \tag{10.33}$$

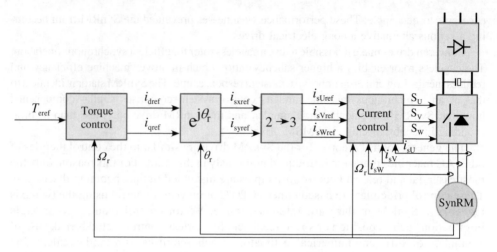

Figure 10.11 Schematic block diagram of the vector control system for a SynRM. S_U, S_V, and S_W are the switching commands for the converter. The current controller may need the speed feedback for possible feedforward voltage calculations.

If i_q is plugged into Equation (10.2), the expression for torque can be written as follows.

$$T_e = \frac{3}{2}p(L_d - L_q)i_d i_q = \frac{3}{2}\frac{p}{L_q}(L_d - L_q)i_d\sqrt{\psi_s^2 - (L_d i_d)^2} = \frac{3}{2}p\left(\frac{L_d}{L_q} - 1\right)i_d\sqrt{\psi_s^2 - (L_d i_d)^2}$$

$$(10.34)$$

Solving $\partial T_e/\partial i_d = 0$ using the above expression gives the constant i_d value to be maintained

$$i_d = \frac{\psi_s}{\sqrt{2}L_d} \tag{10.35}$$

The maximum torque at a given stator flux linkage magnitude ψ_s is

$$T_{emax} = \frac{3}{4}p\frac{L_d - L_q}{L_d L_q}\psi_s^2 = \frac{3}{4}p\left(\frac{1}{L_q} - \frac{1}{L_d}\right)\psi_s^2 \approx \frac{3}{4}\frac{p}{L_q}\psi_s^2 \tag{10.36}$$

The last part of Equation (10.36) reveals that SynRM q-axis inductance must be minimized to produce adequate torque. For i_d = constant control and at speeds below the rated speed, the value of i_d must be kept at

$$i_{dref} = \frac{\psi_{smaxref}}{\sqrt{2}L_d} \tag{10.37}$$

where the maximum value for the reference stator flux linkage is

$$\psi_{smaxref} = \sqrt{\frac{4|T_{eref}|L_d L_q}{3p(L_d - L_q)}} \tag{10.38}$$

EXAMPLE 10.5: Derive (10.37).

SOLUTION: Differentiating the torque equation

$$T_e = \frac{3}{2}p\left(\frac{L_d}{L_q} - 1\right)i_d\sqrt{\psi_s^2 - (L_d i_d)^2}$$

with respect to i_d, results in

$$\frac{\partial T_e}{\partial i_d} = \frac{3}{2}p\left(\frac{L_d}{L_q} - 1\right)\left[\sqrt{\psi_s^2 - (L_d i_d)^2} + i_d\frac{-2L_d^2 i_d}{2\sqrt{\psi_s^2 - (L_d i_d)^2}}\right].$$

The zero for the differential can be found by setting the bracketed term to zero.

$$\left[\sqrt{\psi_s^2 - (L_d i_d)^2} + i_d\frac{-2L_d^2 i_d}{2\sqrt{\psi_s^2 - (L_d i_d)^2}}\right] = 0.$$

This results in

$$\left[\frac{\psi_s^2 - 2(L_d i_d)^2}{\sqrt{\psi_s^2 - (L_d i_d)^2}}\right] = 0.$$

The numerator becomes 0 when

$$i_d = \frac{\psi_s}{\sqrt{2L_d}}.$$

In field weakening, the i_{dref} component drops with speed as follows.

$$i_{dref} = \frac{\psi_{maxref}}{\sqrt{2L_d}}\frac{\omega_N}{\omega_s} \tag{10.39}$$

The reference value i_{qref} for the q-axis stator current component is

$$i_{qref} = \frac{2T_{eref}}{3p(L_d - L_q)i_{dref}} \tag{10.40}$$

When i_d varies, L_d also changes, and therefore torque is no longer directly proportional to the q-axis current component i_q. In this case, more advanced methods must be applied such as self-tuning, and model reference adaptive controllers must be used to produce the current reference i_{qref} (Vas 1998).

10.3.3   Constant angle κ control

There are three methods used to keep the angle κ constant between the stator current space vector and the d-axis of the rotor:

- fastest torque control (gives the fastest torque response),

- maximum torque/current control, and

- maximum power factor control.

All the above control strategies depend on $\tan \kappa$.

Fastest torque control

As was true in the previous case for i_d, the solution of the extreme value problem for the q-axis current component, Equation (10.34), yields

$$i_q = \frac{\psi_s}{\sqrt{2L_q}} \tag{10.41}$$

Dividing both sides of Equation (10.41) by Equation (10.35) results in the following.

$$\frac{i_q}{i_d} = \frac{\psi_s/\sqrt{2L_q}}{\psi_s/\sqrt{2L_d}} = \frac{L_d}{L_q} \quad (= \tan \kappa) \tag{10.42}$$

which can be interpreted as $\tan \kappa$. Compare with Figure 10.4. The highest rate of change for torque results when the angle between the stator current space vector and the d-axis is

$$\kappa = \arctan \frac{L_d}{L_q} \tag{10.43}$$

The d-axis current reference i_{dref} is obtained by plugging the magnitude of the flux linkage solved from Equation (10.35) into Equation (10.36) and substituting the real torque T_{emax} with its reference.

$$\psi_{smaxref} = i_{dref}\sqrt{2L_d} \tag{10.44}$$

$$i_{dref} = \frac{\sqrt{\dfrac{4|T_{eref}|L_d L_q}{3p(L_d - L_q)}}}{\sqrt{2L_d}} = \sqrt{\frac{4|T_{eref}|L_d L_q}{3p(L_d - L_q)2L_d^2}} = \sqrt{\frac{2|T_{eref}|L_q}{3p(L_d - L_q)L_d}} \tag{10.45}$$

Finally, by taking Equation (10.43) into account

$$i_{dref} = \sqrt{\frac{2|T_{eref}|}{3p(L_d - L_q)\tan \kappa}} \tag{10.46}$$

The reference for q-axis current comes from Equations (10.43) and (10.44). Because it was ignored in the derivation, the sign of the reference torque must be taken into account.

The q-axis current reference becomes

$$i_{qref} = \frac{i_{dref}\,\text{sgn}(T_{eref})}{\tan \kappa}$$ (10.47)

Maximum torque/current and maximum power factor control

In the maximum torque/current control strategy, it is obvious that the maximum value for the T_e/i_s ratio results when the angle κ between the current and the d-axis of the rotor is $\pi/4$. This is accounted for by the load angle equation of the SynRm.

As shown previously, maximum power factor is achieved when the angle of the stator current with respect to the d-axis is as follows. See the derivation of Equation (10.28).

$$\kappa = \arctan \sqrt{\frac{L_d}{L_q}}$$ (10.48)

The current references in the dq reference frame are determined as was done for maximum torque/current control.

10.3.4   Combined current-voltage vector control

Combined current-voltage vector control is appropriate for high rotational speeds. The control scheme avoids drawbacks resulting from delays in the AC current references and saturation. The monitored rotor speed information is fed to a function generator, which outputs the d-axis stator flux linkage reference ψ_{dref}. During field weakening, the reference signal is a function of the rotor speed. The actual values of the stator flux linkage components or estimates of their values are obtained with the measured phase currents and the direct and q-axis inductances in the stator flux linkage estimation circuit. The flux linkage controller (a PI controller) produces a d-axis stator current reference i_{dref} from the difference of the stator flux linkage component reference value and the actual value. The difference of the d-axis stator current reference and the measured stator current component, fed to the current controller (PI controller), yields the d-axis voltage component reference. Finally, the rotational voltage induced by the q-axis stator flux linkage must be subtracted from the voltage reference obtained with the current controller, as shown by the equivalent circuit of Figure 10.3.

A speed controller, usually also PI-based, introduces the torque reference T_{eref}. The q-axis current reference is

$$i_{qref} = \frac{T_{eref} + \psi_q i_d}{\psi_d}$$ (10.49)

The error reference Δi_q obtained from the difference of i_{qref} and i_q is fed into the PI-based electric current controller, as was the d-axis component to obtain the q-axis voltage component reference. The rotational voltage $\omega_s \psi_d$ is added to the output signal as a feedforward to get the final u_{qref}. This feedforward serves as a simple means of decoupling of the d- and q-axis phenomena. Using coordinate transformations and two-to-three-phase transformations, phase voltage references are generated from the dq voltage references. Control of the PWM inverter is based on these obtained references. Figure 10.12 shows the block diagram of the control.

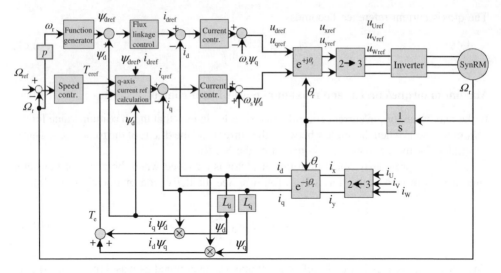

Figure 10.12 Block diagram of the combined current-voltage vector control. The notation 1/s indicates integration with respect to time. The feedforward voltages $\omega_s\psi_d$, $\omega_s\psi_q$ help in removing the cross coupling.

10.3.5   Direct torque and flux linkage control

Direct torque control can also be used to operate synchronous reluctance as well as other rotating-field machines. The SynRM, however, has a special feature that makes the application of DTC somewhat complicated—the stator flux of the machine can collapse when DTC tries to increase the torque by accelerating the stator voltage space vector. Since the maximum acceptable static load angle in a SynRM motor drive is only 45°, its load angle must be carefully monitored to keep the drive from exceeding peak torque. Since the peak torque is also limited compared to, for example, induction machine peak torque, the controller must be accurately monitored during periods of high-power SynRM drive operation.

Adding correction based on an electric current model, as is done for a separately excited synchronous machine, results in a more suitable DTC control method for a SynRM; however, rotor position sensing is required. When a rotor position feedback system based on a pulse encoder is integrated with a SynRM drive, the initial rotor angle can be easily determined because of the large difference between the d-axis and q-axis reluctance. In this case, the DTC should be supplemented with references inherent to this machine type, for instance, with respect to the power factor. Any of the previously discussed control schemes can be applied also to the case of a DTC inverter. In addition, a position-sensorless drive applying drift correction can be adapted for the purpose. If a damper winding is also included, the d- and q-axis transient inductances will be more similar, which facilitates finding the appropriate drift direction of the stator flux linkage. With no damper winding and high saliency, the vector dot product result is so strongly affected by the inductance variation that a dedicated version of this correction method is needed.

Moreover, a very low q-axis inductance may prove problematic for DTC. The low inductance may result in excessively high q-axis current ripple, or it may cause measurement problems.

10.4   Further development of SynRM drives – PMaSynRM

Although the characteristics of a SynRM drive improve with increasing saliency ratio, as is the case with an induction machine, constructing a SynRM machine with a very large number of poles is difficult. This is because increasing the number of poles increases stator flux leakage, which degrades the performance characteristics of the drive, whether induction machine or SynRM. When constructing multipole machines, separately excited synchronous machines and PMSMs are superior alternatives. The practically ideal pole-pair number for a SynRM motor is $p = 2$.

The lack of rotor current in a SynRM provides both benefits and detriments. The benefit is that no fundamental Joule losses are present in the rotor. However, when compared to the induction motor there is no rotor current to compensate for the armature reaction in the q-axis, which easily degrades the SynRM drive power factor. Figure 10.13 compares induction machine and SynRM drives at their rated operating points for both machines, where, $L_d/L_q = 7$ with per-unit values $L_q = 0.42$, $L_d = 3$ for SynRM and $L_{s\sigma} = 0.1$. These values can easily be reached using the rotor geometry that was previously introduced as Figure 10.1f, a flux-barrier rotor.

There is a simple but costly way of improving the power factor of a SynRM drive. Q-axis PMs can be added to compensate for the q-axis armature reaction. The rotor geometry that was depicted in Figure 10.1f could be equipped with q-axis PMs, which would increase the power factor significantly. This raises the question, "Is the result after adding the PMs a reluctance synchronous machine, or is it a PMSM?" If the torque produced by the magnets is smaller than the reluctance torque, the machine is normally referred to as a PM-compensated or PM-assisted SynRM (PMaSynRM). If vice versa, the machine is a PMSM with embedded magnets and some reluctance torque. Figure 10.14 illustrates the compensating effect of q-axis PMs.

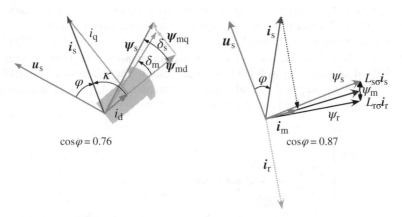

$$\cos\varphi = 0.76 \qquad\qquad\qquad \cos\varphi = 0.87$$

Figure 10.13 Comparison of SynRM and induction machine from the point of view of power factor. For both machines, $L_{dSynRM} = L_{mIM} = 3$, $L_{qSynRM} = 0.32$, and $L_{s\sigma} = 0.1$. $L_{r\sigma} = 0.1$ for the induction machine. In the induction machine rotor, electric current compensates for the stator current reaction, and only i_m produces significant flux linkage. Stator and rotor currents have their leakage components, but they are limited because both stator and rotor leakage inductances are small. In the SynRM vector diagram, the quadrature armature reaction ($i_q L_{mq} = 0.91 \times 0.32 \approx 0.3$) is significant and noncompensated, which makes the SynRM power factor smaller than that of an induction machine with similar inductances.

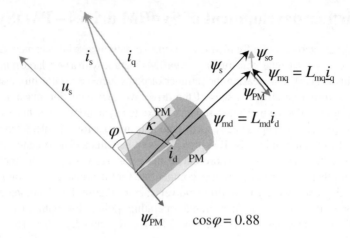

Figure 10.14 Improving the power factor of a SynRM by adding q-axis PMs to compensate for the q-axis armature reaction. All machine parameters are the same as in the previous example, however, the q-axis armature reaction is compensated for with PM flux linkage. The power factor is significantly improved from 0.76 to 0.88.

The control of a PM-compensated SynRM resembles the control of a PMSM, and it can be implemented according to the same principles. The dq axes are now different, however, which must be taken into account in the control. Figure 10.15 illustrates torque production as a function of the current angle κ.

Relatively inexpensive PMs, such as ferrites, can be used in the PM-compensated SynRM to produce good machine characteristics. However, the added PM mass to the rotor and the resulting higher mechanical stresses will require a more robust rotor structure.

As stated previously, the SynRM with PMs may also be regarded as a PMSM with saliency. In that case, PMSM control methods may be applied. The d- and q-axes switch places with the d-axis aligning with the PM magnetic axis. The saliency of this architecture, however, will be big compared to traditional PM machines with embedded magnets, and therefore the proportion of reluctance torque will be big.

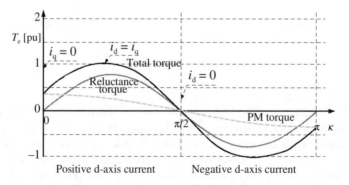

Figure 10.15 Torque production as a function of current angle with constant current ($i_s = 1$) for a PM-compensated SynRM. PM torque has its maximum at $\kappa = 0$. The reluctance torque maximum is at $\kappa = \pi/4$. The maximum torque is at the current angle $\kappa < \pi/4$.

In SynRM control research, a phenomenon called cross saturation is attracting interest. This effect has proven to induce large errors in rotor position estimation. The effect is based on the differential anisotropy of the magnetic circuit. To evaluate it properly, the magnetic behaviour of the machine must be understood. Magnetically, both the d- and q- axis linkage fluxes exhibit nonlinear behaviours, but this is more evident in the d-axis. The d- and q- axis magnetic linkage fluxes are functions of both the d- and q- axis electric currents as follows.

$$\psi_d = \psi_d\left(i_d, i_q\right) \tag{10.50}$$

$$\psi_q = \psi_q\left(i_d, i_q\right) \tag{10.51}$$

The reciprocity condition also holds.

$$\frac{\partial \psi_d}{\partial i_q} = \frac{\partial \psi_q}{\partial i_d} \tag{10.52}$$

Differential inductances depend heavily on the working point. The effect is seen mainly at current levels higher than rated. Therefore, this phenomenon must be taken into account mainly in drives where overloading of the machine can occur. As pointed out by Vagati et al. (2000), the effect of cross saturation on maximum torque is low at rated current, but gets continually higher at elevated currents (7 % at rated current and 15 % at 3 times rated current). This strong effect of the cross saturation on the sensorless control of the SynRM drive has been proven by analytical investigation as well as by experiment.

The future of the SynRM seems promising. Big companies like ABB have taken this machine type into wide production in recent years. ABB offers a full industrial range of SynRM drives. Figure 10.16 shows a typical ABB SynRM drive. The figure illustrates how the rotor of the motor is lace-like and has large flux barriers in the laminations and only thin

Figure 10.16 ABB's industrial SynRM drive components. A synchronous motor with a flux-barrier rotor (Figure 10.1f) and ABB's frequency converter tuned to control the SynRM. *Source:* Reproduced with permission from ABB.

bridges to maintain rotor integrity. Even so, the motor drive offers reliable performance and is a serious competitor to corresponding induction-machine drives.

References

Betz, R. E. (1992). Theoretical aspects of control of SynRMs. *IEE Proceedings-B*, 139(4), 355–364.

Betz, R. E., Lagerquist, R., Jovanovic, M., Miller, T. J. E., & Middleton, R. H. (1993). Control of SynRMs. *IEEE Transactions on Industry Applications*, 29(6), 1110–1122.

Boldea, I. (1996). *Reluctance synchronous machines and drives*. Oxford, UK: Clarendon Press.

Boldea, I., Fu, Z. X., & Nasar, S. A. (1993). High-performance reluctance generator. *IEE Proceedings-B*, 140(2), 124–130.

Guglielmi, P., Pastorelli, M., & Vagati, A. (2006). Impact of cross-saturation in sensorless control of transverse-laminated synchronous reluctance motors. *IEEE Transactions on Industrial Electronics*, 53(2), 429.

Haataja, J. (2003). A comparative performance study of four-pole induction motors and synchronous reluctance motors in variable speed drives. Acta Universitatis Lappeenrantaensis. Dissertation, Lappeenranta University of Technology.

Matsuo, T., & Lipo, T. A. (1994). Rotor design optimization of SynRM. *IEEE Transactions on Energy Conversion*, 9(2), 359–365.

Moghaddam, R. (2011). SynRM (SynRM) in variable speed drives (VSD) applications. Dissertation Royal University of Technology, Stockholm, Sweden. 1653–5146, ISBN 978-91-7415-972-1.

Moncada, R. (2012). Effects of saliency on the performance of permanent magnet machine drives. Dissertation University of Concepcion, Concepcion Chile.

Pyrhönen, J., Jokinen, T., & Hrabovcova, V. (2014). *Design of rotating electric machines*. Chichester, UK: John Wiley & Sons, Ltd. ISBN: 978-1-118-58157-5,

Staton, D. A., Miller, T. J. E., & Wood, S. E. (1993). Maximising the saliency ratio of the synchronous reluctance motor. *IEE Proceedings-B*, 140(4), 249–259.

Vagati, A., Pastorelli, M., Scapino, F., & Franceschini, G. (2000). Impact of cross saturation in synchronous reluctance motors of the transverse-laminated type. *IEEE Transactions on Industry Applications*, 36(4).

Vas, P. (1998). *Sensorless vector and direct torque control*. Oxford, UK: Oxford University Press.

11

Asynchronous electrical machine drives

This chapter offers an overview beginning with a review of the working principle for the alternating current (AC) induction motor, that is, the asynchronous motor. Induction motors are well standardized, and a review of the applicable standards is given. One section has been dedicated to pulse-width-modulated (PWM)-supplied induction motor losses, followed by a more detailed discussion of the scalar control, vector control, and direct torque control (DTC) for asynchronous motors.

11.1 The working principle of the induction motor – direct online drives

The induction motor is the most common motor type used by industry. Industrial motors are usually three-phase configurations with squirrel-cage rotors. The machines are rugged, reliable, and inexpensive to manufacture. Industrial versions are also well standardized, so motors from different manufacturers are interchangeable. Figure 11.1 illustrates an industrial induction motor manufactured by ABB.

Today, however, the induction motor is increasingly facing competition from other motor types, and significantly improving its performance is difficult. The induction motor relies on slip to produce rotor current. Although the effect is slight, slip lowers efficiency. Asynchronous motor slip is analogous to the mechanical slip that occurs in the torque converter or clutch of an automobile drivetrain. In the torque converter, slip in the fluid coupling results in a loss of efficiency and increased cooling load. Slip in a clutch generates heat that reduces clutch life. In asynchronous machine drives, slip makes accurate speed control without rotor speed feedback challenging. A measurement of speed is often needed.

Electrical Machine Drives Control: An Introduction, First Edition. Juha Pyrhönen, Valéria Hrabovcová and R. Scott Semken.
© 2016 John Wiley & Sons, Ltd. Published 2016 by John Wiley & Sons, Ltd.

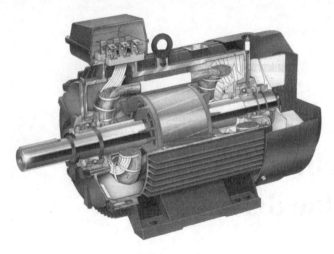

Figure 11.1 Totally enclosed ABB M2BA 280 fan-cooled motor (shaft height 280 mm), 75 kW, 3000 rpm. The rotor squirrel-cage bars are barely visible. The squirrel-cage end rings are equipped with cooling fan blades. The stator features a standard distributed three-phase winding with all six ends brought to the connection box enabling delta or star connection. *Source:* Reproduced with permission from ABB.

The induction motor is stator fed. The stator winding must excite the machine, and together with the short-circuited rotor winding, must take care of energy conversion. If the machine is supplied with three-phase current, the stator and rotor winding work together to produce a rotating field in the air gap, the flux lines of which intersect the rotor bars when the rotor runs with slip. In an asynchronous machine, slip is necessary for torque production. When slip is present, an electromotive force develops and induces rotor current in the rotor conductor bars. According to the Lorentz force, the force effect between the current and the rotating magnetic field produces the torque of the machine. The motor starts to rotate, when the electrical torque is higher than the torque of the load braking the rotor. As the rotor speed increases, the speed between the rotor bars and the field decreases, and consequently, the rotor voltage reduces and its frequency decreases. The rotor rotates slower than the magnetic field, and therefore its speed deviates from the synchronous speed. That is why an induction motor is called an asynchronous motor.

The per-unit slip s of the motor expresses the difference between the synchronous speed n_s determined by the line frequency and the rotor speed n_r.

$$s = \frac{n_s - n_r}{n_s} = \frac{\omega_s - \omega_r}{\omega_s} = \frac{\dfrac{\omega_s}{p} - \Omega_r}{\dfrac{\omega_s}{p}} \tag{11.1}$$

As Equation (11.1) shows, the slip can be defined also by the electrical angular frequencies ω_s and ω_r or the mechanical angular frequency of the rotor Ω_r.

The rotation speed n_r [1/s] of an asynchronous motor is

$$n_r = \frac{f_s}{p}(1 - s) \tag{11.2}$$

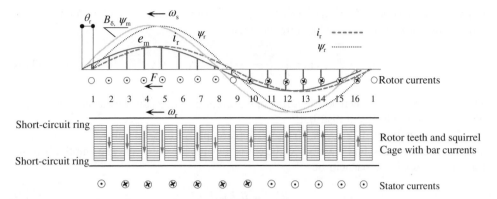

Figure 11.2 Operating principle of an asynchronous machine. The rotor is described with laminated teeth, squirrel cage (in plane), and bar currents. The air-gap flux density distribution $B_\delta(x)$ moves to the left at ω_s, while the rotor travels at slightly lower electrical angular velocity ω_r in the same direction. This results in a small internal rotor slip angular frequency ω_{slip}. ψ_r describes the virtual rotor flux-linkage distribution.

and the synchronous speed n_s in revolutions per second is

$$n_s = \frac{f_s}{p} \tag{11.3}$$

The frequency f_s is the line frequency and p is the number of pole pairs in the machine.

Figure 11.2 illustrates the asynchronous machine operating principle. The representation shows the machine rotor unwrapped into a plane. The rotor induced voltages and currents and corresponding fluxes are shown.

In the air gap of the machine, the air-gap flux Φ_m (surface integral of B_δ) and the corresponding flux linkage ψ_m travel at the electrical angular velocity ω_s of the stator input voltage. As the rotor slips, the rotor rotation electrical angular velocity is

$$\omega_r = \omega_s(1 - s) \tag{11.4}$$

The electrical angular velocity caused by the slip in the rotor will therefore be

$$\omega_{slip} = \omega_s - \omega_s(1 - s) = s\omega_s \tag{11.5}$$

The rotor radius is r_r, and the length of the bars is l. The peripheral slip speed of a rotor bar is therefore

$$v_{r,bar} = s\omega_s r_r \tag{11.6}$$

Voltages will be induced in the rotor bars according to the Lorentz force.

$$e_{r,bar}(x) = s\omega_s r_r l B_\delta(x) \tag{11.7}$$

These voltages are shown in Figure 11.2 as bars and as a fundamental wave e_m of the air-gap voltage. The air-gap voltage affects the rotor and electric current is induced. Because the rotor has both resistance R_r and leakage inductance $L_{r\sigma}$, the rotor current distribution fundamental i_r will lag the air-gap induced voltage by the angle θ_r.

$$i_r = \frac{e_r}{R_r + j\omega_{slip}L_{r\sigma}} = i_r | \theta_r \tag{11.8}$$

The rotor currents are shown both as arrow tips and tails and as vectors on the unwrapped rotor plane and the fundamental distribution i_r.

The Lorentz force effect works to move, for example, rotor bar number 5 to the left. Flux density B_δ is upwards, and the electric current is towards the observer. The amplitude of the force acting upon the bar will be

$$F_{bar} = B_\delta i_{r,bar} l \tag{11.9}$$

This bar contributes to the torque of the machine as follows.

$$T_{bar} = F_{bar} r_r \tag{11.10}$$

Figure 11.2 reveals the electric current in bars 1 through 8 to be downward. There happens to be zero current in bar 1. Current in bars 9 through 16 is upward, and bar 9 has zero current. Positive rotor torque develops, because the resulting flux density at bars 1 through 8 is opposite in sign to the flux density at bars 9 through 16. Stator torque is the reaction to the rotor torque and has the opposite sign.

The rotor current in Figure 11.2 contributes to the rotor flux Φ_r and corresponding rotor flux linkage ψ_r. This flux linkage ψ_r is in phase with the rotor current. The air-gap flux density B_δ can be integrated into the air-gap flux Φ_m, which produces, again with the stator winding, the air-gap flux linkage ψ_m.

Figure 11.3 illustrates an induction motor with its space vectors and the same distributions as Figure 11.2. The figure shows how the space vectors coincide with the peak values of the real machine flux distributions. The stator flux linkage ψ_s leads the air-gap flux linkage ψ_m, because the stator produces leakage flux linkage $i_s L_{s\sigma}$ just as rotor current produces its own leakage flux linkage $i_r L_{r\sigma}$. Serving as examples, the Lorentz forces producing torque are marked at rotor bars 5 and 13.

The slip s also brings some challenges. Rotor loss is proportional to the per-unit slip s, and therefore maximum theoretical rotor efficiency can be expressed as $\eta_{rot} \approx 1 - |s|$. In other words, as per-unit slip increases motor efficiency decreases. The absolute value of $|s|$ is selected so the machine can also operate in generating mode at negative slip values. Because slip is expressed as a per-unit value, its effect is weaker at higher machine speeds. Torque is produced with the absolute angular velocity $s\omega_s$ of the rotor currents, but the efficiency is limited by the per-unit slip. For a 2 Hz rotor frequency, for example, the maximum rotor efficiency of a 50 Hz two-pole machine $1 - 2/50 = 0.96$. For a 500 Hz case, the maximum rotor efficiency will be $1 - 2/500 = 0.996$.

Part of the electrical power taken by the motor is converted into heat in the stator, yet most of the power is transferred as air-gap power through the rotating magnetic field to the rotor. The mechanical power P_m generated by the motor at the rotation speed n_r is

$$P_m = 2\pi n_r T_{em} = \frac{\omega_s}{p}(1 - s)T_{em} = \frac{\omega_r}{p}T_{em} = \Omega_r T_{em} \tag{11.11}$$

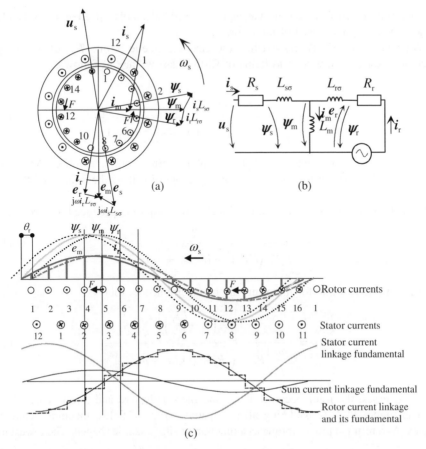

(a)   (b)

(c)

Figure 11.3 Operating principle of an induction motor showing (a) a cross-sectional view of a squirrel-cage configuration with a space-vector diagram and corresponding currents, (b) a space-vector equivalent diagram in the stator coordinate system with corresponding motor voltages, currents, and flux linkages, and (c) the stator and rotor currents, flux-linkage space vectors, and stator and rotor current linkages and their sum. In (b), all quantities are referred to the stator, but the traditional prime is not used to designate the referred quantities, because the real rotor currents must also be known for machine design cases. The stator current linkage peak value is between stator slots 12 and 1, which also includes the stator current vector. The rotor corresponding current linkage peak value is between rotor slots 8 and 10, which includes the rotor current vector. The sum of the stator and rotor current linkages has its maximum between stator slots 2 and 3, which is where the corresponding sum current vector i_m and the air-gap flux-linkage vector ψ_m can be found. The figure also shows flux-linkage distributions corresponding to vectors ψ_s, ψ_m, and ψ_r. Only ψ_m is built based on the real air-gap flux-density distribution. In reality, the components of ψ_s and ψ_r are distributed around the machine with their leakage flux-linkage components. However, these flux linkages induce real back emfs in the stator and rotor windings.

where T_{em} is the electromagnetic torque, ω_s is the electrical angular speed of the grid and Ω_r is the mechanical angular speed of the rotor.

Pyrhönen et al. (2014) showed that the electromagnetic torque T_{em} of a three-phase induction motor can be written in terms of RMS phase voltage as follows.

$$T_{em} = \frac{3\left[U_{s,ph}\left(1 - \frac{L_{s\sigma}}{L_m}\right)\right]^2 \frac{R_r'}{s}}{\frac{\omega_s}{p}\left[\left(R_s + R_r'/s\right)^2 + \left(\omega_s L_{s\sigma} + \omega_s L_{r\sigma}'\right)^2\right]} \tag{11.12}$$

The primed parameters here (e.g., R_r') are stator-referred rotor quantities. According to Equation (11.12) the electromagnetic torque is proportional to the square of the supply voltage.

The pull-out torque T_b (maximum or breakdown torque) is produced at slip value s_b.

$$s_b = \pm\frac{R_r'}{\sqrt{(R_s)^2 + \left(\omega_s L_{s\sigma} + \omega_s L_{r\sigma}'\right)^2}} \tag{11.13}$$

The pull-out torque value for motoring can be determined as follows.

$$T_b = \frac{3\left[U_s\left(1 - \frac{L_{s\sigma}}{L_m}\right)\right]^2}{2\frac{\omega_s}{p}\left[R_s + \sqrt{R_s^2 + \left(\omega_s L_{s\sigma} + \omega_s L_{r\sigma}'\right)^2}\right]} \tag{11.14}$$

As slip increases, the rotor becomes more and more inductive and torque takes the form shown in Figure 11.4. Considering all the spatial harmonic fields, machine behaviour is complex. As a result, a plot of torque as a function of slip is saddle shaped. The lowest torque

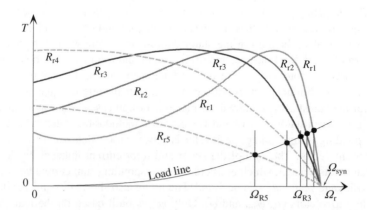

Figure 11.4 Adjusting the speed of a slip-ring induction motor by increasing the rotor resistance (R_{r1} is the smallest rotor resistance). The peak torque speed is lowered by increasing rotor resistance. Pull-out torque slip occurs at approximately $s_b \approx R_r/(\omega_s L_k)$ with rotor resistance referred to the stator voltage level and L_k being the short-circuit inductance ($L_k \approx L_{s\sigma} + L_{r\sigma}$). If stator resistance can be neglected, the pull-out torque can be considered constant.

is produced at about 1/7th of the rated speed, which is where the seventh stator harmonic brakes the rotor most significantly. This has been studied in more detail by, for example, Pyrhönen et al. (2014).

An asynchronous machine is well adapted for low-accuracy direct online (DOL) drives. DOL drives offer little in the way of machine control. Historically, slip-ring asynchronous machines were a common choice. By adjusting external rotor resistance, the static torque curve of the DOL asynchronous machine could be changed, making it possible to reduce the rotation speed of the drive when necessary. See Figure 11.4.

However, the method of changing speed by adjusting external rotor resistance is lossy, and therefore no longer widely applied in industry. The method may, however, be implemented also with power electronics so that rotor currents are rectified and slip energy is fed back to the network. A modern version of this approach has been developed, the so-called doubly fed drive, where the rotor of a slip ring induction machine is controlled by a four-quadrant frequency converter. This drive technology is now dominant in wind power drives.

The speed control of an asynchronous machine cannot normally be implemented by just controlling the stator voltage amplitude and keeping the frequency unchanged. To carry out this type of speed adjustment, a deliberately high rotor resistance must be selected to achieve a torque curve similar to the curve R_{r4} in Figure 11.4. Otherwise, the operating point of the drive may not be stable. See Figure 11.5.

When the stator voltage is lowered from the rated voltage U_{s1}, rotor speed drops stably but only slightly at first, which means that motor power does not drop significantly. The per-unit slip, however, increases rotor losses heavily, bringing the motor into thermal danger. Since peak torque is proportional to the square of the voltage, it drops rapidly with voltage. The operating point with U_{s3} moves closer to the peak torque of the machine, and the operating point with U_{s4} is no longer stable, because the induction motor torque and the load line differentials have the same sign. Any increase in the load torque curve should reduce rotor speed dramatically. This method of voltage speed control works if the rotor resistance is high (see R_{r5} in Figure 11.4), and if the motor is designed so the heat of its large rotor losses may be safely removed.

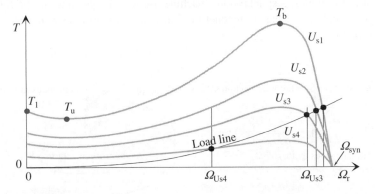

Figure 11.5 Adjusting the speed of a fan or pump drive with a low-resistance-rotor induction motor by decreasing stator voltage amplitude and keeping frequency unchanged. The illustrated starting torque T_1, minimum torque T_u, and breakdown torque T_b for voltage U_{s1} are typical induction motor torque-producing behaviours.

When frequency converters were first brought to market, engineers dreamt of replacing DC-motors with regular squirrel-cage induction motors in speed controlled drives without modifying the motor for a frequency converter drive system. A squirrel-cage asynchronous machine is the simplest machine type for a speed-controlled drive if its modest dynamics and poor speed accuracy can be accepted. The machine operates well with scalar control as the inverter takes care only of the ratio of the supply voltage and the frequency in compliance with the requirements of the induction law.

However, for accurate vector control, an asynchronous machine presents a challenge. In particular, the typical slip makes it difficult to accurately control speed without feedback. Despite their complex structures, the synchronous machines discussed in the previous chapters are, after all, easier to control accurately than asynchronous machines. In practice, accurate speed control of induction motors needs accurate rotor speed feedback or modified rotors with some saliency. For saliency, signal injection can be used to detect the rotor speed (Tüysüz et al., 2012).

A squirrel-cage motor designed for DOL-applications is not ideally suited for use in a frequency converter drive, and therefore the original dream of using regular DOL motors in speed-controlled applications cannot be fully satisfied.

11.1.1   Industrial motor types – the role of the induction machine

Electric motors account for nearly 45% of total global electricity consumption. Industry still mainly utilizes three conventional electric machine types despite the more recently introduced synchronous reluctance machine and the permanent magnet synchronous machine types. In number, induction motors make up the largest group. DC machines are common for speed-controlled drives, and synchronous machines are popular for large constant-speed DOL drive applications. The large synchronous machines used in wood grinders employed by the forest products industry is a typical example. Figure 11.6 illustrates the distribution of electric motors in a typical forest industry plant in Finland.

Therefore, asynchronous machines clearly comprise the largest industrial energy consumption group. Their share also among control drives is constantly increasing along with new control technologies. Asynchronous machine use may be peaking. In the future, other machine types, such as permanent magnet machines and synchronous reluctance machines, may gain ground.

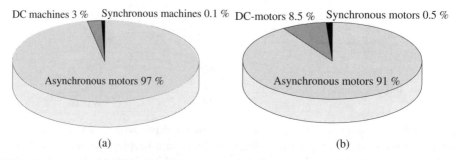

Figure 11.6 The distribution of motors (a) by quantity and (b) by installed power in a large Finnish wood-processing plant (Kaukonen, 1994).

11.2   Asynchronous machine structures and the main norms

The structures and characteristics of electric machines are regulated by various international and national norms. Induction machines in particular are regulated by several standards. The most important international standards have been published by IEC (International Electrotechnical Commission), CENELEC (Comité Européen de Normalisation Électrotechnique), and NEMA (National Electrical Manufacturers Association). The German VDE (Verband Deutscher Elektroingenieure) standards are largely congruent with the IEC (International Electrotechnical Commission) norms. The British Standard norms are widely adhered to in former British Empire areas, whereas throughout North America, the adhered to norms are the NEMA motor standards.

Totally enclosed fan-cooled induction motors constitute the most significant motor group. The structure and behavioural characteristics of this motor type are briefly discussed in the following text. According to ABB manufacturer information, totally enclosed induction motors comply, for example, with the following norms.

- The structure, rating, and performance are compliant with IEC/EN 60034-1.

- Methods for determining losses and efficiency IEC/EN 60034-2-1 and 60034-2-2

- Terminal markings and rotation direction IEC/EN 60034-8

- Thermal protection IEC/EN 60034-11

- Starting performance IEC/EN 60034-12

- Induction motors fed with converters IEC/EN 60034-17

- Motors designed especially for converter supply IEC/EN 60034-25

- Effects of unbalanced supply IEC/EN 60034-26

- Efficiency – IE codes IEC/EN 60034-30

- Standard voltages IEC 60038

- The enclosure class complies with IEC/EN 60034-5. Totally enclosed machines in the enclosure class IP55 are dust tight and water resistant.

- In totally enclosed fan-cooled machines, the type of cooling is IC 01 41, in compliance with IEC/EN 60034-6. Frame-surface cooling is accomplished with an external shaft-mounted blower. Machines are also supplied with a separate blower.

- The mounting and terminal box positioning are compliant with IEC/EN 60034-7. In the standard, mounting position is indicated by two codes. Code I covers only the motor and its bearing end shield and a single shaft extension. Code II is a general code. Motor mounting positions are identified by an induction motor code. There are dozens of different mounting position codes for foot-mounted motors and for flange-mounted motors with either small or large flanges. There are a number of different code variants that cover mounting orientation (i.e., whether or not the motors are mounted horizontally or vertically) and shaft configuration (i.e., the number and type of free shaft ends). An ordinary foot-mounted motor for normal floor mounting is designated IM B3 or IM 1001. A horizontal flange-mounted machine, large flange, is designated IM B5 or IM 3001.

- D-end and N-end: Motor ends in compliance with IEC/EN 60034-7 are designated by the codes D (drive end) and N (nondrive end).

- The positive rotation direction for the motors is the clockwise direction when looking at the D-end from the front of the motor.

- Noise limits IEC/EN 60034-9

- Mechanical vibration IEC/EN 60034-14

- Dimensions and output series IEC/EN 60072-1

- Degrees of protection by enclosures EN 50102

- General purpose three-phase induction motors with standard dimensions and outputs – Frame sizes 36-315 and flange numbers 65 to 740 EN 50347

- Insulation and insulation classes: Insulation materials are categorized into insulation classes in compliance with IEC 60085. The insulation class expresses the upper limit of the operating ambient temperature range for the insulation material under normal operating circumstances. Motor dimensioning is for an ambient temperature of 40 °C. And, the motors may be operated at their rated power at maximum elevations up to 1000 m above sea level. See Table 11.1.

Insulating materials are classified according to their ability to resist high temperatures without failure. Table 11.2 shows the temperatures according to the current IEC standard, and the previous, although still commonly used, NEMA standard temperature classes with letter codes. The hot spot allowance gives the highest permitted temperature the warmest part of the insulation may reach. Temperature rise allowance denotes the highest permitted temperature rise of the winding at rated load.

The most commonly used temperature class in industrial electric machines is 155 (F). Also, the 130 (B) and 180 (H) classes are common. A machine may be designed according to insulation class 130, but make use of materials belonging to class 155. Such an approach should naturally offer a long lifetime for the insulation in practice.

As insulation ages, its long-term thermal resistance degrades, that is, its temperature rise allowance drops. The concept of thermal index is used when evaluating the long-term thermal

Table 11.1 Effect of temperature and mounting level on the load capacity

Ambient temperature [°C]	30	40	45	50	55	60	70	80
Allowed power in per cent of the rated power	107	100	96.5	93	90	86.5	79	70
Altitude above sea level [m]	1000	1500	2000	2500	3000	3500	4000	
Allowed power in per cent of the rated power	100	97	94.5	92	89	86.5	83.5	

Table 11.2 Temperature (insulation) classes of insulating materials (IEC 60034-1)

Insulation Class (previous designation)	Allowed Hot Spot Temperature [°C]	Permitted Temperature Rise for a 40 °C Ambient Temperature [°C]	Permitted Average Winding Temperature Determined by Resistance Measurement [°C]
90 (Y)	90		
105 (A)	105	60	
120 (E)	120	75	
130 (B)	130	80	120
155 (F)	155	100	140
180 (H)	180	125	165
200	200		
220	220		
250	250		

resistance of a single insulator. The thermal index is the maximum temperature an insulator can be safely exposed to in operation for an average life of 20,000 hours based on specified controlled test conditions (e.g., IEEE). This time is short. In practice, therefore, the rated temperature should be marginally reduced to extend the operating life of the insulation. A 10 °C drop in operating temperature approximately doubles insulation life. Short-term thermal resistance refers to thermal stress durations of a few hours at maximum. Exposed to short-term thermal stresses, an insulator may melt, bubble, shrink, or become charred. The insulation material should not be damaged in any of the aforementioned ways if the safe operating temperature is only moderately exceeded during normal operation. In Table 11.2, temperature rise refers to the permitted temperature rise of a winding at rated load. This temperature rise is not sufficient to cause premature aging of the insulator. An insulator exposed to excessive temperature fluctuation may become brittle and/or develop cracks. In certain operating situations, the frost resistance of the insulation material may become important.

The following tolerances are specified by IEC 60034-1.

Power factor is measured at rated power. The tolerance limit compliant with the standard is $(1/6)(1 - \cos \varphi)$, however, the error must be in the range of [0.02, 0.07].

Voltage and frequency-The tabular values hold for rated power and frequency. The motor may be constantly loaded at its rated power if the voltage deviates from the rated value ±5% at maximum. In that case, the temperature rise may be 10 K. The voltage may deviate only temporarily 5–10% from the rated value.

Slip-The tolerance for slip is ±20%.

Efficiency-The tolerance for efficiency is $-10\%\ (1 - \eta)$ when the motor power is greater than 50 kW and $-15\%\ (1 - \eta)$ when the motor power is less than or equal to 50 kW.

Rotation speed-AC motors must withstand at least 1.2 times the rated speed.

Table 11.3 Torque requirements of a three-phase induction motor producing normal torque (N-type) in compliance with IEC 60034-12. See the definitions of T_1, T_u, and T_b in Figure 11.5

Power Range [kW]	Number of Poles											
	2			4			6			8		
	T_1	T_u	T_b	T_1	T_u	T_b	T_1	T_u	T_b	T_1	T_u	T_b
$0.4 < P_N \le 0.63$	1.9	1.3	2.0	2.0	1.4	2.0	1.7	1.2	1.7	1.5	1.1	1.6
$0.63 < P_N \le 1.0$	1.8	1.2	2.0	1.9	1.3	2.0	1.7	1.2	1.8	1.5	1.1	1.7
$1.0 < P_N \le 1.6$	1.8	1.2	2.0	1.9	1.3	2.0	1.6	1.1	1.9	1.4	1.0	1.8
$1.6 < P_N \le 2.5$	1.7	1.1	2.0	1.8	1.2	2.0	1.6	1.1	1.9	1.4	1.0	1.8
$2.5 < P_N \le 4.0$	1.6	1.1	2.0	1.7	1.2	2.0	1.5	1.1	1.9	1.3	1.0	1.8
$4.0 < P_N \le 6.3$	1.5	1.0	2.0	1.6	1.1	2.0	1.5	1.1	1.9	1.3	1.0	1.8
$6.3 < P_N \le 10$	1.5	1.0	2.0	1.6	1.1	2.0	1.5	1.1	1.8	1.3	1.0	1.7
$10 < P_N \le 16$	1.4	1.0	2.0	1.5	1.1	2.0	1.4	1.0	1.8	1.2	0.9	1.7
$16 < P_N \le 25$	1.3	0.9	1.9	1.4	1.0	1.9	1.4	1.0	1.8	1.2	0.9	1.7
$25 < P_N \le 40$	1.2	0.9	1.9	1.3	1.0	1.9	1.3	1.0	1.8	1.2	0.9	1.7
$40 < P_N \le 63$	1.1	0.8	1.8	1.2	0.9	1.8	1.2	0.9	1.7	1.1	0.8	1.7
$63 < P_N \le 100$	1.0	0.7	1.8	1.1	0.8	1.8	1.1	0.8	1.7	1.0	0.7	1.6
$100 < P_N \le 160$	0.9	0.7	1.7	1.0	0.8	1.7	1.0	0.8	1.7	0.9	0.7	1.6
$160 < P_N \le 250$	0.8	0.6	1.7	0.9	0.7	1.7	0.9	0.7	1.6	0.9	0.7	1.6
$250 < P_N \le 400$	0.75	0.6	1.6	0.75	0.6	1.6	0.75	0.6	1.6	0.75	0.6	1.6
$400 < P_N \le 630$	0.65	0.5	1.6	0.65	0.5	1.6	0.65	0.5	1.6	0.65	0.5	1.6

Instantaneous overcurrent-AC generators must withstand 1.5 times the rated current for a duration of 30 s (15 s for generators above 1200 MVA). Three-phase AC generators below 1 kV and 315 kVA must withstand 1.5 times the rated current for a duration of 2 min. The capability of withstanding instantaneous overcurrent for three-phase motors larger than this has not been specified. Neither has a specification been defined for any single-phase motors.

Torque is the most important motor parameter. IEC 60034-12 defines the torque requirements for various induction motors within normal and high torque ranges and for star-delta starting. In all, there are four motor types: N (normal torque), H (high torque), and NY (normal torque, star-delta starting) and HY (high torque, star-delta starting). The locked rotor torque T_1, the saddle torque T_u, and the pull-out torque T_b are given for machines producing normal torque as a relative value of the rated torque according to Table 11.3. These values are minimum values at the rated voltage ($U_N \le 690$ V). The torque values required of the H-type machines are notably higher. Relative starting torque must vary from 2 for large machines to 3 for small machines. Correspondingly, saddle torque varies between 1.4 and 2.1. In NY-motor start-up, T_1 and T_u

must be 25% of the values of equivalent N-motors. In DOL applications, the starting of large induction motors may be difficult for high torque loads. The minimum torque produced in DOL drives, for example by a 630 kW N-type motor, is only 50% of the rated torque. If higher starting torque is needed, a frequency converter drive offers a straightforward solution that also stresses the supply network significantly less than the DOL drive.

Unless otherwise determined, the breakdown torque T_b of the machine should be at least 160% of the rated torque under rated circumstances. At the rated voltage, the minimum torque T_u for three-phase machines below 100 kW must be at least 0.5 times the rated torque, and 0.5 times the locked-rotor torque. Correspondingly, the minimum torque must be at least 0.3 times the rated torque and 0.5 times the locked-rotor torque for motors ≥ 100 kW. For single-phase motors and three-phase multi-speed motors, the minimum torque must be at least 0.3 times the rated torque.

Irrespective of duty type, polyphase asynchronous motors and DC motors must withstand torques that exceed their rated values by 60% for a duration of 15 s without a sudden change in rotation speed. Machine frequency and voltage must remain at the rated values.

An asynchronous motor with a wound rotor (i.e., a slip-ring machine) must withstand a torque that exceeds its rated value by 35% for a duration of 15 s. Correspondingly a nonsalient-pole synchronous motor must withstand 35% overtorque, and a salient-pole machine must withstand 50% overtorque without pulling out of synchronism when the excitation remains at the rated value.

Starting current and starting time in DOL applications

When starting, DC braking, or reversing the direction of rotation in an induction motor operating direct on line, the ratio of actual-to-rated current increases by orders of magnitude, because at stall, rotor squirrel-cage currents heavily oppose stator-current–induced flux. The rotor currents force the flux to travel mostly around the rotor, and therefore the machine exhibits transient inductance that largely corresponds to the sum of the stator and rotor leakage inductances. Current magnitude also affects line voltage drop and temperature rise. Voltage drop also reduces the starting torque of the motor and may disturb other devices operating direct on line, particularly in a frail grid. The excessive temperature rise of the motor, in turn, weakens and reduces the service life of the electrical insulation materials.

International standards delimit the starting torque of induction motors by defining a maximum apparent power or a maximum current for a motor at stall. Table 11.4 presents the maximum permitted starting apparent powers proportioned to the motor shaft output power in compliance with the IEC standard.

Table 11.4 Maximum permitted starting apparent powers S_{st} in compliance with the IEC when the motor is at stall supplied at its rated voltage

Motor Rated Power P_N [kW]	S_{st}/P_N
$0.4 < P \leq 6.3$	13
$6.3 < P \leq 25$	12
$25 < P \leq 100$	11
$100 < P \leq 630$	10

For instance the starting apparent power of a 4 kW motor may be 52 kVA, which at 400 V yields the permitted starting current of 75 A [52000 VA/(400 V $\times \sqrt{3}$) = 75 A]. The rated current for a 400 V and 4 kW motor is typically 7.5 A, and therefore the maximum allowed starting current could be 10 times the rated motor current. A current demand this high can easily burden the supply network resulting in a significant network voltage drop. The motor would have to start with lowered voltage, which results in a significantly reduced starting torque, because torque capability is proportional to the square of the terminal voltage.

Traditionally, DOL motors have used starting currents that are below the limiting values. However, reducing losses and increasing efficiency often results also in increasing starting current level, so the highest efficiency motors may have starting currents that reach the limiting values.

The allowed maximum DOL starting time for motors at normal operating temperature varies between 10 and 40 s depending on motor size and number of poles. With star-delta starting, these times triple. If the motor is at room temperature prior to starting, the starting times may double.

Industrial DOL induction motors and their losses

In industry, electric motors account for more than 80% of total electricity consumption. Of that 80%, about 75–80% goes to asynchronous motors. On average, 6% of the electricity consumption of the asynchronous machines can be attributed to losses in the machine.

In the wood-processing industry, for example, asynchronous motors are by far the largest electrical energy consumer group. A single plant may operate more than 10,000 asynchronous motors, and 80–90% of the asynchronous motors run at powers of 37 kW or below. Previously, no special attention was paid to selecting small induction motors best adapted to the purpose or to the efficiency of the motors. Figure 11.7 presents the power balance of a

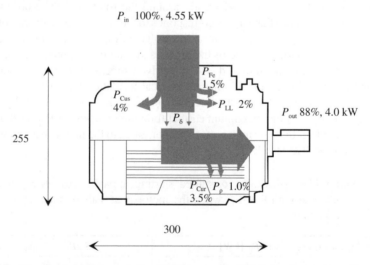

Figure 11.7 The Sankey diagram of a 4 kW two-pole IE3 induction motor. P_{Fe} represents the iron losses, P_{Cus} is the resistive stator losses, and P_{LL} represents additional load losses. P_δ is air-gap power, P_{Cur} is resistive rotor losses, and P_ρ represents friction losses. The sum of these losses (550 W total) must be removed from the machine using the available temperature difference with respect to ambient temperature. *Source:* Pyrhönen et al. (2014). Reproduced with permission of John Wiley & Sons Ltd.

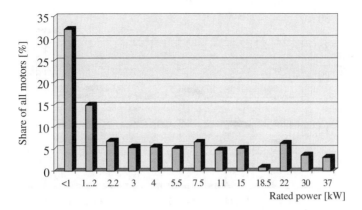

Figure 11.8 Size distribution of induction motors with power of 37 kW or below in a large wood-processing plant with more than 10,000 installed induction motors (Kaukonen, J. (1994). *Epätahtimoottoreiden sähkönkulutus ja häviöt puunjalostusteollisuudessa.* M.Sc. thesis, Lappeenranta University of Technology).

typical IE3 4 kW induction motor. Operating at rated power, 12% of the supplied electrical energy is lost as heat. The proportion of the copper (load) losses is quite large – 62.5% of the total losses. Because of the high cost of copper, these asynchronous machines are designed for general-purpose use, and therefore they are not necessarily dimensioned for the demands of the drives operating constantly close to their rated power. In fact, drives of this kind would require a machine type of their own.

Figure 11.8 illustrates the distribution of different motor sizes in a wood-processing plant with more than 10,000 installed motors.

An induction motor is well adapted to industrial drives thanks to its simple and durable configuration and good loading characteristics. Asynchronous motors account typically for 70–80% of the total industrial electrical energy consumption. Small-scale motors (≤37 kW) account for 15–20% of the total number of asynchronous motors. Figure 11.9 shows the percentage distribution of the electrical energy consumption for asynchronous motors in a typical wood-processing plant.

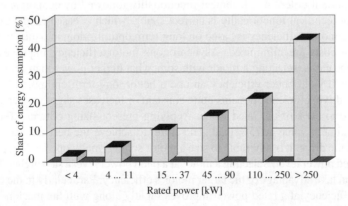

Figure 11.9 Percentage distribution of electrical energy consumption for asynchronous motors at a large wood-processing plant (Kaukonen, J. (1994). *Epätahtimoottoreiden sähkönkulutus ja häviöt puunjalostusteollisuudessa.* M.Sc. thesis, Lappeenranta University of Technology).

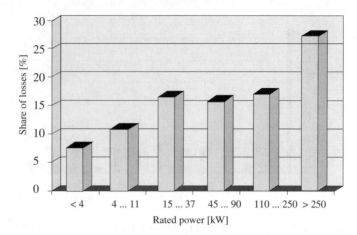

Figure 11.10   Distribution of losses in asynchronous motors by power class at a large wood-processing plant (Kaukonen, J. (1994). *Epätahtimoottoreiden sähkönkulutus ja häviöt puunjalostusteollisuudessa*. M.Sc. thesis, Lappeenranta University of Technology).

At the subject wood-processing plant, losses account for about 6% of the electrical energy consumed by asynchronous motors. Motors with a rated power of 37 kW or below account for 35% of all losses. Comparing the ratio of losses to electricity consumed by the motor group as a whole reveals that the relative amount of energy consumed in losses in the motors with a rated power of 37 kW or below is 2.5 times as high (37%/15%, share of losses/share of energy consumption) as in the motors above 37 kW.

Appropriate DOL motor selection

Recently, the EU has established new regulations concerning electric motor efficiencies. The IE-classes (IEC/EN 60034-30) dictate the minimum efficiencies of induction motors for different applications. Figure 11.11 illustrates the efficiency limits for efficiency classes IE1–IE4.

An asynchronous motor drive must be based on a motor best adapted to the target purpose. The most important criterion is technical functionality followed by economic efficiency. A prerequisite of technical functionality is correct sizing, which is based on the expected load conditions. Economic efficiency is based on long-term optimization, in which purchase price and operating costs are minimized. Asynchronous motor efficiency improves with rated power. In some cases, selecting a motor with somewhat higher power than actually required for the duty will yield better efficiency and be a better long-term optimum.

In a DOL drive, efficiency remains quite constant or increases slightly when power drops from the rated to 50% of the rated value. Applying present sizing criteria, efficiency often reaches a maximum value at 75% load. Therefore, to achieve the best efficiency within the limitations of the drive, it is advisable to size a motor so it normally operates at approximately 75% of its rated power level. Selecting a motor at a rated power level a step higher than required by the load can take advantage of the larger motor's efficiency. Particularly in the case of small motors, the efficiency at a rated power improves rapidly along with the machine size.

Figure 11.12 illustrates the efficiency of some DOL motors as a function of both relative and absolute power.

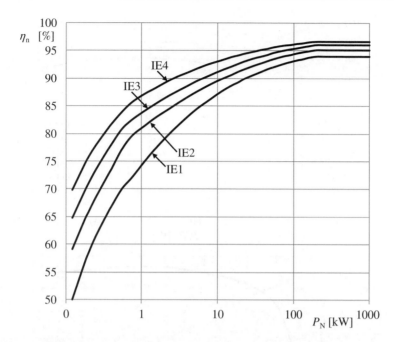

Figure 11.11 Minimum nominal efficiencies η_n for four-pole asynchronous motors as a function of rated power P_N. Class IE1 motors can no longer be sold. IE2 motors may be sold for frequency converter drives. DOL motors must be at least class IE3. With an induction motor, it is difficult to achieve a class IE4 efficiency level and simultaneously fulfil all the boundary conditions set by standards.

The figure clearly shows which machine should be chosen for a certain DOL duty to maximize efficiency. The motor with the topmost efficiency curve at the shaft output power in question should always be selected. The figure supports the advisability of sizing the motor so it operates at about ¾ power, which means choosing a motor that is rated at the next higher power level. Selecting a larger or smaller motor results in lower efficiency.

Consider what happens if the selected motor is too large. A 5.5 kW motor will operate at about 90% efficiency if the actual power required is 5 kW. A more appropriately sized 7.5 kW motor will operate at about 91.5% under the same load conditions. This represents a substantial improvement in efficiency. However, if a 15 kW motor is used, it will be operating at only 33% of its rated power, and its efficiency at this point will be about 89.5%.

Motor cooling, the selection of motor size, and duty type in DOL applications

In electric machines, part of the supplied energy is always converted into heat. For instance, when the efficiency of a 4 kW standard IE3 induction machine at rated power is 88%, there are 4% and 1.5% heat losses in the stator copper and stator iron, respectively, and 3.5% heat losses in the rotor. Friction and additional losses amount to about 3% of the total. The remaining 88% of the supplied energy is converted to the mechanical energy of the shaft. The internal heat being produced in the machine must be removed by transferring it to the surrounding environmental medium. Electric machine cooling methods are defined in IEC 60034-6. Enclosure classes are defined in IEC 60034-5. The enclosure class depends on the

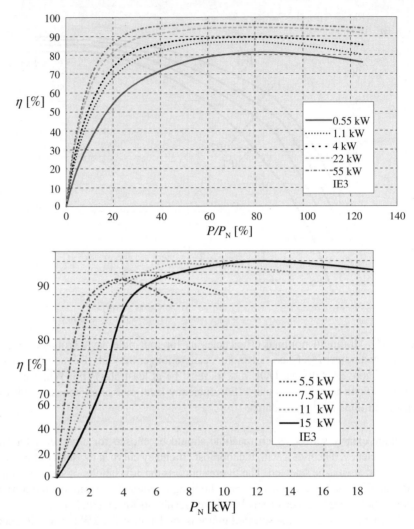

Figure 11.12 Class IE3 three-phase four-pole DOL induction motor efficiencies as a function of load at rated voltage.

cooling method. For instance, the enclosure class IP 44 designates a good mechanical and moisture protection that are not compatible with the cooling method IC 01, since it requires an open machine. Table 11.5 introduces the most common IC classes.

Electric machine duty types are designated according to IEC 60034-1 (2004) as S1, S2, S3 . . . S10.

Duty type S1 – continuous running duty

Operation at constant load maintained for sufficient time to allow the machine to reach thermal equilibrium - A machine for this kind of operation is stamped with the abbreviation

Table 11.5   Most common IC classes of electric machines

Code	Definition
IC 00	The coolant surrounding the machine cools the inner parts of the machine. The ventilating effect of the rotor is insignificant. The coolant is transferred by free convection.
IC 01	Like IC 00, but with an integral fan mounted on the shaft or the rotor to circulate the coolant, a common cooling method for open induction motors
IC 03	A method similar to IC 01, but with a separate motor-mounted blower having same power source with the machine to be cooled
IC 06	A method similar to IC 01, but the coolant is circulated with separate motor mounted blower with different power source. There can also be a single extensive blower system supplying the coolant for several machines.
IC 11	The coolant enters the machine via a ventilating duct and passes freely to the surrounding environment. The circulation of the coolant is carried out with a motor or shaft mounted blower.
IC 31	The rotating machine is inlet and outlet pipe ventilated. The circulation of the coolant is carried out with a motor or shaft mounted blower.
IC 00 41	Totally enclosed internal circulation of the coolant by convection and cooling through the frame with no separate blower.
IC 01 41	Like IC 00 41, but the frame-surface cooling takes place with a separate shaft mounted blower causing the circulation of the coolant. *This is the cooling method used in ordinary, totally enclosed fan-cooled induction motors.*
IC 01 51	Totally enclosed internal cooling by convection - Heat is transferred through an internal air-to-air heat exchanger to the surrounding medium, which is circulated by a shaft mounted blower.
IC 01 61	Like IC 01 51, but the heat exchanger is mounted on the machine.
IC W37 A71	Totally enclosed internal cooling by convection - The heat is transferred through an internal water-to-air heat exchanger to the cooling water, which is circulated either by supply pressure or an auxiliary pump.
IC W37 A81	Like IC W37 A71, but the heat exchanger is mounted on the machine.

S1. This is the most common duty type and normally stamped on the nameplate if no special requirements are set to a machine.

Duty type S2 – short time duty

Operation at constant load for a given time, less than that required to reach thermal equilibrium - Each operation period is followed by a time at deenergized rest of sufficient duration to re-establish the temperature of the surrounding air. For machines of short time duty, the recommended durations of the duty are 10, 30, 60, and 90 minutes. The appropriate abbreviation is S2, followed by an indication of the duration of the duty, stamping, for example, S2 60 min. The same motor that is stamped 1.1 kW S1 may be stamped, for example, 1.5 kW S2 30 min.

Duty type S3 – intermittent periodic duty

A sequence of identical duty cycles, each including a time of operation at constant load and a time at de-energized rest. Thermal equilibrium is not reached during a duty cycle. Starting currents do not significantly affect the temperature rise. The cyclic duration factor is 15%, 25%, 40%, or 60% of the 10-minute duration of the duty. The appropriate abbreviation is S3, followed by the cyclic duration factor, stamping, for example, S3 25%.

Duty type S4 – intermittent periodic duty with starting

A sequence of identical duty cycles, each cycle including a significant starting time, a time of operation at constant load and a time at deenergized rest. Thermal equilibrium is not reached during a duty cycle. The motor stops by naturally decelerating, and therefore the motor is not thermally stressed. The appropriate abbreviation for stamping is S4, followed by the cyclic duration factor, the number of cycles in an hour (c/h), the moment of inertia of the motor (J_M), the moment of inertia of the load (J_{ext}) referred to the motor shaft, and the permitted average counter torque T_v during a change of speed given by means of the rated torque. Stamping, for example, S4 - 15% - 120 c/h - $J_M = 0.1$ kgm^2 - $J_{ext} = 0.1$ kgm^2 - $T_v = 0.5$ T_N.

Duty type S5 – intermittent periodic duty with electrical braking

A sequence of identical duty cycles, each cycle consisting of a starting time, a time of operation at constant load, a time of braking, and a time at deenergized rest. Thermal equilibrium is not reached during a duty cycle. In this duty type, the motor is decelerated with electrical braking, for example, counter-current braking. The appropriate abbreviation for stamping is S5, followed by the cyclic duration factor, the number of cycles per hour (c/h), the moment of inertia of the motor J_M, the moment of inertia of the load J_{ext}, and the permitted counter torque T_v. Stamping, for example, S5 - 60% - 120 c/h-$J_M = 1.62$ kgm^2 - $J_{ext} = 3.2$ kgm^2 - $T_v = 0.35$ T_N.

Duty type S6 – continuous-operation periodic duty

A sequence of identical duty cycles, each cycle consisting of a time of operation at constant load and a time of operation at no-load. Thermal equilibrium is not reached during a duty cycle. The cyclic duration factor is 15%, 25%, 40%, or 60% and the duration of the duty is 10 min. Stamping, for example, S6 60%.

Duty type S7 – continuous-operation periodic duty with electrical braking

A sequence of identical duty cycles, each cycle consisting of a starting time, a time of operation at constant load and a time of electrical braking. The motor is decelerated by countercurrent braking. Thermal equilibrium is not reached during a duty cycle. The appropriate abbreviation is S7, followed by the moment of inertia of the motor, the moment of inertia of the load and the permitted counter torque (cf. S4). Stamping, for example, S7 - 500 c/h - $J_M = 0.06$ kgm^2 - $T_v = 0.25$ T_N.

Duty type S8 – continuous-operation periodic duty with related load/speed changes

A sequence of identical duty cycles, each cycle consisting of a time of operation at constant load corresponding to a predetermined speed of rotation, followed by one or more times of operation at other constant loads corresponding to different speeds of rotation (carried out, for example, by means of a change in the number of poles in the case of induction motors). There is no time at deenergized rest. Thermal equilibrium is not reached during a duty cycle. The appropriate abbreviation is S8, followed by the moment of inertia of the motor, the moment of inertia of the load, and the number of duty cycles in an hour. In addition, a permitted counter torque and the cyclic duration factor must be given. Stamping, for example, S8 - $J_M = 2.3$ kgm^2 - J_{ext} 35 kgm^2, 30 c/h - $T_v = T_N$ - 24 kW - 740 r/min - 30%, 30 c/h - $T_v = 0.5\ T_N$ - 60 kW - 1460 r/min - 30%, 30 c/h - $T_v = 0.5\ T_N$ - 45 kW - 980 r/min - 40%

Duty type S9 – duty with nonperiodic load and speed variation

Duty type S10 – duty with discrete constant loads

Loads and combinations of rotation speeds are stamped in the order they occur in the duty cycle.

When choosing the motor, the torque profile of the load and the thermal time constant of the machine must be known. When the duty comprises periods that are clearly shorter than the thermal time constant of the motor, it is possible to overload the motor for short periods if the average losses are smaller than the rated losses of the motor. The thermal time constants of a totally enclosed induction motor vary depending on machine size from tens of minutes to an hour or even to several hours. For instance, according to LUT measurements, a standard IE3 four-pole 15 kW TEFC induction motor, operating at the rated load, reaches its end temperature in about five hours, and its thermal time constant is about 45 minutes (Kärkkäinen 2015). In the dimensioning of the motor, an RMS value must be determined for the effective power P_{ef}, and the motor rated power P_N must be at least this equivalent power. According to the definition of the RMS value, the following statement can be written.

$$P_N \geq P_{ef} = \sqrt{\frac{1}{t_{cef}} \int_0^{t_o} P^2(t)\mathrm{d}t} \tag{11.15}$$

In Equation (11.15), $P(t)$ is the motor operating power as a function of time, t_o is the operating period (motor current on), t_{cef} is the effective cooling period depending on the motor running or standing. Of the total time the motor is standing, only 20% can be counted in the effective cooling time.

The duty types S1 (continuous running duty) may be defined with Equation (11.15). Also S3 (intermittent periodic duty), and S6 (continuous-operation periodic duty) allow dimensioning based on Equation (11.15) as long as the cycle time t_j is short compared to the thermal time constant of the machine. Motor selection in compliance with duty type S2 requires specific knowledge of the machine. In-depth technoeconomical dimensioning analysis must be carried out if heavy starting and braking duties are expected.

In rated duty, rated losses are removed by the cooling system. From the thermal point of view, the motor selection in the periodic duties is based for instance on the definition of dissipation energy balance. In that case, a dissipation power corresponding to the rated duty is removed from the machine during the effective cooling time t_{cef} of the cycle. Therefore, it is possible to base the dimensioning of the motor on the determination of the rated current corresponding to the load. The rated current I_N of the motor corresponding to the effective I_{ef}

load current must be

$$I_N \geq I_{ef} = \sqrt{\frac{1}{t_{cef}} \int_0^{t_o} I^2(t)dt} \tag{11.16}$$

In dimensioning the motor rated current, the operating current may be divided into sub-currents according to corresponding operating cycle sub-periods, which are then used to calculate the effective motor current.

$$I_N \geq I_{ef} = \sqrt{\frac{1}{t_{cef}} \left[\int_0^{t_{o1}} I_1^2(t)dt + \int_{t_{o1}}^{t_{o2}} I_2^2(t)dt + \int_{t_{o2}}^{t_{o3}} I_3^2(t)dt + \ldots + \int_{t_{o(n-1)}}^{t_{on}} I_n^2(t)dt \right]}. \tag{11.17}$$

The stator current of an asynchronous machine also includes inductive current, and therefore the power factor of the motor is included in the calculation. In the case of partial loads, the power factor of the motor must be employed. Actually, Equation (11.15) is not accurate for induction motors as they have different efficiencies and different power factors at different loads. Therefore, current-based estimations may give results that are more reliable.

Determining equivalent cooling time is problematic, since the cooling properties of the motor are usually highly dependent on rotation speed. As a rule of thumb, the cooling power of a machine at rest and de-energized is about 20% of the rated cooling power. Therefore, for example, for duty type S3, 20% of the standing time is included in the effective cooling time. See Example 11.1.

EXAMPLE 11.1: Check whether a totally enclosed fan cooled (TEFC) motor, with S1 compliant motor power that is exceeded by 30%, can be applied to S3-25% duty in 10-minute cycles. The power factor at 130% is $\cos \varphi = 0.87$ and $\cos \varphi = 0.85$ at rated power. The effect of changing efficiency can be ignored.

SOLUTION: If the time constant of the subject machine is long enough relative to the 10-minute cycle, the following approach can be used. The thermal time constants of TEFC induction motors are typically tens of minutes. In the case of larger motors, the time constants can be even hours. The effective cooling time is $t_{cef} = 0.25 + 0.20 \times 0.75 = 0.4$ of the operating period. The motor operating current during the load period is $I = 1.30$ $I_N \times 0.85/0.87 = 1.27$ I_N. These values are used in (11.17) to determine the effective current of the machine.

$$I_{ef} = \sqrt{\frac{1}{0.4} \left(\int_0^{0.25} 1.27^2 I_N^2 dt + \int_{0.25}^{1.0} 0^2 dt \right)} = 1.0 I_N$$

Because the effective current corresponds to the motor rated current, and the operating period is short, it can be assumed that the motor windings reach the same temperature in this S3-25% 10-minute 30% overload compared to S1 cycle or S1 rated operating.

The dimensioning is checked with Equation (11.15).

$$P_{ef} = \sqrt{\frac{1}{0.4} \int_0^{0.25} 1.3^2(t)dt} = 1.03P_N$$

The result obtained from Equation (11.15) ignores the improved power factor, and therefore indicates that a slightly larger machine than the initially dimensioned one is needed. However, if the ambient temperature of the motor is lower than 40 °C, and the mounting altitude is less than 1000 m, this motor can most probably be used safely.

However, if a motor is intended for high performance industrial use, it should probably be over dimensioned to ensure a longer operating life. According to Kärkkäinen (2015), the thermal time constant for a typical 15 kW four-pole 400 V asynchronous motor is approximately 54 minutes. For such an example, a 10 minute period can be regarded as "short" with respect to the thermal time constant of the motor.

11.3 Frequency converter drives – losses in a PWM inverter drive

Induction motor inverter drives are common. The sine-triangle comparison, commonly used in the PWM technique, produces PWM pulse trains for the voltage references between points U and N and V and N as was discussed in Chapter 7. Sidebands of the motor-supply voltage harmonics occur in regions near the multiples of the inverter switching frequency. Therefore, the lowest harmonics of the PWM inverter are found near the switching frequency. When the target is to obtain current waveform that is as close to sinusoidal as possible, the motor switching frequency must be raised. State-of-the-art motor control systems (e.g., the DTC) do not necessarily have fixed modulation frequencies, and therefore the frequency spectrum does not remain constant. However, harmonics occur in a manner that is somewhat similar to the occurrence of harmonics in a traditional PWM. The difference is that the frequency spectrum is spread over a wider range, and the peaks are not as clear as in the sine modulation.

The rate of current change in an inverter-driven induction motor at the inverter switching frequency is limited by the transient inductance L'_s of the machine. See Figure 11.34. The transient inductance characterizes the inductance experienced by the fast pulse that is fed to the stator. Only via its leakage inductance does the rotor manage to react to the fast pulse. This transient inductance can be determined, for instance, by employing the short pulses obtained from the inverter. In principle, the transient inductance of an induction motor corresponds somewhat accurately to the short circuit inductance L_k of the machine. In practice, $L'_s \neq L_k$ because of the different magnetic states in the regions where these inductances are measured.

In modern vector-controlled inverters, the pulse pattern does not necessarily follow a fixed modulation scheme. However, in steady state, the pulse pattern is selected to introduce to the inductive load as sinusoidal a current as possible.

Controlling an induction motor with an inverter drive is challenging. When inverters were first introduced, they seemed to offer effective induction motor control. However, thyristor PWM inverters operating at low switching frequencies increased losses to such a degree that continuous-duty rated powers had to be lowered by 10–20%. In thyristor inverters, the switching frequencies were in the range of only a few hundred Hertz. The low switching frequencies and the resulting slow voltage-rise rates were an advantage from the insulation point of view. With low du/dt rates, insulation stresses caused by the PWM are minimized. However, motor cables must be kept short. High du/dt rates may cause severe overvoltages in the windings that can damage the motor insulations. The switching frequencies of inverters that use, for example, IGC-Thyristors, are low, which distorts the current waveforms so the motors can only accommodate 90–95% loading at rated speed. With today's IGBTs or SiC MOSFETs, the switching frequency can be several kHz. du/dt filters are used in the inverters to protect motor insulation.

Irrespective of the switching frequency of the inverter, distortion is caused in the inverter-driven motor by the distortion voltages to the phase current of the motor. The high-frequency phenomena experience the transient inductance L'_s of the motor. The amplitudes of flux components of the switching frequency occurring in the air gap is quite small, and their penetration to the iron circuits is quite limited. Nevertheless, the inverter's switching-frequency loss components manifest in the induction motor. Higher switching frequencies result in better (nearly sinusoidal) motor current. Lower switching frequencies result in more distorted current and greater inverter-induced losses.

An inverter drive increases induction motor losses in the following ways.

- High-frequency current components increase skin effects in the windings. The stator winding of small machines, however, is usually made of thin enough wire that skin-effect losses are insignificant. If the winding of the machine is made of solid preformed copper, the losses caused by the skin effect are greater.

- Despite the damping effect of the leakage inductances, the high-frequency current components produce small high-frequency components in the air-gap flux of the machine, thus increasing the iron losses of the machine.

- The rotor surface of the machine is particularly susceptible to losses in the inverter drive. Depending on the ratio of the rotor flux leakage and the magnetizing inductance, a portion of the fast transients pass through the magnetizing inductance, producing high-frequency components in the air-gap flux. When current harmonics cause high-frequency air-gap time harmonics, the flux components tend to be damped by the eddy currents induced in the induction motor cage winding. Consequently, losses occur in the rotor conductors and especially in the iron near the rotor slot opening. See Figure 11.13.

To reduce the losses in the rotor slot opening, the slot opening type illustrated in Figure 11.14 was introduced for inverter motors. Since the saturating slot wedge is rather long, a slot opening of this type provides a smaller leakage flux in the rotor slot opening than traditional solutions. Simultaneously, the losses caused by eddy currents in the iron of the rotor slot opening are reduced.

An open rotor slot best prevents surface-loss problems in an inverter motor. Therefore, the rotor windings of motors intended for inverter drive are often manufactured of preformed copper. The copper is shaped to leave the rotor slot opening slightly open. On the other hand, this reduces the transient inductance of the machine and may cancel the effect of the open rotor slot.

Figure 11.15 depicts the largest voltage and current harmonic components shown as a frequency spectrum for a DTC converter fed induction motor. The phase voltage is

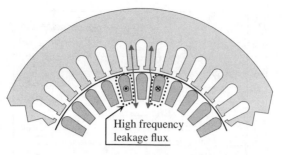

Figure 11.13 Because the air-gap flux includes harmonic components at the inverter switching frequencies, air-gap flux has corresponding high-frequency components. Currents that resist high frequency fluxes are produced in the rotor winding. These currents produce high-frequency leakage flux components that pass through the rotor slot opening. In the slot-opening region, there are considerable losses in the saturating iron, particularly in the case of closed slots.

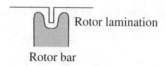

Figure 11.14 Example of slot-opening shape applied to inverter motors when using an aluminium pressure-cast rotor. This configuration provides a path for the leakage flux, and therefore the losses in the long iron bridge remain lower than in the case of ordinary rotor bars.

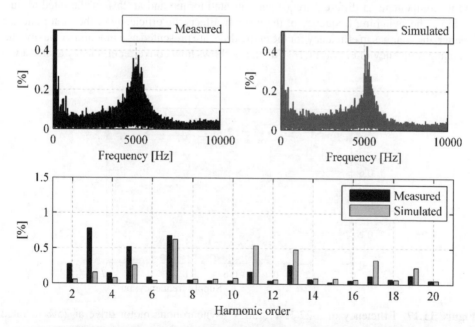

Figure 11.15 Measured and simulated spectra of phase currents from a DTC converter with an average switching frequency of 3 kHz with a 25 Hz fundamental frequency scaled out of the figure. *Source:* Reproduced with permission from Dr Aarniovuori.

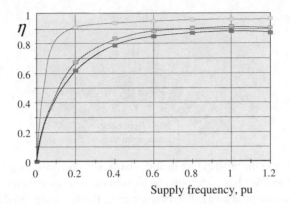

Figure 11.16 Efficiency of a 22 KW, 400 V asynchronous motor drive at rated torque. The uppermost curve corresponds to converter efficiency, the next curve down is motor efficiency, and the lowermost curve is measured drive efficiency. The peak system efficiency is 89%, and the motor rated efficiency at the sinusoidal supply is 93%. In this case, motor efficiency is reduced by about 1% in the converter supply.

considerably distorted; however, the phase current is almost sinusoidal including only small-amplitude harmonic components when applying an average switching frequency of 3 kHz.

To examine the efficiency behaviour of an IGBT inverter and a 22 KW, 400 V induction motor, Figures 11.16 and 11.17 present the measured efficiencies of an electrical motor drive and its components at different frequencies at rated torque and at 25% of the rated torque.

Since the switching frequency of the inverter affects the efficiency of the motor and the inverter, the changes in efficiency for the entire drive are presented as a function of the inverter switching frequency in Figure 11.18. The figure shows how converter efficiency decreases as

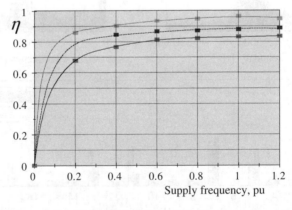

Figure 11.17 Efficiency of a 22 KW, 400 V asynchronous motor drive at 25% of rated torque. The uppermost curve corresponds to converter efficiency, the next curve down is motor efficiency, and the lowermost curve is the measured drive efficiency. These efficiencies are clearly lower than at rated torque.

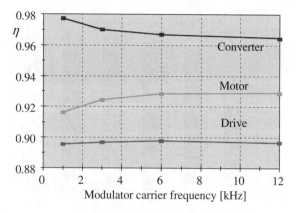

Figure 11.18 The effect of modulator carrier frequency, which is naturally close to the average switching frequency, on the efficiency of a converter drive at 75% torque and a rated supply frequency of 50 Hz. The maximum carrier frequency of the inverter is 12 kHz. Motor current is nearly sinusoidal at this carrier frequency, and motor efficiency is almost as high as when driven by a sinusoidal line voltage. These measurements were carried out at the highest motor efficiency at about 75% power and at the rated frequency. The peak efficiency of the drive occurs at approximately 6-kHz carrier frequency.

switching frequency increases, and how the efficiency of the motor improves. The decrease in inverter efficiency is clearly due to increased switching losses. In the example case, as the modulator carrier frequency increases, motor losses decrease faster than inverter losses, and therefore the efficiency of the whole drive improves with the increased carrier frequency until reaching a 6-kHz carrier frequency. At 12 kHz, the motor efficiency increase is smaller than the inverter efficiency decrease, and therefore total drive efficiency begins to decrease.

Finally, Figure 11.19 sums up the total efficiencies as a function of supply frequency with modulator carrier frequency as a parameter when the motor is running at its rated torque. The figure shows that the efficiency of the drive does not significantly depend on the carrier frequency of the output stage of the inverter. In this example case, the changes in the inverter and motor losses nearly cancel. In general, the PWM converter and induction motor combination can provide high efficiency over different frequency and torque ranges.

As speed decreases, the cooling effectiveness of the totally enclosed fan-cooled machine starts to decrease rapidly, and its torque must be limited. Figure 11.20 illustrates how much load the machine can tolerate when supplied with different switching frequencies and different modulation principles.

Figure 11.21 depicts the estimated load capacity of a totally enclosed fan-cooled induction motor in corresponding circumstances.

11.4 Frequency converter control methods for an induction motor

In chronological order, the control methods for an induction motor are scalar control, vector control, and DTC. The first inverters were based on the scalar control principle. Blaschke (1971)

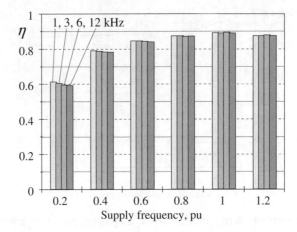

Figure 11.19 The efficiency of the example induction motor drive as a function of supply frequency at the rated torque with modulator carrier frequency as a parameter. The bars are arranged in the following order: 1 kHz, 3 kHz, 6 kHz, and 12 kHz. Carrier frequency does not significantly influence drive efficiency. As machine efficiency improves with increased carrier frequency, inverter efficiency drops. In general, a carrier frequency of 3 kHz will result in nearly sinusoidal motor current and a machine efficiency that cannot be significantly improved by further increasing carrier frequency.

introduced the idea of vector control for rotating field machines in the early 1970s, and vector control was more widely adopted in the 1980s. In 1986, direct flux-linkage control (DFLC) was introduced, and it was followed by direct torque control when ABB introduced its DTC inverter in 1994.

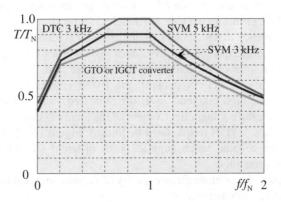

Figure 11.20 Load capacity of a TEFC (IC 01 41) motor driven by various inverter types. For DTC inverters with a 3 kHz average switching frequency or space-vector–modulated inverters with a 5 kHz switching frequency, the machine can be operated at rated torque and at its rated speed without overheating. The middle curve relates to space-vector–modulated inverters at 3 kHz switching frequency. With GTO converters using switching frequencies lower than 1 kHz, the lowest curve must be followed. At speeds above rated, torque is restricted by field weakening (based on public data published by ABB).

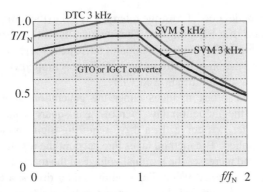

Figure 11.21 Load capacity of a totally enclosed inverter-driven induction motor with an independent motor-shaft-driven fan. Exterior cooling is effective, but rotor heat exchange is impaired, because the integral blower is dependent on the rotation speed of the machine. The effect of switching frequency is clear. GTO inverters typically apply switching frequencies below 1 kHz, which results in motor current with high harmonic content that increases losses (based on public information from ABB).

11.4.1 Scalar control

Scalar control is based chiefly on steady-state motor information. The scalar control parameters are motor frequency, motor voltage, and corrections (adjustments) made in response to current measurements. Figure 11.22 shows a schematic of inverter scalar control.

The external reference of a scalar-controlled inverter drive system is either a frequency, a torque, or a speed reference. The frequency reference is fed directly through the voltage

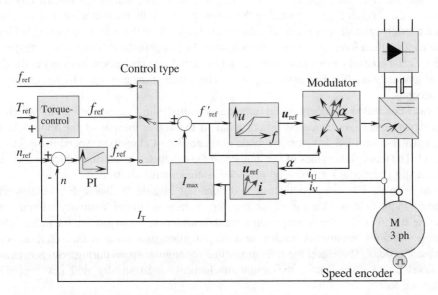

Figure 11.22 Schematic of the control system for a scalar-controlled inverter.

reference block to the modulator. The torque reference is fed to the torque controller, which is usually a digital version of a PI controller. The inverters typically also include a digital PI controller for speed control. The implementation of a frequency reference f_{ref} might cause an inverter overcurrent, and therefore the frequency reference is limited via a block that monitors current and limits it to a maximum allowed value.

The voltage reference block includes either the generation of the constant u/f ratio obtained straight from the induction law, a root-mean-square voltage curve, or an IR-compensated voltage reference curve. In a drive requiring high torque at low speeds, IR-compensation is also applied. The method takes into account the resistive voltage drop of the inverter and the motor and increases the terminal voltage of the machine at low speeds to keep the stator flux linkage at its rated value. For slightly lighter drives, a constant ratio for the voltage and frequency can be applied. To save energy in blower drives, motor voltage can be reduced at low frequencies and a quadratic voltage curve can be used.

The inverter modulator is often a unit that digitally implements sine-triangle comparison and generates the references for the inverter changeover switches. As its references, the modulator requires a frequency and amplitude modulation stage that corresponds to the desired voltage. The modulators of modern scalar inverters implement asynchronous pulse width modulation. Conventional thyristor inverters use a synchronous modulation technique. In certain cases, a third harmonic is added to the voltage reference to increase the RMS value of the voltage. However, in Figure 11.22, a space vector modulator is used, which enables more sophisticated drive system control. With space vector modulation, for example, estimation of torque producing current in steady state becomes possible.

Torque control is possible by estimating the torque producing current I_T of the motor. When the space vector modulator produces voltage vectors, an instantaneous average (the average calculated at a point in time over a very short time period) can be computed for the vectors to estimate the exact position α of the voltage vector. A corresponding current vector is determined with the measured phase currents. The current vector has amplitude and phase angle, and therefore the angle between the current and voltage can be calculated. The active current of the inverter is proportional to the torque produced by the motor, so it is possible to obtain a rough estimate for the actual value of the torque from the active current. This kind of scalar control, therefore, can estimate the torque of the motor in the absence of an actual motor model. Scalar control cannot react immediately to rapid step changes in torque, but the drive moves gradually into the new operating state. The settling time can typically be hundreds of milliseconds.

Scalar-controlled inverters often include so-called slip compensation, in which the frequency reference is increased proportional to the active current of the motor. In this way, a scalar-controlled drive can be made to rotate almost at constant speed irrespective of the load. The nameplate values of the motor and the calculated rated slip, which is needed to perform slip compensation, must be provided to the control electronics.

The simplest U/f speed-control method does not include feedback. Particularly when operating in the field-weakening region, the stepped increase in the frequency reference may cause the slip to exceed the corresponding breakdown torque value moving the motor into an unstable operating regime. A similar unstable situation may arise if the reference value decreases in steps. Therefore, the reference value for angular speed during both acceleration and deceleration must follow the actual mechanical speed so slip will not exceed the breakdown torque.

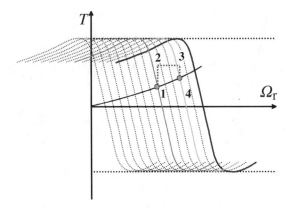

Figure 11.23 Stable increase in rotation speed in scalar control from operating point 1 to operating point 4. The figure depicts the static torque curves of an asynchronous machine at different supply frequencies.

Figure 11.23 illustrates stable rotor acceleration using scalar control. When the control system receives a stepwise change in the rotation speed reference and raises the frequency reference, the motor slip will increase until the stator current reaches the set limit. The situation is illustrated by the range 1–2 in the torque-angular-speed graph. Next, the frequency is increased under current limitation in the constant torque range (2–3). After this, the stator current decreases until a new point 4 continuous state is reached.

If the U/f ratio cannot be held constant, the air-gap flux linkage produced may be saturated or remain below the desired value. The fluctuation of the stator circuit parameters resulting from temperature and saturation may cause fluctuations in the air-gap flux linkage. When the machine's air-gap flux linkage decreases, slip must increase (an increase in the stator current) to reach the same torque. As a result, the transient stability of the motor will be reduced.

Field weakening in an induction motor

Field weakening is necessary when operating the motor at higher than rated speed is the goal. The field weakening behaviour of an induction motor is complex. Care must be taken not to overload it, especially in the case of scalar control, which lacks a motor model. In vector control, similar overloading problems can be avoided, because the motor model predicts motor behaviours.

In field weakening, motor voltage cannot be increased along with rotation speed, since the available voltage reserve is already being used, that is, the voltage has reached its rated value. If the voltage is kept at its rated value while frequency increases, the U/f ratio decreases and flux linkage diminishes. The breakdown torque of an asynchronous machine is reached when the angle γ between the stator and rotor flux linkage is about 45°. If the angle increases, the magnitude of the air-gap flux linkage drops because of the stator and rotor leakage inductances. When operating at high torque in the deep field-weakening region, the magnitudes of stator and rotor current and the angle between them increase. Therefore, the magnetising current i_m and corresponding air-gap flux linkage ψ_m decrease. The magnetizing

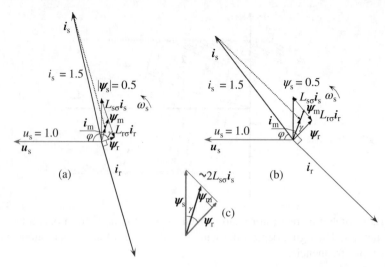

Figure 11.24 Vector diagrams of an induction machine in the deep field-weakening region at about twice the rated speed (a) operating above its breakdown torque slip and (b) operating near the breakdown point. Because $\omega_s = 2$, the magnitude of the stator flux linkage is about 0.5. The per-unit values of the stator and rotor leakage inductances are 0.15. In (a), the rotor flux-linkage magnitude is 0.16, and the magnitude of the rotor current-producing torque is about 1.48. The per-unit torque in (a) is about 0.24. Correspondingly, the torque in (b) is about 0.44. At the breakdown point, the vector diagram is approximately equal to the simplified small illustration (c).

current required by the asynchronous machine is a component of the stator current i_s, as shown in Figure 11.24.

The simplified vector diagram (Figure 11.24c) shows clearly that approximate breakdown torque is reached as if the stator and rotor flux linkage are in an angle of about 45°. Assuming the stator and rotor leakage inductances are equal, the approximately right-angled flux-linkage triangle is obtained. Since torque is proportional to the cross product $\psi_s \times \psi_r$, as was previously shown, in theory, the highest torque will be reached within the set boundary conditions when these flux linkages are perpendicular. This, however, is not possible in an induction motor, as can be seen in Figure 11.24. If the power factor angle φ increases excessively, the air-gap flux linkage and the rotor flux linkage start to decrease, and consequently, the torque starts to decrease.

Field weakening enables an asynchronous machine to run, for example, at twice rated speed. However, its torque production capacity must be carefully considered as was revealed in the previous discussion. Figures 11.25 and 11.26 show the behaviour of an asynchronous machine at different speeds. The operating ranges can be divided into constant torque, constant power, and high-speed. In the constant torque range, the air-gap flux linkage is held constant, and the machine is capable of producing a constant torque at the constant slip frequency f_{slip}. Relative slip decreases as speed increases. In field weakening, the ratio of terminal voltage and angular frequency u_s/ω_s decreases, and therefore flux must also decrease. In the constant power range, the drop in machine torque is inversely proportional to increasing frequency, because the air-gap flux linkage must be reduced proportionally to increasing frequency.

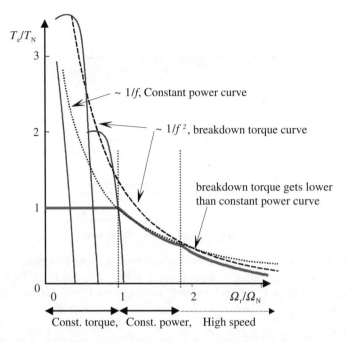

T_e/T_N

3

~ 1/f, Constant power curve

2

~ 1/f², breakdown torque curve

breakdown torque gets lower
than constant power curve

1

0

0   1   2 Ω_r/Ω_N

Const. torque, Const. power, High speed

Figure 11.25 Motor torque producing capability and limitations as a function of the per-unit rotation angular velocity Ω_r. The constant power curve, induction motor breakdown torque behaviour, and motor operating regions are illustrated. In the constant torque and constant power regions, torque is limited by machine cooling. However, if the current resources of the converter allow, the motor may temporarily produce very high torques at low speeds. In the high-speed area, torque is limited by the breakdown torque capability of the motor.

In the high-speed range, machine torque can be expressed in terms of rotor current and air-gap flux linkage, which are approximately perpendicular.

$$T_e = \left| \frac{3}{2} p \psi_m \times i_r \right| \cong \psi_m i_r \tag{11.18}$$

The electromotive force RMS value induced to the rotor is given by

$$E_r = k_E \psi_m f_{slip} \tag{11.19}$$

In the rotor, this emf generates an RMS current phasor $\bar{I}_r$.

$$\bar{I}_r = \frac{\bar{E}_r}{R_r + j\omega_{slip} L_{r\sigma}} \approx \frac{k_E \psi_m f_{slip}}{R_r} \cong \psi_m f_{slip} \tag{11.20}$$

Combining Equations (11.18) and (11.20) yields the following.

$$T_e \cong \psi_m^2 f_{slip} \tag{11.21}$$

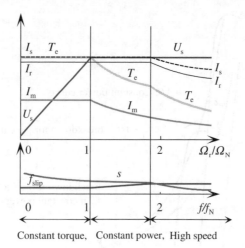

Figure 11.26 Performance of an asynchronous machine in normal operation as a function of rotor speed Ω_r in field weakening, in the constant power range, and in the high-speed range. Plots are given for voltage U_s, stator current I_s, rotor current I_r, magnetizing current I_m, and available torque T_e.

Assuming that $U_s \approx \omega_s \psi_m$, Equation (11.21) becomes

$$T_e \cong \frac{U_s^2}{\omega_s^2} f_{slip} \tag{11.22}$$

In field weakening in the constant power range, the rotor current must be at the rated value. This corresponds to constant per-unit slip, which is shown by Equation (11.20), the information $\psi_m \approx U_s/\omega_s$, and the definition of the slip frequency $f_{slip} = sf$.

$$I_r \cong U_s s = \text{constant} \tag{11.23}$$

In the constant power range (subscript cp), supply frequency f increases, and slip frequency increases with this increasing supply frequency $f_{slip} = sf$. The per-unit slip, however, remains constant. When U_s and f_{slip}/f remain constant in the constant power range, and the magnitude of the flux linkage decreases inversely proportional to frequency, pull-out torque $T_{b,cp}$ can be easily determined from rated frequency f_N and torque T_N.

$$T_{b,cp} = \frac{f_N}{f} T_N \tag{11.24}$$

In practice, the motor can produce slightly more than rated power in the constant power range, because an increased current proportion can be used for torque production as magnetization current and air-gap flux are reduced. The reduction of air-gap flux linkage decreases iron losses. The resulting higher speed also improves cooling.

In field weakening, the air-gap flux linkage of the machine decreases constantly as the speed of the machine increases if terminal voltage is held constant. Depending on the structure of the machine – typically at the relative speeds of 1.5–2—air-gap flux linkage drops until the motor approaches its breakdown torque. Some margin must be left with respect to the breakdown situation, and torque must be further decreased to further increase speed and move into the high-speed range. In this range, the slip frequency of the rotor cannot be increased, and maximum torque is inversely proportional to the square of supply frequency.

$$T_{b,cp} \cong \frac{1}{f^2} \tag{11.25}$$

To move to the high-speed range, both motor current and torque must decrease, because motor breakdown torque, not thermal load capacity, limits the drive.

The rapid reduction of breakdown torque in an asynchronous machine results from the increase in slip frequency and from the fact that the rotor becomes more and more inductive in field weakening. Figure 11.26 plots voltage U_s, stator current I_s, rotor current I_r, magnetizing current I_m, and available torque T_e for an asynchronous machine in normal duty as a function of rotor speed Ω_r in field weakening, in the constant power range, and in the high-speed range.

The per-unit pull-out torque of the motor also defines the constant power range length. The frequency at which the pull-out torque equals the constant power producing torque is the theoretical maximum frequency in the constant power region. The decrease in constant power producing torque is inversely proportional to frequency, and the decrease in pull-out torque is inversely proportional to frequency squared. These two torque expressions are set equal to find the constant power range.

$$\frac{f_N}{f_{s,max,cp}} T_N = \left(\frac{f_N}{f_{s,max,cp}} \right)^2 T_b \tag{11.26}$$

where f_N is the rated frequency, $f_{s,max,cp}$ is the maximum constant power frequency, T_N is the rated torque, and T_b is the pull-out torque. The maximum constant power stator frequency is as follows.

$$f_{s,max,cp} = f_N \frac{T_b}{T_N} \tag{11.27}$$

The expression reveals that per-unit pull-out torque directly defines the length of the constant power range of an induction motor. For example, an induction motor having a per-unit pull-out torque of 3.0 can produce rated power at three times the rated speed.

When increasing frequency, sometimes the rated voltage of a motor can be exceeded. An inverter-driven motor can be run at values above the rated voltage and angular speed without going into field weakening. If the machine is wound for 230/400 V, 50 Hz with Δ/Y connections, it can operate (with the consent of the manufacturer) using a delta connection at 86.6 Hz frequency and 400 V voltage at the rated air-gap flux linkage. Therefore, the rated power increases, in principle, with a ratio of $\sqrt{3} : 1$.

Figure 11.27 shows operating point A of the motor at the rated values. The rated line-to-line voltage in the Δ connection is 230 V, the frequency is 50 Hz, and the synchronous rotation speed is 1500 rpm. At operating point B, the motor runs at 1.74 times the voltage of 400 V at a frequency of 86.6 Hz. By increasing voltage and frequency in the 50 to 86.6 Hz range (operating point B), the constant torque range can be expanded from the original. Torque production capacity remains stable up to the rotating speed of 2598 rpm.

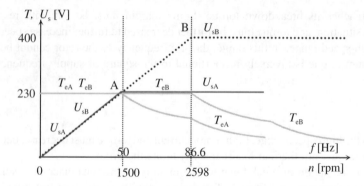

Figure 11.27 The voltage and frequency of an induction motor at rated values (operating point A) and at 174% of rated values (operating point B). By using the voltage and frequency values at point B, the constant torque range of the motor can be expanded from its original. Higher current is required of the inverter than in the original 50 Hz 400 V drive ($\sqrt{3} : 1$).

11.4.2   The vector control of an induction machine

Vector control is a magnetic-field-oriented current control method for an induction motor. Vector control is a form of torque control, for which the rotation speed controller gives a reference value. In vector-control computation, measured stator current is divided into flux producing and torque producing components using a two-axis model of the asynchronous machine. The excitation state of the machine is derived from the flux-axis current component $i_{s\psi}$, and it gives the actual value for the flux linkage. The torque-axis component i_{sT} provides machine torque, and it gives the actual value for the torque. Figure 11.28 depicts the schematic diagram for conventional vector control.

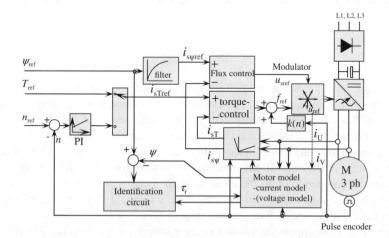

Figure 11.28   Schematic diagram for conventional induction motor vector control. Feedback from the space-vector modulator can be added, and the stator flux linkage can be integrated from the machine voltages offering enhanced vector control. Typically two motor-phase currents i_U and i_V are measured assuming that $i_W = -(i_U + i_V)$. The block $k(n)$ is used as a modulator stabilizing feedforward term, for example, at low speeds to compensate for the slip effects.

Figure 11.28 reveals that vector control does include a motor model. The motor model is normally based on the two-axis equivalent motor circuit, which may be fixed to the rotor-flux-linkage reference frame. The required control parameters are calculated using this motor model and feedbacks. In the identification block, the rotor time constant τ_r is computed. This time constant is important for controlling the dynamic state of the motor. Vector control is notably better than scalar control and managing motor dynamics.

Good alternatives for implementing the vector control of an asynchronous machine include the stator reference frame (the general reference frame is fixed with the stator and $\omega_g = 0$), the rotor reference frame ($\omega_g = p\Omega_r$), or a flux-linkage reference frame. The flux-linkage equations based on space-vector theory are

$$\psi_s = L_s i_s + L_m i_r = L_{s\sigma} i_s + \psi_m \tag{11.28}$$

$$\psi_r = L_r i_r + L_m i_s = L_{r\sigma} i_r + \psi_m \tag{11.29}$$

$$\psi_m = L_m(i_s + i_r) = L_m i_m \tag{11.30}$$

All the flux linkages depend on both the stator and rotor currents. However, in a squirrel-cage induction machine, measuring rotor current is practically impossible. To implement vector control, one of the flux linkages must be estimated using measurable parameters.

Figure 11.29 shows an asynchronous machine presented in the rotor reference frame. A conventionally applied method in vector control uses the rotor reference frame and rotor position—in practice, measured with a pulse encoder.

The rotor voltage equation in the rotor reference frame is as follows.

$$R_r i_r = -\frac{d\psi_r}{dt} \tag{11.31}$$

From Equation (11.29), rotor current is

$$i_r = \frac{\psi_r - L_m i_s}{L_r} \tag{11.32}$$

Substitution leads to

$$\frac{d\psi_r}{dt} = -\frac{R_r}{L_r}\psi_r + \frac{R_r}{L_r}L_m i_s \tag{11.33}$$

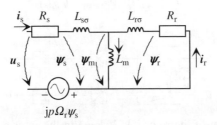

Figure 11.29 Equivalent circuit of an induction motor based on space-vector theory presented in the rotor reference frame.

The no-load time constant of the rotor is

$$\tau_r = \frac{L_r}{R_r} \tag{11.34}$$

This term is often referred simply as the rotor time constant. Inserting Equation (11.34) into Equation (11.33) yields

$$\tau_r \frac{d\psi_r}{dt} + \psi_r = L_m i_s \tag{11.35}$$

which can be transformed into the Laplace domain.

$$\tau_r s \psi_r(s) + \psi_r(s) = L_m i_s \tag{11.36}$$

from which the rotor flux linkage is

$$\psi_r(s) = \frac{L_m i_s}{1 + \tau_r s} \tag{11.37}$$

The above statement corresponds to a first-order low-pass filter, and therefore the rotor flux linkage can be found by filtering the stator current multiplied by the magnetizing inductance. In practice, this estimate is done by transforming the measured stator current vector to the rotor reference frame, multiplying by the magnetizing inductance, and by filtering the result with a low-pass filter based on the rotor time constant. See Figure 11.30.

Since the rotor is magnetically isotropic in the rotation plane, relative information suffices and no absolute rotor angle information is required; therefore, a pulse encoder is well adapted for the purpose. The d- and q-axes of the machine are not obvious, but may be freely selected. In practice, when commissioning the drive, an arbitrary rotor position is chosen for the rotor d- and q-axes, and thereafter the pulse encoder measures the rotor position with reference to this initial position.

In practice, the method is problematic, since none of the above-mentioned parameters are constant. Rotor resistance changes with skin effect and temperature, and magnetic saturation, a function of both flux level and torque, has an effect on magnetizing inductance. Consequently, a saturation model and a rotor temperature estimator should be constructed for the machine. Torque can be calculated using the following expressions.

$$T_e = \frac{3}{2} p \psi_s \times i_s = \frac{3}{2} p \left(L_s i_s + \frac{L_m}{L_r} (\psi_r - L_m i_s) \right) \times i_s = \frac{3}{2} p \frac{L_m}{L_r} \psi_r \times i_s,$$

$$T_e = \frac{3}{2} p \frac{L_m}{L_r} \left(\psi_{rd} i_{sq} - \psi_{rq} i_{sd} \right) \tag{11.38}$$

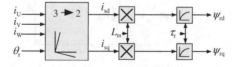

Figure 11.30 Estimation of the rotor flux linkage in the rotor reference frame. Measured rotor position feedback is needed. The stator current dq components are presented in the rotor reference frame.

The latter applies especially in the rotor reference frame with its dq-components. Therefore, when machine parameters are known, the actual control values can be estimated. Both the d and q current components affect torque calculated in the dq-reference frame. Therefore, it might be more convenient to use a coordinate system oriented to the rotor flux linkage, so rotor current only has a torque-producing component and the machine is excited by stator current.

Since an asynchronous machine is usually voltage fed, how the stator voltage control is constructed to implement the rotor-flux–oriented current vector control must be determined. In a system based on rotor flux linkage, the machine equations can conveniently be transformed into the rotor-flux-linkage reference frame ψ_rT. When transforming from the rotor reference frame (dq) to the rotor-flux-linkage reference frame (ψ_rT) with the rotor flux linkage $\boldsymbol{\psi}_r = \psi_{rd} + j\psi_{rq}$ in angle $\theta_{\psi d}$, the required trigonometric functions are obtained from the known parameters.

$$\sin \theta_{\psi d} = \frac{\psi_{rq}}{\sqrt{\psi_{rd}^2 + \psi_{rq}^2}} = \frac{\psi_{rq}}{\psi_r}, \cos \theta_{\psi d} = \frac{\psi_{rd}}{\psi_r} \tag{11.39}$$

Further, the torque can be expressed based on the cross-field principle in the rotor flux linkage–oriented (ψ_rT) coordinate system.

$$T_e = \frac{3}{2}p\frac{L_m}{L_r}\boldsymbol{\psi}_r \times i_s,$$

$$T_e = \frac{3}{2}p\frac{L_m}{L_r}\left(\psi_{r\psi}i_{sT} - \psi_{rT}i_{s\psi}\right) = \frac{3}{2}p\frac{L_m}{L_r}\psi_{r\psi}i_{sT}. \tag{11.40}$$

This simple result is explained by the fact that the rotor flux linkage does not have a quadrature-axis component in the rotor-flux-linkage reference frame. Torque control is performed by the component i_{sT} perpendicular to the rotor flux linkage. Its reference value is obtained directly from Equation (11.37). The control of the flux linkage is not quite as simple as the torque control.

This is now a general reference frame (familiar from Chapter 4, Figure 4.10) with its real axis fixed to the rotor-flux-linkage vector. In the general reference frame, the stator current has two components—the rotor-flux-linkage–producing component and torque-producing component.

The rotor voltage (4.47) is written in the rotor-flux-linkage reference frame (ψ_rT) using components with rotor flux linkage angular velocity $\omega_{\psi r}$T and rotor mechanical angular velocity Ω_r.

$$\boldsymbol{u}_r^{\psi_r T} = u_{r\psi} + ju_{rT} = R_r\boldsymbol{i}_r^{\psi_r T} + \frac{d\boldsymbol{\psi}_r^{\psi_r T}}{dt} + j\left(\omega_{\psi_r T} - \omega_r\right)\boldsymbol{\psi}_r^{\psi_r T},$$

$$u_{r\psi} + ju_{rT} = R_r\left(i_{r\psi} + ji_{rT}\right) + \frac{d\left(\psi_{r\psi} + j\psi_{rT}\right)}{dt} + j\left(\omega_{\psi_r T} - \omega_r\right)\left(\psi_{r\psi} + j\psi_{rT}\right)$$

$$u_{r\psi} = R_r i_{r\psi} + \frac{d\psi_{r\psi}}{dt} - \left(\omega_{\psi_r T} - p\Omega_r\right)\psi_{rT} = 0,$$

$$u_{rT} = R_r i_{rT} + \frac{d\psi_{rT}}{dt} + \left(\omega_{\psi_r T} - p\Omega_r\right)\psi_{r\psi} = 0. \tag{11.41}$$

In the rotor-flux-linkage reference frame, the result is simplified further since there exists in the rotor flux-linkage coordinate system a ψ-axis component of the flux ($\psi_{r\psi} \neq 0$ and $\psi_{rT} = 0$).

$$u_{r\psi} = R_r i_{r\psi} + \frac{d\psi_{r\psi}}{dt} = 0,$$

$$u_{rT} = R_r i_{rT} + (\omega_{\psi_r T} - p\Omega_r)\psi_{r\psi} = 0. \tag{11.42}$$

Equation (11.42) yields a low-pass-filter type equation. See also Equation (11.35).

$$\tau_r \frac{d\psi_{r\psi}}{dt} + \psi_{r\psi} = L_m i_{s\psi} \tag{11.43}$$

The magnitude of the rotor flux linkage follows the stator current component $i_{s\psi}$ aligned with the rotor linkage flux at the rotor no-load time constant, which varies typically between 0.15 and 1.5 s depending on the machine size. The rotor flux linkage remains constant as the stator current component $i_{s\psi}$ is constant at $i_{s\psi} = \psi_r/L_m$. When operating in the field-weakening region, the differential term of Equation (11.44) must be taken into account when speed varies rapidly. If the rotor flux linkage ψ_r is successfully held constant as torque changes, $i_{r\psi} = 0$ according to Equation (11.42). The flux linkages behave as shown in Figure 11.31.

If the magnetizing inductance does not saturate, the tips of the stator and air-gap flux-linkage vectors plot the loci indicated by dashed lines.

The stator voltage reference can be determined from the stator voltage equation in the rotor-flux-linkage reference frame (superscript $\psi_r T$).

$$\begin{aligned}
u_s^{\psi_r T} &= R_s i_s^{\psi_r T} + \frac{d\psi_s^{\psi_r T}}{dt} + j\omega_{\psi_r T}\psi_s^{\psi_r T} \\
&= R_s i_s^{\psi_r T} + L_{s\sigma}\frac{di_s^{\psi_r T}}{dt} + \frac{d\psi_m^{\psi_r T}}{dt} + j\omega_{\psi_r T}\left(L_{s\sigma}i_s^{\psi_r T} + \psi_m^{\psi_r T}\right).
\end{aligned} \tag{11.44}$$

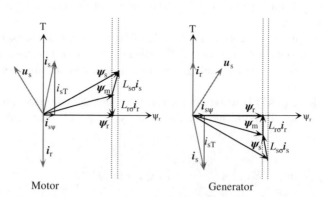

Motor   Generator

Figure 11.31 Behaviour of flux linkages in the rotor-flux-linkage reference frame for different machine load states. The vector diagrams are for steady state at about 65% speed ($u_s = 0.65$, $\psi_s = 1$) and at partial load ($i_s = 0.65$). Stator resistance effect are ignored. Integrating stator voltage yields the stator flux linkage. The current components are drawn with respect to the rotor flux linkage–oriented coordinate system. Stator current is divided into rotor flux–linkage and torque-producing components. Rotor current is only torque producing.

Remembering that $\boldsymbol{\psi}_m = \boldsymbol{\psi}_r - L_{r\sigma}\boldsymbol{i}_r$, the actual value for the air-gap flux linkage is obtained by inserting the rotor current $\boldsymbol{i}_r = (\boldsymbol{\psi}_r - L_m\boldsymbol{i}_s)/L_r$ from Equation (11.42).

$$\boldsymbol{\psi}_m^{\psi_r\mathrm{T}} = \frac{\boldsymbol{\psi}_r^{\psi_r\mathrm{T}}L_r - L_{r\sigma}\boldsymbol{\psi}_r^{\psi_r\mathrm{T}} + L_{r\sigma}L_m\boldsymbol{i}_s^{\psi_r\mathrm{T}}}{L_r},$$

$$\boldsymbol{\psi}_m^{\psi_r\mathrm{T}} = \frac{\boldsymbol{\psi}_r^{\psi_r\mathrm{T}}L_m + L_{r\sigma}L_m\boldsymbol{i}_s^{\psi_r\mathrm{T}}}{L_r}, \qquad (11.45)$$

$$\boldsymbol{\psi}_m^{\psi_r\mathrm{T}} = \frac{L_m}{L_r}\left(\boldsymbol{\psi}_r^{\psi_r\mathrm{T}} + L_{r\sigma}\boldsymbol{i}_s^{\psi_r\mathrm{T}}\right)$$

Substituting this into Equation (11.44) results in the following equation.

$$\boldsymbol{u}_s^{\psi_r\mathrm{T}} = R_s\boldsymbol{i}_s^{\psi_r\mathrm{T}} + \left(L_{s\sigma} + \frac{L_m}{L_r}L_{r\sigma}\right)\frac{d\boldsymbol{i}_s^{\psi_r\mathrm{T}}}{dt} + \frac{L_m}{L_r}\frac{d\boldsymbol{\psi}_r^{\psi_r\mathrm{T}}}{dt} + j\omega_{\psi_r\mathrm{T}}\left(\left(L_{s\sigma} + \frac{L_m}{L_r}L_{r\sigma}\right)\boldsymbol{i}_s^{\psi_r\mathrm{T}} + \frac{L_m}{L_r}\boldsymbol{\psi}_r^{\psi_r\mathrm{T}}\right)$$

$$(11.46)$$

Inserting $\boldsymbol{i}_r = (\boldsymbol{\psi}_r - L_m\boldsymbol{i}_s)/L_r$ into the rotor voltage Equation (11.42) gives a statement for the differential of the rotor flux linkage.

$$\frac{d\boldsymbol{\psi}_r^{\psi_r\mathrm{T}}}{dt} = -\frac{R_r}{L_r}\boldsymbol{\psi}_r^{\psi_r\mathrm{T}} + \frac{L_m}{L_r}R_r\boldsymbol{i}_s^{\psi_r\mathrm{T}} - j\left(\omega_{\psi_r\mathrm{T}} - p\Omega_r\right)\boldsymbol{\psi}_r^{\psi_r\mathrm{T}} \qquad (11.47)$$

This expression can be inserted into the previous equation to produce

$$\boldsymbol{u}_s^{\psi_r\mathrm{T}} = \left(R_s + \left(\frac{L_m}{L_r}\right)^2 R_r\right)\boldsymbol{i}_s^{\psi_r\mathrm{T}} + \left(L_{s\sigma} + \frac{L_m}{L_r}L_{r\sigma}\right)\frac{d\boldsymbol{i}_s^{\psi_r\mathrm{T}}}{dt} - \frac{L_m}{L_r^2}R_r\boldsymbol{\psi}_r^{\psi_r\mathrm{T}}$$

$$+ j\left(\omega_{\psi_r\mathrm{T}}\left(L_{s\sigma} + \frac{L_m}{L_r}L_{r\sigma}\right)\boldsymbol{i}_s^{\psi_r\mathrm{T}} + \Omega_r p\frac{L_m}{L_r}\boldsymbol{\psi}_r^{\psi_r\mathrm{T}}\right).$$

$$(11.48)$$

Usually, when operating with processors, a Cartesian representation is easier to use than a polar representation. To accomplish this, the voltage vector can be decomposed next into its components by inserting $\boldsymbol{i}_s^{\psi_r\mathrm{T}} = i_{s\psi} + ji_{s\mathrm{T}}$.

$$u_{s\mathrm{T}} = \left(R_s + \left(\frac{L_m}{L_r}\right)^2 R_r\right)i_{s\mathrm{T}} + \left(L_{s\sigma} + \frac{L_m}{L_r}L_{r\sigma}\right)\frac{di_{s\mathrm{T}}}{dt} + \omega_{\psi_r\mathrm{T}}\left(L_{s\sigma} + \frac{L_m}{L_r}L_{r\sigma}\right)i_{s\psi} + \Omega_r p\frac{L_m}{L_r}\psi_{r\psi}$$

$$(11.49)$$

$$u_{s\psi} = \left(R_s + \left(\frac{L_m}{L_r}\right)^2 R_r\right)i_{s\psi} + \left(L_{s\sigma} + \frac{L_m}{L_r}L_{r\sigma}\right)\frac{di_{s\psi}}{dt} - \omega_{\psi_r\mathrm{T}}\left(L_{s\sigma} + \frac{L_m}{L_r}L_{r\sigma}\right)i_{s\mathrm{T}} - \frac{L_m}{L_r^2}R_r\psi_{r\psi}$$

$$(11.50)$$

The angular speed $\omega_{\psi_r T}$ of the rotor-flux-linkage reference frame can be determined from Equation (11.42) by inserting the rotor current $i_r = (\psi_r - L_m i_s)/L_r$.

$$\omega_{\psi_r T} = \frac{R_r}{L_r} \frac{L_m}{\psi_{r\psi}} i_{sT} + p\Omega_r = \frac{L_m}{\tau_r \psi_{r\psi}} i_{sT} + p\Omega_r \tag{11.51}$$

EXAMPLE 11.2: Derive angular speed $\omega_{\psi_r T}$ of the rotor-flux-linkage reference frame Equation (11.51) using Equation (11.42).

SOLUTION: The equation for angular speed from Equation (11.42) is $u_{rT} = R_r i_{rT} + (\omega_{\psi_r T} - p\Omega_r)\psi_{r\psi} = 0$. Solving it for the angular velocity yields the following.

$$\omega_{\psi_r T}\psi_{r\psi} = p\Omega_r \psi_{r\psi} - R_r i_{rT}$$

Rotor current written in terms of its components is

$$i_r = i_{r\psi} + ji_{rT} = (\psi_r - L_m i_s)/L_r = (\psi_{r\psi} + j\psi_{rT} - L_m(i_{s\psi} + ji_{sT}))/L_r$$
$$\Rightarrow i_{rT} = (\psi_{rT} - L_m i_{sT})/L_r$$

Because $\psi_{rT} = 0$ in the rotor-flux-linkage reference frame, $i_{rT} = -L_m i_{sT}/L_r$.

Substitution results in this equation for angular velocity.

$$\omega_{\psi_r T} = \frac{p\Omega_r \psi_{r\psi} + R_r i_{sT} L_m/L_r}{\psi_{r\psi}} = p\Omega_r + \frac{R_r L_m i_{sT}}{L_r \psi_{r\psi}} = p\Omega_r + \frac{R_r}{L_r} \frac{L_m}{\psi_{r\psi}} i_{sT} = p\Omega_r + \frac{1}{\tau_r} \frac{L_m}{\psi_{r\psi}} i_{sT}$$

The above is identical to Equation (11.51).

In steady state, the angular speed of the flux-linkage reference frame is equal to the electrical angular speed ω_s supplying the machine. Solving i_{sT} and inserting the result into the torque equation gives

$$T_e = \frac{3}{2} p \frac{L_m}{L_r} \psi_{r\psi} \frac{\omega_{r\psi} - p\Omega_r}{L_m} \psi_{r\psi} \tau_r = \frac{3}{2} p \frac{\omega_r \psi_{r\psi}^2}{R_r}. \tag{11.52}$$

This is a typical torque equation used in scalar control. ω_r is the electrical angular frequency of the rotor $(p\Omega_r)$.

The voltage Equations (11.49) and (11.50) resemble those used to control a DC machine, because one of the voltage components gives the desired rotor flux linkage and the other gives the current that produces torque. These voltage components must be transformed into the stator reference frame to enable actual motor control. To facilitate the coordinate transformations, the sine and cosine values of the rotor angle are determined from the calculated flux-linkage components according to Equation (11.39).

The above vector control method for an induction motor was originally introduced by Blaschke (1971). It is known as rotor-flux–oriented control. The system functions mainly as described above in a reference frame fixed to the rotor flux-linkage vector.

Inserting the rotor flux-linkage expression into the rotor flux-linkage differential Equation (11.31), and then writing the derivatives as delta parameters will help to explain why the rotor-flux-linkage–oriented control method is a justified choice.

$$\Delta\boldsymbol{\psi}_r = L_m\Delta\boldsymbol{i}_s + L_r\Delta\boldsymbol{i}_r = -R_r\boldsymbol{i}_r\Delta t \tag{11.53}$$

The limit value for the above as transition time approaches zero is as follows.

$$\lim_{\Delta t\to 0}\Delta\boldsymbol{\psi}_r = \lim_{\Delta t\to 0}(L_m\Delta\boldsymbol{i}_s + L_r\Delta\boldsymbol{i}_r) = \lim_{\Delta t\to 0}-R_r\boldsymbol{i}_r\Delta t = 0 \tag{11.54}$$

The equation clearly reveals that changes to rotor flux linkage cannot be instantaneous. For the change in the rotor current

$$\Delta\boldsymbol{i}_r = -\frac{L_m}{L_r}\Delta\boldsymbol{i}_s \tag{11.55}$$

The changes in the flux linkage resulting from the current step are obtained from the electric current equations of the flux linkages.

$$\Delta\boldsymbol{\psi}_s = \frac{L_sL_r - L_m^2}{L_r}\Delta\boldsymbol{i}_s \tag{11.56}$$

$$\Delta\boldsymbol{\psi}_m = \frac{L_m}{L_r}L_{r\sigma}\Delta\boldsymbol{i}_s \tag{11.57}$$

EXAMPLE 11.3: Show that fast changes are dictated by the effect of leakage inductances.

SOLUTION: In the machine to be investigated, assume that $L_s = L_r = 2.5$, $L_m = 2.4$, and $L_{s\sigma} = L_{r\sigma} = 0.1$. These are typical per-unit values for an induction machine. Now Equation (11.56) may be reduced as follows.

$$\Delta\boldsymbol{\psi}_s = \frac{L_sL_r - L_m^2}{L_r}\Delta\boldsymbol{i}_s = \frac{(L_m + L_{s\sigma})(L_m + L_{r\sigma}) - L_m^2}{L_r}\Delta\boldsymbol{i}_s = \frac{L_m^2 + L_mL_{s\sigma} + L_mL_{r\sigma} + L_{s\sigma}L_{r\sigma} - L_m^2}{L_r}\Delta\boldsymbol{i}_s$$

$$= \frac{L_mL_{s\sigma} + L_mL_{r\sigma} + L_{s\sigma}L_{r\sigma}}{L_r}\Delta\boldsymbol{i}_s \approx \frac{L_m(L_{s\sigma} + L_{r\sigma})}{L_r}\Delta\boldsymbol{i}_s \approx (L_{s\sigma} + L_{r\sigma})\Delta\boldsymbol{i}_s$$

$\Delta\boldsymbol{\psi}_s \approx (L_{s\sigma} + L_{r\sigma})\Delta\boldsymbol{i}_s = 0.2\Delta\boldsymbol{i}_s$, while the more accurate value is $\Delta\boldsymbol{\psi}_s = 0.196\Delta\boldsymbol{i}_s$, and correspondingly, $\Delta\boldsymbol{\psi}_m \approx L_{r\sigma}\Delta\boldsymbol{i}_s = 0.1\Delta\boldsymbol{i}_s$ while the more accurate value is $\Delta\boldsymbol{\psi}_m = 0.096\Delta\boldsymbol{i}_s$. A rapid change in stator current, therefore, passes chiefly through leakage inductances.

Rotor flux linkage is inherently most stable; however, stator flux linkage can be changed most rapidly. Therefore, it is advisable to base vector control on rotor flux linkage, because rotor flux-linkage magnitude and angular velocity can be held constant with least difficulty. The magnitude of stator flux-linkage changes with changes in loading. Therefore, stator voltage magnitude should be adjusted as a function of load. Adjusting stator voltage magnitude is straightforward at lower speed, but can become problematic as speeds increase and field weakening becomes an issue. To guarantee sufficient voltage, the selected magnitude for the rotor flux linkage should be low enough to maintain a suitable voltage reserve.

Finally, the pull-out torque of an asynchronous machine and the corresponding slip angular frequency with stator resistance neglected can be expressed as follows.

$$T_b = \frac{3}{2} p \frac{L_m^2}{2\omega_s^2 \left(L_r L_s^2 - L_s L_m^2\right)} |u_s|^2 = \frac{3}{2} p \frac{L_m^2}{2\omega_s^2 \left(L_r L_s^2 - L_s L_m^2\right)} |\psi_s|^2 \qquad (11.58)$$

$$\omega_b = \pm \frac{L_s R_r}{L_r L_s - L_m^2} \qquad (11.59)$$

Improving vector control using the voltage model of the motor

Naturally, introducing the voltage integral can also benefit vector control. The principles of DFLC can be applied, if the embedded processor of the control system has sufficient capacity, which makes it possible to use the voltage model as a corrective means for the motor control. In DFLC, the stator flux linkage is integrated from the converter voltages $\psi_{s,u} = \int (u_s - R_s i_s) dt$. In vector control, the same value is calculated as $\psi_{s,i} = L_{s\sigma} i_s + L_m i_m$. The difference is $\Delta\psi_s = \psi_{s,u} - \psi_{s,i}$. Using a suitable weighting scheme, this difference can be used to correct the stator flux-linkage calculation. When both voltage and current models are used, superior drive performance can be achieved irrespective of the control principle. In practice, therefore, there may be little difference between vector control and DTC. Previously in Figure 11.28, this option was added by bringing the voltage information from the space vector modulator to the motor model.

11.4.3   Direct torque control

Direct torque control was originally developed for induction motors. In previous chapters, the application of direct flux linkage and DTC methods to various synchronous machine types was discussed. As the name implies, the DTC of an induction motor controls its flux linkage and torque directly. ABB introduced DTC technology in 1988 and has launched the ACS600/800/850/880 and ACS1000/2000/6000 inverters based on DTC technology. Figure 11.32 shows a simplified schematic diagram of the DTC. The most important elements of the DTC system include direct hysteresis control of the stator flux linkage and torque, optimal switching logic, and an adaptive rotor model.

As shown in the figure, the DTC measures two motor-phase currents and the DC voltage of the DC link. Unlike in induction-motor vector control, there is no rotor position measurement needed. Current and voltage information as well as the switch position information of the switches S_U, S_V, and S_W are fed to the motor model. The motor model gives estimate values for torque, flux linkage, and rotation speed. The optimal switching logic selects optimum instantaneous switch positions based on torque and the state of the flux linkage. All switch control decisions for S_U, S_V, and S_W are based on the electromagnetic state of the machine. In ABB's classic DTC converters, switch control decisions are made at the intervals of 25 μs, which also dictates the minimum DTC pulse width. In many applications, this is fast enough, but there are also applications where shorter voltage pulses would be more acceptable. Compared to vector control low-voltage converters, the minimum pulse width is

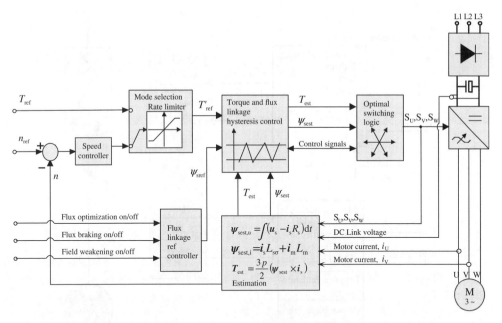

Figure 11.32   Block diagram of DTC.

long, because a vector modulator or a sine triangle comparison modulator will typically pulse in the range of 3 µs.

In an asynchronous machine, the inductance-based current model can be used to stabilize the behaviour of the induction machine without rotor position feedback. Naturally, this is based on the isotropy of an asynchronous machine. The entire analysis can be performed in the stator xy-reference frame, and a dq-reference frame fixed to the rotor is not required. This asynchronous machine control may seem simpler than typical synchronous machine control; however, quite the opposite is true. The difficulty arises because estimating rotation speed in isotropic motors is extremely challenging. Synchronous machines, at least, rotate synchronously. For asynchronous machines, many tricks have been introduced to determine rotor speed without encoders. So far, however, a speed sensor is still needed for the accurate speed control of induction motor drives.

Figure 11.33 illustrates DTC implementation at the hardware level. Optimal switching logic was originally implemented in the ABB ACS600/800 motor control system using ASICs (Application Specified Integrated Circuit). The logic always selects the best switch position based on the computed needs for changes in torque and flux linkage. The calculation interval between switchings, 25 µs, is faster, in practice, than the motor's electrical time constants. Consequently, fast torque control response is achieved. With sufficient voltage reserve, rated torque response can be reached in 2–3 ms. Therefore, with respect to torque commands; the DTC is faster than vector control.

The motor model in a DTC drive

Unlike in DTC for asynchronous machines where accurate rotor position feedback is needed, the motor current model in induction-motor DTC drives does not need additional input. As a

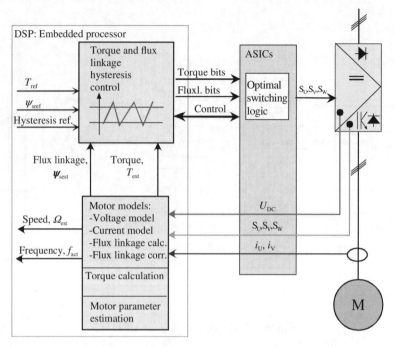

Figure 11.33 Schematic diagram of an example DTC (ABB ACS600/800). The core of the DTC includes torque control, flux-linkage control, a motor model, and optimal switching logic. It is executed using a DS processor. Processor operations are divided by time level. Switch control is computed from the motor model at the fastest 25 µs time level. Other time levels in order of increasing duration include, e.g., 100 µs, 200 µs, 500 µs, and 1 ms. Tasks are prioritized. Rotor speed is computed at 1 ms intervals. The two motor phase currents are measured via current measurement sensors based on the Hall effect. After A/D conversion, the signal is sent to the DSP. The DSP uses the motor model for switch overload protection; however, actual short-circuit protection is implemented via the ASICs alongside the switch. For data transfer outside the processor card, optical fibres provide galvanic separation and isolate the control system from environmental disturbance.

result, the motor-parameter–based current model has always been included for the stabilization of induction motor DTC drives. Using both the voltage model and the current model results in strong stabilization for induction-motor DTC drives.

The models generate corrected estimates for the flux-linkage vectors and torque to control flux linkage and torque hysteresis. Furthermore, estimates for rotation speed and frequency are obtained to satisfy the needs of other DTC blocks. Estimation is based on identifying the motor parameters and on electric current measurement. The currents of two motor phases and the voltage of the DC link are measured, and the state of the inverter switches is investigated. Simple sensors measuring the on state of switches are used to get better knowledge of the voltage vectors supplied to the motor.

The big advantage that DTC offers over vector control is that it works well in the absence of an accurate knowledge of the motor parameters. In contrast, the performance of vector control bases solely on the accuracy of the motor model. In DTC, the current model is used more for stabilization than to provide accurate flux-linkage estimation. Naturally, a more accurate motor model results in control that is more accurate.

The DTC controller identifies the most important motor parameters and then computes the stator flux linkage of the motor. The motor model also includes a temperature compensation model, which is essential for the static speed accuracy.

The DTC motor model also computes motor shaft speed from the differential of the rotor flux linkage. It is not necessary to measure the speed, if a static accuracy of 0.5% is sufficient, as is the case for most industrial applications. However, for most accurate drive control, speed sensing is required. This is particularly true for induction motor drives, which are characterized by unpredictable rotor slip that makes it even more difficult to estimate rotor speed.

Compared to a motor model based on nameplate values, using measured parameter values taken during the commissioning identification run to build the motor model provides more control accuracy. The most important parameters determined in the identification run are stator resistance R_s, stator leakage inductance $L_{s\sigma}$, and magnetizing inductance L_m. The accuracy of the motor model can also be improved by taking into account the voltage and torque saturation behaviour of the machine. Using an accurate electric current model with current feedback notably improves the estimation of stator flux linkage particularly at low speeds. High starting torque is achieved, which becomes linear over the entire speed range.

Once an accurate estimate of stator flux linkage has been made, torque estimation is simple. The motor model also estimates shaft speed and frequency. Rotor frequency determination is based on knowledge of the rotor flux-linkage vector, which is computed every millisecond making it practical to estimate rotor speeds up to 400 Hz.

Vector control or DTC are not actually necessary for high-speed drives. At high speeds, scalar control is normally adequate even for nonisotropic machines.

For an asynchronous machine, good practice suggests the electric current model be determined using measured parameter values taken during the commissioning identification run. The first task is to determine transient inductance. The equivalent circuit corresponding to the transient inductance is illustrated in Figure 11.34.

According to the illustration, the transient inductance is

$$L'_s = L_{s\sigma} + \frac{L_{r\sigma}L_m}{L_{r\sigma} + L_m} \tag{11.60}$$

Adding and subtracting L_m leads to the following equations.

$$L'_s = L_{s\sigma} + L_m + \frac{L_{r\sigma}L_m}{L_{r\sigma} + L_m} - \frac{L_r L_m}{L_r} \quad \text{and} \quad L'_s = L_s + \frac{L_{r\sigma}L_m}{L_r} - \frac{(L_{r\sigma} + L_m)L_m}{L_r}$$

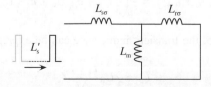

Figure 11.34 Equivalent circuit corresponding to the transient inductance of an asynchronous machine. The transients pass mainly through the stator and rotor leakage inductances bypassing the rotor, because the rotor cage opposes any transient fluxes.

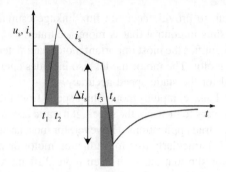

Figure 11.35 Stator current behaviour in response to two voltage pulses in a transient inductance test.

These lead to the conventional representation for transient inductance.

$$L'_s = L_s - \frac{L_m^2}{L_r}.$$
(11.61)

Transient inductance characterizes the inductance experienced by a fast pulse fed to the stator. In other words, the rotor manages to react to a fast pulse only through its leakage inductance. There is little flux through the magnetizing inductance. Transient inductance can be determined using a short pulse from an inverter as shown in Figure 11.35. When a voltage pulse is fed to the machine, there is a rapid change in the machine current.

If the opposite voltage vectors, for example, u_1 and u_4 are used, the resulting transient inductance can be calculated as

$$L'_s = u_4 \frac{t_4 - t_3}{\Delta i_s}$$
(11.62)

u_4 is the length of the stator voltage vector u_4 (not shown in the figure).

The hysteresis state of the machine rotor may affect the determination of the dynamic inductance, and therefore the effect of hysteresis should be eliminated from the measured parameters by selecting the information at an appropriate position. The measurement can be carried out using only one pulse, if a suitable data acquisition period can be selected. See Figure 11.36.

$|u_4|$ is the length of the stator voltage vector.

$$u_s = \frac{2}{3}\left(u_{sU} + au_{sV} + a^2u_{sW}\right) = \frac{2}{3}\left(\frac{2}{3}U_d + a\frac{1}{3}U_d + a^2\frac{1}{3}U_d\right) = \frac{2}{3}U_d$$
(11.63)

According to Figure 11.35, the transient inductance can be expressed as follows.

$$L'_s = \frac{2}{3}U_d\frac{t_4 - t_3}{\Delta i_s}$$
(11.64)

A corresponding expression can be written for the test illustrated in Figure 11.36. In general, stator leakage inductance builds up to about 60–70% of the transient inductance. When

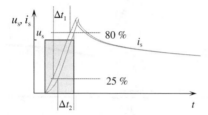

Figure 11.36 Stator transient inductance can be measured in one pulse, in which case the excitation state of the machine before the measurement pulse must be taken into account. The shape of the first part of the current curve depends on the hysteresis state of the machine. When the change in current is for instance between 25% and 80%, the measurement result is almost independent of any hysteresis.

measuring transient inductance, iron bridge saturation discharge causes an interesting phenomenon to develop in the region of declining current if there are closed slots in the rotor. See Figure 11.37.

Another important parameter value to determine is the magnetizing inductance. Unfortunately, a no-load measurement does not correctly depict magnetizing inductance behaviour, because inductance saturates not only as a function of voltage, but also as a function of torque. Machine flux under load stretches into the air gap and increasingly across the slots. This may be observed as saturation of the magnetizing inductance. The behaviour of the magnetizing inductance can be modelled both as a function of voltage and torque as shown in Figure 11.38.

Stator resistance is also an essential parameter, which can be determined easily with a DTC inverter. The stator resistance may be estimated using an iterative algorithm by supplying the motor at standstill with DC current. The flux linkage of the motor will be

$$\boldsymbol{\psi}_s = L_s \boldsymbol{i}_s + \boldsymbol{\psi}_{s0} \qquad (11.65)$$

The initial value of the motor flux linkage is either zero or based on remanent flux density of the rotor. The same algorithm can be used also for a PMSM assuming that current can be supplied in the d-axis. In that case, the initial flux linkage equals the PM flux linkage $\boldsymbol{\psi}_{s0} \approx \boldsymbol{\psi}_{PM}$. A suitable initial value for the stator resistance estimation is R_{sest}. It is also possible to set $R_{sest} = 0$. When electric current is held constant, the differential of the flux

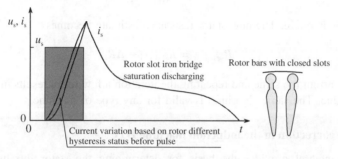

Figure 11.37 Rotor slot closing bridge saturation on the current decline in a transient inductance test.

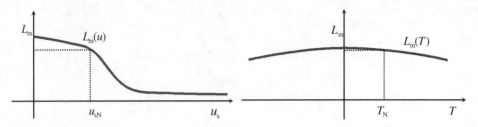

Figure 11.38 Magnetizing inductance as a function of air-gap voltage and torque. The rated points are marked with subscript N.

linkage will be zero, and the estimated stator flux linkage will be

$$\boldsymbol{\psi}_{sest} = \boldsymbol{\psi}_{s0} + \int (\boldsymbol{u}_s - \boldsymbol{i}_s R_{sest})dt \tag{11.66}$$

The estimated stator resistance can be expressed in terms of its value and an error correction.

$$R_{sest} = R_s + \Delta R_s \tag{11.67}$$

If the current and voltage information is assumed sufficiently correct, there will be a difference in the actual and estimated flux linkages.

$$\boldsymbol{\psi}_{sest} - \boldsymbol{\psi}_{s0} = \boldsymbol{\psi}_{s0} + \int_{t_0}^{t_1} (\boldsymbol{u}_s - \boldsymbol{i}_s R_{sest})dt - \boldsymbol{\psi}_{s0} + \int_{t_0}^{t_1} (\boldsymbol{u}_s - \boldsymbol{i}_s R_s)dt = -\Delta R_s \boldsymbol{i}_s (t_1 - t_0) \tag{11.68}$$

The error can be calculated using Equation (11.68). Since the actual value of the stator flux linkage is not known, two stator flux-linkage estimates $\boldsymbol{\psi}_{s1}$ and $\boldsymbol{\psi}_{s2}$ are needed.

$$\begin{aligned} \boldsymbol{\psi}_{sest1} - \boldsymbol{\psi}_{s1} &= \Delta R_s \boldsymbol{i}_s (t_0 - t_1), \\ \boldsymbol{\psi}_{sest2} - \boldsymbol{\psi}_{s2} &= \Delta R_s \boldsymbol{i}_s (t_0 - t_2). \end{aligned} \tag{11.69}$$

Subtracting the latter from the former and knowing that with constant current $\boldsymbol{\psi}_{s1} = \boldsymbol{\psi}_{s2}$ yields the following error correction equation.

$$\Delta R_s = \frac{\boldsymbol{\psi}_{sest2} - \boldsymbol{\psi}_{sest1}}{\boldsymbol{i}_s (t_2 - t_1)} \tag{11.70}$$

The process is iterative. The new stator resistance estimate becomes

$$R_{sest\ new} = R_{sest\ old} - \Delta R_s \tag{11.71}$$

Keeping the current the same and repeating the iteration a few times results in a sufficiently small ΔR_s value. The same algorithm is valid for any type of machine.

Flux linkage correction in the induction machine DTC

The voltage integral provides the basis for determining the stator flux-linkage vector. However, the flux-linkage integration introduces some error, as was shown in the introduction

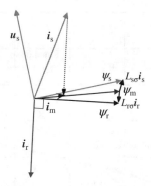

Figure 11.39 Space-vector diagram of an asynchronous machine at the rated operating point in steady state.

to the DTC. For an asynchronous machine, there are methods available to correct this error. Figure 11.39 shows again the space-vector diagram of an asynchronous machine.

According to the figure, the rotor current vector should be perpendicular to the rotor flux linkage in steady state. For an induction motor drive, the error correction (stabilization) method for the stator flux linkage can be based on this information. The rotor current vector is determined next. The following equations can be written based on the space-vector diagram in the figure.

$$\psi_r = \psi_s - L_{s\sigma}i_s + L_{r\sigma}i_r \tag{11.72}$$

$$\psi_s = L_s i_s + L_m i_r \Rightarrow i_r = \frac{1}{L_m}(\psi_s - L_s i_s) \tag{11.73}$$

Since the rotor current vector should be perpendicular to the rotor flux linkage, the correction term can be calculated by imposing the perpendicularity condition.

$$(\psi_{r,est}) \cdot (\psi_s - L_s i_s) = \varepsilon \tag{11.74}$$

When the dot product is low-pass filtered before it is supplied to the controller, ripple caused by the noise of the current measurement and the errors of the flux linkage can be filtered. Constant corrections $(k_d + jk_q)$ are then made to the stator flux linkage. These corrections hold the dot product Equation (11.74) at zero.

$$\psi_{s,est\ new} = \psi_{s,est\ old} + \varepsilon(k_d + jk_q)\psi_{s,est\ old} \tag{11.75}$$

The perpendicularity conditions do not hold during transients; however, the equation can be applied also during a transient if and extra term is added to the dot product.

$$k\tau_r \frac{d\psi_r}{dt} \tag{11.76}$$

After adding the additional term, the result must still be zero. Accomplishing this is difficult, since the rotor time constant varies.

Another possible means of stabilization is to construct a steady-state $I_s T$ curve based on nameplate values derived using stator resistance, stator leakage inductance, magnetizing

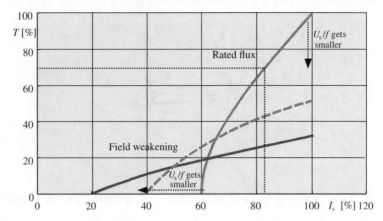

Figure 11.40 The torque production [%] of an asynchronous machine as a function of stator current with flux as a parameter.

inductance, rotation speed, and voltage as illustrated in Figure 11.40. This curve depicts the current-to-torque ratio of the machine at a given flux level in steady state. The curve becomes input for the stabilization algorithm. Since the motor currents can be measured, and since voltage level and speed dictate the amplitude of the flux-linkage, motor torque at a given current should match the look-up table curves in steady state. If it does not match, the flux-linkage estimate must be erroneous and must be corrected accordingly.

The two just introduced methods of stabilizing the instantaneous flux-linkage value provide the basis for stator flux-linkage correction in the controller; because the rotor current vector and the excitation flux linkage are always perpendicular in steady state. Further, the torque/current curve is known, and it can be used to compare the torque obtained by equation $T_e = 3/2p\psi_s \times i_s$ from the measured stator current vector with the integrated stator flux-linkage vector. If the result does not fit the current-torque curve of Figure 11.40, the stator flux linkage must be corrected.

The operator determines, during the commissioning identification run of the motor; the rated current i_N, the rated voltage u_N, the rated frequency f_N, the rated rotation speed n_N, and the power factor $\cos\varphi_N$. The stator resistance R_s is automatically identified during startup. Therefore, the operator does not have to manually determine a suitable value for the so-called IR compensation of the stator voltage drop. The change in stator resistance during the temperature rise of the machine is estimated by the thermal model of the machine.

The auxiliary control systems of the DTC

The core of the DTC efficiently produces the torque estimate and supplies the motor parameters. A DTC-controlled inverter unit requires several auxiliary controllers. Figure 11.41 illustrates the total DTC motor control system of an asynchronous machine.

Torque control

The reference for the torque control is either the torque reference provided by the speed controller or an external torque reference. The torque reference must be moderated from both

external torque reference

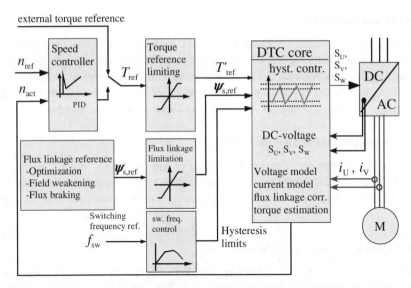

Figure 11.41 Schematic diagram of the DTC motor control system.

the drive system mechanics and current limiting points of view. DFLC is so fast that damage may result if mechanical system reacts anomalously.

Limiting the torque reference is important also in braking the motor so the permitted DC link voltage is not exceeded.

Speed control

Speed is controlled via a PID controller. In many applications, speed control is the most important drive system task. The quality of the overall control process often depends on the accuracy of rotation speed control. The output of the speed controller acts directly as a reference to the torque controller. This enables the rotation speed controller to take several requirements of the process into account. The controller also includes an acceleration compensator, which provides special advantages when it is necessary to minimize control deviations in the acceleration and deceleration of inertial loads. Once the mechanical time constant of the drive has been identified, PID speed control can be tuned to act more as a load compensator.

Flux linkage reference

The magnitude of the stator flux linkage can be fed to the DTC block as a reference value. This stator flux-linkage reference may be selected based on several criteria. One practical criterion is, for example, to minimize stator current for a given torque and improve drive system energy efficiency. Flux linkage optimization offers good energy economy, especially in drives that can run at normal speed with weakened machine flux. Examples include pump and blower drives.

Flux optimization improves the total efficiency of the motor drive. At small loads, the losses of the machine can be reduced even by 60% by suitably modifying the flux density. Flux optimization also reduces motor noise.

Switching frequency reference

The typical average switching frequency for IGBT switches in DTC drives is in the 3 kHz range. Thermal stress in the switch semiconductor material is the main constraint. In practice, switching frequency is controlled via the hysteresis bands in the optimal switching logic circuits. Switching frequency is lower at the lower speeds, because at low speeds there is higher voltage reserve stressing the switches than there is at higher speeds.

Performance of the induction motor DTC

DTC drive dynamic behaviour makes it possible to use AC drives to carry out duties that traditionally required DC machines. Examples include drives for cranes, lifts, presses, and reel-ups. In some cases, speed sensorless DTC drives can be used in place of flux-oriented vector controls.

Torque response and control accuracy

The DTC can provide the fastest torque response. When operating at rated flux, the minimum rise time for torque response is typically below 2 ms for a non–pulse-encoder control, while the torque step rise time is about 20 ms for typical DC and PWM-AC drives equipped with a speed-encoder. For a speed-sensorless vector controlled drive, the torque response time can be hundreds of milliseconds.

Figure 11.42 illustrates the performance of the DTC. The figure shows the step response of a 15 kW induction machine in a 70% torque step without pulse encoder feedback. Achieving a shorter response time is difficult, because the leakage inductances of the motor constitute a limiting factor for the required current changes. One could say that DTC is capable of using all the resources a machine can offer.

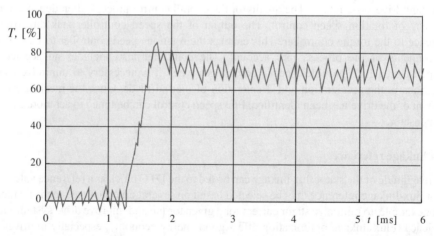

Figure 11.42 Typical fastest possible torque step in a DTC drive. The torque step height is 70% of the rated torque at 25 Hz operating frequency. A speed encoder was not used. The fast torque ripple shown in the plot cannot be seen, but produces audible noise at all but the highest switching frequencies. *Source:* Adapted from Aaltonen et al. (1995). Reproduced with permission from Finnish Electrical Engineers Association.

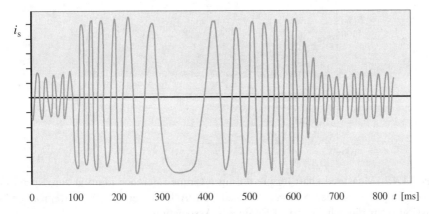

Figure 11.43 Phase current during a fast reverse at 20% torque in the absence of a pulse encoder. *Source:* Adapted from Aaltonen et al. (1995). Reproduced with permission from Finnish Electrical Engineers Association.

DTC can also guarantee linear and fast torque control at low speeds. Figure 11.43 depicts the motor phase current behaviour during a fast reverse with 20% load torque.

Figure 11.44 illustrates 80% shaft torque level during the slow reversing. Linearity is good for a drive without a pulse encoder.

Figure 11.45 represents a slow torque ramp at zero speed. The actual shaft torque is measured and the linearity error stays below 10%.

Accuracy of the speed control

The accuracy of the shaft speed estimate is good over the entire speed range. The static speed accuracy of the DTC is 10% of the rated slip of the motor, typically being between +0.1% and 0.5% of the rated speed. This accuracy usually suffices for industrial drives, to which speed-sensor–based vector control has conventionally been applied. Should this accuracy be insufficient, a pulse encoder feedback can be employed instead of the speed estimate, in which case the speed accuracy depends on the encoder characteristics.

Since the DTC is capable of fast torque response, the speed fluctuation in a sudden load change is notably smaller than in the traditional open-loop control drives. The gain of the speed controller can also be increased considerably from the previous values.

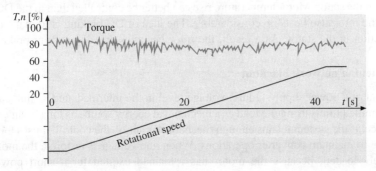

Figure 11.44 Estimated torque and measured shaft speed for a slow speed reversal with 80% torque. *Source:* Adapted from Aaltonen et al. (1995). Reproduced with permission from Finnish Electrical Engineers Association.

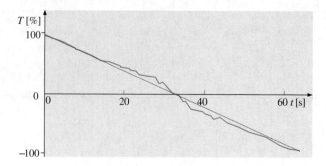

Figure 11.45 Measured shaft torque during a torque reference ramp at zero speed in the absence of a pulse encoder. *Source:* Adapted from Aaltonen et al. (1995). Reproduced with permission from Finnish Electrical Engineers Association.

Starting characteristics

With the DTC, a motor can be fast started in all electromagnetic states. A machine at rest and de-energized is started by first exciting it with direct current. Using one active voltage vector and the corresponding zero vector, the rated motor magnetizing current is first fed in by the converter. Thereafter, the machine can be started at its rated flux level by rotating the stator flux in the desired direction based on DFLC principles. If the shaft is rotating, the synchronous speed can be determined in some milliseconds using an algorithm similar to that described for permanent magnet synchronous machines, after which the DTC is immediately synchronized. See Chapter 9. In addition, starting can be performed rapidly when the rotor flux has already disappeared by first injecting an exciting pulse and then observing the machines possible rotation as is done in the case of a PMSM.

Braking characteristics

Because it is straightforward to control the amplitude of the stator flux linkage, this feature can also be utilized to brake a drive with just a diode bridge in the network terminals (i.e., without supplying power back to the network). The DTC braking feature includes so-called flux braking. In flux braking, the stator flux-linkage amplitude is raised from its rated value, and as a result, copper and iron losses in the machine increase. With this braking method, most motor losses are in the stator. Motor temperature rise can better be controlled than in the DC braking, in which the rotor also heats up considerably. The idea of DC braking is to make use of DC magnetization in the stator. As a result, the rotor becomes an eddy-current brake.

Management of network blackout

In the event of a power supply failure, voltage level in the intermediate DC link can be kept within permitted limits using the load as a mechanical energy source as long as any rotation is taking place. If the system retains enough mechanical energy, the controller uses the motor as a generator to maintain converter operations. When line voltage is restored, the motor can be immediately loaded, because the motor has remained excited for a short power failure duration. Power failures may vary from seconds to minutes depending on the inertia of the driven system.

11.5   A summary of industrial induction motor drives

Induction motors are suited DOL operation. They can also be operated in scalar control, in vector control and with DTC. DOL drives cannot be controlled, except by switching the drive on and off. Such drives stress both network and machinery and may be energy inefficient. Often simple scalar control suffices, which can also offer energy savings, especially in pump and fan applications. Vector control was originally developed for demanding drives and usually requires rotor speed feedback (a speed sensor).

An induction motor DTC drive is a general-purpose solution, and an induction motor DTC is in principle quite simple. Such drives are suitable for a number of applications. They can be used in less demanding roles, such as in static pumps and blower drives, or in extremely demanding drive applications. Both DTC and vector control are accurate and substantially faster than scalar control, and they are easier to use, at least in principle. For example, unlike scalar control, more sophisticated controllers do not require tuned IR or slip compensation. They offer automatic starting, improved braking characteristics, improved reliability, fast DC voltage control, avoidance of overcurrent releases, extremely fast reaction to sudden impacts on the load side, and improved operation during line-side power failures.

Table 11.6 compares the performance of different drives. The table affirms that DTC performs well in comparison to more traditional motor control techniques.

Current versions of DTC and vector control are similar. DTC uses the motor current model, which means it includes, in principle, all the elements of vector control except the current component controllers and a fixed modulator carrier frequency. In addition, improved stability and easier model tuning results if the voltage model is used to integrate the stator flux linkage, thereby improving classic vector control. A clear advantage of vector control is that it uses a predetermined switching pattern, and therefore the thermal stresses of the switching components can be better controlled than in DFLC where no accurate switching frequency is

Table 11.6 Asynchronous motor drive performance comparison with typical parameters

	TORQUE CONTROL METHOD				
	Speed Sensorless Scalar Controlled PWM	Vector Controlled PWM + Speed Sensor	DC Drive and a Digital Speed Sensor	Speed Sensorless DTC	DTC + Speed Sensor
Linearity	12%	4%	3%	4%	3%
Repeatability	4%	1%	–	1%	1%
Response time	150 ms	10 . . . 20 ms	10 . . . 20 ms	1 . . . 2 ms	1 . . . 2 ms
SPEED CONTROL					
Static accuracy	1 . . . 3%	0.01%	0.01%	0.1 . . . 0.5%	0.01%
Dynamic accuracy	3%s*	0.3%s	0.3%s	0.4%s	0.1%s

* %s is the time integral of speed deviation from the reference value in a load transient.

found. In the most demanding dynamic drives, DTC reacts faster than vector control, which always includes the current controller dynamics.

11.6   Doubly fed induction machine drives

Doubly fed induction machine (DFIM) drives are currently the most popular drive type for wind turbines. The DFIM has a wound rotor with multiphase winding sets on the rotor and stator. DFIMs can be referred to as wound rotor slip-ring induction machines. When this motor type was used as an industrial motor, the rotor winding set was connected to resistors via slip rings for smooth starting.

Before the era of static frequency converters, wound rotor slip-ring induction machines were also used as rotating frequency converters. The technology was used again in hydroelectric pump-storage power plants, where high-power variable-speed drives were required, operating both in motoring and generating modes at variable speeds. Currently, slip-ring machines are being used in new applications, especially as generators in wind turbines.

A DFIM drive can be adapted easily for use in wind turbines based on classic gearbox drivetrains. Moreover, DFIM efficiencies are good, so only a partial-power converter is needed to control the rotor circuitry. Figure 11.46 offers a schematic representation of a wind turbine drive train based on a doubly fed induction generator (DFIG).

A doubly fed machine drive can work as either a motor or a generator. Normally, however, the rotor can only turn in one direction without modifying the phase order of the stator. In principle, the DFIM can be operated in both rotor directions independent of stator phase order, but only by changing rotor and rotor converter dimensions. The rotor converter must be a four-quadrant (4Q) converter. Figure 11.47 illustrates the power flow options of a doubly fed

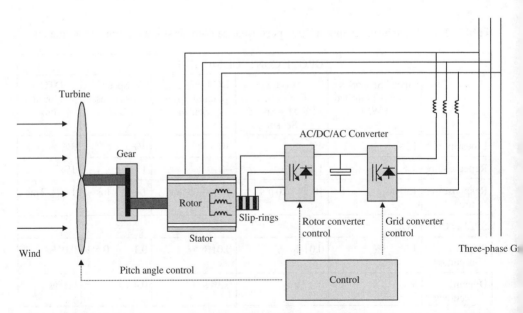

Figure 11.46   Schematic representation of a wind turbine drive train based on a DFIG.

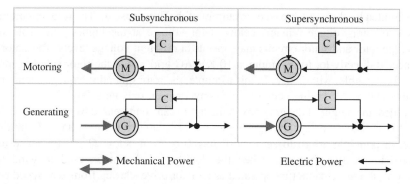

Figure 11.47 Practical operation modes for a DFIM. The converter must operate in both flux-rotating directions and in both power-flow directions.

system used as a motor and as a generator with the same rotating direction determined by the stator phase sequence.

The DFIM of the figure, therefore, can operate in four different modes in the $T\Omega$-plane as Figure 11.48 illustrates.

Figure 11.48 illustrates how the rotation speed of the DFIM drive is always in the same direction. Opposite rotation could be realized by changing the stator phase sequence. The height of the gray box in the figure is limited by the voltage capabilities of the rotor system. Its

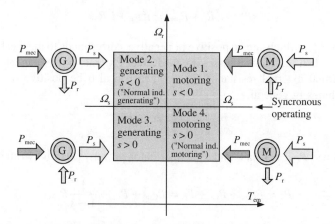

Figure 11.48 Practical operation modes of a DFIM in the $T\Omega$-plane. In principle, modes 2 and 4 could be realized passively using normal induction machines without any power electronics in the rotor circuit. Modes 1 and 3 are modes where the rotor must be supplied actively with power electronics that can realize "negative rotor circuit external resistance". The doubly fed induction machine can also act as a nonsalient-pole synchronous machine when DC is supplied to the rotor. In that case, one rotor phase is connected in series with two other phases connected in parallel to produce a DC-excitation winding. The phrases "normal induction machine generating" or "normal induction machine motoring" are modes included in the rotor shorted areas. When the rotor is shorted, the magnitude of slip must be as low as in squirrel-cage machines, where slip is typically in the range of 1% or less. Mechanical power is P_{mec}, stator power is P_s, and rotor power is P_r.

width is limited by the available electric current of the drive system. The rotor circuit must be designed so rotor voltage remains below stator voltage at the highest allowable slip. In practice, the rotor and stator circuits may operate at different voltage levels. The stator could be connected directly to, for example, a 10 KV AC grid, while the rotor side system could operate, for example, with 690 V. In this example, the rotor operating voltage should remain below this 690 V AC so the network-side converter can operate safely.

If needed, the height of the gray box in the figure can be increased by specifying less turns in the rotor winding or by specifying a higher rotor system maximum voltage. Therefore, it is possible, in principle, to go to zero speed $\Omega_r = 0$ or even below $\Omega_r < 0$ (opposite rotation direction). In practice, however when the doubly fed system is used in wind turbine applications, the drive cannot be operated as a motor drive starting from zero speed because of the voltage limitations in the rotor circuit converter.

The power balance of a wound-rotor machine is the same, in principle, as that of the cage-rotor machine. The electromechanical power P_{mec}, including output and mechanical losses, is dependent on air-gap power P_δ via slip s.

$$P_{mec} = T_{em}\Omega_r = (1 - s)P_\delta \tag{11.77}$$

In a DFIM, power sP_δ is the sum of the Joule losses and the real power of the rotor external circuit, that is, the power with phasors $m\mathrm{Re}(\underline{U}_r \, \underline{I}_2^*)$ and with m the number of phases. sP_δ can be large for a DFIM, because it also includes the external circuit power P_{rext} of the system. The equation for sP_δ is

$$sP_\delta = i_r^2 R_r + P_{rext} = i_r^2 R_r + i_r^2 R_{rext} \tag{11.78}$$

External power P_{rext}, which can be positive or negative, corresponds to the mechanical power of the machine when mechanical losses are neglected. External power can take place in the presence of virtual external resistance R_{rext}, which can also be positive or negative. For induction machines in general,

$$P_{rext} = \frac{s}{1 - s}P_\delta - P_{Cur} \tag{11.79}$$

In large machines, P_{Cur} is small, and $P_r \approx sP_\delta$. Stator power P_s can be written

$$P_s = P_{Cus} + P_{Fe} + P_\delta = P_{Cus} + P_{Fe} + \frac{1}{1 - s}P_{em} \tag{11.80}$$

The (to or from) grid power is

$$P_{grid} = P_s - P_{rext} \tag{11.81}$$

Neglecting losses yields

$$P_{rext} = \frac{s}{1 - s}P_\delta \tag{11.82}$$

$$P_s = \frac{1}{1 - s}P_{em} \tag{11.83}$$

$$P_{grid} = P_s - P_{rext} = (1 - s)P_\delta = P_{mec} = T_{em}\Omega_r \tag{11.84}$$

According to Equation (11.84), mechanical power is the difference between the stator input power P_s and the power passing from the rotor to the brushes via the slip-rings P_{rext}. Equation (11.84) is written with positive P_s, P_{rext}, and P_{mec} in accordance with induction motor tradition for subsynchronous operation.

Since the effective number of turns may be different, the transformation ratio K_{rs} must be defined between the stator and rotor; common in wound rotor applications. Assuming the number of phases m is the same for both stator and rotor, with $k_{w1s}N_s$ and $k_{w1r}N_r$ effective turns in series per phase, the transformation ratio becomes

$$K_{sr} = \frac{k_{w1s}N_s}{k_{w1r}N_r} \tag{11.85}$$

The ratio between the air-gap–induced voltages of the machine is as follows.

$$E_{mr} = sE_{ms}\frac{1}{K_{sr}} \tag{11.86}$$

Using the equivalent circuit of Figure 11.49 leads to an expression for the rotor RMS current phasor.

$$\underline{I}_r = \frac{\underline{E}_{mr} - \underline{U}_{slipring}}{R_r + js\omega_s L_{r\sigma}} \tag{11.87}$$

This current may be referred to the stator side via the ratio

$$\underline{I}'_r = \frac{\underline{I}_r}{K_{sr}} \tag{11.88}$$

The referred current flows in the equivalent rotor winding producing the same current linkage as the real rotor current. The expression for referred current can be rewritten as

$$\underline{I}'_r = \frac{1}{K_{sr}}\frac{sE_{ms}\frac{1}{K_{sr}} - \underline{U}_{slipring}}{R_r + js\omega_s L_{r\sigma}} = \frac{E_{ms} - \dfrac{U'_{slipring}}{s}}{\dfrac{R'_r}{s} + j\omega_s L'_{r\sigma}} \tag{11.89}$$

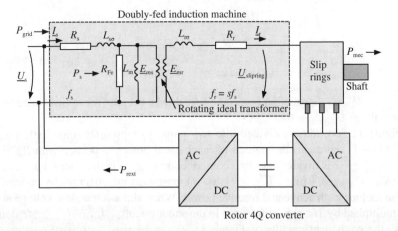

Rotor 4Q converter

Figure 11.49 DFIM drive system equivalent circuit. Power flow given for subsynchronous motoring. The rotating ideal transformer transforms both the voltage and frequency (Miller, T. J. E. (2010). Theory of the doubly fed induction machine in the steady state. *2010 XIX International Conference on lectric Machines (ICEM)*, pp. 1–6, 6–8 September 2010).

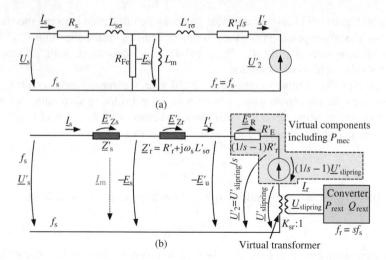

(a)

(b)   Virtual transformer

Figure 11.50 (a) Equivalent circuit of DFIM in the stator voltage level with rotor quantities referred (') to the stator and (b) solving the behaviour of the DFIM-drive ((Miller, T. J. E. (2010). Theory of the doubly fed induction machine in the steady state. *2010 XIX International Conference on lectric Machines (ICEM)*, pp. 1–6, 6–8 September 2010).

The right-side expression in Equation (11.89) results when the expression to its left has been expanded by K_{sr}/s. It is equivalent to the traditional presentation of an IM at the stator voltage level. With Equation (11.89), it is possible to use the traditional RMS equivalent circuit of an induction motor. The only difference is the extra voltage source in the rotor circuit representing the external voltage arrangement of the DFIM. See Figure 11.50.

In Figure 11.50,

$$\underline{U}'_2 = \frac{\underline{U}'_{slipring}}{s},$$

$$\underline{U}'_{slipring} = K_{sr}\,\underline{U}_{slipring},$$

$$R'_r = K_{sr}^2 R_r, \text{ and} \qquad (11.90)$$

$$L'_{r\sigma} = K_{sr}^2 L_{r\sigma}.$$

The $\underline{U}'_2 = \underline{U}'_{slipring}/s$ equation is specific for DFIMs, because for squirrel-cage machines, the rotor voltage $U_{slipring} = 0$.

In Figure 11.50b, R'_r/s is divided into two parts. R'_r represents rotor Joule losses, and $R'_E = R'_r(1/s - 1)$ represents the portion of mechanical power associated with R_r. Similarly, voltage $\underline{U}'_{slipring}/s$ is divided into two components. $\underline{U}'_{slipring}$ is the referred actual slip-ring voltage ($K_{sr}\,\underline{U}_{slipring}$), while $\underline{U}'_{slipring}(1/s - 1)$ represents the effect of the external circuit, which can include both active and reactive power. When slip s is low, low voltage at the slip rings is multiplied by $1/s$, appearing large in the stator circuit. $\underline{U}'_2 = \underline{U}'_{slipring}/s$ dominates.

Now, the equivalent circuits of Figure 11.50 can be used to solve Equation (11.89). Figure 11.49 is first reduced to Figure 11.50a. The ideal rotating transformer disappears, and the equations are solved entirely at the stator frequency. The task is simplified further by making use of $R_s + j\omega_s L_{s\sigma} = \underline{Z}_s$ and the magnetization path to a single admittance $\underline{Y}_m$, and by reducing the stator part to its Thévenin equivalent.

$$\underline{U}'_s = \frac{\underline{U}_s}{1 + \underline{Z}_s \underline{Y}_m},$$

(11.91)

$$\underline{Z}'_s = \frac{\underline{Z}_s}{1 + \underline{Z}_s \underline{Y}_m}.$$

(11.92)

The stator current is now solved via Figure 11.50b.

$$\underline{I}'_s = \frac{\underline{U}'_s - \underline{U}'_2}{\underline{Z}'_s + R'_r/s + j\omega_s L'_{r\sigma}}.$$

(11.93)

The equations for voltages (see Figure 11.50), magnetizing branch current, and stator current are as follows.

$$\underline{E}'_u = \underline{E}'_R + \underline{U}'_2,$$

(11.94)

$$\underline{E}_s = \underline{E}'_u + \underline{U}'_{Zr},$$

(11.95)

$$\underline{I}_m = \underline{E}_s Y_m,$$

(11.96)

$$\underline{I}_s = \underline{I}_m + \underline{I}'_r.$$

(11.97)

The power balance analysis of the system starts with rotor copper losses. The real rotor Joule loss is given by

$$P_{Cur} = m I_r^2 R_r$$

(11.98)

In Figure 11.50 the rotor power associated with R_r is

$$P_{Rr} = m I_r'^2 \frac{R'_r}{s} = m \left(\frac{I_r}{K_{sr}}\right)^2 \frac{K_{sr}^2 R_r}{s} = m I_r^2 \frac{R_r}{s}.$$

(11.99)

This obviously includes much larger power than just the Joule losses, which should be only $m I_r^2 R_r$. Obviously, there is additional power in the stator

$$m I_r'^2 \left(\frac{1}{s} - 1\right) R'_r = P_{Cur}\left(\frac{1}{s} - 1\right)$$

(11.100)

Electromechanical power conversion includes this term (11.100). This is the only power component associated with power conversion for a normal squirrel-cage motor. In a DFIM, however, there is another component included in $m\,\underline{U}'_r\,\underline{I}'^*_r$ (* stands for complex conjugate), which is associated with $\underline{U}'_{slipring}/s$. The apparent power $m\,\underline{U}'_{slipring}\,\underline{I}'^*_r = m\,\underline{U}_{slipring}\,\underline{I}^*_r$ is the actual apparent power on the slip rings and converter terminals. However, the current and voltage frequencies are equal to the rotor slip frequency. The virtual component $m\,\underline{U}'_{slipring}\,\underline{I}'^*_r(1/s - 1)$ is analogous to $\underline{I}'^2_r R'_E$ or $\underline{I}'^2_r R'_r(1/s - 1)$. Although $\underline{I}'^2_r R'_E$ represents pure active power, both stator and rotor external circuits can have both active and reactive power. Therefore, machine excitation can be shared by the stator and rotor currents if so desired.

The following phasor equations can be written for the DFIM drive.

$$P_{\text{rext}} = m\text{Re}\left(\underline{U}'_{\text{slipring}}\ \underline{I}'^{*}_{\text{r}}\right) = m\text{Re}\left(\underline{U}_{\text{slipring}}\ \underline{I}^{*}_{\text{r}}\right), \qquad (11.101)$$

$$P_{\text{Cur}} = m\ \underline{I}'^{2}_{\text{r}}R'_{\text{r}} = m\ \underline{I}^{2}_{\text{r}}R_{\text{r}}, \qquad (11.102)$$

$$P_{\text{mec}} = (1 - s)\frac{P_{\text{Cur}} + P_{\text{rext}}}{s} = m\underline{I}'^{2}_{\text{r}}R'_{\text{E}} + m\text{Re}\left[(1 - s)\ \underline{U}'_{2}\ \underline{I}'^{*}_{\text{r}}\right] \qquad (11.103)$$

P_{rext} is actual converter power. It is regarded positive when the power is flowing from the rotor via the slip rings to the converter and to the network. Establishing this flow direction as positive conforms to the sign convention for the power components of an asynchronous motor.

P_{mec} is the mechanical power of the machine. It is positive for motoring. This power is entirely represented by the virtual power gray box in Figure 11.50b. The mechanical power $P_{\text{mec}} = (1 - s)P_{\delta}$, and the air-gap power is given by the following equation.

$$P_{\delta} = \frac{P_{\text{Cur}} + P_{\text{rext}}}{s} = m\underline{I}'^{2}_{\text{r}}R'_{\text{r}} + m\text{Re}\left[\underline{U}'_{2}\ \underline{I}'^{*}_{\text{r}}\right]. \qquad (11.104)$$

Air gap power can also be written as

$$P_{\delta} = m\text{Re}\left[\underline{E}_{s}\ \underline{I}'^{*}_{\text{r}}\right]. \qquad (11.104)$$

The electrical analysis above provides evaluation of the component powers, especially the slip-ring power P_{rext}, which does not appear in the case of the squirrel-cage motor.

11.6.1   Controlling the DFIM

To analyse the basic principles of DFIM control, its equivalent circuit can be simplified by ignoring inessential stator components. See Figure 11.51.

Following Figure 11.51 and multiplying both sides of the equation by the per-unit slip s, the rotor voltage equation can be written.

$$\underline{u}'_{\text{r}} = R'_{\text{r}}\underline{i}'_{\text{r}} + js\omega_{\text{s}}L'_{\text{r}\sigma}\underline{i}'_{\text{r}} + s\underline{u}_{\text{s}} \qquad (11.105)$$

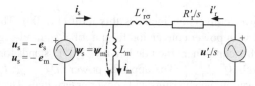

Figure 11.51   Simplified equivalent circuit for the doubly fed induction machine. The rotor voltage $\underline{u}'_{\text{r}}/s$ describes external voltage $\underline{u}_{\text{r}}$, which should be zero in squirrel-cage machines. In this simplified presentation, air-gap voltage is equivalent to stator terminal voltage.

This results if the following definition of the steady-state rotor current vector.

$$i''_r = \frac{u'_r - su_s}{R'_r + js\omega_s L'_{r\sigma}}$$
(11.106)

The main idea for speed and torque control is to apply a suitable external voltage vector u_r to the rotor terminals so the rotor is driven towards new slip and torque values. Different speeds (i.e., different slips) represent different no-load rotor voltages. A new value can be defined, the **base slip** value s_0 for different rotor voltages, by setting the rotor current to zero at the respective rotor voltage.

$$i''_r = 0 = \frac{u'_{r0} - su_s}{R'_r + js\omega_s L'_{r\sigma}} \Rightarrow u'_{r0} = su_s \Rightarrow s_0 = \frac{u'_{r0}}{u_s}$$
(11.107)

At s_0 and u_{r0}, the DFIM runs at no load at a speed corresponding to s_0. Therefore, with $u_r = 0$, the base slip value $s_0 = 0$ indicates that no-load speed corresponds to stator frequency, and the machine operates like a normal squirrel-cage machine, that is, like a short-circuited wound-rotor machine. Correspondingly for $u_r = u_s$, for example, a base slip value of $s_0 = 1$ results at the no-load speed of zero (like a wound rotor with an open-circuit winding). If negative values are used for u_r, the machine no-load speed is on the super synchronous side where $\omega_r > \omega_s$.

In practice, if per-unit stator voltage $u_s = 1$ is used, the maximum rotor slip variation in a wind power generator can typically be in the range of ±30%. This range can be achieved for base slip as follows.

$$s_0 = \pm 0.3 = \frac{u'_{r0}}{u_s} \Rightarrow u'_{r0} = \pm 0.3 \text{ pu}$$
(11.108)

The base slip operational parameter s_0 is important to the understanding of DFIM rotor behaviour. It establishes a virtual synchronous no-load operation point for the DFIM. If external torque causes DFIM speed to increase from this base-slip speed value, there will be extra voltages induced in the rotor winding driving rotor currents and the machine operates as a generator. This corresponds to the rotor behaviour of a normal induction motor at super synchronous speed. If the external load torque slows machine speed to below the base-slip speed, voltages will again be induced in the rotor, and the resulting currents will induce electromagnetic torque that opposes the change of speed. The equation for base-slip speed is

$$\omega_r = (1 - s_0)\omega_s.$$
(11.109)

If s_0 varies ±0.3 pu, the real electromagnetic rotating speed of the machine also varies.

$$\omega_r \in [0.7, 1.3]\omega_s.$$
(11.110)

In addition, mechanical angular velocity varies according this the following equation.

$$\Omega_r = \frac{\omega_r}{p} \in [0.7, 1.3]\frac{\omega_s}{p}.$$
(11.111)

The base-slip speed forms a **virtual synchronous operating point** for the machine, and the torque producing slip can be regarded as a Δs slip change from this base.

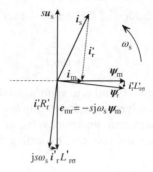

Figure 11.52 Simplified (no stator leakage) steady-state space-vector diagram of an induction motor with $s = 0.01$ and real-rotor–induced quantities corresponding to a base slip of $s_0 = 0$. The diagram corresponds to a squirrel-cage induction motor with $u_r = 0$. The voltage induced in the rotor winding is defined by the slip $s = 0.01 = \Delta s$ and is $e_m = -\Delta s j\omega_s\psi_m$. Since the rotor terminal voltage is zero, the induced voltage $0.01u_s$ is dropped totally from the expression for rotor impedance $R_r + js\omega_sL_{r\sigma}$. $R_r = 0.01$ pu. $L_{r\sigma} = 0.1$ pu.

A slightly modified version of the space-vector diagram shown in Figure 11.3 can be used to model the doubly salient machine. For an external rotor voltage supply, the rotor circuit current is driven only by the air-gap flux-linkage–induced voltage in the rotor winding. The equation for this voltage in terms of base slip s_0 is

$$e_{mr} = -j\Delta s\omega_s\psi_m = -j(s - s_0)\omega_s\psi_m \qquad (11.112)$$

Instead of values referred to the stator voltage level, absolute values for the rotor-side voltage vectors should be given. Therefore, being directly per-unit slip-dependent, rotor-induced voltage may be low. Figure 11.52 illustrates the modified space-vector diagram of an induction motor with shorted rotor terminals $(u_r = 0)$.

Figure 11.53 illustrates the DFIG at no load for two different base slips and corresponding rotor voltage supplies. Voltages and base slips are ± 0.3. When the induced voltage and rotor terminal voltage are equal, there is no rotor current, and the machine runs at no load.

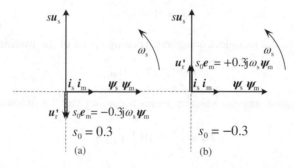

Figure 11.53 DFIG at no load with ± 0.3 base slip and corresponding voltages. Graph (a) represents positive base slip and graph (b) represents negative base slip.

EXAMPLE 11.4: Determine how DFIM behaviour changes as rotor voltage is reduced or increased by 1% unit from the no-load slip value of 0.25 pu. The per-unit rotor resistance is $R_r = 0.01$, and the leakage inductance is $L_{r\sigma} = 0.1$.

SOLUTION: The base slip and rotor voltage at DFIM no-load are $s_0 = u_r' = s_0 u_s = 0.25$. Vector direction can be assumed according to Figure 11.53. Then, rotor voltage can be dropped to $u_r' = 0.24$. Based on Equation (11.106), rotor current will be

$$i_r' = \frac{u_r' - s_0 u_s}{R_r' + j s \omega_s L_{r\sigma}'} = \frac{0.24 e^{j3\pi/2} - 0.25 \cdot 1 e^{j3\pi/2}}{0.01 + j0.025 \cdot 1 \cdot 0.1} = \frac{-0.01 e^{j3\pi/2}}{0.01 + j0.025} = -0.37 e^{j3\pi/2 - 0.37 j\pi}$$

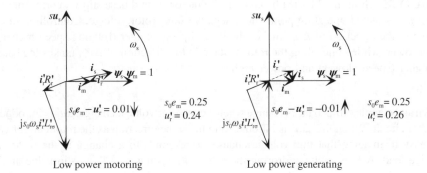

Low power motoring   Low power generating

Figure 11.54 Vector diagrams of machine behaviour for small voltage changes from the no-load value to a value nearer to the 0.25 base slip.

Increasing rotor voltage to $u_r' = 0.26$ in the original vector direction results in the following.

$$i_r' = \frac{u_r' - s_0 u_s}{R_r' + j s \omega_s L_{r\sigma}'} = \frac{0.26 e^{j3\pi/2} - 0.25 \cdot 1 e^{j3\pi/2}}{0.01 + j0.25 \cdot 1 \cdot 0.1} = \frac{0.01 e^{j3\pi/2}}{0.01 + j0.025} = 0.37 e^{j3\pi/2 - 0.37 j\pi}$$

The voltage that produces current in the rotor winding is the difference between $u_{r0}' = s_0 u_s$ and the real rotor voltage u_r'. The rotor winding angular frequency is dictated by the base slip frequency resulting in $s_0 \omega_s$. Figure 11.54 illustrates the vector diagrams of the two example scenarios.

The vector diagram introduced in Figure 11.55 illustrates the effect of rotor terminal voltage variation using three rotor circuit voltage change values $\Delta u_{r1}'$, $\Delta u_{r2}'$, and $\Delta u_{r3}'$. Each of these has been selected to be in phase with the original rotor current. Such voltage changes emulate voltage drops over a virtual—positive or negative—external rotor circuit resistance. The rotor induced voltage $s_{01} e_m$ at base slip s_{01} is represented by a gray vector, while the rotor-induced voltages at other slips caused by changes in the rotor circuit voltage in the doubly fed machine are represented by dashed lines. The voltage drops over the virtual rotor complex impedances are depicted for each of these values.

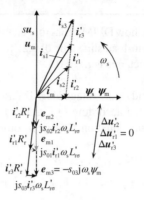

Figure 11.55 Simplified DFIM behaviour of a motor around base slip s_{01} corresponding to an *a priori* selected operating point and a negative $s_{01}u_s$ rotor voltage. Controlling the rotor voltage in the direction of $\Delta u'_{r2}$ makes the base slip s_{02} smaller (rotating speed higher) and power lower, while controlling the rotor voltage in the direction of $\Delta u'_{r3}$ makes the base slip s_{03} higher (speed slower) and power higher.

The virtual complex impedances change when a converter voltage change $\Delta u'_r$ is introduced in the rotor circuit. When this $\Delta u'_r$ is introduced in the same direction as the rotor current vector, it is easy to imagine that this voltage change corresponds to a change in the rotor circuit resistive load. A large voltage drop in the rotor corresponds to a large slip value and high torque, while a small voltage drop in the rotor corresponds to a small slip value and lower torque.

Looking at Figure 11.55, a positive $\Delta u'_{r2}$ change in rotor voltage forces the rotor current to a new smaller value i'_{r2} according to Equation (11.106). The stator current moves to i_{s2}, and the base slip moves to s_{02}. The result is an increase in speed and a reduced voltage $s_{02}e_m$ induced in the rotor circuit. Similarly, a negative change $\Delta u'_{r3}$ in the rotor terminal voltage forces the rotor and stator currents to i'_{r3} and i_{s3} and the base slip to s_{03}. Speed is reduced and the induced voltage on the rotor side increases to $s_{03}e_m$. Since the changes in rotor voltage and current are in phase, the external circuit converter handles only active power in this case.

Naturally, the rotor circuit can also be supplied with a voltage change that reverses the rotor current, and therefore changes the torque direction. The DFIM begins to operate as a generator. Figure 11.56 illustrates a DFIG in operation.

The simplified equivalent circuit in Figure 11.51 can also be used to study DFIM cases involving zero stator resistance and leakage.

The stator terminals power factor for a DFIM system can be freely selected by controlling the rotor voltage vector and, therefore, the rotor current. The rotor winding can be used as the field winding and share the excitation task between the stator and the rotor. Figure 11.57 illustrates the possibilities in motoring mode.

Figure 11.48 shows the DFIM vector diagrams for four operating modes accounting for stator leakage inductance. The equivalent circuit is presented in Figure 11.58.

Figure 11.59 describes mode 1 operation, that is, supersynchronous motoring with unity power factor in the stator side.

The vector diagrams for modes 2 through 4 are shown in Figures 11.60–11.62.

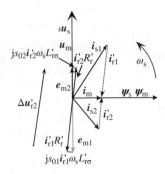

Figure 11.56 DFIM operation changes from motoring to generating with increased external rotor voltage $\Delta u'_{r2}$.

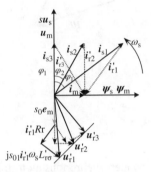

Figure 11.57 Controlling the stator-side power factor using rotor voltage and current at base slip s_0 with the active power being constant.

DFIG control can be realized as vector control based on Figure 11.63. It is modelled in the stator flux-linkage–oriented reference frame. Since grid voltage dictates the stator flux linkage anyway, this is natural.

DFIG torque and active power control is carried out with the q-component of current. Excitation and reactive power control is carried out with the current d-component. Rotor speed control is accomplished by manipulating the rotor currents.

$$\Omega_m p = \omega_s - \omega_r = \omega_s - 2\pi f_r. \tag{11.113}$$

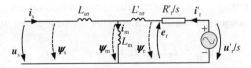

Figure 11.58 Equivalent circuit of DFIM accounting for stator leakage.

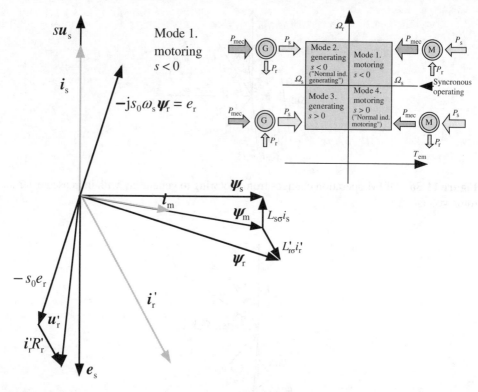

Figure 11.59 Vector diagram of supersynchronous motoring at negative slip. Magnetization comes via the rotor current. The stator has a power factor of one. The diagram is drawn in the traditional manner with rotor quantities referred to the stator. However, base slip is taken into account in the rotor back emf. Both rotor and stator powers are positive, so the converter must actively supply the rotor to achieve motor super synchronous operation.

The general Park model of the induction machine is used.

$$u_s = R_s i_s + \frac{d\psi_s}{dt} \tag{11.114}$$

$$u_r = R_r i_r + \frac{d\psi_r}{dt} - j\omega_s \psi_r \tag{11.115}$$

$$\psi_s = L_s i_s + L_m i_r \tag{11.116}$$

$$\psi_r = L_m i_s + L_r i_r \tag{11.117}$$

$$L_s = L_m + L_{s\sigma}, L_r = L_m + L_{r\sigma} \tag{11.118}$$

The stator flux-linkage components, assuming the stator flux-linkage orientation, are as follows. The flux-linkage– and torque-producing components are designated with subscripts ψ or T.

$$\psi_{s\psi} = L_s i_{s\psi} + L_m i_{r\psi} = \psi_s = L_m i_m \tag{11.119}$$

$$\psi_{sT} = 0 \tag{11.120}$$

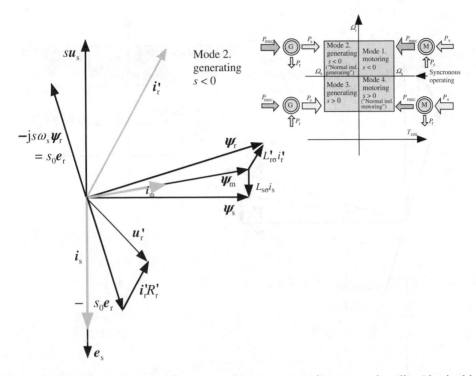

Figure 11.60 Vector diagram of super-synchronous generating at negative slip. Also in this case, magnetization comes via the rotor current. The stator has a power factor of one. Base slip is taken into account in the rotor back emf. Both rotor and stator powers are negative, so the converter must actively rectify rotor power and supply the power to the network.

The following express leakage factor σ and equivalent inductance L_o.

$$\sigma = 1 - \frac{L_m^2}{L_s L_r} \tag{11.121}$$

$$L_o = \frac{L_m^2}{L_s} \tag{11.122}$$

The rotor side voltage and flux-linkage equations are

$$u_{r\psi} = R_r i_{r\psi} + \sigma L_r \frac{di_{r\psi}}{dt} - \omega_r \sigma L_r i_{rT} \tag{11.123}$$

$$u_{rT} = R_r i_{rT} + \sigma L_r \frac{di_{rT}}{dt} + \omega_r \left(L_o i_m + \sigma L_r i_{r\psi} \right) \tag{11.124}$$

$$\psi_{r\psi} = \frac{L_m^2}{L_s} i_m + \sigma L_r i_{r\psi} \tag{11.125}$$

$$\psi_{rT} = \sigma L_r i_{rT} \tag{11.126}$$

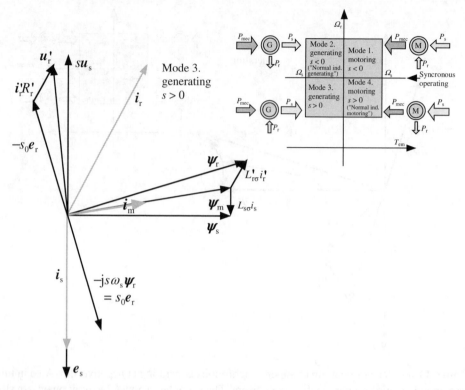

Figure 11.61 Vector diagram of subsynchronous generating at positive slip. Also in this case, magnetization comes via the rotor current. The stator has a power factor of one. Base slip is taken into account in the rotor voltages. Rotor power is positive while stator power is negative, so the converter must actively supply rotor power from the network.

Because of the stator flux-linkage orientation, $\psi_{s\psi} = \psi_s$, and $\psi_{sT} = 0$. Active and reactive power are

$$P_s = u_{s\psi}i_{s\psi} + u_{sT}i_{sT} \tag{11.127}$$

$$Q_s = u_{s\psi}i_{sT} + u_{sT}i_{s\psi} \tag{11.128}$$

where

$$i_{sT} = -\frac{L_m}{L_s}i_{rT} \tag{11.129}$$

$$i_{s\psi} = i_m - \frac{L_m}{L_s}i_{r\psi} \tag{11.130}$$

Currently, the DFIM is dominant for wind power applications, at least in the 3 to 5 MW range for onshore applications. It will be interesting to see which machine architectures prevail as wind turbine nameplate power ratings continue to increase. The largest commercial wind turbine in operation today is manufactured by Vestas, and its 9 MW drivetrain uses a semi-geared medium-speed permanent magnet generator interfacing with a full power converter. Enercon offers a 7 MW direct-drive synchronous generator system.

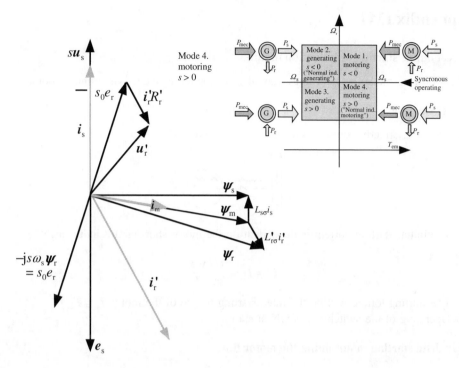

Figure 11.62 Vector diagram of subsynchronous motoring at positive slip. Also in this case, magnetization comes via the rotor current. The stator has a power factor of one. Base slip is taken into account in the rotor voltages. Rotor power is negative, while stator power is positive so the converter must actively absorb the rotor power and supply it to the network.

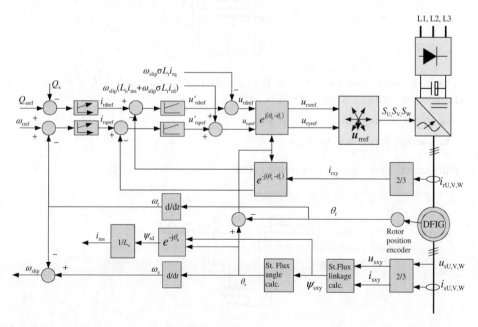

Figure 11.63 Vector control of a DFIG.

Appendix IM1

Direct online starting of induction motors

The following methods and values may also be applied to DOL synchronous motors.

Direct starting

The motor is directly connected to the grid via switch S1.

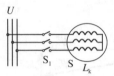

The initial starting current is dependent on the motor short circuit inductance L_k.

$$I_s = I_1 \approx \frac{U/\sqrt{3}}{2\pi\omega_s L_k}$$

The starting torque will be the rated starting torque of the motor $T_s = T_1$.
Operating of the switch: S_1 is ON at start.

Star-delta starting maintaining the motor flux

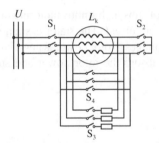

Operating sequence of the switches: 1) S_1 and S_2 is ON at start. 2) S_3 is ON after a suitable acceleration time. 3) S_2 is OFF. 4) S_4 is ON.

The initial starting current is $I_s = \frac{I_1}{3}$.

The starting torque will be $T_s = \frac{T_1}{3}$.

Starting transformer with a star point switch

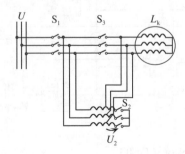

Operating sequence of the switches: 1) S_2 is ON, 2) S_1 is ON 3) S_2 is OFF, and S_3 is ON.

The initial starting current is $I_s = \left(\frac{U_2}{U}\right)^2 I_1$.

The starting torque will be $T_s = \left(\frac{U_2}{U}\right)^2 T_1 = \frac{I_s}{I_1} T_1$.

Starting transformer without a star point switch

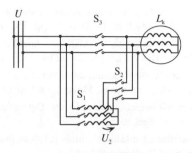

Operating sequence of the switches: 1) S_1 and S_2 are ON at start. 2) S_1 and S_2 are fast-switched OFF, then S_3 is immediately switched ON.

The initial starting current is $I_s = \left(\frac{U_2}{U}\right)^2 I_1$.

The starting torque will be $T_s = \left(\frac{U_2}{U}\right)^2 T_1 = \frac{I_s}{I_1} T_1$.

Block transformer effect

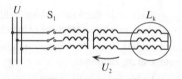

Operating sequence of the switches: 1) S_1 is ON at start.

The initial starting current is $I_s = \left(\frac{U_2}{U}\right)^2 I_1$.

The starting torque will be $T_s = \left(\frac{U_2}{U}\right)^2 T_1 = \frac{I_s}{I_1} T_1$.

Star point choke limiting the starting current

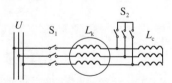

Operating sequence of the switches: 1) S_1 is ON at start, and 2) S_2 is ON.

The initial starting current is $I_s = \frac{L_k}{L_k + L_c} I_1$.

The starting torque will be $T_s = \left(\frac{I_s}{I_1}\right)^2 T_1$.

References

Aaltonen, M., Heikkilä, S., Lalu, J., & Tiitinen, P. (1995). Direct torque control (DTC). *Sähkö & Tele*, 68, 33–38 [in Finnish].

Aarniovuori, L. (2010). Induction motor drive energy efficiency – simulation and analysis. Dissertation Lappeenranta University of Technology, 2010, ISBN 978-952-214-962-6 ISBN 978-952-214-963-3 (PDF), available at http://urn.fi/URN:ISBN:978-952-214-963-3

Blaschke, F. (1971). *Das Prinzip der Feldorientierung, die Grundlage für die Transvector-Regelung von Drehfeldmaschinen*. Siemens-Zeitschrift 45, Heft 10. pp. 757–760.

Depenbrock, M. (1985). *Direckte Selbstregelung (DSR) für hochdynamische Drehfeldantriebe mit Stromrichterspeisung*. Etz Archiv BD 7, H.7. pp. 211–218.

Gonzalo, A. (2008). Predictive direct control techniques of the doubly fed induction machine for wind energy generation applications. Doctoral thesis, Mondragon University, 3 July 2008.

Kärkkäinen, H. (2015). Converter-fed induction motor losses: Determination with IEC methods. M.Sc. thesis, Lappeenranta University of Technology.

Kaukonen, J. (1994). Epätahtimoottoreiden sähkönkulutus ja häviöt puunjalostusteollisuudessa. M.Sc. thesis, Lappeenranta University of Technology.

Miller, T. J. E. (2010). Theory of the doubly fed induction machine in the steady state. *2010 XIX International Conference on lectric Machines (ICEM)*, pp. 1–6, 6–8 September 2010.

Mohan, N., Undeland, T. M., & Robbins, W. P. *Power electronics: Converters, applications and design*. Chichester, UK: John Wiley and Sons.

Niiranen, J. (2000). Sähkömoottorikäytön digitaalinen ohjaus. Otatieto (in Finnish).

Pena, R. Clare, J. C., & Asher, G. M. (1996). Doubly fed induction generator using back-to-back PWM converters and its application to variable-speed wind-energy generation. *Electrical Power Applications, IEE Proceedings*, 143(3), 231–241.

Pohjalainen, P. (1987). Vuon ja vääntömomentin kaksipistesäätöihin perustuva invertterin välitön ohjaus oikosulkumoottorikäytössä. Master's thesis, Helsinki University of Technology [in Finnish].

Pohjalainen, P., Tiitinen P., & Lalu J. (1994). The next generation motor control method - Direct torque control, DTC. EPE Chapter Symposium, Lausanne.

Rabelo, B., & Hofmann, W. (2010). Doubly fed induction generator drives for wind power plants, wind power. In S. M. Muyeen (Ed.). ISBN: 978-953-7619-81-7. InTech. Available from: http://www. intechopen.com/articles/show/title/doublyfed-induction-generator-drives-for-wind-power-plants

Tuysuz, A., Schoni, M., & Kolar, J. W. (2012). Novel signal injection methods for high speed self-sensing electrical drives. Energy Conversion Congress and Exposition (ECCE), 2012 IEEE, pp. 4663–4670, 15–20 September 2012.

12

Switched reluctance
machine drives

Modern signal processing capabilities and advanced control algorithms enable effective management of even strongly nonlinear electromechanical phenomena in electrically produced driving force architectures. The switched reluctance (SR) motor is simple in construction, arguably the simplest of rotating machines, but complicated to control. Its working principle basics are easy to understand but, for example, SR torque production calculations are more complicated than with alternating-current (AC) or direct-current (DC) motors. This motor type does not have a simple equivalent circuit model (a current model) suitable for engineering analysis. Figure 12.1 shows where the SR machine drive falls with respect to other general machine drive types in terms of construction and control complexity.

As the figure shows, the mechanical construction of the traditional fully compensated DC machine is quite complex, but its control is simple. The motor drive can even be accurately realized using simple analog control electronics. Synchronous machines, with their numerous windings, are not only mechanically complicated, but they are also remarkably more complex to control. Squirrel-cage induction motors (IMs) are rugged and moderately complex in construction. If sensorless positioning is not needed, scalar control of the squirrel-cage IM is relatively simple. However, if accurate sensorless positioning is needed, then more complex vector control is required, because of IM slip. Brushless DC machines (BLDCs) and permanent magnet synchronous motors (PMSMs) are both moderately simple in construction. PMSM control, which is demanding but relatively straightforward, is more difficult than BLDC control, which makes use of position feedback. All DC machines use a simple parameter-based (electric current) control model, while all AC machines use a voltage-integral–based model (the voltage model).

Similar equivalent-circuit control models that are available for DC or AC machines do not exist for SR machines. These drives, which demand dedicated control electronics, operate autonomously with neither a DC or an AC supply, but smooth torque production is challenging.

Electrical Machine Drives Control: An Introduction, First Edition. Juha Pyrhönen, Valéria Hrabovcová and R. Scott Semken.
© 2016 John Wiley & Sons, Ltd. Published 2016 by John Wiley & Sons, Ltd.

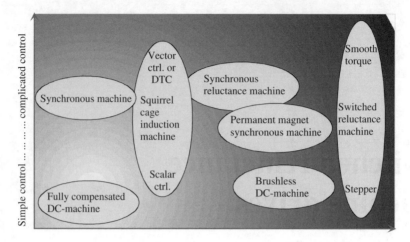

Figure 12.1 Control complexity as a function of machine construction.

An SR machine may be used as a stepper motor, which can be operated using simple step control; however, the discrete torque pulsation of the output is not suitable for all applications. Actually, the stepper motor can be considered the predecessor of the SR motor. The stepper can have the same geometry as the SR machine; however, it may have a solid rotor, which is subject to high losses but steps reliably from one excitation state to the next without excessive oscillation. If smooth torque is required, a low-loss high-performance SR machine coupled with intelligently controlled power electronics is the better choice. The rotor of a stepper motor rotates in discrete steps while the rotor of an SR motor rotates continuously. In an SRM drive, the ON and OFF switching angles must be determined in advance. That is, each next phase must be switched on before the previous phase reaches its aligned position. Within their load limits, stepper motors can operate well without rotor position feedback. However, this open-loop method of control is not stable for a low-loss high-performance SR motor.

The principle of an SR motor with a mechanical chopper was introduced as early as 1838. The motor was used to drive the targeting and firing of a cannon. However, the SR motor principle could not be more generally applied without today's greater knowledge and more advanced power electronics and control components. In the 1970s, electronic commutation based on rotor position was patented, and subsequently the performance of the SR motor became more competitive with DC and AC drives, at least for some special applications. The structure and theory of the SR motor had been well documented by the end of the 1970s, and development has been steady ever since.

To avoid any possible confusion, keep in mind that a salient-pole reluctance machine may be either a doubly salient switched or singly-salient synchronous reluctance AC machine. It is important to understand the difference. The SynRM belongs to the family of rotating field AC machines. It was discussed previously in Chapter 10. The SR machine is not a rotating field machine, an exception in electromechanics. In principle, all other motor types discussed here are rotating field machines; even the DC machine has a rotating field in its armature.

Switched reluctance refers to an electronically commuted motor type, the stator and rotor of which are both salient-pole constructions. The poles of the machine are designed so the maximum saliency ratio (the ratio of the highest and lowest inductance $L_{\max}/L_{\min}$) at two

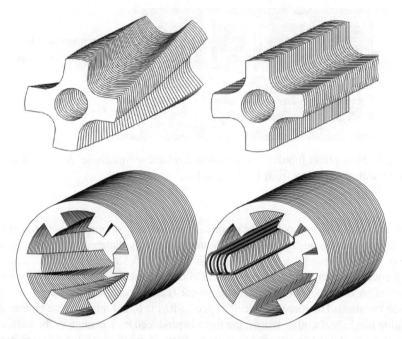

Figure 12.2 Stator and rotor laminations of 6/4-pole SR machines. A skewed rotor may have, at least in principle, its own secondary working function. For example, it can act as its own cooling air impeller. The lower right image in the figure also shows a simple representation of a single stator tooth coil. *Source:* Adapted and reproduced with permission from Lic.Sc Kimmo Tolsa.

different rotor angles is reached, however, in the motor phases, with certain restrictions. There is no winding in the rotor, and shaft excluded, the rotor is usually made of laminated electrical steel. An ideal switched reluctance machine for industrial or mobile applications is a fully undamped machine. The stator is also made of laminated iron, and each stator pole has an independent winding coil. Figure 12.2 depicts some parts of the magnetic circuits of three-phase reluctance machines.

In English, the term switched reluctance motor (SRM) is used of a doubly salient reluctance motor; the term "switched" refers to the commutation method rather than the reluctance of the machine. In the United States, the machine is commonly known as a variable reluctance motor (VRM), which is also a term for a certain stepper motor type. In the German language, the term is *Reluktanzmaschine mit beidseitig ausgeprägten Polen* ("a reluctance machine with saliency on both sides").

An SRM is an electric machine that produces torque because the rotor has the tendency to move into a position where the energy of the magnetic circuit is lowest and the inductance of the circuit is highest. In practice, controlling the motor to produce torque evenly is challenging, because the inductance of the circuit changes nonlinearly as the rotor rotates. Because of its structure, an SRM will function, to some degree, as a stepper motor. When employing a stepper motor as a position servo, rotor position feedback is not necessarily required, as long as the stepped motion of the rotor is somehow damped. An SRM without

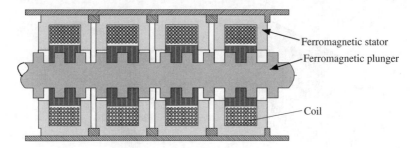

Figure 12.3 Four-phase tubular linear switched reluctance machine. *Source:* Adapted and reproduced with permission from Dr Jussi Salo.

feedback and with low losses is only marginally stable as a stepper motor. It is difficult to drive at high speed, because oscillations are not sufficiently damped between control steps.

The oscillation tendency can be avoided in an SRM position servo with rotor position feedback control, because the torque of the SRM can be reset at every instant appropriate for the situation. A linear switched reluctance motor (LSRM) can also be based on the same reluctance variation. The force produced by the LSRM is based on the same phenomenon as in a rotating machine. In other words, the force is produced by a change in the inductance of the magnetic circuit due to a change in mover position. In many cases, the linear motor is well suited as a position servo. An LSRM can be more compact than a rotating SRM as it directly converts electrical energy to linear motion.

The advantages of an LSRM servo drive are as follows.

- direct conversion of electrical energy to linear motion

- accurate positioning (no clearance, no notable friction)

- maintenance-free (does not include parts subject to wear)

- high holding force is achieved with low losses

- applicable to hazardous environments

The larger size of the LSRM servo drive compared to the achieved mechanical force can be considered a drawback of the drive. However, the comparison is unfavourable only if the drive is compared to a hydraulic cylinder, the power production capacity of which is superior to a magnetic device. Figure 12.3 illustrates the structure of a cylindrical four-phase LSRM.

12.1   The torque or force of an SR machine

The stator and rotor of a reluctance motor constitute a magnetic circuit, in which the stator winding generates a magnetising current linkage, and the flux linkage ψ penetrates the rotor and air gap. Figure 12.4 depicts the cross section of a three-phase 6/4-pole SR motor, in which the coils of the pole winding belonging to the same phase are connected in series. The inductance L of the magnetic circuit depends strongly on the pole angle γ of the rotor as it is shown in Figure 12.6. The magnetic force effects tend to minimize the magnetic resistance of

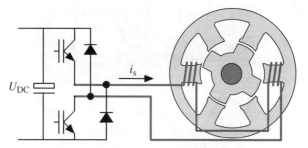

Figure 12.4 Three-phase, 6/4-pole reluctance machine, where i_s is the current of a single phase. The same number of power electronic switching components are needed as in a two-level AC motor inverter. However, the connections are different. See Chapter 7.

the magnetic circuit, that is, the reluctance, and the torque exerted on the rotor tends to align the rotor poles with the stator poles.

A straightforward way to predict the torque of an SR motor is to apply d'Alembert's principle. However, this method is most applicable for machine designers. The approach is not of much help to a practical expert in drives engineering. Utilization of the principle of virtual work presupposes that hysteresis and eddy current losses can be neglected, in which case, the energy W of the magnetic field or the co-energy W^* can be expressed with the rotor angle (reference to the phase of the motor is omitted).

$$T(\psi_s, \gamma) = -\left(\frac{\partial W}{\partial \gamma}\right)_{\psi_s} = \left[-\frac{\partial}{\partial \gamma}\int_0^{\psi} i_s(\psi_s, \gamma)\, d\psi_s\right]_{\psi_s = \text{constant}},$$

$$T(i_s, \gamma) = \left(\frac{\partial W^*}{\partial \gamma}\right)_{i_s} = \left[\frac{\partial}{\partial \gamma}\int_0^{i} \psi_s(i_s, \gamma)\, di_s\right]_{i_s = \text{constant}}.$$

(12.1)

Similarly, in case of a linear motor, force F is produced as a function of the flux linkage ψ and movement, for example, in the x-direction as

$$F(\psi_s, x) = -\left(\frac{\partial W}{\partial x}\right)_{\psi_s} = \left[-\frac{\partial}{\partial x}\int_0^{\psi} i_s(\psi_s, \gamma)\, d\psi_s\right]_{\psi_s = \text{constant}},$$

$$F(i, x) = \left(\frac{\partial W^*}{\partial x}\right)_{i_s} = \left[\frac{\partial}{\partial x}\int_0^{i} \psi_s(i_s, \gamma) di_s\right]_{i_s = \text{constant}}.$$

(12.2)

If the torque or force of an SRM were alternatively defined using numerical methods, the advantage of employing the virtual work principle would be that the distribution of forces in the SRM could be determined simultaneously in the same way that flux density distribution can be determined by the finite element method. The advantage would be that in the design of the machine structure, both the electromagnetic and mechanical stresses could be taken into account.

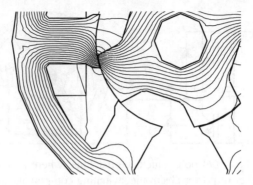

Figure 12.5 Flux solution of a three-phase four-rotor-pole reluctance machine in the case of overlapping stator and rotor poles. *Source:* Adapted and reproduced with permission from Dr Jussi Salo.

A general method to calculate the mechanical forces exerted by the magnetic field on a steel object is to employ the Maxwell stress equations. Figure 12.5 shows the flux solution for a 6/4 machine in the case of overlapping poles.

When the magnetic field strength vector H is decomposed into the components orthogonal (n) and tangential (t) to the surface Π as follows,

$$H = H_n\,n + H_t\,t, \quad |n| = |t| = 1 \tag{12.3}$$

the Maxwell stress equations can be written in the following form.

$$
\begin{aligned}
\sigma_n &= \tfrac{1}{2}\,\mu\!\left(H_n^2 - H_t^2\right), \\
\sigma_t &= \mu\,H_n H_t.
\end{aligned}
\tag{12.4}
$$

For instance, if the surface Π is selected to be a cylinder of radius r, located in the air gap of the stator and rotor of the SRM, the magnitude and direction of the total force exerted on the rotor can be determined by integrating the equation.

$$F = \oint_{\Pi} \left[\tfrac{1}{2}\,\mu\,(H_n^2 - H_t^2)n + \mu H_n H_t t \right] \mathrm{d}\Pi = \oint_{\Pi} \left[\mu H_n H - \tfrac{1}{2}\,\mu\,H^2 n \right] \mathrm{d}\Pi, \tag{12.5}$$

the result of which should be zero, if the rotor is centered. Therefore, the magnitude of torque is obtained from the following equation.

$$T = \oint_{\Pi} \mu\,(H \cdot n)(r \times H)\,\mathrm{d}\Pi, \quad r = rn \tag{12.6}$$

where r is a vector from the rotation axis of the rotor to the integration point of the surface Π. In the motor air gap, if the magnetic field strength H, the flux density B, or the magnetic vector potential A have been determined in numerical form via, for example, the finite element method, then Equation (12.6) applies.

The machine inductance of a reluctance machine changes as a constant function of machine rotation angle. Because the machine saturates near the pole edges, determining

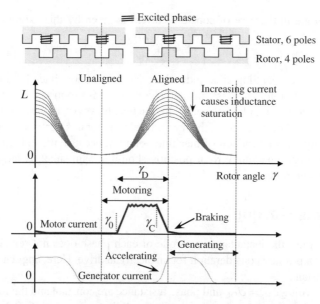

Figure 12.6 Inductance of a saturated switched reluctance motor as a function of angle (or position) with electric current as a parameter and the current pulses for both motoring and generating operation. As a motor, the phase voltage is switched on at ignition angle γ_0 and after the torque stroke is commutated with the angle γ_C. Motor current does not quite reach zero before reaching the aligned position. Some slight braking occurs as the machine shifts away from the aligned position. The unaligned and aligned positions of a 6/4 machine are shown at the top of the figure. Note: The pitch of the schematic motor figure and the inductance figure are different. They coincide only at the marked minimum or maximum inductance points (Miller, T. J. E. [1993]. *Switched reluctance motors and their control.* Oxford, UK: Magna Physics Publications, Oxford University Press. ISBN 1-881855-02-3).

inductances is difficult at different machine positions and current levels. Average torque grows with the inductance difference between the aligned and unaligned (direct and quadrature) positions. As shown previously, the instantaneous torque of the reluctance machine can be computed from the change in the co-energy of the machine as a function of rotation angle. Operating as a motor, the electric current of the machine is often kept constant with a switch-mode power supply. See Figure 12.6.

If the magnetization curves and phase currents are known at a sufficient number of rotor angles, the magnetic co-energy W^* can be calculated for each rotor position. Moreover, the change in W^* with respect to the rotor angle can be examined using the difference equations. Instantaneous torque can be approximated in this manner.

In the (nonpractical) special case in which the motor does not saturate, the magnetization curve is linear, and therefore the magnetic co-energy W^* and the stored energy W_e are equal.

$$W^* = W_e = \frac{1}{2}Li_s^2 \qquad (12.7)$$

Instantaneous torque in the case of constant current is given by this expression.

$$T = \frac{1}{2}i_s^2 \frac{dL}{d\gamma} \qquad (12.8)$$

In the unsaturated operating area, inductance changes almost linearly, and therefore its change with respect to rotor angle is constant. Here, torque is proportional only to the square of the current, and torque regulation is easy. In practice, however, because it would require a large air gap and overdimensioning of the machine, aiming for unsaturation is not advisable. It would reduce torque (as a function of machine volume), because the inductance difference between the direct and quadrature rotor position should remain small. A larger air gap also requires a larger controller.

12.2 Average torque

As rotor angle varies, the instantaneous torque of each phase does not remain constant. Its average torque is a primary consideration for an electrical drive. Here, angular rotor speed Ω_r is presumed constant.

When supply voltage $u_s = U_{DC}$ and phase resistance are constant and the resistive voltage loss remains low, there is a linear increase in the flux linkage ψ_s as an integral of voltage after the voltage is switched on.

$$\psi_s(t) = \int (u_s - R_s i_s)dt = \frac{1}{\Omega_r}\int(U_{DC} - R_s i_s)d\gamma \qquad (12.9)$$

Figure 12.7 shows three plots of flux linkage versus electric current that reveal the mechanical work regions for the transistor conduction period, diode conduction period, and energy

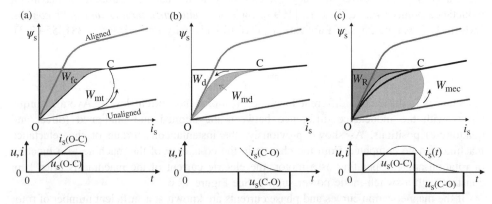

Figure 12.7 Energy conversion during single voltage pulse operation at a higher speed. (a) Transistor conduction period, (b) diode conduction period, and (c) energy conversion loop. W_{fc} is the energy stored in the magnetic field at the commutation point. W_{mt} is the magnetic energy converted into mechanical energy before commutation. W_d is the energy stored in the magnetic circuit along the current-switching-off path. It corresponds to W_R, which is finally returned to the DC-link as electric energy. W_{md} is the magnetic energy converted into mechanical following commutation. W_{mec} is the magnetic energy converted into mechanical work.

conversion loop. The loop is built with rotor position dependent flux-linkage curves as function of current. The curves have to be built between two boundaries: unaligned and aligned positions. In the unaligned position the inductance is small and the flux-linkage curve behaves almost linearly as a function of current. However, in the aligned position the flux-linkage curve saturates heavily when the motor current increases. The energy conversion loop starts from the origin when the motor is in the unaligned position and the exciting voltage is turned on.

In Figure 12.7a, the voltage is applied to the phase at an ignition angle γ_0 at the unaligned position. This angle varies depending on the speed of the motor. Phase current i_s first increases linearly as machine inductance L remains low near the unaligned position. As the poles start overlapping and approach alignment, inductance increases rapidly, and the resulting back emf $(-\mathrm{d}\psi_s/\mathrm{d}t = -i_s\mathrm{d}L/\mathrm{d}t - L\mathrm{d}i_s/\mathrm{d}t)$ restricts the electric current. This movement corresponds to the O-C curve. With the rotor angle γ_C at point C, the phase in question is commutated. Now, the energy brought to the system is $W_{mt} + W_{fc}$ (the white and shaded areas). W_{fc} is the energy stored in the magnetic field, and W_{mt} is the energy converted into mechanical work when the transistor is conducting. In this period, mechanical work is approximately equal to the energy stored in the magnetic circuit.

After commutation (Figure 12.7b), the polarity of the voltage changes and energy W_d is returned through the diode to the voltage source. The remaining energy W_{md} is the mechanical work produced along curve C-O. The complete working stroke is illustrated in Figure 12.7c. For the complete working stroke, the mechanical work is thus $W_{mec} = W_{mt} + W_{md}$, and the energy returning to the voltage source is $W_R = W_d$. In this example, mechanical energy is about 65% of the total energy, a percentage analogous to the power factor of AC drives. The remaining 35% is the *reactive energy* of the reluctance machine, which can be stored either in the electric field of the DC-link capacitor or in the magnetic field of the motor magnetic circuit.

A useful parameter for an SR machine is the energy ratio Γ. This ratio reveals how much mechanical energy is produced from a given amount of electrical energy over the course of the energy conversion loop.

$$\Gamma = \frac{W_{mec}}{W_{mec} + W_R} = \frac{W_{mec}}{W_{el}} \tag{12.10}$$

This energy ratio is to some degree analogous to the power factor in AC machines. In the example introduced previously in Figure 12.7, the energy ratio is approximately $\Gamma = 0.65$. Because the number of strokes per revolution is known, the average torque can be determined. In one revolution, all poles N_r of the rotor must interact with all the stator phases, and therefore the number of strokes per revolution is mN_r. The average electromagnetic torque over one revolution thus becomes

$$T = \frac{mN_r}{2\pi} W_{mec} \tag{12.11}$$

Due to inaccuracies in manufacturing, the energy conversion loops for each current phase may differ slightly in practice, and therefore produce different torques. In SR machines, the air gap is manufactured as small as possible and therefore even smallest deviations from the desired geometry may cause asymmetry in the machine.

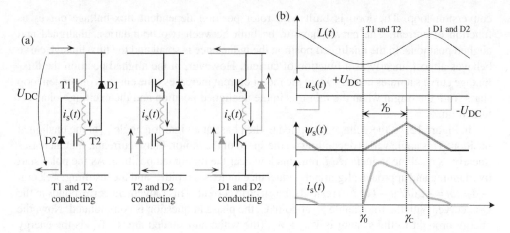

Figure 12.8 (a) The circuit controlling one phase of a reluctance motor and the current flow during the working stroke (b) The electric current waveform in single-pulse operation for the idealized inductance case (Miller, T. J. E. [1993]. *Switched reluctance motors and their control*. Oxford, UK: Magna Physics Publications, Oxford University Press. ISBN 1-881855-02-3).

The original energy W_{el} supplied by the power electronics can be expressed as a fraction k of the product $i_{sC}\psi_{sC}$, where ψ_{sC} is the value of the stator flux linkage at the instant of commutation, and i_{sC} is the value of the current at the same moment. If the flux linkage increases linearly during the flux formation period, shown as O-C previously in Figure 12.7, then

$$\psi_{sC} = U_{DC}\gamma_D/\Omega_r \qquad (12.12)$$

Here γ_D is the dwell angle during which the power stage supplies power to the machine. See Figure 12.8. Electrical work can therefore be expressed as follows.

$$W_{el} = \frac{W_{mec}}{\Gamma} = \frac{kU_{DC}\gamma_D i_{sC}}{\Omega_r} \qquad (12.13)$$

and since i_C is the peak value of the current yielded by the power electronics, the required power processing ability of the output stage in an m-phase system is

$$S_{PE} = mU_{DC}i_{sC} = \frac{mW_{mec}\Omega_r}{\Gamma k\gamma_D} = \frac{2\pi T_e\Omega_r}{N_r\Gamma k\gamma_D} \qquad (12.14)$$

The product of torque and angular speed corresponds to the air gap power P_δ, and the product $N_r\gamma_D$ is constant with a maximum value of about $\pi/2$ at the base rotation speed of the machine. The power processing ability of the power electronic bridge must therefore be as follows.

$$S_{PE} = \frac{4P_\delta}{k\Gamma} \qquad (12.15)$$

This required power is independent of phase number and the number of poles. It is inversely proportional to the energy conversion ratio Γ and the utilization ratio k. Both Γ and k depend strongly on the static magnetizing curves of the machine and on the curves of the aligned and unaligned positions in particular. These curves, however, are in practice highly dependent on the pole number N_r, which has a strong indirect effect on the dimensioning of the power stage of the converter. An approximation can be used to compare the power processing ability of the power stage to the shaft output power. Assuming that $k = 0.7$ and $\Gamma = 0.6$, then $S_{PE}/P_\delta \approx 10$. This value is typical of SR motor drives. Inverter power bridges of the same scale are required also in induction motor drives.

12.3   Control systems for a SR machine

SR machines always require an individual control system. Its performance decides the overall machine characteristics. The control system comprises voltage choppers and the control and measurement circuits that control their behaviours. The direction of the torque of a reluctance machine does not depend on the direction of the phase current, and therefore unidirectional switches can be employed in the chopper.

12.3.1   The power electronic circuits

In an SR machine operating as a motor at higher speed, and therefore with single voltage pulse per energy conversion stroke, the magnitude of the stator flux linkage ψ_s will change from zero to its maximum within half of the period of movement from the unaligned position with minimum inductance to the aligned position where circuit inductance spikes. This is accomplished by switching on the supply voltage $u_s = U_{DC}$ at rotor angle γ_0 and switching $u_s = -U_{DC}$ at commutation angle γ_C. Figure 12.8a shows a single-phase circuit applicable to the control of a reluctance motor.

The illustrated switches can be, for example, power MOSFETs or IGB transistors. At low rotation speeds, the switches are controlled so the upper transistor T1 controls the magnitude of current with PWM. The lower transistor T2 is needed for commutation. When the rotor angle reaches position γ_0, both transistors are switched to a conducting state that allows current to pass through the transistors and the phase winding. As the current reaches its upper limit, the upper transistor T1 is turned off and the phase winding current passes through transistor T2 and diode D2, converting the energy stored in the magnetic circuit into mechanical work.

At commutation angle γ_C, the lower transistor state is reversed, and the rest of the energy of the winding discharges through the diodes towards the DC voltage source or DC link capacitor, because the polarity of the voltage across the winding changes. The flux linkage must drop to zero before the rotor passes the aligned position. A nonzero flux linkage beyond that point would result in a negative i.e. braking torque. Figure 12.8b illustrates the flux linkage and the current under the control of a single control pulse (Miller, T. J. E. [1993]. *Switched reluctance motors and their control*. Oxford, UK: Magna Physics Publications, Oxford University Press. ISBN 1-881855-02-3, pp. 53–55).

At high rotation speeds, both transistors are typically controlled simultaneously, which results in added current and torque ripple. This is known as hard chopping. Hard chopping is used at high speeds where there is insufficient time for soft chopping. Both hard and soft chopping are described in the following paragraphs.

12.3.2   Current control

Figure 12.8b showed that a single DC pulse during the working stroke produces an indefinite current waveform, and therefore, uneven torque. The current increases first linearly, but then the back-emf caused by the increasing inductance restricts the current. At the commutation point, the voltage direction changes, which causes a sudden decrease in the phase current. Because dL/dt has a negative sign making $-dL/dt$ positive, it tries to increase current even though there is a negative supply voltage after commutation. Therefore, in the aligned position, the direction of the back emf changes as circuit inductance begins to decrease and the rate of current drop slows. When this happens, back emf can exceed supply voltage, and electric current may begin to rise. Therefore, in single-pulse operation, the commutation angle must precede the aligned position by several degrees. As speed increases, the commutation must be advanced further. Similarly, the turn-on switching angle γ_0 may be advanced well ahead of the unaligned position.

Supply voltage must be chopped to control electric current at low speed. This can be most easily accomplished by holding transistor T2 (Figure 12.8a) in the conductive state as the rotor angle changes from γ_0 to γ_C and switching the transistor T1 on and off at sufficiently high frequency. This is referred to as soft chopping. Hard chopping, where both transistors are switched together at high frequency, can also be used but it is harder on the transistors and results in higher current pulsation. The advantage of this pulse width modulation of voltage is that torque range can be increased by delaying the moment of commutation, because of the lower magnetic energy stored in the magnetic circuit. The peak value of stator flux linkage is limited to a suitable level. At lower speeds single-pulse operation saturates the machine and makes commutation difficult. To reduce acoustic noise, the switching frequency must be kept high.

Considerable current ripple results if the voltage switching frequency does not change throughout the rotor's working stroke. This electrical ripple can be moderated by switching the power transistors on and off as the phase current becomes either greater or less than a reference current. Figure 12.9 illustrates the voltage, flux linkage, and electric current

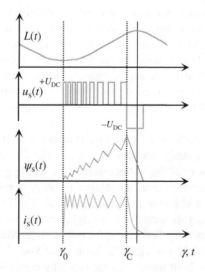

Figure 12.9   The inductance, voltage, flux linkage, and current waveforms that result when electric current magnitude is controlled via PWM. In hysteresis control, soft chopping is used until the rotor angle passes γ_C and flux linkage must be removed (Miller, T. J. E. [1993]. *Switched reluctance motors and their control*. Oxford, UK: Magna Physics Publications, Oxford University Press. ISBN 1-881855-02-3).

waveforms produced by a hysteresis-type current regulator. A simple hysteresis controller maintains the current waveform between the upper and lower limit—within the hysteresis band—when the supply voltage is switched on. The switching frequency decreases as the inductance of the magnetic circuit increases. The example illustrated soft chopping. With soft chopping, less filtering is needed in the DC link.

12.4   The general controller structure

Digital control enables versatile and reliable parameter setting and efficient programming of operating modes. Since they depend strongly on the requirements set by the load, control design and implementation must be carried out independently for each drive case.

To use a switched reluctance motor as a servo drive, it must produce minimal torque ripple and offer rapid dynamic response. In addition, it must be capable of operating at zero speed and reversing smoothly. Even if the motor will not be used for a servo drive, optimal operation as a simple variable-speed drive requires continuous control of the switching angles. Four-quadrant operation, in other words the ability to operate in both rotation directions at positive and negative torques, presupposes fast real-time controllers that directly control phase current and voltage. With ratios between torque, current, speed, and switching angle that are often nonlinear and variable as functions of speed and load, achieving this level of control with an SR motor can be difficult.

Figure 12.10 depicts the general architecture of a switched reluctance machine controller. The design illustrated by the block diagram will not offer servo-quality control, because it cannot dynamically profile the electric current waveform to eliminate torque ripple. However, it uses the stator voltage integral to estimate stator flux linkage as is also done with direct flux-linkage control (DFLC), which should make the operation more accurate.

In Figure 12.10, torque is the output, and the current demand signal i_{ref} and turn-on and turn-off angles γ_0 and γ_C (the switching and commutation angles) are the inputs to the controlled power electronic circuit. The design includes a current hysteresis regulator that can keep electric current within desired limits, *i.e.*, approximately constant. Similarly to DFLC,

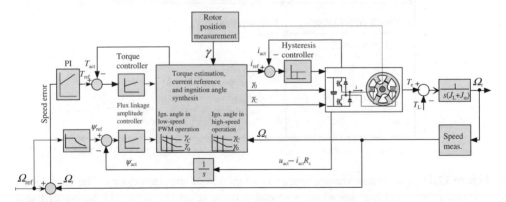

Figure 12.10 The structure of a controller capable of torque control and four-quadrant operation. Speed is controlled via speed sensor feedback. i_{ref} is the current reference signal, T_m is the motor torque, T_L is load torque, Ω_r is angular speed, and Ω_{ref} is its reference value. J_m and J_L are the moments of inertia for the motor and the load. The ignition angle γ_0 and the commutation angle γ_C are used to toggle the electric current controlling switches.

the estimated or measured stator voltage is integrated into stator flux linkage. The stability problems common in DFLC do not manifest, because stator flux linkage always begins and ends at zero value within one stroke. The motor controller must include a flux-linkage current to rotor position to torque chart as a motor model.

The block diagram also includes a notional conversion from torque to angular speed and further to rotor angle. According to Newton's second law of motion, the difference between the electrical and load torques (T_e and T_L) is a product of angular acceleration and the sum of the moments of inertia of the motor and the load ($J_m + J_L$). Therefore, the block after the error element integrates the difference of the torques, and then divides it by the common moment of inertia. The result is the angular speed Ω_r. The next block integrates the angular speed yielding the rotor position angle γ. In practice, rotor position is sensed with an encoder, which generates a digital pulse train. Speed and position are estimated from this pulse train using a suitable digital algorithm. Initial rotor position must be measured or estimated before commissioning a drive. An absolute rotor position encoder can also be used.

The digital speed estimate is compared to the reference speed Ω_{ref}. The error determined is input to the PI controller, which generates the torque demand signal T_{ref}. If the speed error increases, that is, the speed lags behind the reference, the proportional P-controller and the integral I-controller increase the current reference. A current limiter is needed to set the current limit to prevent damage to the current circuit. If the speed exceeds the reference value, four-quadrant operation may even call for braking torque from the motor. In many ordinary variable-speed applications, this is not necessary, because the load torque provides sufficient deceleration. To produce braking torque, electric current must be fed to the circuit as switching angles are delayed. As with motoring torque, braking torque magnitude is a nonlinear function of current and the switching angles.

Figure 12.11 depicts the average torque of an SR motor as a function of rotation speed. From zero to a base speed Ω_b, motor current is held constant by chopping windings voltage,

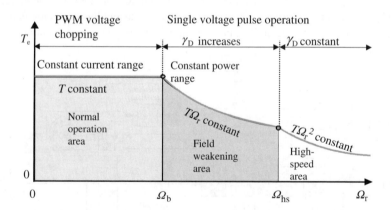

Figure 12.11 Maximum average torque as a function of angular velocity. The base speed, constant power, and high-speed areas resemble those of an IM drive. The high-speed area starts at speed Ω_{hs}. γ_D is the angle of rotation in which a constant voltage pulse acts on the stator winding, and the flux linkage is increasing. At higher speeds the ignition γ_0 and commutation angle γ_C are both advanced suitably to get desired operation. It is no longer possible to increase the angle in the highest speed area. Therefore, torque drops rapidly with additional increasing speed.

and the motor operates at nearly constant torque. As speed increases further, the switch passes current throughout the working stroke, and the motor operates at maximum power. As the power holds constant at its maximum value, the rotation speed of the rotor is controlled based on the switching angles. This behaviour is exactly analogous to the behaviour of AC motors. Also as speed increases, the SR drive must move into field weakening, and then into the high-speed region. The magnitude of flux linkage is not constant in an SR drive, but its maximum value must be limited as speed increases. There is insufficient time to integrate and disintegrate the voltage flux linkage, and field weakening must be enabled.

12.4.1   Determination of rotor position

A closed-loop control system activates the chopper switches based on rotor position and rotation speed. Phase current level is set as a function of rotor position. Rotor position can be measured with optical or magnetic position sensors. An optical encoder comprises a slotted pulse disc that rotates with the rotor, a stationary reading mask, and a light detector connected to the logic unit. Figure 12.12 illustrates the architecture of a simple SR motor drive and Figure 12.13 the position measuring principle. The drive system has three optical encoders to determine rotor position. Since at high speeds the current waveform is dependent on switching angle, it must be measured with high precision.

The difference between the desired and measured speed as processed by the speed error amplifier controls the chopping frequency of the external power transistors. As this speed error increases, chopping frequency is reduced. Measured electric current is compared to the set current limiter value. If the measured value exceeds the permitted value, the transistors are

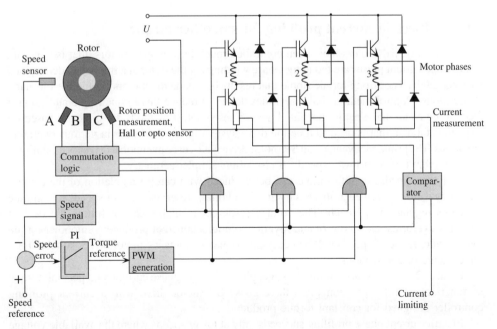

Figure 12.12 Simple SR motor drive system. Optical or Hall effect sensors report rotor position and speed. The motor phases are current controlled based on rotor position.

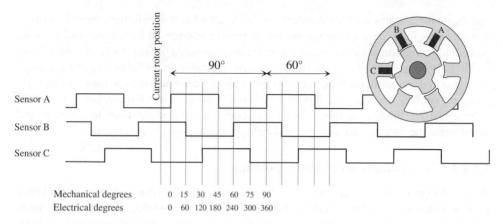

Figure 12.13 Sensor positioning and sensor pulses for a 6/4 SR machine; 60 represents the mechanical phase shift between phases. The sensors are illustrated schematically. In practice, they are placed in the non-driving end of the motor, and they have a similar rotor wheel to measure position.

turned off. Switching angles are selected by the logic controller based on the A, B, and C shaft sensor data. Figure 12.13 illustrates the sensing signals for a 6/4 machine.

Controllers have also been developed that operate without rotor position sensors. This is normally done to minimize cost or improve reliability, particularly in drive subject to extraordinary and difficult operating conditions.

12.4.2   Electric current profiling for smoother torque

Conventionally, SR motors have been controlled much like stepper motors, that is, constant voltage and current pulses are fed to the phase windings of the stator at a frequency determined by rotor angle. The strong torque ripple that results has made the motors unsuitable for many applications. Control methods have been developed to reduce this torque ripple and bring it close to conventional electric machine levels. These control methods are typically based on experimental data that is used to tune the drive system for optimal duty in various operating situations. The primary advantage an SR motor drive has over conventional drives is its ability to produce more torque at lower speeds. It also offers simpler geometry.

At lower speeds, smooth torque can be produced with careful regulation of the electric current waveform. However, there is insufficient voltage reserve at higher speeds to continue with this regulation approach. The voltage resources available cannot follow the desired current waveforms. New control and inverter solutions aimed at producing smoother torque over a wider range of speeds will be necessary to make reluctance motor drives suitable for a wider range of applications.

Figure 12.14 offers plots of flux linkage, phase voltage, phase current, and per-unit torque as functions of rotor position for a three-phase SR motor drive with a current profiling controller designed for constant torque production.

Electric current phase profiling succeeds only at lower speeds where the available voltage reserve needed by the algorithm is sufficient. Based on motor properties and operating conditions, current profiling must be carried out on a case-by-case basis. The torque

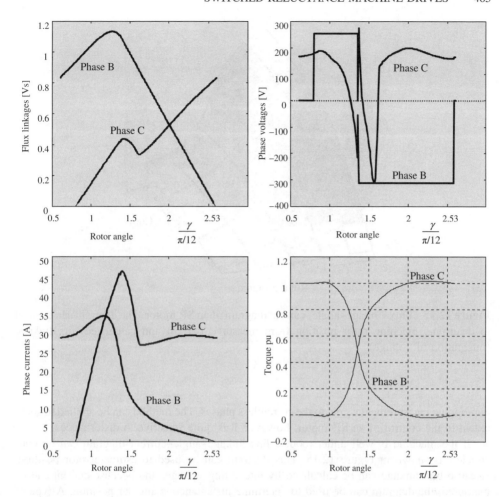

Figure 12.14 Producing constant torque with the limited voltage available to profile the phase currents in a three-phase SR machine. Smooth torque results, but the profiling consumes all the available voltage. *Source:* Adapted and reproduced with permission from Lic.Sc Kimmo Tolsa.

production of the 6/4 motor referenced in Figure 12.13 is slightly low at the commutation point, and therefore the constant phase current must be increased as shown in Figure 12.14.

12.5   The position sensorless operation of an SR machine

SR machine control requires accurate rotor position information. Only an angle sensor can accurately provide this information. However, sensorless rotor position determination may be adequate if drive requirements are not too demanding. For example, inductance measurements of the currentless phases of an SR machine can be used to determine rotor position. This approach, which applies to multiphase machines with at least four phases, feeds an

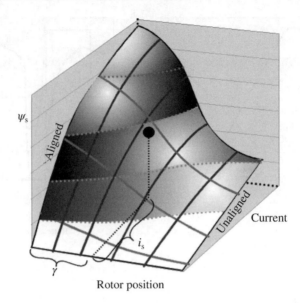

Figure 12.15   Flux-linkage–angle–current diagram of an SR motor. The diagram can be used to determine machine rotor position from measured current and calculated flux linkage (integration of voltage).

appropriate measurement signal to the currentless phases. The method can be applied to some extent to the control of an SR motor; however, it is quite sensitive to disturbances.

If the machine is well understood, a flux-linkage–angle–current diagram can be constructed as shown in Figure 12.15. This diagram can be used to estimate rotor position. Because flux linkage can be calculated by integrating voltage, and because current can be measured, the diagram can be used to determine instantaneous angular position. A type of DFLC, stator flux linkage does not drift with this approach as it does in the DFLC of a rotating field machine, because the flux linkage of the phase increases in value at every stroke, and therefore the possible offset error can always be eliminated before the working stroke of the phase.

Stator flux linkage is integrated from voltage as follows.

$$\psi_s(i_s, \gamma, t) = \int (U_{DC}(t) - R_s i_s(t)) \mathrm{d}t \tag{12.16}$$

Using the values illustrated in Figure 12.15, flux linkage is

$$\psi_s(i_s, \gamma, t) = L(i_s, \gamma) i_s(t) \tag{12.17}$$

To summarize, by integrating voltage to determine flux linkage (Equation 12.16) and by measuring current, it is possible to define rotor position γ given the sensor information shown in Figure 12.13. With these values, an SR motor can be controlled based on a DFLC-like approach.

The method should be reliable, because unlike AC drives, the flux linkage in an SR drive always starts from zero at the beginning of each working stroke so the voltage integral is not subject to substantial drift.

12.6   A summary of SR drives

The SR drive offers the following advantages.

- No winding is required so rotor construction is simple and easy to manufacture.

- The mass moment of inertia of the rotor is low resulting in improved dynamic performance.

- The stator windings are easy to produce, and end-winding losses are less than in a corresponding IM.

- Most losses are in the stator so motor cooling is easier, making possible higher load capacity.

- Large open spaces in the rotor promote efficient ventilation.

- Machine torque is independent of current direction so inverter and controller have more degrees of freedom.

- The machine can produce rated torque at lower rotation speeds and stopped with fairly low current.

- The machine constant of an SR machine is higher than the machine constant of an IM.

- In the event of a failure, both open-loop voltage and short-circuit current are low.

- The power electronic circuits do not have a so-called shoot-through path facilitating implementation of the control system.

- Machine reliability is high, because each phase of the SR motor drive is physically, magnetically, and electrically independent.

- Because the rotor lacks conductors or magnets, higher speeds can be achieved making the machine suitable for high-power-density aviation purposes.

A disadvantage of a reluctance motor is, for instance, discontinuous torque that can cause vibration and noise. In the low-speed range, torque ripple can be restricted to levels comparable to AC motor drives. In the high-speed operation range, reducing torque ripple becomes impossible. Mechanical filtering makes this less of a problem. Already at present, the best drives produce negligible torque ripple at low speeds, which is generally where loads are most vulnerable to the harmful effects of torque ripple and torque production must be smoothest. In small motors, noise can be mitigated at high speeds by selecting a switching frequency above the range of audibility.

In torque control, power pulses are taken from the DC link, and therefore efficient filtering is necessary. In this sense, the SR drive does not differ substantially from the inverter drive of an induction motor. The small air gap required to maximize inductance ratio and improve motor operation increases production costs.

Despite several notable advantages, wider acceptance of SR drives has been restricted, because smooth torque production is not available over a sufficiently wide speed range. New inverter and control solutions are needed to overcome this restriction.

Currently, high-torque, slow-operating motors of 30–200 kW are being used in heavy four-quadrant drives in a number of industries including, for example, the mining industry. General-purpose SR motors are manufactured for various purposes. For example, they can replace old DC and AC drives in applications that require accurate rotation speed control.

Rotation speed control has become more common in pump and blower drives to improve efficiency. So far, SR motors are not competitive with frequency converter drive induction motors in this area, because of their more complicated control requirements. SR motor torque is higher at lower rotation speeds, and therefore the motor type may become more popular in applications that demand high startup torque.

The applicability of reluctance motors to electric tools has also been investigated. In these applications, the size and the torque properties of reluctance motors are most beneficial. A disadvantage of the mixed current motor, which is commonly used at the present, is wear of the mechanical commutator and the commutation-induced electromagnetic disturbances. These problems could be solved with an SR drive. Thanks to its durability and other favourable qualities, an SR motor is also a suitable mechanical power source for electric vehicles where drive system cost must be minimized. A number of automobile manufacturers are currently testing passenger cars powered by an SR motor drive.

References

Becerra, R., Ehsani, M., & Miller, T. J. E. (1993). Commutation of SR motors. *IEEE Trans. Power Electronics*, 8, 257–262.

Carpenter, C. (1959). Surface-integral methods of calculating forces on magnetized iron parts. *IEE Monograph*, No 342.

DiRenzo, D. (2000). Switched reluctance motor control–Basic operation and example using the TMS320F240. Texas Instruments Application Report, SPRA420A.

Husain, I., & Ehsani, M, (1994). Torque ripple minimization in switched reluctance motor drives by PWM current control. Proc. APEC'94, 1994, pp. 72–77.

Ilic-Spong, M., Miller, T. J. E., MacMinn, S. R., & Thorp, J. S. (1987). Instantaneous torque control of electrical motor drives. *IEEE Trans. Power Electronics*, 2, 55–61.

Kjaer, P. C., Gribble, J., & Miller, T. J. E. (1997). High-grade control of switched reluctance machines. *IEEE Trans. Industry Electronics*, 33, 1585–1593.

Miller, T. J. E. (1993). *Switched reluctance motors and their control*. Oxford, UK: Magna Physics Publications, Oxford University Press. ISBN 1-881855-02-3.

Salo, J. (1996). Molemmin puolin avonapainen reluktanssimoottori. Lappeenranta. Lisensiaatintutkimus. Lappeenranan teknillinen korkeakoulu energiatekniikan osasto sähkötekniikan laitos [in Finnish].

Tolsa, K. (1997). Licentiate's thesis, Lappeenranta University of Technology.

13

Other considerations: the motor cable, voltage stresses, and bearing currents

Basic electrical engineering often assumes that voltage pulses travel with infinite speed. For example, it is usually assumed that motor terminals experience full input voltage immediately following activation of the converter switch. In reality, however, the speed of travelling voltage pulses is finite and must be taken into account in some cases. The wave nature of a transmission line means that on the line, voltages travel as waves. A motor cable must be regarded as a transmission line where the voltage pulses travel at a finite speed—typically about 150 m per microsecond, which is half the speed of light. When an electric machine in a direct online (DOL) drive is supplied with typical sinusoidal machinery frequency voltages, this speed seems infinite, and there is no need to evaluate motor cable transmission line properties. Compared to a DOL drive, an inverter-fed electrical drive, however, sets special requirements for the cabling and motor insulation.

Pulse width modulation produces high du/dt differential mode and common-mode voltage pulses by sequentially switching the direct current (DC)-link voltage to the motor windings. The common potential of the phases in relation to ground is called common-mode voltage (subscript cm). In a balanced three-phase system, common-mode voltage is zero. In converters, however, it is nonzero. Common-mode voltage depends primarily on the type of converter used, but it is also a function of modulation. The most common two-level converters often have diode bridges connected to network terminals. Figure 13.1 reviews the topology of a two-level converter.

Across the DC link, voltages in the two-level converter vary as shown in Figure 13.2.

Figure 13.3 illustrates a typical PWM modulation sequence for one switching period. The Figure suggests that the topology schemes can be the same for the time intervals 1 and 7, 2 and

Electrical Machine Drives Control: An Introduction, First Edition. Juha Pyrhönen, Valéria Hrabovcová and R. Scott Semken.
© 2016 John Wiley & Sons, Ltd. Published 2016 by John Wiley & Sons, Ltd.

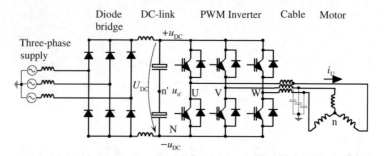

Figure 13.1   Two-level converter topology. The motor cable is described as an LC circuit with parasitic ground capacitances.

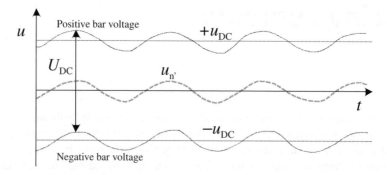

Figure 13.2   Two-level converter DC link positive and negative bar voltages and link capacitor middle-point voltage $u_{n'}$. The figure makes it clear that grounding any part of the DC link is impossible without galvanic separation from the network.

6, and 3 and 5. Therefore, there are four distinct topology schemes possible from 1 to 7. For the intervals 2 and 6 and 3 and 5, the system acts as a simple voltage divider. The voltage at motor-winding star point n is $-U_{DC}/3$ and $+U_{DC}/3$. For intervals 1, 4, and 7, all phases are connected in parallel, and the potential at star point n equals the potential of the connected

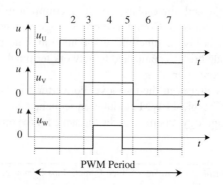

Figure 13.3   Switching positions over one switching period for a two-level converter.

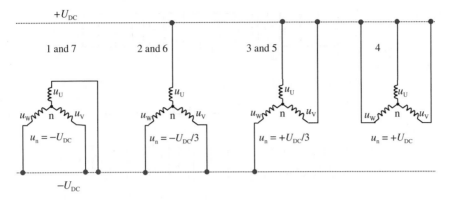

Figure 13.4 Two-level converter motor star-point potentials for time intervals 1 through 7 given the switching positions presented in Figure 13.3.

DC-link terminal. During modulation, the potential of the star point is nonzero and variable. This potential in relation to ground is called the common-mode voltage.

Figure 13.4 illustrates the four distinct equivalent circuits and common-mode voltages that a star-connected motor experiences in the course of a single switching period.

Given the modulation pattern of Figure 13.3, the common-mode voltage (star-point voltage) will conform to the pattern of Figure 13.5.

A motor control arranges sinusoidal currents with respect to the motor phases, and therefore PWM pulse widths are not constant within each period. As Figure 13.6 shows, the common-mode voltage fundamental changes at three times the frequency of the modulated sinusoidal signal.

The instantaneous value of the common-mode voltage fundamental can be calculated by

$$u_{cm}(t) = \frac{u_U(t) + u_V(t) + u_W(t)}{3} \tag{13.1}$$

Motor PWM is characterized by a star-point varying voltage potential that has an amplitude equal to half of the DC-link voltage U_{DC}. This potential produces high-frequency harmonics. Because motors have high ground capacitances, high-frequency currents can pass between the neutral point and ground. Therefore, stray capacitances and associated currents inside the motor must be taken into account.

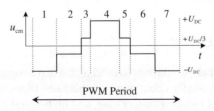

Figure 13.5 Common mode voltage over one switching period for a two-level converter.

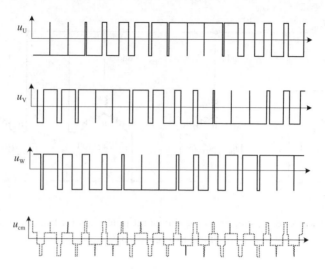

Figure 13.6 Two-level converter motor star-point potentials for time intervals 1 through 7 over one voltage fundamental wavelength given the switching positions presented in Figure 13.3.

This example of a star connection has been presented to help explain how common-mode voltage is determined, which does not depend on the connection of phases or the number of motor phases. It is easier to use the star connection when describing the phenomenon. However, in a delta connection, the situation is practically the same from the point of view of parasitic currents.

As mentioned previously, an inverter-fed electrical drive sets special requirements for the cabling and motor insulation, but it also can produce significant bearing currents leading to bearing erosion. The use of IGBTs (insulated gate bipolar transistors) or MOSFETs (metal oxide semiconductor field-effect transistors) as switching semiconductor devices has led to pulse rise and fall times of approximately 50 ns or less. Prospective silicon carbide (SiC) or gallium nitride (GaN) switching components will be even faster. Because of these shorter switching times, switching losses decrease and less cooling surface area is required. Shorter switching time enables higher switching frequencies, which result in improved current waveforms being fed to the motor. If switching frequencies are adjusted to avoid resonance, the higher frequencies also reduce noise. On the other hand, higher switching frequencies bring down IGBT inverter efficiencies, while lower switching frequencies increase motor losses. Therefore, an optimum switching frequency should be determined when implementing an inverter-fed electrical drive.

13.1   Cable modelling

In many high-power applications, the electromechanical drive components are separated from the converter by a motor supply cable that runs between them. In practice, a low-voltage motor cable can be several hundred meters long, and high-voltage motor cables can be even longer. A motor cable behaves like an *RL* component at low frequencies. However, because

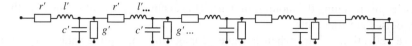

Figure 13.7 Transmission line distributed parameter modelling.

power electronic switch rise times are typically nanoseconds, the current pulses in the cable have high-frequency components. There is, therefore, capacitive coupling between the power cores of the cable and ground, which must be taken into account. A cable transmitting high-speed pulses—a transmission line—must be modelled using distributed parameters. Figure 13.7 illustrates a distributed parameter system. In the figure, r' is the distributed resistance (Ω/m), l' is the inductance (H/m), c' is the capacitance (F/m), and g' is the conductance (S/m).

In cables commonly used in motor drive systems, the speed of electromagnetic wave propagation is typically 150 m/μs. In comparison, with IGBT current rise times that are well below 100 ns, modern inverter switching frequencies are high. Therefore, with frequencies notably higher than those for a normal electrical network, inverter-to-motor cables cannot be described using concentrated cable parameters. Instead, distributed constants must be used to spread the resistance, inductance, conductance, and capacitance over the length of the cable. Resistance, conductance, inductance, and capacitance per unit length determine the characteristic cable impedance Z_0 for each cable. Z_0 also defines the relation between the amplitudes of the corresponding voltage and the current waves on the transmission line at position x.

$$Z_0 = \sqrt{\frac{r' + j\omega l'}{g' + j\omega c'}} \approx \sqrt{\frac{l'}{c'}} \approx \frac{\hat{u}(x)}{\hat{i}(x)} \tag{13.2}$$

Therefore, the cable's geometry and materials determine the magnitude of the characteristic impedance Z_0, which is independent of the cable length. Characteristic cable impedance is typically about 100 Ω. Wave propagation velocity through the cable depends on the materials surrounding its conductors. The highest speed possible is the speed of light C, which is attainable only in pure vacuum. The wave propagation velocity v can be determined as a function of cable characteristics and the speed of light as follows.

$$v = \frac{C}{\sqrt{\mu_r \varepsilon_r}} = \frac{1}{\sqrt{l'c'}} \tag{13.3}$$

where C is the speed of light, ε_r is the relative permittivity, and μ_r is the relative permeability. For example, if the relative permittivity of the motor cable sheathing is $\varepsilon_r = 4$, and its relative permeability is $\mu_r = 1$, then its maximum wave propagation velocity is half of the speed of light or $v \approx 150$ m/μs. In case of a 450 m length motor cable, a voltage pulse needs three microseconds to travel from the converter to the motor terminals. Three microseconds is typically a minimum pulse length for two-level low-voltage IGBT converters. Therefore, the converter is already turning the voltage off when the edge of the pulse is just reaching the motor terminals. This example shows how "slowly" the actual voltage pulses are travelling.

Wave reflection occurs when a wave propagating through a cable reaches a step change in characteristic impedance, that is, a point of discontinuity. The reflection off a point discontinuity that transitions from a smaller to a larger cable impedance can be cursorily explained by the current decrease that results in the larger impedance region. "Extra" charge begins to accumulate at the discontinuity, which increases voltage and produces both reflected and transmitted components of a new wave.

13.2   Reflected voltage at a cable impedance point of discontinuity

Voltage reflection also occurs at the cable terminus, which can be shorted or open.

- If the cable terminus is shorted, the amplitude of the reflected wave is equal in magnitude but opposite in sign to the incident wave, in which case terminus voltage is zero.

- If the cable terminus is open, the amplitude of the reflected wave is equal in magnitude with the same sign, resulting in a terminus voltage that is twice the magnitude of the incident voltage.

Normally, motor impedance is notably higher than the characteristic impedance of the cable. For the motor-incident (incoming) wave of an inverter-fed electrical drive, there is inevitably a reflection. The ratio of the reflected pulse and the incident pulse is expressed by the reflection coefficient ρ. This coefficient depends on the characteristic impedance Z_0 of the motor cable and the characteristic impedance Z_m of the motor (winding) "experienced" by the wave.

$$\rho = \frac{Z_m - Z_0}{Z_m + Z_0} \tag{13.4}$$

As Equation (13.4) shows, the value of the reflection coefficient varies between $0 \le \rho \le 1$ when $Z_m \ge Z_0$. The reflection coefficient at the cable-to-motor junction is commonly between 0.6 and 0.9. The reflected voltage u_r can be expressed in terms of the incoming voltage u and the reflection coefficient ρ as follows.

$$u_r = \rho u \tag{13.5}$$

The reflected component of the incoming voltage pulse wave propagates back through the cable to the inverter. At the inverter, the wave encounters the large DC-link capacitor or a short circuit via transistors or flyback (freewheeling) diodes. The characteristic impedance of the capacitor is close to zero for the high-frequency returning wave, and therefore the wave is reflected back as negative. The reflection coefficient is approximately $\rho = -1$. See Figure 13.8. A voltage surge occurs between the motor and inverter that attenuates according to the reflection coefficient. The attenuation also strongly depends on cable losses.

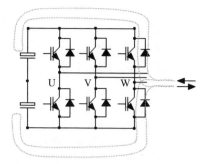

Figure 13.8 Alternative reflection paths at the converter in phase W. A pulse from the motor is transmitted to the DC-link capacitor via a transistor or freewheeling diode. It reflects back from the DC link, which the pulse sees as a short circuit. If the upper transistor of phase W is conducting, the reflected current superposes with the positive current, and the reflected wave may propagate to the DC link via the transistor.

13.3   Continuing voltage at a cable impedance point of discontinuity

At a cable impedance point of discontinuity, continuing wave propagation is affected by the type of cable terminus as follows.

- If the cable terminus is shorted, the amplitude of the continuing voltage surge is 0.

- If the cable terminus is open, the amplitude of the continuing voltage surge is twice the magnitude of the incident voltage.

The ratio of the forward-travelling and incoming backward-travelling pulses is described by the transmission coefficient τ. This factor, like the reflection coefficient, is a function of the motor cable characteristic impedance Z_0 and the motor (winding) characteristic impedance Z_m.

$$\tau = \frac{2Z_m}{Z_m + Z_0} \tag{13.6}$$

As Equation (13.6) shows, the value of the transmission coefficient varies between $1 \leq \tau \leq 2$ when $Z_m \geq Z_0$. The forward-travelling voltage u_2 can be expressed in terms of the incoming voltage u and the transmission coefficient τ.

$$u_2 = \tau u \tag{13.7}$$

Figure 13.9 illustrates the behaviour of the first reflections for a motor cable made of low-loss cable with ideal reflections.

Figure 13.10 illustrates how different current pulses travel through the cable as a function of time assuming a finite rise time t_r for the schematic pulses.

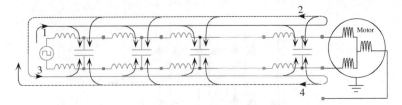

Figure 13.9 Reflections in a motor cable assuming the motor has infinite characteristic impedance and the converter (left) has zero characteristic impedance including (1) initial converter pulse, (2) motor terminal reflection of the initial pulse, (3) converter terminal reflection of the motor-terminal reflected pulse, and (4) second reflection of the pulse at the motor terminals.

During the initial period t_r, the converter pulse builds in amplitude. For period t_p, the pulse traverses the cable and arrives at the motor terminals. The rise time of the pulse reflected back towards the converter is the same as that of the original pulse. Motor terminal voltage doubles. After another period t_p, the pulse arrives at the converter terminals. It is then reflected back with negative amplitude at time instant $2t_p$. At time instant $3t_p$, the negative pulse arrives at the motor terminals and begins to lower the voltage. Simultaneously, terminal voltage moves towards zero. Oscillation is produced.

If the motor cable is "short", that is, if $t_p < t_r$, the positive reflection at the motor terminals and the negative reflection at the converter terminals partly mitigate one another, and the

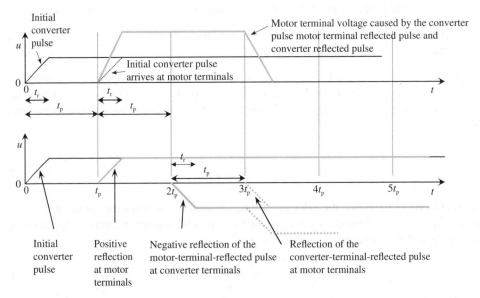

Figure 13.10 Reflections in a "long" motor cable assuming infinite characteristic motor impedance and a converter with zero characteristic impedance. The reflections result in voltage oscillation at the motor terminals. Here t_r is the pulse rise time and t_p is the time needed to travel from the converter to the motor terminals.

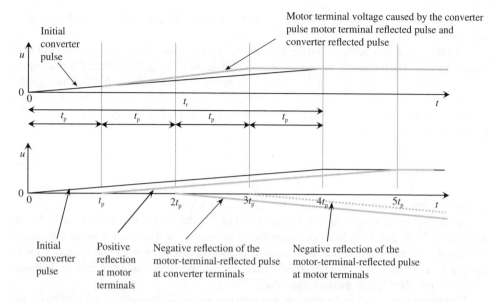

Figure 13.11 Reflections in a "short" motor cable assuming infinite characteristic motor impedance and a converter with zero characteristic impedance. There is insufficient time for the motor terminal voltage to reach dangerous levels, because the negative reflection arriving from the converter supresses voltage. Between time instants t_p and $3t_p$, motor terminal reflection causes the voltage to slightly exceed the amplitude of the converter pulse.

resulting voltage stress at the motor terminals is smaller. Figure 13.11 illustrates the short-cable case.

13.4 Motor overvoltage

If maximum motor overvoltage is twice the amplitude of the inverter pulse, the increase in voltage can be explained in terms of the waveguide theory just presented. Reflected voltage magnitude depends on cable length and the relative impedances of the motor and the cable. For a total mismatch between the motor and cable, the motor can be expressed as an open cable terminus that produces perfect reflection from the motor terminals. Voltage may double.

Voltage doubling can only occur if the cable length is above the critical cable length, that is, long enough to keep the negative wave returning from the inverter from suppressing the overvoltage. If the pulse reaches its peak value before the suppressing wave arrives, in other words if the pulse rise time is shorter than the time required for the wave to travel the distance to and from the inverter, then voltage doubles. Critical cable length l_{cr} can be determined using the following expression.

$$2l_{cr} = t_r v \Leftrightarrow l_{cr} = \frac{t_r v}{2}$$

(13.8)

where t_r is the rise time for the voltage pulse or the pulse rise time, and v is its propagation speed.

The spectral width of the signal depends mainly on the pulse rise time t_r and the switching frequency f_{sw}. The theoretical pulse Fourier spectrum is flat until f_{sw} is reached. Then, it begins to attenuate by 20 dB/decade. After reaching the upper frequency of the bandwidth f_{BW}, the spectrum attenuates by 40 dB/decade. Therefore, f_{BW} can be used as an approximation of the spectral width for the inverter output (Skibinski et al., 1999). The bandwidth frequency f_{BW} can be defined as follows.

$$f_{BW} = \frac{1}{\pi t_r} \tag{13.9}$$

For a rise time of $t_r = 50$ ns, the spectrum reaches ca 8 MHz.

Assuming $v = 150$ m/μs and $t_r = 50$ ns (IGBT) for the wave, the critical cable length becomes 3.75 m. So, to avoid reflections and overvoltages at the motor terminals for these conditions, the converter should be positioned next to the motor. Another approach is to lower the du/dt values to make "long" cables "short". Naturally, cable dielectric and Joule losses also filter out the highest pulse frequencies.

Under certain conditions, motor voltage may rise above the theoretical value of twice the DC-link voltage. If the oscillation of the latest pulse has not attenuated by the time a new pulse arrives, the combined effect of the two pulses may lead to overvoltage magnitudes of three or even four times the pulse voltage. Therefore, switching frequency and modulation significantly influence the magnitude of overvoltage. The rate of pulse rise has less influence. The characteristic oscillation frequency f of the cable significantly affects cable-induced damping, which in turn affects the magnitude of any residual charge. So far, no materials have been found that provide effective damping of high-frequency waves.

The oscillation frequency at the motor terminals is a function of wave propagation velocity v and cable length l. Therefore, oscillation frequency is independent of inverter or motor characteristic values, such as the pulse edge rise time of the inverter or the power of the motor. Wave propagation velocity, in turn, is a function of the characteristic values of the cable according to Equation (13.2). At the critical cable length, pulse rise time and propagation time are equal, that is, $t_r = t_p$. For this special case, the wave travels the length of the cable $l = l_{cr}$ four times in one oscillation period. Figure 13.12 shows the voltage response of a motor to an inverter pulse for a "long" cable, that is, for a cable with a pulse rise time equalling the pulse propagation time from the converter to the motor or $t_r = t_p$.

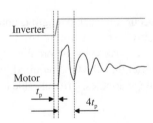

Figure 13.12 Voltage response of a motor to an inverter pulse for a "long" cable, that is, for a cable with $t_r = t_p$.

Cycle time T is now 4 times the wave propagation time t_p. Therefore, the oscillation frequency f can be written

$$f = \frac{1}{T} = \frac{1}{4t_p} = \frac{v}{4l} = \frac{1}{4l\sqrt{l'c'}} \tag{13.10}$$

where c' is capacitance per unit length, l' is inductance per unit length, l is cable length, t_p is wave propagation time, T is oscillation cycle time, and v is wave propagation velocity.

Therefore, the characteristic oscillation frequency of the cable has an indirect effect on wave attenuation. Conductor resistance, and therefore attenuation, increases due to the skin effect. The skin effect, in turn, is a function of oscillation frequency. As a result, an increase in oscillation frequency increases wave attenuation, and to avoid overvoltages that exceed the double DC-link voltage, the oscillation of the reflected wave must be attenuated before the next pulse arrives.

13.4.1 Double pulsing

Figure 13.13 illustrates the so-called double-pulsing phenomenon, which results in large overvoltages. Double pulsing occurs, if the transient (oscillation) caused by the pulse has not attenuated before the next pulse arrives, in other words, if the dwell time is not long enough. Initially, inverter and cable voltages are equal. The falling edge of the pulse is reflected away from the motor terminals based on progressive wave theory, and the reflected wave continues to the inverter, where it is reflected back with a reflection coefficient $\rho = -1$. Coincident with the wave being reflected from the inverter, the rising edge of the inverter control pulse arrives at the motor. The combined effect of these two waves results in the situation illustrated in the figure.

The magnitude of the voltage produced at the motor terminals depends on cable damping, how long the pulse voltage remains at zero, cable length, and inverter switching frequency. In practice, double pulsing should be avoided.

13.4.2 Double switching

To establish the required output frequency, the inverter turns each phase on or off according to its modulation pattern. Considering the case as a whole, the voltage between different phases can be as much as twice the DC-link voltage. See Figure 13.14. This can happen if the states of two inverter branches are switched simultaneously.

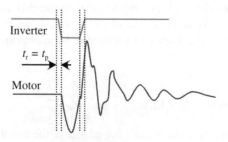

Figure 13.13 The double-pulsing phenomenon.

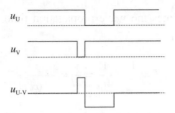

Figure 13.14   The change in line-to-line voltage that occurs when a pair of phase voltages are switched simultaneously.

Double switching causes overvoltages that are even higher than those caused by double pulsing. At worst, these overvoltages can exceed the design limits of the electric machine. Modulator operation should be designed to prohibit double switching, which does not feature among the properties of a space vector modulator, for example.

13.5   Limiting overvoltages with impedance matching

Overvoltages can be reduced by filtering or by inverter programming. The most common way to implement filtering is to add du/dt chokes to the motor cable. They suppress the high-frequency components and therefore round out the pulse edges. The du/dt choke solution naturally increases inductance and voltage drop, which in turn reduces motor voltage and torque. Cable inductance, the inductance of the series inductors, and stator leakage inductance combine to reduce machine breakdown torque.

With small machines ($P < 20 \, \text{kW}$), the characteristic impedance of the motor can be from 10 to 100 times the impedance of the cable, because of the high characteristic impedance of the motor winding. Therefore, the incident wave is reflected back from the motor according to Equations (13.4) and (13.5). However, if the cable is terminated with an impedance corresponding to the characteristic cable impedance, no reflection takes place.

As machine size increases, characteristic impedance typically decreases, so the difference between motor and cable impedances is less for larger machines. As a result, they exhibit less wave reflection, and fewer impedance matching measures are required. However, this does not imply that voltage rise is necessarily less significant in larger machines. On the contrary, large machines typically present more inverter drive problems despite this lower reflection.

Many methods of impedance matching are based on a filters constructed of passive filtering components with impedance Z_s mounted in parallel with the motor. The characteristic impedance of the cable Z_0 is notably lower than the characteristic impedance of the motor Z_m, and therefore Z_0 is decisive. The objective is to bring impedance Z_s closer to Z_0. The parallel connection of the filter and motor can be approximated with the following equation.

$$\rho = \frac{(Z_s \| Z_m) - Z_0}{(Z_s \| Z_m) + Z_0} \approx \frac{Z_s - Z_0}{Z_s + Z_0} \tag{13.11}$$

According to the equation, the characteristic impedance of the motor can be neglected, so the filter should be designed to match the cable characteristic impedance.

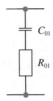

Figure 13.15 An *RC* filter used to mitigate voltage oscillations arising from motor terminal reflections.

The cable could be terminated using a resistance $R = Z_0$ connected in parallel with the motor.

$$R = Z_0 = \sqrt{\frac{l'}{c'}} \tag{13.12}$$

However, the significant amount of power that would be lost through the resistor keeps this from being a real impedance matching approach alternative. At 460 V, for example, resistance losses in a 30 m cable with 190 Ω characteristic impedance would be approximately 1 kW.

Figure 13.15 depicts an impedance matching filter comprising the series connection of a capacitor C_{01} and a resistor R_{01}. The resonant frequency of the filter is selected empirically to be five times the inverter switching frequency. The angular frequency is determined by the operating frequency, which is set close to the resonant frequency.

The impedance of the filter Z_{S01} can be calculated as follows.

$$Z_{S01} = \sqrt{R_{01}^2 + \left(\frac{1}{\omega\,C_{01}}\right)^2} \tag{13.13}$$

where ω is the angular frequency. To achieve an overdamped condition, the value of the filter resistor R_{01} should be selected so that

$$R_{01} > \sqrt{\frac{4L}{C_{01}}} \tag{13.14}$$

where L is the inductance of the cable. Power losses for this impedance matching approach, under the same conditions as in the previous case (460 V, 30 m, and 190 Ω) are 150 W.

Figure 13.16 shows a second-order impedance matching filter comprising a capacitance C_{02}, an inductance L_{02}, and a resistance R_{02}.

The impedance Z_{S02} of the filter in question can be determined as follows.

$$Z_{S02} = \sqrt{\left(\frac{R_{02}\omega^2 L_{02}^2}{R_{02}^2 + \omega^2 L_{02}^2}\right)^2 + \left(\frac{R_{02}^2\omega L_{02}}{R_{02}^2 + \omega^2 L_{02}^2} - \frac{1}{\omega C_{02}}\right)^2} \tag{13.15}$$

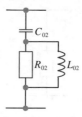

Figure 13.16   A second-order *RCL* filter used to mitigate voltage oscillations arising from motor terminal reflections.

To again produce an overdamped condition, the value of the filter resistor R_{02} should be selected so that

$$R_{02} < \frac{\sqrt{L_{02}C_{02}}}{2C_{02}} \tag{13.16}$$

where L_{02} is the inductance. Power losses in this case are about the same as for the *RC* filter, that is, about 150 W.

Because of these incurred losses, motor-to-cable impedance matching is usually not considered. If filtering is used at all, the target is mainly to modify the steep rising edges of the pulses, which can be accomplished using less lossy filters. In practice, small du/dt filters are applied extensively, particularly in industrial inverter drives. The simplest way of implementing a du/dt filter is to install ferrite rings on each of the motor cable phase conductors at the converter output. These ferrite rings produce extra inductance, which limits the voltage rise in the cable.

Overvoltages that result from double pulsing and double switching can be addressed with converter programming. Double pulsing can be prevented by eliminating pulses that are below a certain duration. A function of cable damping, cable length, and inverter switching frequency, the preset uncharged time is defined individually for each system. Appropriate modulator design prevents double switching.

The output voltage of an inverter consists of approximately rectangular pulses. Because of the inverter's fast rise and fall times, each pulse includes high-frequency harmonics. These harmonics induce "extra" currents in the windings, which increase losses and, consequently, winding temperatures. The higher temperatures can be avoided by limiting power, or the reduction in machine life brought about by operating at higher temperatures can be accepted.

The effect of steep-edged inverter voltage pulses on machine temperature was discussed previously. Even though the inverter controller does not directly affect winding temperature, a poorly designed control may result in higher temperatures that lead to cooling problems. In most electric machines, the coolant impellers and cooling channel geometries have been designed for normal direct-on-line operation. With inverter control, traditional machine cooling approaches may be inadequate at lower rotation speeds and additional cooling may be required. Typically, an external blower is used to deliver this additional cooling.

Ström (2009) proposed a switching pattern to avoid oscillations caused by the steep-edged pulses. The motor cable can be roughly modelled as an LC circuit. Steep-edged DC pulses fed to the circuit produce high-amplitude AC oscillations. However, oscillation can be mitigated if the LC circuit is first prepulsed to charge the circuit capacitor to half of the supply voltage

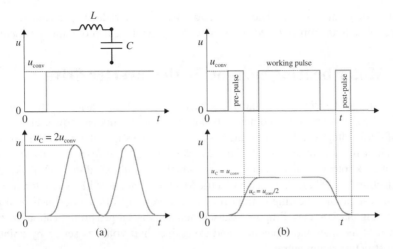

Figure 13.17   (a) When an ideal LC circuit is supplied with a single voltage pulse, voltage doubles, and the oscillation angular velocity is $(LC)^{-0.5}$. (b) If inductor charging is halted when the capacitor voltage reaches half the supply voltage, oscillation will be mitigated. A charging pre-pulse and a discharging post-pulse are necessary.

level and then, after a short delay, fed a single pulse. Another postpulse is needed to avoid oscillations at switch off. Figure 13.17 illustrates this behaviour.

This method may become practical as switches get faster. With IGBTs, the extra switching operations needed may be an overburden that leads to significant derating of power handling capability. With the higher speed of MOSFETs, however, this method becomes practical.

In a DOL machine, the low-frequency (50/60 Hz) supply voltage is evenly distributed throughout the windings, so voltage loss is also evenly distributed. With inverter control, this voltage distribution is uneven and includes steep-edged voltage pulses that have been shown to produce notably larger voltage stresses in the motor windings than the stresses produced by conventional sinusoidal line voltages.

Moreover, the first and last winding coil turns may be adjacent in low-voltage motors using round enamelled wires in the winding. These wires are not deterministically positioned and, in the worst case, may result in the first and last turn of a coil close to one another. If so, the voltage acting upon the entire coil is also the voltage between these outmost coil turns. The insulation of individual conductors in each coil has been designed to prevent a current path between conductors at a normal-frequency line voltage. The uneven distribution of voltage over the winding and between different turns leads to significant overvoltages both to ground and between the conductors. In the worst case, 80% of the phase voltage in a multiple-turn winding is across the active length of the first turn. In this event, the field strengths could exceed the breakdown voltage of the conductor-to-conductor insulation. Partial discharges do occur in high-voltage machines, and if inverter driven, the voltage-driven discharges can prematurely age conductor-to-conductor insulation.

To summarize Section 13.5, waveguide theory explains how overvoltage at the motor terminals is caused by the reflection of voltage waves. The effect of characteristic impedance is obvious. Overvoltages can be categorized according to magnitude with respect to the double-incident voltage. Critical cable length can be determined for overvoltages less than double the incident voltage. There are two methods to reduce overvoltages including impedance matching implemented by passive components and programming methods.

13.6   Motor bearing currents in the inverter drive

Bearing currents in electric machinery has been a topic of discussion for the past hundred years. The bearing current phenomenon is not exclusive to modern power electronics. The effects of shaft voltages and bearing currents have been investigated, for example, by F. Punga and W. Hess in their article titled *"Eine Erscheinung an Wechsel- und Drehstromgeneratoren"* in Elektrotechnik und Maschinenbau, which was published in 1907. P. Alger and H. Samson lectured on "Shaft Currents in Electric Machines" at the A.I.R.E. Conference in 1924. In early investigations, bearing current problems were most often associated with machine manufacturing inaccuracies. Proposed causes included discharges brought about by electrical breakdown of the insulation materials and excessive shaft voltages set up by unintentional magnetic structure asymmetries.

Although modern power electronics has provided numerous operational and control benefits, they have also introduced new bearing current scenarios. At first, it was not clear that power electronics control was the cause of the observed problems. The most important bearing-current triggers associated with today's inverter drives include common-mode voltages fed by the PWM-type inverter to the motor, higher switching frequencies, poor cabling between the inverter and motor, and parasitic windings capacitances. Figure 13.18 offers a 7× micrograph to illustrate the condition of a bearing race that has been subjected to bearing current discharges. The photograph illustrates fluting. Electrostatic discharge between the balls and the race has moved steel and fluted the race surface. This behaviour is analogous to the sand fluting that occurs on a shoreline when light waves of water lap against it and produce fluted patterns in the sand.

Shaft Grounding Systems, a U.S. company, analysed 1000 inverter-driven and 150 DOL AC motors. The analysis revealed that a full 25% of the inverter driven motors experienced

Figure 13.18 The photomicrograph of a damaged bearing race showing fluting brought about by discharge bearing currents (magnified 7 times). The term *fluting* refers to a regular pattern similar to the pattern eroded into lakeshore sand by wave motion. In this case, steel erosion is brought about by continual electrical discharge. In the photograph, the dark areas show undamaged steel. The lighter flutes represent damage. *Source:* Reproduced with permission from ABB.

some type of bearing failure within the first 18 months of operation. Within an average of 2 years' operation, 65% suffered bearing failure. Only 1% of the DOL motors experienced bearing failure in that period (Boyanton, 1995).

13.6.1   Bearing damage caused by electric current—failure mechanism

There are several bearing current triggering mechanisms. However, they all produce equivalent currents. A ball or a roller bearing comprises two cylindrical surfaces (races) with a number of metal balls or cylinders captured in-between. Typically, to reduce friction, an oil-based lubricant coats the metal surfaces. When relative motion in the bearing slows or comes to a halt, the electrical resistance between races drops to a minimum as the metal bearing components move closer together or make contact. This drop in electrical resistance is accompanied by an increase in race-to-race electric current, which is brought about by and proportional to the voltage difference that exists between the races. At higher rotation speeds, more lubricant flows between the bearing components pushing them farther apart. As a result, race-to-race impedance increases and electric current drops abruptly. The change in bearing impedance is a highly nonlinear function of speed. It is low at slower rotation speeds, but quickly increases to the megaohm (MΩ) range as rotation builds to rated speed.

Bearing impedance is also directly proportional to the quality and temperature of the bearing lubricant and the surface roughness of the bearing components. When the lubricant film between bearing elements builds up and effectively breaks the path of conductivity, capacitance develops across the lubricant film and begins to accumulate charge. Figure 13.19 illustrates the geometry and shows the equivalent circuit of a typical motor bearing. If the voltage level that develops between the bearing races becomes sufficiently high, the capacitors discharge, and the substantial discharge current passes through the very small interface areas between the bearing rolling elements and the races. The current density at the interface can be high enough to erode the interface surfaces. The fluting of the bearing race shown previously in Figure 13.18 is an example. The mechanism of erosion is similar to that of electrical discharge machining (EDM).

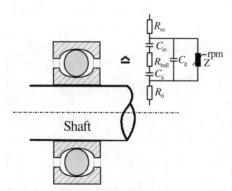

Figure 13.19 A ball bearing and its equivalent circuit. $C_{io,I}$ is the outer or inner capacitance between the bearing balls and races, C_g is the capacitance between the bearing races, R_{ball} is the bearing balls resistance, R_{ri} is the resistance of the inner race, R_{ro} is the resistance of the outer race, and Z is the overall nonlinear impedance of the ball bearing.

The bearing discharge current problem grows with voltage across the bearing, and at some voltage level, it becomes destructive. A cold bearing can withstand PWM inverter pulses of about 35 V. The limit for a warm bearing is 6–10 V. If the voltage is sinusoidal, warm-bearing levels below 1 V are safe (Skibinski, 1997).

Bearing current damage may result in machine failure within a few weeks of operation, or it may take several years. For a machine rotating at constant speed, the damage manifests as a uniform circular array of flutes and channels aligned perpendicular to the direction of motion for the bearing rolling elements as it was shown in Figure 13.18. The array is uniform, because the dielectric breakdowns occur at regular intervals. Bearing current damage for variable speed rotation manifests differently. Because there is no regular breakdown pattern, current discharges are more random, and they erode the entire raceway path more or less evenly.

All converter-driven motors are subject to bearing currents, and motor manufacturing asymmetries are a contributing factor. Manufacturing asymmetries grow with size, and the probability of there being harmful bearing currents in a motor grows with its diameter. Induction motors can be divided into the following machine size groups listed in order of increasing bearing current risk.

1. Industrial induction motors with shaft heights below 280 mm usually have lower risk of failure attributed to bearing current phenomena.

2. Motors with shaft heights over 280 mm have significant risk of failure attributed to bearing currents. Measures must be taken to mitigate bearing current effects. The PWM modulation pattern also may be changed to minimize risk. Converter output du/dt filters are definitely recommended.

3. In industrial induction motors having shaft heights greater than 400 mm, the risk is high.

Stray capacitances develop in an electric machine at higher frequencies, which must be considered in its equivalent circuit (Busse et al., 1995). Figure 13.20 illustrates the three

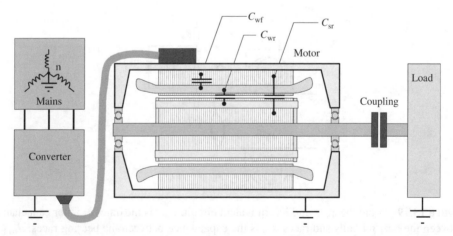

Figure 13.20   The most important stray capacitances develop across the three obvious stator-to-rotor paths for currents produced by the high-frequency voltage harmonics of PWM pulses.

obvious paths for currents produced by the high-frequency voltage harmonics of PWM pulses. They include the capacitance between the stator winding and the stator frame C_{wf}, the capacitance between the stator winding and the rotor core C_{wr}, and the capacitance between the stator and the rotor cores C_{sr}. These stray capacitances usually increase with motor size.

When capacitance-enabled currents run from stator to rotor to ground, they must pass through the rotor bearings. These bearing currents can be roughly divided into the following categories.

- capacitive discharge currents, that is, noncirculating bearing currents

- circulating currents

- shaft grounding currents

13.6.2   Noncirculating bearing currents

Noncirculating currents are produced by an increase in the potential differences between the stator windings and the rotor, between the stator windings and the frame, and between the frame and ground. Common-mode current charges the stator-rotor frame capacitance C_{sr} and the bearing capacitances $C_{io,i}$. Figure 13.21 shows a simplified equivalent circuit for an example motor. The switch S describes galvanic contact through the bearing rolling elements. The discharge current depends on the common-mode voltage rise time and the timing of switch S.

In general, $C_{wf} > C_{sr} > C_{wr}$, and the shaft potential is not sufficient to break down the film of lubrication (Gambica, 2002). Capacitive discharge currents dominate the bearing failure mechanism in smaller motors.

13.6.3   Circulating bearing currents

Circulating currents in large machines drive the most harmful bearing currents. They are produced by PWM voltage pulses acting on the capacitance between the stator windings and the stator frame C_{wf} and between the stator windings and the rotor core C_{wr}. Figure 13.22

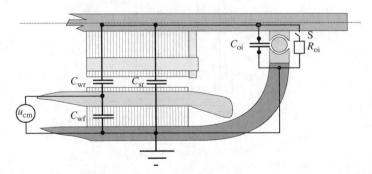

Figure 13.21 Simplified common-mode equivalent circuit of a motor with capacitive discharge currents. The large capacitance between the stator winding and frame, C_{wf}, filters the common-mode voltage. However, some of the current produced by the common-mode voltage passes through the bearings.

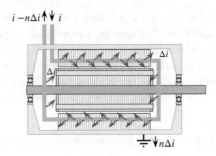

Figure 13.22 Schematic view of current passing through one turn of one phase. The currents leak similarly from all winding turns, and supply all phases with similar voltage as a common mode. n is an empirically determined number describing the amount of leakage.

illustrates the currents via these capacitors and the preceding Figure 13.21 shows the capacitors. Because the capacitances are distributed along the winding, there is charge leakage along its entire length. This current leakage is a loss of winding current, so the level of incoming winding current is higher than the level of outgoing winding current.

Ideally, incoming and outgoing winding currents are equal. The total sum current is zero, and no additional magnetic flux is produced. However, because of the parasitic capacitances, this is generally not the case. Figure 13.23 is a representation of the actual winding current situation for one phase of an example motor. In the figure, i is the current incoming to terminal U1, and $i - n\Delta i$ is the current outgoing from U2. Parasitic capacitance is represented by the columns of capacitive leakage currents of magnitude Δi. The sum of the currents at a given machine cross section A–A is not equal to zero.

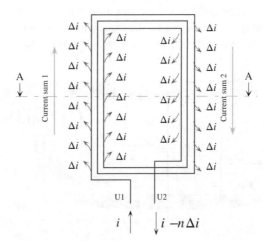

Figure 13.23 Phase U winding currents for an example motor. There is capacitive current leakage along the entire length of the winding. As a result, current sum 1 is greater than current sum 2. In addition, the current travelling back to the converter via the phase line is smaller than the current travelling towards the machine.

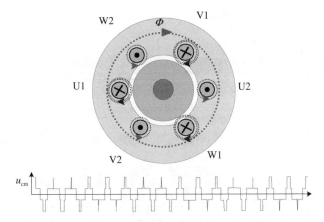

Figure 13.24 Cross section A–A for all three phases of the motor represented in Figure 13.22. Fluxes are produced in the three-phase winding due to the common-mode voltage u_{cm}. In all phases, a sum total current moves in the same direction from one end of the machine to the other producing a circulating magnetic flux.

Figure 13.24 depicts the situation at cross section A–A for all three phases of the example motor of Figure 13.23. Because of the current imbalance brought about by the parasitic capacitances, the motor develops a net flux Φ (Shaotang, 1996).

According to Faraday's induction law, any change in magnetic flux produces a circulating electric field. In this case, the field sets up a voltage difference between the N and D ends of the motor shaft. The only current path available to relieve this potential comprises the motor frame, the bearings, and the shaft. Figure 13.25 illustrates.

High-frequency current in the stator windings induces a voltage across the motor shaft, the magnitude of which may be 20 times the voltage of a motor operating direct on line. It produces a large common mode current from the machine's terminal end to its nonterminal end. An opposite rotor current must compensate (Chen et al., 1998; Ollila et al., 1997). Because the shaft has electrical impedance, an axial shaft voltage develops. Shaft voltages exceeding 300 mV are harmful to metallic bearings.

13.6.4   Shaft grounding current

A drive system offers several paths to ground for common mode currents. See Figure 13.26. These include the protective earthing (PE) conductor of the motor cable

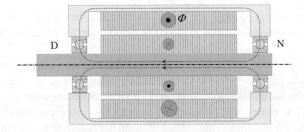

Figure 13.25 Circuit of balancing circulating currents travelling via both motor bearings.

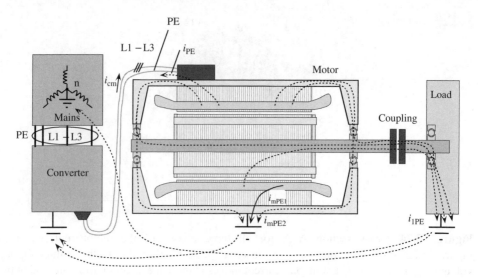

Figure 13.26   The various paths for the grounding currents.

(current i_{PE}), the grounded parts of the frame (currents i_{mPE1} and i_{mPE2}), and the load ground (current i_{lPE}).

The magnitudes of the different grounding currents depend on the impedances of the paths at the different PWM modulation frequencies. The inductances of the grounding paths become more important at higher frequencies. If the load ground path impedance is small enough, current via the motor and the load bearings can result in failures in the load machine.

If the high-frequency rotor–bearing–frame–ground impedance is low, there can be common-mode currents through the motor bearings. The currents I_{mPE1} and I_{PE} illustrated in Figure 13.26 are harmless to bearings but can cause other EMC problems.

13.6.5   The motor cable and capacitive currents

The motor supply cable transfers power from the inverter to the motor with minimal power loss. According to EMC regulations, the cable cannot induce external electromagnetic interference (EMI). Furthermore, it must resist disturbances coming from the environment. Electric safety regulations also set limits on the cross-sectional geometries of the conductors. To eliminate EMI, the motor supply cable must be shielded with a conductive material. This shielding is necessary for both AC and DC drives.

If asymmetric supply cables are used, against the recommendations of frequency converter manufacturers, substantial voltage can develop on the PE conductor. Figure 13.27 depicts asymmetric motor supply cable configurations.

Voltage can be induced on the PE conductor by the common-mode voltage coming from the inverter and the phase conductor voltages, which can include the high-frequency du/dt differential mode and common-mode voltage pulses if no filters are used. For instance, the rise time for fast IGBT switches may be below 50 ns, which implies spectrum frequencies above 20 MHz. When the PE conductor and cable armouring is connected to the inverter frame, the voltage potential of the motor relative to ground increases.

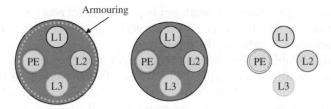

Figure 13.27 Asymmetric supply cable configurations. The cable illustrated on the left comprises one protective earth (PE) and three live conductors separated by plastic insulation and jacketed with conductive armouring. The similar cable shown in the middle does not include the armouring. The supply cable on the right uses just four separate insulated conductors. In all three examples, the asymmetric positioning of the live conductors within the cable assembly can lead to unbalanced electromagnetic behaviours and therefore to EMC problems.

Figure 13.28 shows a block diagram of an inverter-fed drive and attached power tool. In this system, if the impedance of the cable armouring is too large, some of the high-frequency current can divert through the rotor shaft and power tool bearings to ground.

As stated, a tool in an inverter drive and power tool system can be subjected to unwanted bearing currents. This situation can develop in paper machines, roller mill drives, and in other drives that compose solid metal structures. The constantly varying shaft grounding impedance in roller mill drives presents another problem, as the machined work piece connects and disconnects the shafts of the motor to and from the ground potential.

The bearings of an inverter drive power tool can be damaged if the bearing lubricant in the tool is a better insulator than the bearing lubricant used in the motor bearings. The higher impedance of the tool lubricant causes higher voltages to develop, so dielectric breakdown across the tool bearings results in higher discharge current densities. Paradoxically, bearings exposed to large bearing currents will last longer using a poorer quality lubricant. Figure 13.28 shows that

$$i_{PE} = i_{arm} + i_g \tag{13.17}$$

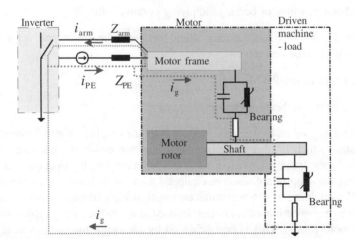

Figure 13.28 Block diagram of an inverter-fed drive and driven power tool illustrating the non-circulating capacitive-discharge bearing-current path when the cable used has asymmetric construction that induces voltages in the PE conductor.

If the cable armouring provides a sufficiently low impedance return path for i_{PE}, then $i_{PE} \approx i_{arm}$. The cable construction should provide a suction transformer for the common mode pulses. Symmetric construction is necessary for this purpose.

There are also notable quality differences between symmetrical cables. Analyses have shown that poor-quality motor supply cables induce 13-fold voltage in the armouring compared to good-quality supply cables. Correspondingly, there can be a 56-fold difference in cable-to-cable noise conduction between poor- and good-quality cables. Clearly, cable quality can be a significant factor in the development of bearing currents (Bentley, 1996).

13.7   Reducing bearing currents

While bearing currents in inverter drives cannot be eliminated, good system design can keep bearing currents below damaging levels. There are several ways to accomplish this, and the best results are achieved by combining alternatives. In addition, the construction of the bearings themselves plays an important role in mitigating the problem. Active magnetic bearings, for example, have a big air gap between the bearing rotor and stator that effectively breaks the bearing current conduction path.

In general, bearing current mitigation methods can be divided into the following four categories:

1. Effective grounding solutions inside the drive to bypass harmful currents coming from the bearings

   - Effective electrical installation of the drive

   - Applying a shaft grounding system

2. Creating high-impedance paths inside the motor for bearing currents

   - Insulated bearings or bearings applying ceramic balls

   - Applying a grounded Faraday shield between the stator and rotor to mitigate capacitive rotor currents

3. Using filters in the PWM output

4. Using conductive bearing crease

A properly designed and assembled electric machine exhibits minimal impedance in the grounding paths, which in turn minimizes stray currents inside the machine. The voltage potential of the motor frame relative to ground can be reduced with proper cabling. In addition to its armouring, the supply cable must be equipped with a solid layer of electromagnetic radio frequency (RF) shielding to provide protection against high-frequency interference.

Low induction connection methods must be used to attach the supply cable armour and RF shield to the motor frame. This is best achieved by enclosing the cable with a conductive sleeve that is galvanically connected to the cable shielding. The sleeve itself must be fixed to the converter and motor using a conductive bushing. Experts talk about 360° grounding, which means the shield must completely envelop the live lines. Finally, the cable armouring

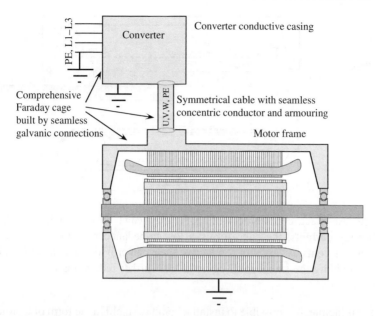

Figure 13.29 Faraday-cage methodology applied to a typical converter drive. The machine and controller housings and the cable protection shield form a seamless cage that mitigates bearing currents and other EMC problems. The entire drive system is contained inside a solid "metal housing". The 360° connections at the motor and converter terminals are essential for the PE-line.

and shielding must be wired to the PE buss as directly as possible to establish a Faraday cage around the cable conductors all the way from the inverter to the machine.

Figure 13.29 illustrates the Faraday cage methodology applied to a typical converter drive. The machine and controller housings and the cable protection shield form a seamless cage to mitigate bearing currents and other EMC problems.

It is also possible to mitigate bearing currents by grounding the rotor to the motor housing using conductive "brushes". Electrically, these brushes and the rotor bearings are connected in parallel. However, it makes little sense to use grounding brushes in AC drives, because AC drives were initially introduced to eliminate the maintenance problems associated with the commutator brushes used in DC motors.

Special insulated bearings are recommended for large motors drives. Insulated bearings provide a strong remedy against circulating and shaft grounding currents. The simplest approach is to add an aluminium-oxide-based 50 to 300 μm insulating coating to the outside of the outer race. Another approach is to insulate the bearing from the frame using a significantly thicker insulation embedded appropriately in the bearing shield.

Using hybrid bearings (steel races and ceramic rolling elements) will also eliminate bearing currents. However, the silicon nitride ceramic balls in hybrid bearings are much stronger and stiffer than steel. As a result, the stresses in their steel races are higher, and load capacity must be reduced. Moreover, hybrid ball bearings are expensive and not as well characterized as steel ball bearings.

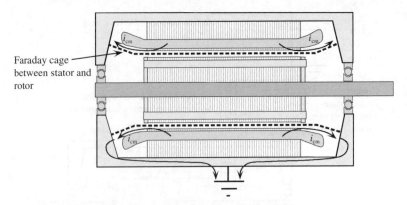

Figure 13.30 An electric machine with a Faraday cage positioned in the air gap. To eliminate bearing currents, the cage forms a low-impedance path directly to ground for capacitive stator currents. A very thin conductive sheet can provide the path while minimizing Joule losses. Dividing the shield into segments can further reduce losses.

At least in principle, it is possible to install a Faraday shield in the form of a grounded can between the stator and the rotor. The grounded can would give capacitive currents a path directly to the PE conductor. The primary electromagnetic disadvantage comes from the eddy currents and subsequent losses that develop in a conductive material positioned in the moving magnetic field of an electric machine. Figure 13.30 depicts an electric machine with a Faraday cage positioned in the air gap.

A universal solution for differential- and common-mode problems is to ensure that only sinusoidal voltages are applied to the motor terminals. PWM pulses can be filtered in a number of ways to soften the PWM output and achieve a more sinusoidal shape. Output inductors, du/dt filters, and sinus filters are typically used. Output inductors add to the drive system's leakage inductance and lower system performance. du/dt filters round the PWM pulses and decrease the voltage change rate. Sinus filters have a low cut-off frequency, and therefore they filter the inverter waveform making the output voltage almost sinusoidal.

13.7.1 PWM inverter output filters or chokes

The purpose of output filters or chokes is to reduce output voltage du/dt values, thereby eliminating the higher frequencies of the output voltage spectrum. This significantly decreases the current through the parasitic capacitances of the motor, bringing the current levels to and from the windings more into balance and consequently decreasing rotor shaft voltage. Cable induced voltages also decrease. An output choke, therefore, affects both circulating and noncirculating bearing currents. A properly designed filter comprising the appropriate inductances and capacitances can shape the output voltage to approximate a sine wave. Given a sinusoidal input voltage, the bearing current levels of an inverter driven motor fall into the same range as those of a DOL motor.

Specialty filters have been designed, in particular, to suppress common-mode voltages. Applying these filters can lead to considerably lower bearing current levels. Surrounding phase conductors with a lossy magnetic core is also an effective approach to filtering common-mode currents; however, this method results in higher losses.

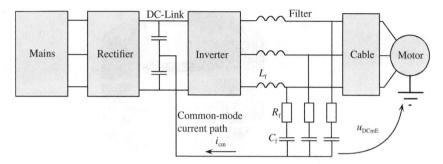

Figure 13.31 The schematic for a conventional inverter output filter comprising inductances, capacitances, and resistances. Voltage u_{DCmE} is from the DC-link midpoint to the PE conductor.

Filters can be either active or passive. Passive filters are more reliable and less expensive. Figure 13.31 is a schematic for a conventional inverter output filter comprising inductances, capacitances, and resistances. The most significant drawback to this filter design is its inability to effectively filter common-mode voltages if the DC-link midpoint connection is missing. Some converters will not work using this connection configuration, flagging it as a fault condition.

In the ideal case where the filter eliminates reflections in the system, common-mode voltage at the motor terminals can be expressed as follows.

$$u_{cm} = \frac{1}{3} R_f i_{cm} + \frac{1}{C_f} \int i_{cm} dt + u_{DCmE} \tag{13.18}$$

The common-mode voltage at the motor terminals is proportional to R_f, inversely proportional to C_f, and proportional to the voltage u_{DCmE} between the DC-link midpoint and the PE conductor. If $C_f \to \infty$ and $R_f \to 0$, then $u_{cm} \approx u_{DCmE}$ and the common-mode potential is close to the potential of the DC-link midpoint. The shape of the common-mode voltage becomes considerably smoother without harmful high-frequency harmonics. If the DC-link midpoint is not available, the filter elements can be duplicated and the capacitor star points can be connected to both the positive and negative terminals of the DC link.

There are numerous other filters available of varying types. Dzhankhotov (2007) proposed an air-core hybrid LC filter comprising a triple layer foil choke. The result combines both filtering and capacitance. See Figure 13.32.

13.7.2   Using a conductive bearing lubricant

Using a conductive bearing lubricant may reduce or even eliminate the damaging effects of bearing currents. However, this possibility has not been thoroughly investigated, and there is little practical experience with using conductive lubricants to mitigate bearing current damage and little documentation of the long-term effects. In his publication, Chen mentions that using conductive bearing lubricant may be one way to avoid using brushed to ground the rotor shaft (Shaotang, 1996).

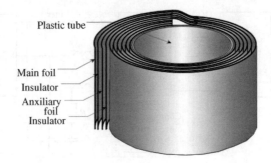

Figure 13.32 Configuration of an air-core hybrid LC filter built of aluminium foil layers separated by insulation. The filter combines both filtering and capacitance. (Dzhankhotov, 2011).

Thus far, EMC and bearing current problems have not been totally eliminated. However, because power electronic drives offer a number of important advantages, these problems are tolerated in general and have been successfully addressed in practice. Newer and faster switches will aggravate EMC and bearing current problems, but techniques made possible by their improved switching performance, such as the one suggested by Ström (2009), can be used to mitigate at least part of the remaining issues.

References

Akagi, H., & Tamura, S. (2005). A passive EMI filter for eliminating both bearing current and ground current from an inverter-driven motor. In Power Electronics Specialists Conference, PESC'2005, Recife, Brasil, pp. 2442–2450.

Bentley, J. M. (1997). Evaluation of motor power cables for PWM AC drives. *IEEE Transactions on Industry Applications*, 33(2), 342–358.

Binder, A., & Muetze, A. (2007). Scaling effects of inverter-induced bearing currents in AC machines. In *Proceedings of IEEE International Electric Machines & Drives Conference, IEMDC'2007*, Antalya, Turkey, May 3–5, 2007, pp. 1477–1483.

Boyanton, H. (1995). Bearing damage due to electric discharge. *Shaft Grounding Systems*, 1–29.

Busse, D., Erdman, J., Kerkman, R., Schlegel, D., & Skibinski, G. (1995). Bearing currents and their relationship to PWM drives. In Proceedings of International Conference on Industrial Electronics, IECON'1995, Orlando, FL, November 6–10, Vol. 1, pp. 698–705.

Chen, S., Lipo, T. A., & Novotny, D. W. (1998). Circulating type motor bearing current in inverter drives. *Industry Applications Magazine*, 4(1), 32–38.

Dzhankhotov, V. (2009). Hybrid LC filter for power electronic drives: Theory and implementation. Dissertation LUT, ISBN 978-952-214-826-1 ISBN 978-952-214-827-8 (PDF), available at http://urn.fi/URN:ISBN:978-952-214-827-8

Esmaeli, A. (2006). Mitigation of the adverse effects of PWM inverter through passive cancellation Method. In Proceedings of International Symposium on Systems and Control in Aerospace and Astronautics, ISSCAA'2006, January 19–21, 2006, pp. 47–751.

Esmaeli, A., Sun, Y., & Sun, L. (2006). Mitigation of the adverse effects of PWM inverter through active filter technique. In Proceedings of International Symposium on Systems and Control in Aerospace and Astronautics, ISSCAA'2006, January 19–21, 2006, pp. 770–774.

Finlayson, P. (1998). Output filters for PWM drives with induction motors. *IEEE Industry Applications Magazine*, 4(1), 46–52.

Gambica Association. (2002). Variable speed drives and motors: Motor shaft voltages and bearing currents under PWM inverter operation, Report No. 2 (2nd. ed.). Available at http://www.rema.uk.com/pdfs/Report%20No%202.pdf

Hanigovszki, N., Poulsen, J., & Blaabjerg, F. (2004). A novel output filter topology to reduce motor overvoltage. *IEEE Transactions on Industrial Applications*, 40(3), 845–852.

Hongfei, M., Dianguo, X., & Lijie, M. (2004). Suppression techniques of common-mode voltage generated by voltage source PWM inverter. In Proceedings of Power Electronics and Motion Control Conference, IPEMC'2004, August 14–16, Vol. 3, 1533–1538.

Kerkman, R. J., Leggate, D., & Skibinski, G. L. (1997). Interaction of drive modulation and cable parameters on AC motor transients. *IEEE Transactions on Industry Applications*, 33(3), 722–731.

Kuisma, M., Dzhankhotov, V., Pyrhönen, J., & Silventoinen, P. (2009). Air-cored common-mode DC filter with integrated X and Y capacitors. In Proceedings of 13th Conference on Power Electronics and Applications, EPE'2009, September 8–10, 2009.

Mbaye, A., Bellomo, J. P., Lebey, T., Oraison, J. M., & Peltier, F. (1997). Electrical stresses applied to stator insulation in low voltage induction motors fed by PWM drives. *IEE Proceedings Electr. Power Applications*, 144(3), 191–198.

Muetze, A., & Binder, A. (2003). Experimental evaluation of mitigation techniques for bearing currents in inverter-supplied drive-systems – Investigations on induction motors up to 500 kW. In *IEEE International Electric Machines and Drives Conference, IEMDC'2003*, June 1–4, 2003, pp. 1859–1865.

Mei, C., Balda, J. C., Waite, W. P., & Carr, K. (2003). Minimization and cancellation of common-mode currents, shaft voltages and bearing currents for induction motor drives. In *Proceedings of Power Electronics Specialist Conference, PESC'2003*, June 15–19, 2003, Vol. 3, pp. 1127–1132.

Ollila, J., Hammar, T., Iisakkala, J., & Tuusa, H. (1997). On the bearing currents in medium power variable speed AC drives. In *IEEE International Electric Machines and Drives Conference Record*, 1997, Milwaukee, WI, May 18–21, MD1/1.1–MD1/1.3.

Punga, F., & Hess, W. (1907). Eine Erscheinung an Wechsel- und Drehstromgeneratoren. *Elektrotechnik und Maschinenbau*, 25, 615–618.

Rendusara, D. A., & Enjeti, P. N. (1998). An improved inverter output filter configuration reduces common and differential modes dv/dt at the motor terminals in PWM drive systems. *IEEE Transactions on Power Electronics*, 13(6), 1135–1143.

Shaotang, C. (1996a). Source of induction motor bearing currents caused by PWM inverters. *IEEE Transactions on Energy Conversion*, 11(1), 25–32.

Shaotang, C. (1996b). Circulating type motor bearing current in inverter drives. IEEE-IAS Annual Meeting 1996, Vol. 1, pp. 162–167.

Skibinski, G. (1997). Bearing currents and their relationship to PWM drives. *IEEE Transactions on Power Electronics*, 12(2), 243–251.

Skibinski, G. (1996). Effect of PWM inverters on AC motor bearing currents and shaft voltages. *IEEE Transactions on Industry Applications*, 32(2), 250–259.

Ström, J.-P. (2009). Active du/dt filtering for variable-speed AC drives, dissertation LUT, available at http://urn.fi/URN:ISBN:978-952-214-889-6

von Jouanne, A., & Enjeti, P. (1997). Design considerations for an inverter output filter to mitigate the effects of long motor leads in ASD applications. *IEEE Transactions on Industrial Applications*, 33(5), 1138–1145.

von Jouanne, A., Rendusara, D., Enjeti, P., & Gray, W. (1996). Filtering techniques to minimize the effect of long motor leads on PWM inverter-fed AC motor drive systems. *IEEE Transactions on Industry Applications*, 32(4), 919–926.

Index

Electrical Machine Drives Control: An Introduction, First Edition. Juha Pyrhönen, Valéria Hrabovcová
and R. Scott Semken.
© 2016 John Wiley & Sons, Ltd. Published 2016 by John Wiley & Sons, Ltd.